Applied Finite Element Analysis for Engineers

FRANK L. STASA
Florida Institute of Technology

Holt, Rinehart and Winston
New York Chicago San Francisco Philadelphia
Montreal Toronto London Sydney Tokyo
Mexico City Rio de Janeiro Madrid

To Frank W. and Lena C. Stasa,
Ann R. Holt,
and to the memory of
John A. Holt

HRW
Series in
Mechanical
Engineering

L. S. Fletcher, Series Editor

F. L. Stasa   APPLIED FINITE ELEMENT ANALYSIS FOR ENGINEERS
B. J. Torby   ADVANCED DYNAMICS FOR ENGINEERS

Copyright © 1985 CBS Publishing
All rights reserved

Address correspondence to
383 Madison Avenue, New York, NY 10017

Library of Congress Cataloging in Publication Data

Stasa, Frank L.
   Applied finite element analysis for engineers.

   (HRW series in mechanical engineering)
   Includes bibliographies and index.
   1. Finite element method. I. Title. II. Series.
TA347.F5S72   1985   620'.001'515353   85-742
ISBN 0-03-062737-0

Printed in the United States of America

5678   038   987654321

CBS COLLEGE PUBLISHING
Holt, Rinehart and Winston
The Dryden Press
Saunders College Publishing

Preface

Written for senior-year undergraduates and first-year graduate students with solid backgrounds in differential and integral calculus, this book is oriented toward engineers and applied mathematicians. A course in linear algebra is helpful but not essential. Courses in elasticity, heat transfer, and fluid mechanics should facilitate the student's understanding of the applications emphasized in this text. Overall, the author's approach represents a compromise between the purely mathematical and the purely applied developments.

Stress or structural analysis is given about the same level of treatment as thermal and fluid flow analysis. However, this book is structured so that it may be used in courses in which the application area is strictly stress analysis or strictly thermal analysis. Moreover, it is also possible to use this book without covering any of the material related to variational calculus and variational formulations. This strategy is made possible by emphasizing the Galerkin weighted-residual method in thermal (and fluid flow) analysis and the principle of virtual displacements in stress analysis. Consequently, this book should be useful to instructors teaching a course on the finite element method to senior undergraduate students. If the nonvariational path is desired, the instructor should omit the following sections: the last part of Sec. 4-2, Sec. 4-3, parts of Secs. 4-4, and Secs. 4-8, 5-3, and 8-6.

The first two chapters are introductory in nature. Chapter 1 contains a brief survey of what the finite element method is, as well as a brief history of the method. Chapter 2 contains a review of the necessary mathematical concepts of matrices, vectors, and determinants. The reason for including this review in the text proper was to establish a common ground from which the finite element method could be developed, regardless of the specific background of the engineering student.

All aspects of the finite element are explored in Chapter 3 by way of one of the simplest of all engineering applications—the truss. The so-called *direct approach* is adopted for this purpose and each step in the finite element solution process is given in full detail. For this reason, all students must be exposed to (and indeed should master) Chapter 3, which is the only structural analysis chapter that must be covered by all users of this book. Both two- and three-dimensional trusses are covered, but the former is given broader, more comprehensive treatment. The student is also introduced to computer programming concepts and a two-dimensional truss program.

Chapter 4 provides the general framework for the development of nearly all (nonstructural) finite element models. Here the student is introduced to the concepts of globally based approximations to the true solution of simple ordinary differential equations. Among the methods covered are the Ritz, Rayleigh-Ritz (variational), point collocation, subdomain collocation, least squares, and Galerkin methods. The last four of these methods comprise the class of approximate solution methods known as the weighted-residual methods. Both the Rayleigh-Ritz (variational) and the Galerkin (weighted-residual) methods are extended to piecewise continuous approximations and, hence, to the finite element method itself. The student is introduced, of course, to the concept of shape functions at this point. Chapter 4 concludes with an application in the thermal analysis area: a pin fin (a type of extended surface). Instructors may wish to have their students review Chapter 1 before proceeding with the next chapter.

Chapter 5 is devoted specifically to the development of finite element models in the stress (or structural) analysis area. A brief review of some of the more important concepts of elasticity is provided and may be skipped by those who have had a prior graduate course on this subject. Depending on the instructor's preference, either of two developments may be used: the principle of minimum potential energy (variational) or the principle of virtual displacement with a simple one-dimensional application: the uniaxial stress member. The results from Chapter 5 are used throughout Chapter 7, the primary stress application chapter.

Chapter 6 is, in effect, a catchall chapter, which contains essential material that has not been covered adequately up to this point. Here the student is introduced more formally to the concept of parameter functions (such as displacement and temperature functions) and to the compatibility and completeness requirements that these functions should satisfy. Shape functions (C^0-continuous only) are derived and presented for the following types of one-, two-, and three-dimensional elements: two-node lineal (1-D), three-node triangular (2-D), four-node rectangular (2-D), four-node tetrahedral (3-D), and eight-node brick (3-D). Local, normalized coordinates, such as length, area, and volume coordinates, as well as serendipity coordinates, are introduced. Axisymmetric elements are also presented. Three simple integration formulas are given in terms of length, area, and volume coordinates for integrations over lineal, triangular, and tetrahedral elements, respectively. Finally, an alternative to the matrix inversion technique is provided, namely, the active zone equation solver. This method is based on triangular decomposition, forward elimination, and backward substitution, and takes advantage of the banded, and often symmetric, nature of the assemblage stiffness matrix. This method requires that the assemblage stiffness matrix be stored as a column vector (instead of a square matrix).

Chapter 7 is the main stress analysis application chapter. Among the topics covered are the following: two-dimensional stress analysis (plane stress and plane strain), axisymmetric stress analysis, three-dimensional stress analysis, and the analysis of beams. The notion of substructuring and condensation is also introduced and working equations are developed. The chapter concludes with a brief description of the development of a two-dimensional stress analysis program. The instructor is referred to the Instructor's Solutions manual for a listing (in FORTRAN) of one version of this program—called program STRESS (the user's manual to the program is also included in the Instructor's Solutions manual).

Chapter 8 is the principal thermal and fluid flow analysis chapter, in which the following topics are covered: one-, two-, and three-dimensional thermal analysis, and axisymmetric thermal analysis. Material on variational formulations in two-dimensional problems is provided, but this also may be skipped by those preferring the nonvariational path. Other application areas in Chapter 8 include convective energy transport, two-dimensional potential flow, and two-dimensional incompressible fluid flow. The chapter concludes with a brief description of the development of a two-dimensional, steady-state thermal analysis program. Again the instructor is referred to the Instructor's Solutions manual for a listing (in FORTRAN) of one version of this program—called HEAT (the user's manual to the program is also included in the Instructor's Solutions manual).

Chapter 9 introduces higher-order elements and numerical integration (quadrature). The one-, two-, and three-dimensional elements introduced in Chapter 6 are extended to quadratic and cubic order. Subparametric, isoparametric, and superparametric elements are also introduced. Two- and three-dimensional isoparametric formulations are developed for the quadrilateral and triangular elements as well as for the brick and tetrahedral elements. Special formulas are presented, which facilitate greatly the evaluation of the integrals that naturally arise.

In conclusion, Chapter 10 is devoted to transient thermal analysis and dynamic structural analysis. The concept of partial discretization is presented and applied to stress analysis and thermal analysis. Lumped and consistent capacitance and mass matrices are discussed. Solution methods are developed based on the finite element method itself (i.e., in time) and on the finite difference method. The result is two- and three-point recurrence schemes for transient thermal analysis and dynamic structural analysis, respectively. The chapter concludes with a brief introduction to modal analysis.

Appendix A contains the material property data to be used in the problems, unless otherwise noted in the problem statements. Appendix B contains a short user's manual to the (two-dimensional) truss program along with the program listing (in FORTRAN). Appendix C contains listings of subroutines ACTCOL and UACTCL (and function DOT), which have been used with the written permission of McGraw-Hill and which appear in Professor Zienkiewicz's third edition of *The Finite Element Method*. The length of the STRESS and HEAT programs and their respective user's manuals precluded their inclusion in this text. Instructors who would like to have copies of all of these programs on floppy disks are encouraged to write to the author. The disk formats available are IBM PC, Apple II series, and 8-inch IBM 3740 standard format for CP/M-based machines. *Other FORTRAN programs that can be obtained on these disk formats include: beam analysis, a TurboPascal** version of the two-dimensional truss program, fin analysis, transient one-dimensional thermal analysis, and transient two-dimensional thermal analysis.

The following suggestions are made to instructors teaching on the quarter and semester system. *For those teaching on the quarter system:* Chapters 1 to 4 can be covered comfortably during the fall quarter. The winter quarter could be devoted to stress analysis with coverage of Chapters 5, 6, 7, 9, and 10 (omit Secs. 10-4, 10-7, and 10-8). The spring quarter could be devoted to thermal and fluid flow analysis with coverage of Chapters 6, 8, 9, and 10 (omit Secs. 10-3, 10-9, and 10-10). For students who take all three courses, some of the material is necessarily repeated. The author has found this repetition not to be a problem, because these students get a firmer grasp of the material the second time around.

For those teaching on the semester system: Chapters 1 to 6 could be covered comfortably during the first semester, and Chapters 7 to 10 during the second semester. This pace allows sufficient class time for the discussion of computer programming techniques. A useful project for the first course is to have the students modify the two-dimensional truss program in Appendix B so that it could be used in three-dimensional truss applications.

The Instructor's Solutions manual contains a listing (in FORTRAN) of a two-dimensional, static stress analysis program (for Chapter 8). The manual is available from the publisher upon proper written request. The instructors may find it useful to distribute copies of these programs to their classes.

The author wishes to thank all of the students who have used the original notes and class-tested the text manuscript. Their comments and suggestions were taken seriously, and their words of encouragement will always be remembered. Deserving of special recognition is Jay A. Buckman, who did an outstanding job of proofreading the page proofs. A special debt of gratitude is owed to General Herbert McChrystal, who first suggested this project. Very special thanks are extended to U. Shripathi Kamath, who had the monumental task of providing the solutions manual to the text, and to Michael Weaver for converting the FORTRAN version of the truss program into Pascal. The author is also thankful to the editors

---

*CP/M is registered trademark of Digital Research, Inc.
**TurboPascal is a trademark of Borland International.

John J. Beck, Lynn Contrucci, and Rachel Hockett for their help and cooperation in publishing the manuscript. Finally, and most important of all, the author is once again deeply indebted to his wife, Donna, who meticulously typed every page of the manuscript and without whose encouragement this book would probably not have been completed, and to Lisa Ann, for helping to put this project into proper perspective and for providing the perpetual light at the end of tunnel.

Frank L. Stasa

CONTENTS

1

General Concepts

1-1 INTRODUCTION

The finite element method of analysis is a very powerful, modern computational tool. The method has been used almost universally during the past 15 years to solve very complex structural engineering problems, particularly in the aircraft industry. It is now gaining wide acceptance in other disciplines such as thermal analysis, fluid mechanics, and electromagnetics. The method requires the use of a digital computer because of the large number of computations involved.

Structures take many different, sometimes very diverse, forms and utilize many different materials. Oftentimes a structure itself is composed of a multitude of materials. For example, bridges and space-deployable antennas are both structures, but they hardly resemble each other in appearance and materials. A space-deployable antenna looks like a gigantic umbrella that is *closed* during launch and *opened*, or deployed, once in earth orbit. Figure 1-1 shows an antenna in the deployed configuration. The structural design and analysis of this antenna was done with large finite element computer programs such as NASTRAN [1] and STARDYNE [2].

In the design of a bridge, engineers are concerned with the stresses developed in the bridge as a result of the weight of the vehicles riding on it, in addition to wind loads and its own weight. These stresses are important because all materials have upper limits on allowable stresses. When these allowable stresses are exceeded, the structure may no longer perform its intended function. In the extreme case, the structure may actually collapse. A tragic example of this is the collapse of two 32-ton skywalks at the Hyatt Regency Hotel in Kansas City on July 27, 1981.

In the case of the space-deployable antenna, excessively large deflections of its carefully contoured umbrellalike surface will deteriorate the antenna perfor-

Figure 1-1   Photograph of a space-deployable antenna in solar thermal vacuum test chamber. (*Courtesy of Harris Corporation and TRW.*)

mance. If the distortions are too large, the antenna may not be able to perform its intended function. These deflections may be caused by the heating effect of the sun. Most materials stretch when heated and shrink when cooled. If part or all of the antenna becomes shaded by the spacecraft, an improperly designed antenna may distort enough to render itself useless.

In a typical thermal analysis application, we may need to know the temperatures within the body being analyzed. For example, for the space-deployable antenna the amount of distortion as a result of the temperature change cannot be ascertained until the temperatures themselves are determined. Sometimes in addition to the temperatures, we need to know how much heat is flowing through a given area. In fluid mechanics problems, the analyst may need to determine the fluid velocities and pressures within or around some device or object. The complex flow pattern in a coal gasifier, for example, may be needed [3]. The finite element method is being used today to solve these and many other more complex problems.

It should be obvious that it is not possible to obtain closed-form, analytical solutions to complex problems such as these. Often a skilled engineer can make simplifying assumptions to obtain a closed-form solution, but the very nature of

Figure 1-2 Three-noded triangular element.

some problems, such as the bridge and space-deployable antenna, may preclude such an approach. Complexities may arise because of the irregular and varied geometry, mixed boundary conditions, nonlinear material behavior, and nonuniform loading conditions. The finite element method is particularly well-suited to handle these and other complications.

1-2 WHAT IS THE FINITE ELEMENT METHOD?

The basic idea behind the finite element method is to divide the structure, body, or region being analyzed into a large number of *finite elements*, or simply elements. These elements may be one, two, or three dimensional. A popular and classical two-dimensional element is the triangle shown in Fig. 1-2. When a *two-dimensional* structure (or heat transfer device, etc.) is divided into hundreds or sometimes thousands of these nonoverlapping triangles, we can see that essentially all planar geometries can be easily accommodated. Note that this particular element has three nodes (i, j, k) appropriately placed at the vertices of the triangle. Before explaining the purpose of these nodes or nodal points, let us see how we may use this type of element to represent an irregularly shaped plate.

Figure 1-3 shows such a plate. By way of illustration, let us say that the temperature distribution within the plate is sought for a given plate material, specified boundary conditions, etc. The field variable of interest here is the temperature. Because there is an infinite number of points in the plate, there is also an infinite number of temperatures to be determined. The problem is therefore said to have an infinite number of degrees of freedom. It should be recalled that when closed-

Figure 1-3 Plate with variable thickness and an irregular shape.

form analytical solutions exist, they theoretically allow us to compute the temperature at any point (and hence at an infinite number of them) in the region because the solution is expressed as a mathematical function of x and y [in the form $T = f(x,y)$]. For the plate in Fig. 1-3, the irregular geometry precludes such a closed-form solution. Therefore, an alternate approach is sought.

Instead of requiring that the temperature be determined at every point in the plate, let us change this unreasonable requirement to a determination of the temperature at only a finite number of points. It would appear that these points could be taken as the vertices (or nodes) of the triangles. Figure 1-4 shows the plate discretized into several triangular elements. Note how the curved boundaries are approximated by the straight-sided triangles. The finite element method will provide the analyst with the temperatures of the plate only at the nodal points. Interpolation functions are used within each element to describe the variation of the field variable (e.g., here the field variable is the temperature) as a function of the global coordinates (x and y in this case). Knowledge of the temperatures at the nodal points allows the analyst to ascertain the general temperature field within the region. These temperatures are often referred to as the *nodal temperatures*. From these, the heat-flows through each element are also easily determined.

As mentioned earlier, a high-speed digital computer is used to perform the large number of computations involved. An analyst must provide the relevant computer program with the coordinates of each node, element definitions, material data, boundary condition data, etc. Generally the accuracy improves as the number of elements increases, but computational time, and hence cost, also increase. There is no substitute for experience in trying to determine the number of elements to be used. The computer program determines the temperature at each node and the heat-flows through each element. The results are usually printed or presented in graphic form. The analyst may then use this information in the design process.

Although the example chosen above was in the heat transfer area, the same basic idea is applicable to the structural and the fluids areas, as well as to others. In a structural analysis application, the field variables may be displacements and/or deflections and slopes. Therefore, in a structural finite element model, we determine the nodal displacements and/or the nodal deflections and slopes. From these, the stresses and/or bending moments within an element are easily derived. In a fluid flow analysis, the field variables are frequently, but not always, the fluid

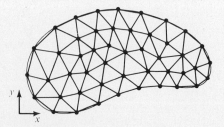

Figure 1-4   Irregularly shaped plate shown discretized into many triangular finite elements.

velocities and pressure. Therefore, the corresponding finite element model enables a determination of these parameters at the nodal points. These results may then be used to obtain the shear stresses within each element if they are needed.

In summarizing, the two key ideas of the finite element method are (1) discretization of the region being analyzed into finite elements and (2) the use of interpolating polynomials to describe the variation of a field variable within an element. Both of these notions are developed in subsequent chapters. However, the power of discretization can be demonstrated by trying to ascertain the value of π as shown in Sec. 1-3.

1-3 DISCRETIZATION

Let us try to compute the circumference of a circle without assuming a value of π. If the circumference can be computed (or at least approximated) in this way, the value of π may be computed (approximately, since the circumference is approximate) from the well-known definition

$$\pi = \frac{C}{d} \qquad \textbf{(1-1)}$$

where C is the circumference of the circle and d is the diameter.

Two scholars were largely responsible for the determination of the value of π. The first, Eudoxus of Cnidus, was born about 2400 years ago in southwest Asia Minor [4]. He developed the method of exhaustion in which he determined the area bounded by one or more closed curves and the volume bounded by a closed surface by replacing the original problem with many simpler ones. He then added up the individual results to obtain the solution to the original problem. For example, to determine the area under the curve shown in Fig. 1-5(a) from point a to point b, Eudoxus would have divided the unknown area into a number of rectangles whose areas are readily calculated. Two ways of accomplishing this are shown in Figs. 1-5(b) and (c). Note that the method illustrated in Fig. 1-5(b) would give a low estimate of the area whereas that in Fig. 1-5(c) would give a high estimate. This gave rise to the notion of bounding an approximate result. As more and more rectangles are used, it seems intuitively obvious that better and better approximations result: this is called *convergence*.

With this method, Eudoxus showed that the area of a circle is proportional to the square of its diameter. He also demonstrated that the volumes of a pyramid and cone are one-third the volumes of the rectangular prisms and cylinders, respectively, with the same heights and bases. The *method of exhaustion* is fully elaborated upon in Book XII of Euclid's *Elements*. This method is the forerunner of integral calculus, which surprisingly was not developed until about 20 centuries later by Leibnitz and Newton.

The scholar who further advanced the notion of discretization was Archimedes, a Greek mathematician, scientist and inventor, born in 287 B.C. in the Greek city-state of Syracuse in Sicily [5]. He extended the method of exhaustion to the point

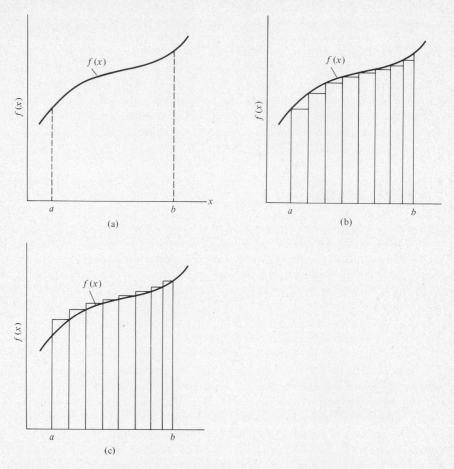

Figure 1-5 Illustration of the method of exhaustion in obtaining the area under a curve. (a) Exact area. (b) Lower bound on estimate by using inscribed rectangles. (c) Upper bound on estimate by using circumscribed rectangles.

where it was actually equivalent to integration in some cases. In one of his extant treatises, *Measurement of a Circle*, Archimedes placed the value of the circumference of a circle between $3\frac{1}{7}$ and $3^{10}/_{71}$ times its diameter. Consequently, firm limits were placed on the value of π. Although others throughout the ages subsequently improved the value of π, it was Eudoxus and Archimedes who were largely responsible for determining the value of this important constant.

By way of illustration, let us determine a sequence of approximate values of π by inscribing and circumscribing regular polygons inside and around a circle as shown in Fig. 1-6 for the case of regular inscribed and circumscribed octagons. The lengths of the sides of the polygons are readily measured (because they are straight and not curved), and the resulting lower and upper bounds of π are readily

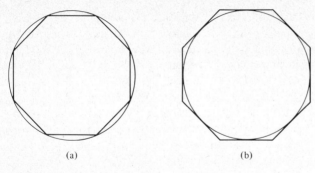

(a)   (b)

Figure 1-6 Approximation to the circumference of a circle by (a) an inscribed octagon and (b) a circumscribed octagon.

calculated from Eq. (1-1). The results are given in Table 1-1, first, for an eight-sided polygon, then, for each successive case doubling the number of sides until a staggering 1,048,576 sides [6] are used—hence the name method of exhaustion! Note that the lower estimate converges more rapidly to the accepted value for π of 3.1415926536 (to 10 decimal places).

These simple and antiquated examples illustrate the concept of *discretization*—the essence of which is the division of a larger, more difficult problem into a number

Table 1-1 Calculation of the Value of π by the Method of Exhaustion Using Inscribed and Circumscribed Regular Polygons

Number of sides	Inscribed polygons	Circumscribed polygons
8	3.061467458921	4.959315235374
16	3.121445152258	3.878006734963
32	3.136548490546	3.477392565633
64	3.140331156955	3.302370812490
128	3.141277250933	3.220307554543
256	3.141513801144	3.180543968220
512	3.141572940367	3.160968277095
1024	3.141587725277	3.151255641334
2048	3.141591421511	3.146417964326
4096	3.141592345570	3.144003766021
8192	3.141592576585	3.142797824426
16384	3.141592634339	3.142195142707
32768	3.141592648777	3.141893874079
65536	3.141592652387	3.141743257818
131072	3.141592653289	3.141667954200
262144	3.141592653515	3.141630303519
524288	3.141592653571	3.141611478460
1048576	3.141592653585	3.141602066002

of smaller, simpler problems such that the sum of the solutions for the simpler problems, in some sense, approximates the solution to the original problem. The example involving the computation of the area under a curve utilized a very common finite element—the rectangular element. The example involving the evaluation of π utilized another finite element—the lineal element. The rectangle, like the triangle in Sec. 1-2, is a two-dimensional element, whereas the lineal element is one-dimensional. Quite appropriately, the two-dimensional element was used to evaluate an area, whereas the one-dimensional element was used to determine the circumference (which is a length and, hence, also one-dimensional). Not surprisingly, three-dimensional elements such as tetrahedra and rectangular prisms (or bricks) may be used in the method of exhaustion to determine the volume of three-dimensional objects with complicated boundaries.

1-4   RELATIONSHIP TO THE FINITE-DIFFERENCE METHOD

The finite-difference method (FDM) is an alternate solution technique that may be used to solve the same types of problems for which the finite element method (FEM) is suited. This is particularly true in the cases of thermal and fluid flow analyses. Both methods require the analyst to discretize the structure, object, or region being analyzed or modeled. However, the way in which the discretization is done is fundamentally different.

For the purpose of this discussion, let us restrict our attention to the two-dimensional case. It should be recalled that the triangular and rectangular elements may be used in FEM. So as not to obscure the basic point that is to be made here, let us further restrict the discussion to the case of the rectangular element. In FEM, the nodes are placed at the corners of the rectangles. Since the rectangle has four corners, there are four nodes. This is quite different from the way in which the discretization is done in FDM.

In FDM, the object or region being analyzed is divided into a finite number of *lumps*. Note that the word "lump," not the word "element," is used. In the basic FDM approach, each of these lumps is assumed to have a constant value of the pertinent field variable (this value, however, may be a function of time in transient problems). For example, in a heat transfer or thermal analysis problem, the relevant field variable is the temperature. Therefore, each lump would be isothermal. Since the entire lump is isothermal, it seems natural that the *node* for the lump should be associated not with any of its *corners*, but rather at its geometric center or centroid. Figure 1-7 shows the two different methods of discretization.

In Fig. 1-7(a), a typical rectangular finite element is shown with four nodes. It should be recalled from the discussion in Sec. 1-2 that each of these four nodes may have a different value of the field variable (displacements, temperatures, velocities, pressures, etc.). It should further be recalled that by using interpolation functions for the field variable within each element, the field variable is generally not constant over an element but varies in some prescribed manner, depending on the interpolation polynomial being used.

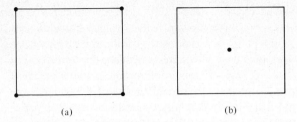

(a)   (b)

Figure 1-7   Two main types of discretization. (a) An element used in finite element analysis. (b) A lump used in finite-difference analysis. Note the four nodes on the element in (a) and the one node on the lump in (b).

Figure 1-7(b) shows a typical lump used in FDM analysis. The one and only node appears at the center of the rectangular lump. In this case, the entire lump is assumed to have a constant value of the field variable. In both FEM and FDM, the nodes represent the locations at which the field variables are to be determined.

To further illustrate the difference between the FEM and FDM approaches, let us discretize a rectangular plate into elements and lumps such that the same number of nodes is used in each case. Figures 1-8(a) and (b) clearly show the difference between the two approaches. The FDM discretization in Fig. 1-8(a) requires that if nodes are to be placed on the plate boundaries (e.g., to impose a prescribed temperature on this boundary), then the surface lump is essentially one-half the size of an interior lump. Corner lumps are seen to be one-fourth the size of interior lumps. This assumes that the nodes are to be equally spaced; if they are not, the situation is even more complicated. Note that for the FEM discretization in Fig. 1-8(b), each element is of the same size for equally spaced nodes. If it is desirable to assign nodes that are not equally spaced, the FEM approach is more easily adapted.

More importantly, in cases in which the geometry of the region being analyzed is irregular, as shown in Fig. 1-9(a), the triangular and quadrilateral elements are routinely used, and it is seen that the plate boundaries are approximated quite well

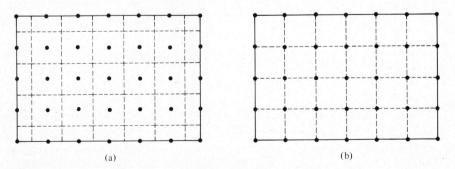

(a)   (b)

Figure 1-8   Rectangular plate discretized into (a) lumps for a finite-difference analysis and (b) rectangular elements for a finite element analysis.

with these elements. Note that different element types may be used in the same model, as illustrated here by the use of the triangular, rectangular, and quadrilateral elements. On the other hand, if the same plate is discretized into lumps for an FDM analysis, as shown in Fig. 1-9(b), then the effective plate boundary becomes jagged as shown. The alternative is to write special finite-difference equations for these nodes that explicitly include the effect of the curved boundary [7]. Unlike the formulation in FEM, the FDM formulation is nontrivial when curved boundaries are present.

The different methods of discretization in FEM and FDM, however, are not the only significant differences between the two types of formulations. In FDM, a basic law is written for each node, such as the first law of thermodynamics (conservation of energy) or Newton's second law of motion. For example, in a heat transfer formulation, an energy balance is made on each lump. This is relatively straightforward (except when curved boundaries are present).

In FEM, although a direct energy balance approach can be used, other more powerful and more popular approaches are taken such as the principle of virtual work, variational methods, and weighted-residual methods. These various methods of FEM formulations are developed in detail throughout this text. For structural and stress analysis problems, the principle of virtual work will usually be used, whereas for nonstructural applications, the method of weighted residuals will be used almost exclusively. The reasons for favoring the method of weighted residuals over variational formulations will become evident in Chapters 4 and 8. As the name implies, the variational formulations require knowledge of variational calculus, an obvious disadvantage.

1-5   ADVANTAGES AND DISADVANTAGES OF THE FINITE ELEMENT METHOD [8]

As mentioned in the previous section, one of the main advantages of FEM over most other approximate solution methods, including the popular finite-difference

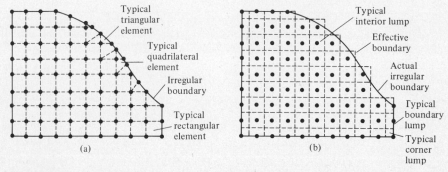

(a)   (b)

Figure 1-9   Irregularly shaped plate shown discretized into (a) rectangular, triangular, and quadrilateral elements and (b) lumps.

method, is the fact that FEM can handle irregular geometries routinely. The triangular element in two-dimensional applications is used with no special considerations. Of course, closed-form analytical solutions are invariably nonexistent when irregular geometries are present.

Another significant advantage of FEM is that a variable spacing of the nodes is also routinely handled. When a body is discretized using finite elements (in FEM) or lumps (in FDM), the nodes are said to form a *mesh*. Typical two-dimensional meshes were shown in Figs. 1-8 and 1-9. When the nodes are not equally spaced, the mesh is said to be *graded*. The finite element method lends itself to the use of graded meshes. It will be seen in later chapters that special subroutines can be written that will automatically or semiautomatically generate unequally spaced nodes, thereby reducing the amount of input required by the user of such a computer program.

Another advantage of FEM, again over FDM and especially over analytical solution techniques (as opposed to numerical techniques) is the ease with which nonhomogeneous and anisotropic materials may be handled. Materials whose properties are not spacially dependent are said to be *homogeneous*, whereas materials with spacially dependent properties are *heterogeneous*. A special case of a heterogeneous material in a thermal analysis is one for which the thermal conductivity is temperature-dependent. The reason for this is simple: the thermal conductivity is a function of temperature, the temperature is a function of spacial coordinates, and, therefore, the thermal conductivity must also be a function of spacial coordinates. Concrete is an example of a heterogeneous material.

Materials may also be classified as isotropic and anisotropic. An *isotropic* material is one whose properties (Young's modulus, thermal conductivity, etc.) do not exhibit a direction sensitivity. For example, even though concrete may be nonhomogeneous, each direction appears to have the same (rather random) variation in thermal conductivity and, therefore, it is isotropic. *Anisotropic* materials, on the other hand, will have one or more properties that are direction-dependent. For example, a laminated metallic structure quite frequently will have different values of certain properties, such as Young's modulus or thermal conductivity, in different directions. Wood is another example of an anisotropic material; it is generally stronger in the direction of the grain and hence would have a higher value of Young's modulus in this direction. Very little extra effort is required in the FEM formulation when heterogeneous and/or anisotropic materials are to be modeled, even when some parts of the structure or body are made of one material and other parts are made of different materials.

All the various types of boundary conditions that we may encounter in a typical FEM application except those that require prescribed values of the field variables themselves, are automatically included in the formulation. Recall that typical field variables are displacements in structural and stress analysis, temperatures in thermal analysis, fluid velocities and pressures in fluid flow analysis, etc. In other words, prescribed displacements, temperatures, velocities, pressures, etc., are not automatically included in the FEM formulation and solution. They are systematically enforced just before the solution for the nodal values of the unknown field variables is obtained, as discussed in subsequent chapters. Among the boundary conditions that are automatically included are the following:

Structural and stress analysis
    Concentrated surface forces
    Distributed surface forces (tractions)
Thermal analysis (heat conduction)
    Convection
    Radiation
    Applied heat fluxes
    Insulation
Fluid flow analysis
    Pressure gradients
    Velocity gradients

In all cases, these conditions need not be constant. When these boundary conditions or other properties are a function of the (unknown) field variable, the problem becomes nonlinear. Special solution techniques, one of which is discussed in Chapter 8, must be applied in these cases. The basic finite element method is applicable, however, for both linear and nonlinear problems.

Another advantage is that higher-order elements may be implemented with relative ease. Several higher-order elements are shown in Fig. 1-10. Higher-order elements require the use of higher-order interpolating polynomials. Note that additional nodes are introduced along the sides of the two-dimensional elements and between the two end nodes of the one-dimensional element. In fact, by using isoparametric elements defined in Chapter 9, curved sides may actually be used thereby allowing very close fits to essentially all irregular geometries. Occasionally, interior nodes are introduced as shown in Fig. 1-10. The use of these nodes requires special considerations as discussed in Chapter 9.

Among the disadvantages of FEM is the necessity for a digital computer and fairly extensive software. In fact, the MacNeal-Schwendler version of NASTRAN [1] has more than 430,000 lines of FORTRAN source code. Of course, such a large general program need not always be used. More specific and significantly smaller programs may be all that are needed in some applications. The programs included in the appendices of this book are provided mainly for educational purposes, but on occasion they may be used to solve small practical problems.

A couple of obvious questions have probably arisen. The first is whether or not the finite element method is more accurate than the finite-difference method. We cannot answer this question without a significant number of qualifications, but FEM seems to be more accurate when curved boundaries are present. By referring to Fig. 1-9 again, we can see why this is not really surprising: the finite element method can usually include the curved boundary more precisely than the finite-difference method can.

The second question that may be asked pertains to the relative execution times of typical and comparable FDM and FEM models. Once again it is difficult to make comparisons, but FEM seems to have longer execution times. However, it

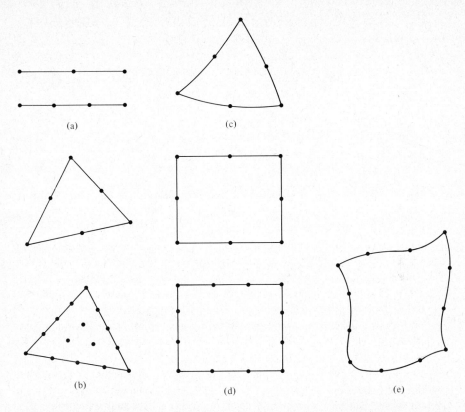

Figure 1-10 Representative higher-order elements. (a) One-dimensional elements. (b) Two-dimensional triangular elements with straight sides. (c) Two-dimensional triangular element with curved sides. (d) Two-dimensional rectangular elements with straight sides. (e) Two-dimensional quadrilateral element with curved sides.

must be said that FDM models will generally require more hours of computer input preparation by people than will FEM. As will be seen in subsequent chapters, many of the tedious calculations in FEM are relegated to the computer. For example, in a thermal application, internal conductances need not be calculated in FEM models because they are inherently included in the computer software. On the other hand, FDM thermal models quite frequently require the user to calculate many of the internal conductances by hand. This will become an even more significant advantage of FEM as engineering labor costs rise and computational costs decline.

1-6 BRIEF HISTORY OF THE FINITE ELEMENT METHOD

An early but crude application of finite elements was presented in Sec. 1-3 where the value of π was determined by the method of exhaustion. However, the finite element method used today was developed to its present state very recently. According to Zienkiewicz [9], the development has occurred along two major paths: one in mathematics and the other in engineering. Somewhere in between these two extremes are the variational and weighted residual methods, both of which require the use of trial functions to effect a solution. The use of these functions dates back almost 200 years.

A *trial function* is an assumed mathematical function that is usually based on physical intuition, is applied globally to the region being analyzed, and approximates the expected behavior of the region to some external forces. These functions are used in various types of integral formulations [10], described in detail in Chapter 4. A qualitative example may help to clarify these abstract points.

Let us consider again the heat conduction problem in a rectangular plate. If the temperature is known to be zero along the boundaries of the plate and a maximum at the center based on physical reasoning, then one or more trial functions could be chosen that exhibit this known trend. Such a method is rather limited for two reasons: (1) the trial functions must satisfy the boundary conditions exactly, and (2) the trial functions must be chosen based on physical reasoning. In other words, it is very difficult to apply this method to any but the simplest of problems.

As indicated earlier in this chapter, the finite element method is predicated along the same lines with an important exception: the trial functions are not applied globally to the entire region being analyzed but only locally (i.e., on an element basis). This point is of paramount importance, since the method becomes applicable to real problems with irregular geometries and unknown solutions. This important step did not occur until 1943 when Courant [11] introduced piecewise continuous trial functions. But this is getting ahead of the story, which begins in 1795.

As mentioned earlier, the use of trial functions is associated with neither the purely mathematical nor the engineering developments. In a paper by Gauss [12] in 1795, trial functions (on a global basis) were used in what is now called the method of weighted residuals. Later, these functions were used in variational methods by Rayleigh [13] in 1870 and by Ritz [14] in 1909. In fact, any variational-based approximate solution method is frequently referred to today as the Rayleigh-Ritz method. In a landmark paper in 1915, Galerkin [15] introduced a particular type of weighted residual method, which to this day bears his name, the Galerkin method. However, all of these early uses of trial functions were done on a global basis.

In 1943, Courant [11] introduced *piecewise continuous trial functions*. As the name suggests, these functions were not applied globally to the entire region being analyzed but rather to many small regions or elements. By using trial functions on a local basis, Courant greatly extended the applicability of trial functions in obtaining approximate solutions to real-world problems. Many feel that the finite element method has its roots in this paper by Courant, although he did not introduce the terminology *finite element*.

Along more mathematical lines, the finite-difference method was introduced by Richardson [16] in 1910 and later improved by Liebman [17] in 1918 and Southwell [18] in 1946. Variational finite differences were developed by Varga

[19] in 1962. These mathematical developments lead directly to the present-day finite element method, and it may now be said that the finite-difference method is a special case of the more general finite element method [20].

In the early 1940s, aircraft engineers were developing and using analysis methods that are now recognized as early forms of the finite element method. The first applications used the so-called *force matrix method* (also known as the redundant force method). In this method, the nodal unknowns are the forces, not the displacements. When the displacements of each node are taken as the unknowns, the method is called the *stiffness method*.

However, the stiffness method was not developed until 1953 in an important paper by Levy [21]. Other key contributors were Schuerch [22] also in 1953; Argyris and Kelsey [23] in 1955; and Turner, Clough, Martin, and Topp [24] in 1956. In a paper in 1960, Clough [25] first introduced the term *finite element*. Readers interested in the historical details of the development of the finite element method along the structural engineering lines are referred to the book by Martin and Carey [26] in which a fascinating account is given.

The application of the finite element method to nonstructural problems has an equally interesting history. Among the first to apply the method to general problems described by Laplace's and Poisson's equations were Zienkiewicz and Cheung [27] and Visser [28] in 1965. The application at this particular time happened to be conduction heat transfer, but it was immediately recognized that the procedure was applicable to all problems that could be stated in a variational form.

Other researchers, such as Szabo and Lee [29], showed how the method of weighted residuals, particularly the Galerkin method, could be used in the study of nonstructural applications to retain the basic finite element process. Zienkiewicz [30] in 1971, in a second edition of an earlier book, was evidently the first to include in one book the general applicability of the finite element method to problems describable by ordinary and partial differential equations, or field problems in general. Field problems include all the problems normally associated with the continuum. Zienkiewicz's third edition [31] of the same book in 1977 is an outstanding example of the level of maturity that the finite element method now enjoys. Since the second edition of Zienkiewicz's book, numerous other books have appeared; those that have not already been cited are included in references 32 to 46.

Interestingly, the present-day finite element method does not have its roots in any one discipline. The mathematicians continue to put the finite element method on firm theoretical ground, and the engineers continue to find interesting applications and extensions in many branches of engineering. These concurrent developments have made the finite element method one of the most powerful of the approximate solution methods.

1-7 REMARKS

In this chapter some of the more important introductory concepts were developed. The basic finite element method was introduced by contrasting it to the more familiar finite-difference method. The notion of discretization, which is critical to the use of the finite element method, was introduced. The advantages and disadvantages of the method were also discussed. The use of trial functions, particularly on a local (or element) basis, was also introduced.

It cannot be emphasized enough that it is the notions of discretization and the use of local trial functions (or interpolating polynomials) that make the finite element method applicable to real-world problems with irregular geometries, heterogeneous and anisotropic materials, mixed boundary conditions, and so forth. Although these concepts may seem abstract to the uninitiated, they will undoubtedly become much more familiar in subsequent chapters.

The reader will be exposed further to discretization beginning with Chapter 3 and to global and local trial functions beginning with Chapter 4, where the variational and weighted residual methods are also formally introduced. Because the direct approach is taken, trial functions are not needed in Chapter 3 in which the two-dimensional truss model is developed. Our study begins in Chapter 2 with a review of some of the more important mathematics needed in the development of the truss model.

REFERENCES

1. *MSC/NASTRAN*, User's Manual, MacNeal-Schwendler Corp., Los Angeles, 1979.
2. *STARDYNE for Scope 3.4 Operating System*, User Information Manual, Cybernet Services, Control Data Corp., Minneapolis, Minn., 1978.
3. Kolbe, R. L., and K. K. Kimble, "The Development and Use of a Two-dimensional Finite Element Program for Incompressible-Viscous Flow Through a Complex Internal Geometry," *Int. Congr. Technol. Technol. Exchange*, Pittsburgh, Pa., May 3–6, 1982.
4. Macropaedia, "Eudoxus of Cnidus," *Encyclopaedia Britannica*, vol. 6, pp. 1021–1022, 1981.
5. Macropaedia, "Archimedes," *Encyclopaedia Britannica*, vol. 1, pp. 1087–1088, 1981.
6. Demonstration Program, *FORTRAN-80 User's Guide, Microsoft FORTRAN-80 for the Apple II Computer*, Microsoft Consumer Products, Bellevue, Wash., © 1980.
7. Myers, G. E., *Analytical Methods in Conduction Heat Transfer*, McGraw-Hill, New York, 1971, pp. 310–313.
8. Zienkiewicz, O. C., *The Finite Element Method*, McGraw-Hill (UK), London, 1977, pp. 430–432.
9. Zienkiewicz, O. C., *The Finite Element Method*, McGraw-Hill (UK), London, 1977, pp. 3–4.
10. Arpaci, V. S., *Conduction Heat Transfer*, Addison-Wesley, Reading, Mass., 1966, pp. 59–83.
11. Courant, R., "Variational Methods for the Solution of Problems of Equilibrium and Vibrations," *Bull. Am. Math. Soc.*, vol. 49, pp. 1–23, 1943.
12. Gauss, C. F., *Carl Friedrich Gauss Werks*, vol. VII, Gottingen, Germany, 1871.
13. Lord Rayleigh, "On the Theory of Resonance," *Trans. Roy. Soc. (London)*, vol. A161, pp. 77–118, 1870.
14. Ritz, W., "Uber eine neue Methode zur Losung gewissen Variations—Probleme der mathematischen Physik," *J. Reine Angew. Math.*, vol. 135, pp. 1–61, 1909.

15. Galerkin, B. G., "Series Solution of Some Problems of Elastic Equilibrium of Rods and Plates" (in Russian), *Vestn. Inzh. Tekh.*, vol. 19, pp. 897–908, 1915.
16. Richardson, L. F., "The Approximate Arithmetical Solution by Finite Differences of Physical Problems," *Trans. Roy. Soc. (London)*, vol. A210, pp. 307–357, 1910.
17. Liebman, H., "Die angenaherte Ermittlung: harmonischen, functionen und konformer Abbildung," *Sitzungsber. Math. Physik Kl. Bayer Akad. Wiss. Muchen*, vol. 3, pp. 65–75, 1918.
18. Southwell, R. V., *Relaxation Methods in Theoretical Physics*, Clarendon Press, Oxford, England, 1946.
19. Varga, R. S., *Matrix Iterative Analysis*, Prentice-Hall, Englewood Cliffs, N.J., 1962.
20. Zienkiewicz, O. C., *The Finite Element Method*, McGraw-Hill (UK), London, p. 569, 1977.
21. Levy, S., "Structural Analysis and Influence Coefficients for Delta Wings," *J. Aeronaut. Sci.*, vol. 20, no. 7, pp. 449–454, July 1953.
22. Schuerch, H. U., "Delta Wing Design Analysis," *Nat. Aeron. Meeting, Soc. Automot. Eng.*, Los Angeles, Paper No. 141, September 29–October 3, 1953.
23. Argyris, J. H., and S. Kelsey, *Energy Theorems and Structural Analysis*, Butterworth, 1960 (reprinted from *Aircraft Eng.*, 1954–1955).
24. Turner, M. J., et al., "Stiffness and Deflection Analysis of Complex Structures," *J. Aeronaut. Sci.*, vol. 23., no. 9, pp. 805–824, September 1956.
25. Clough, R. W., "The Finite Element in Plane Stress Analysis," *Proc. 2nd ASCE Conf. on Electron. Computation*, Pittsburgh, Pa., September 1960.
26. Martin, H. C., and G. F. Carey, *Introduction to Finite Element Analysis: Theory and Application*, McGraw-Hill, New York, 1973, pp. 1–13.
27. Zienkiewicz, O. C., and Y. K. Cheung, "Finite Elements in the Solution of Field Problems," *The Engineer*, 1965, pp. 507–510.
28. Visser, W., "A Finite Element Method for the Determination of Nonstationary Temperature Distribution and Thermal Deformations," *Proc. Conf. Math. Methods Struc. Mech.*, Air Force Institute of Technology, Dayton, Ohio, 1965.
29. Szabo, B. A., and G. C. Lee, "Derivation of Stiffness Matrices for Problems in Plane Slasticity by Galerkin's Method," *Int. J. Numer. Methods Eng.*, vol. 1, pp. 301–310, 1969.
30. Zienkiewicz, O. C., *The Finite Element Method in Engineering Science*, McGraw-Hill (UK), London, 1971.
31. Zienkiewicz, O. C., *The Finite Element Method*, McGraw-Hill (UK), London, 1977.
32. Desai, C. S., and J. Abel, *Introduction to the Finite Element Method*, Van Nostrand-Reinhold, New York, 1971.
33. Oden, J. T., *Finite Elements of Nonlinear Continua*, McGraw-Hill, New York, 1972.
34. Gallagher, R. H., *Finite Element Analysis: Fundamentals*, Prentice-Hall, Englewood Cliffs, N.J., 1975.
35. Norrie, D. H., and G. de Vries, *The Finite Element Method*, Academic, New York, 1973.
36. Robinson, J., *Integrated Theory of Finite Element Methods*, Wiley, London, 1973.
37. Huebner, K. H., *The Finite Element Method for Engineers*, Wiley, New York, 1975.
38. Bathe, K.-J., and E. L. Wilson, *Numerical Methods in Finite Element Analysis*, Prentice-Hall, Englewood Cliffs, N.J., 1976.
39. Segerlind, L. J., *Applied Finite Element Analysis*, Wiley, New York, 1976.
40. Tong, P., and J. N. Rossettos, *Finite-Element Method: Basic Technique and Implementation*, MIT Press, Cambridge, Mass., 1977.

41. Norris, D. H., and G. de Vries, *An Introduction to Finite Element Analysis*, Academic, New York, 1978.

42. Hinton, E., and D. R. J. Owen, *An Introduction to Finite Element Computations*, Pineridge Press Limited, Swansea, UK, 1979.

43. Cheung, Y. K., and M. F. Yeo, *A Practical Introduction to Finite Element Analysis*, Pitman, London, 1979.

44. Desai, C. S., *Elementary Finite Element Method*, Prentice-Hall, Englewood Cliffs, N.J., 1979.

45. Becker, E. B., G. F. Carey, and J. T. Oden, *Finite Elements: An Introduction*, vol. 1, Prentice-Hall, Englewood Cliffs, N.J., 1981.

46. Bathe, K.-J., *Finite Element Procedures in Engineering Analysis*, Prentice-Hall, Englewood Cliffs, N.J., 1982.

2

Mathematical Preliminaries

2-1 INTRODUCTION

Before actually developing the truss model in Chapter 3, a review of the necessary mathematics is given here. In all practical finite element analyses, matrix and vector notation must be used because of the large number of variables and computations involved.

This mathematical review includes the following topics relating to matrices and vectors: the definitions of matrices and vectors; matrix and vector equalities; matrix and vector algebra; transpose, determinant, cofactor, minor, adjoint, and inverse of a matrix; matrix partitioning; length of a vector; direction cosines; and solutions to linear systems of equations. Additional review material is presented at appropriate places throughout the text, usually immediately preceding the application of the topic. The matrices and vectors we discuss are always assumed to be composed of only real elements unless explicitly stated otherwise.

2-2 DEFINITION OF A MATRIX

A *matrix* is a rectangular or square array of elements or numbers. The use of the word "element" should not be confused with its use in the phrase "finite element." The context usually makes the usage clear. An *element* in a matrix is simply an entry in the matrix as explained below. The notion of a matrix is more easily understood by means of a linear system of equations. For example, the system of equations

$$a_{11}x_1 + a_{12}x_2 + a_{13}x_3 = b_1$$
$$a_{21}x_1 + a_{22}x_2 + a_{23}x_3 = b_2$$

(2-1)

could be written in the compact matrix form

$$\mathbf{Ax} = \mathbf{b}$$

(2-2)

where by definition

$$\mathbf{A} = \begin{bmatrix} a_{11} & a_{12} & a_{13} \\ a_{21} & a_{22} & a_{23} \end{bmatrix} \qquad \mathbf{x} = \begin{bmatrix} x_1 \\ x_2 \\ x_3 \end{bmatrix} \qquad \mathbf{b} = \begin{bmatrix} b_1 \\ b_2 \end{bmatrix}$$

The array of elements in the matrix $\mathbf{A}$ is a matrix of coefficients. The size (or dimension) of $\mathbf{A}$ is said to be 2×3 (or 2 by 3). Note that first the number of rows is given, then the number of columns is given. The matrix $\mathbf{x}$ contains three rows and one column, and the matrix $\mathbf{b}$ two rows and one column. Hence, the matrices $\mathbf{x}$ and $\mathbf{b}$ are said to be of size 3×1 and 2×1, respectively.

A *column vector* is defined as a matrix with only one column. Therefore, $\mathbf{x}$ and $\mathbf{b}$ are also referred to as column vectors. A *row vector* is a matrix with only one row. An example of a row vector is given by the 1×4 matrix

$$[1 \quad 3 \quad -2 \quad 0]$$

These notions are readily generalized. An $m \times n$ matrix is a rectangular array of elements with m rows and n columns. If m is equal to n, the matrix is said to be square. Quite frequently, each entry in a matrix may be an algebraic expression, a derivative, or possibly an integral; it need not always be a pure number. A column vector is an $n \times 1$ array of elements, whereas a row vector is a $1 \times n$ array. After the transpose of a matrix is introduced, the row vector will always be written in terms of its column vector representation (see Sec. 2-5).

An element or entry in the ith row and jth column of an $m \times n$ matrix $\mathbf{A}$ is written a_{ij}. The first subscript denotes the ith row and the second subscript the jth column, where $1 \leq i \leq m$ and $1 \leq j \leq n$. The ith entry in an $n \times 1$ column vector $\mathbf{b}$ may be written as b_i, where $1 \leq i \leq n$. Similar remarks hold for a $1 \times n$ row vector. In general, matrices (where neither m nor n is unity) are denoted with uppercase English and Greek boldface letters. Column (and row) vectors are denoted with lowercase English and Greek boldface letters.

Two special types of square matrices are the identity matrix and the null matrix. The *identity matrix* is denoted as $\mathbf{I}$ or $\mathbf{I}_{n \times n}$, and has unity on its main diagonal and zero elsewhere. For example, the 3×3 identity matrix is given by

$$\mathbf{I}_{3 \times 3} = \begin{bmatrix} 1 & 0 & 0 \\ 0 & 1 & 0 \\ 0 & 0 & 1 \end{bmatrix}$$

The main diagonal always runs from the upper left to the lower right. The *null matrix* is denoted as **0** or $\mathbf{0}_{n \times n}$ and, as its name suggests, has only zero elements or entries. The 2×2 null matrix is given by

$$\mathbf{0}_{2 \times 2} = \begin{bmatrix} 0 & 0 \\ 0 & 0 \end{bmatrix}$$

The equation $\mathbf{Ax} = \mathbf{b}$ contains the notions of matrix equality and matrix multiplication. Matrix equality is very easy to define and understand and is reviewed next.

2-3   EQUALITY OF TWO MATRICES

Two $m \times n$ matrices **A** and **B** are said to be equal if and only if the corresponding elements of **A** and **B** are equal. Note that **A** and **B** must be of the same size. Stated differently, two $m \times n$ matrices **A** and **B** are equal if and only if $a_{ij} = b_{ij}$ for $1 \leq i \leq m$ and $1 \leq j \leq n$.

Example 2-1

Consider the four matrices **A**, **B**, **C**, and **D** given below and determine which of the various combinations are equal (taking two at a time):

$$\mathbf{A} = \begin{bmatrix} 3 & 1 & 6 \\ 2 & -2 & 8 \\ 1 & 0 & 7 \end{bmatrix} \qquad \mathbf{B} = \begin{bmatrix} 3 & 1 \\ 2 & -2 \\ 1 & 0 \end{bmatrix}$$

$$\mathbf{C} = \begin{bmatrix} 3 & 1 \\ 2 & -2 \\ 1 & 0 \end{bmatrix} \qquad \mathbf{D} = \begin{bmatrix} 3 & 1 \\ 8 & -2 \\ 1 & 0 \end{bmatrix}$$

Solution

Only **B**, **C**, and **D** can possibly be equal because they are at least of the same size. Each element in **B** is equal to the corresponding element in **C**; hence $\mathbf{B} = \mathbf{C}$. However, $\mathbf{B} \neq \mathbf{D}$ because $b_{21} = 2$ and $d_{21} = 8$. Since these two elements are not equal, the corresponding matrices **B** and **D** cannot be equal. Similarly, $\mathbf{C} \neq \mathbf{D}$. Therefore, only **B** and **C** are equal.   ∎

2-4   MATRIX ALGEBRA

In this section the notions of matrix addition and subtraction are discussed. This is followed by a review of the two types of matrix multiplication and finally a summary of the commutative, distributive, and associative laws for matrices.

Matrix Addition and Subtraction

Matrix addition (and subtraction) is performed by adding (or subtracting) the corresponding elements of two $m \times n$ matrices to form a third $m \times n$ matrix. Let $\mathbf{A}$ and $\mathbf{B}$ be two $m \times n$ matrices and let the $m \times n$ matrices $\mathbf{C}$ and $\mathbf{D}$ be defined by

$$\mathbf{C} = \mathbf{A} + \mathbf{B} \qquad \text{and} \qquad \mathbf{D} = \mathbf{A} - \mathbf{B}$$

The matrices $\mathbf{A}$ and $\mathbf{B}$ are said to be *conformable* for addition (and subtraction) because they both have the same number of rows and columns. From the above definitions, these may be written on an element basis as

and
$$\left. \begin{array}{l} c_{ij} = a_{ij} + b_{ij} \\[2mm] d_{ij} = a_{ij} - b_{ij} \end{array} \right\} \qquad \text{for} \quad 1 \leqslant i \leqslant m \quad \text{and} \quad 1 \leqslant j \leqslant n$$

Example 2-2

For $\mathbf{A}$ and $\mathbf{B}$ given by

$$\mathbf{A} = \begin{bmatrix} 6 & -1 & 2 \\ 0 & 3 & 5 \end{bmatrix} \qquad \text{and} \qquad \mathbf{B} = \begin{bmatrix} 2 & 0 & 5 \\ -8 & 3 & 4 \end{bmatrix}$$

determine $\mathbf{A} + \mathbf{B}$ and $\mathbf{A} - \mathbf{B}$.

Solution

First, both $\mathbf{A}$ and $\mathbf{B}$ are of the same size so the matrices may be added and subtracted. The results are

$$\mathbf{A} + \mathbf{B} = \begin{bmatrix} 6 + 2 & -1 + 0 & 2 + 5 \\ 0 + (-8) & 3 + 3 & 5 + 4 \end{bmatrix} = \begin{bmatrix} 8 & -1 & 7 \\ -8 & 6 & 9 \end{bmatrix}$$

and

$$\mathbf{A} - \mathbf{B} = \begin{bmatrix} 6 - 2 & -1 - 0 & 2 - 5 \\ 0 - (-8) & 3 - 3 & 5 - 4 \end{bmatrix} = \begin{bmatrix} 4 & -1 & -3 \\ 8 & 0 & 1 \end{bmatrix} \qquad \blacksquare$$

Matrix Multiplication

Two types of multiplication may be defined: (1) multiplication of a matrix by a scalar and (2) multiplication of two (or more) matrices. The former is simpler and is defined first.

Multiplication of a Matrix by a Scalar

Let s be any real scalar and $\mathbf{A}$ any $m \times n$ matrix. The product of s and $\mathbf{A}$, written $s\mathbf{A}$, is obtained by multiplying every element of $\mathbf{A}$ by s. In other words, the elements of $s\mathbf{A}$ are sa_{ij}. Note that the product is of size $m \times n$ (such as $\mathbf{A}$ itself). This type of matrix multiplication is commutative; that is, $s\mathbf{A} = \mathbf{A}s$.

Example 2-3

For the matrix $\mathbf{A}$ given in Example 2-2, determine the matrix $\mathbf{C}$ defined by $6\mathbf{A}$.

Solution

$$6\mathbf{A} = 6\begin{bmatrix} 6 & -1 & 2 \\ 0 & 3 & 5 \end{bmatrix} = \begin{bmatrix} 36 & -6 & 12 \\ 0 & 18 & 30 \end{bmatrix}$$ ■

Multiplication of Two Matrices

The product of an $m \times p$ matrix $\mathbf{A}$ and a $q \times n$ matrix $\mathbf{B}$, written $\mathbf{AB}$, is only defined when p equals q. Furthermore, the matrices $\mathbf{A}$ and $\mathbf{B}$ are said to be conformable for multiplication when p equals q. The matrix $\mathbf{A}$ is said to premultiply $\mathbf{B}$, and the matrix $\mathbf{B}$ to postmultiply $\mathbf{A}$.

Let an $m \times r$ matrix $\mathbf{A}$ premultiply an $r \times n$ matrix $\mathbf{B}$ to form a third ($m \times n$) matrix $\mathbf{C}$. The elements of $\mathbf{C}$, or the c_{ij}'s, are given by

$$c_{ij} = \sum_{k=1}^{r} a_{ik} b_{kj} \tag{2-3}$$

for $1 \le i \le m$ and $1 \le j \le n$. It can be shown that this type of matrix multiplication is not commutative in general; that is, $\mathbf{AB} \ne \mathbf{BA}$. In fact, the product $\mathbf{BA}$ may not even be defined if $\mathbf{B}$ and $\mathbf{A}$ are not conformable for multiplication (even though $\mathbf{A}$ and $\mathbf{B}$, in that order, are conformable). Another way of checking whether or not two matrices $\mathbf{A}$ and $\mathbf{B}$ are conformable for multiplication is to check if the number of columns of $\mathbf{A}$ is equal to the number of rows of $\mathbf{B}$; if they are, $\mathbf{AB}$ is defined. A similar test may be stated for the product $\mathbf{BA}$.

Example 2-4

Given the two matrices $\mathbf{A}$ and $\mathbf{B}$, determine if $\mathbf{AB}$ and $\mathbf{BA}$ are defined; if they are, compute them. The matrices $\mathbf{A}$ and $\mathbf{B}$ are

$$\mathbf{A} = \begin{bmatrix} 2 & 0 \\ 6 & 4 \\ -5 & 3 \end{bmatrix} \quad \text{and} \quad \mathbf{B} = \begin{bmatrix} -4 & 7 \\ 5 & 0 \end{bmatrix}$$

Solution

$\mathbf{A}$ is of size 3×2 and $\mathbf{B}$ is 2×2. Therefore, $\mathbf{AB}$ is defined and is of size 3×2 by 2×2 or 3×2. Using Eq. (2-3), we get

$$\mathbf{AB} = \begin{bmatrix} (2)(-4) + (0)(5) & (2)(7) + (0)(0) \\ (6)(-4) + (4)(5) & (6)(7) + (4)(0) \\ (-5)(-4) + (3)(5) & (-5)(7) + (3)(0) \end{bmatrix} = \begin{bmatrix} -8 & 14 \\ -4 & 42 \\ 35 & -35 \end{bmatrix}$$

On the other hand, **BA** is not defined because the number of columns of **B** (in this case two) is not equal to the number of rows of **A** (in this case three). The matrices **A** and **B** are conformable for multiplication, but **B** and **A** (in this order) are not. ∎

The Commutative, Distributive, and Associative Properties

Let **A**, **B**, and **C** be matrices and let s be a scalar. In the laws summarized below, it is assumed that the matrices are conformable for addition and multiplication as the situation warrants.

Commutative laws:

$$\mathbf{A} + \mathbf{B} = \mathbf{B} + \mathbf{A} \tag{2-4}$$

$$s\mathbf{A} = \mathbf{A}s \tag{2-5}$$

Distributive laws:

$$s(\mathbf{A} + \mathbf{B}) = s\mathbf{A} + s\mathbf{B} \tag{2-6}$$

$$\mathbf{A}(\mathbf{B} + \mathbf{C}) = \mathbf{A}\mathbf{B} + \mathbf{A}\mathbf{C} \tag{2-7}$$

$$(\mathbf{A} + \mathbf{B})\mathbf{C} = \mathbf{A}\mathbf{C} + \mathbf{B}\mathbf{C} \tag{2-8}$$

Associative laws:

$$\mathbf{A} - \mathbf{B} = \mathbf{A} + (-\mathbf{B}) \tag{2-9}$$

$$(\mathbf{A} + \mathbf{B}) + \mathbf{C} = \mathbf{A} + (\mathbf{B} + \mathbf{C}) \tag{2-10}$$

$$(\mathbf{A}\mathbf{B})\mathbf{C} = \mathbf{A}(\mathbf{B}\mathbf{C}) \tag{2-11}$$

The proofs of these laws are relatively straightforward and are left to the reader as exercises (see Problems 2-18 to 2-20). It should be remembered that the product of two matrices is not commutative, that is, $\mathbf{AB} \neq \mathbf{BA}$.

2-5 TRANSPOSE OF A MATRIX

The transpose of an $m \times n$ matrix **A** with elements a_{ij} is an $n \times m$ matrix denoted as $\mathbf{A}^T$ (and stated **A** transpose) whose elements are given by a_{ji}. The switching of the subscripts should be noted. The transpose of a product of two or more matrices occurs often and an important property is given by

$$(\mathbf{AB})^T = \mathbf{B}^T\mathbf{A}^T \tag{2-12}$$

$$(\mathbf{ABC})^T = \mathbf{C}^T\mathbf{B}^T\mathbf{A}^T \tag{2-13}$$

and so forth. If $\mathbf{A} = \mathbf{A}^T$ (or $a_{ij} = a_{ji}$), the matrix **A** is said to be *symmetric*. If $\mathbf{A} = -\mathbf{A}^T$ (or $a_{ij} = -a_{ji}$), **A** is *skew-symmetric*. A skew-symmetric matrix must have zeros on its main diagonal. Symmetric and skew-symmetric matrices must be square. The transpose of a column vector is a row vector, and vice versa. From this point on a lower case boldface letter will always represent a column vector. The transpose will always be used to indicate a row vector (see Example 2-10).

Example 2-5

Determine the transpose of

$$\mathbf{A} = \begin{bmatrix} 2 & -1 \\ 1 & 9 \\ 6 & 5 \end{bmatrix}$$

Solution

Since $\mathbf{A}$ is 3×2, the transpose is 2×3 and is given by

$$\mathbf{A}^T = \begin{bmatrix} 2 & 1 & 6 \\ -1 & 9 & 5 \end{bmatrix}$$

■

Example 2-6

Determine if the matrices $\mathbf{A}$, $\mathbf{B}$, $\mathbf{C}$ defined below are symmetric, skew-symmetric, or neither:

$$\mathbf{A} = \begin{bmatrix} 3 & -1 & 6 \\ -1 & 2 & 0 \\ 6 & 0 & 5 \end{bmatrix} \qquad \mathbf{B} = \begin{bmatrix} 0 & -1 & 3 & 4 \\ 1 & 0 & -6 & -7 \\ -2 & 6 & 5 & 5 \\ 4 & 7 & -5 & 0 \end{bmatrix}$$

$$\mathbf{C} = \begin{bmatrix} 0 & 1 & -6 \\ -1 & 0 & -3 \\ 6 & 3 & 0 \end{bmatrix}$$

Solution

Since $a_{ij} = a_{ji}$ for $i = 1$ to 3 and $j = 1$ to 3, the matrix $\mathbf{A}$ is symmetric. Since $b_{31} \neq b_{13}$ and $b_{31} \neq -b_{13}$, the matrix $\mathbf{B}$ is neither symmetric nor skew-symmetric. Finally, since $c_{ij} = -c_{ji}$ for $i = 1$ to 3 and $j = 1$ to 3, the matrix $\mathbf{C}$ is skew-symmetric. Symmetric matrices arise quite frequently in finite element analysis.

2-6 DETERMINANTS, MINORS, AND COFACTORS [1]

Consider an $n \times n$ matrix $\mathbf{A}$. The determinant of $\mathbf{A}$, written det $\mathbf{A}$, is a scalar function of $\mathbf{A}$ which for $n = 1$, 2, and 3 is given by

$$\det \mathbf{A}_{1 \times 1} = a_{11} \tag{2-14a}$$

$$\det \mathbf{A}_{2 \times 2} = a_{11}a_{22} - a_{21}a_{12} \tag{2-14b}$$

$$\det \mathbf{A}_{3 \times 3} = a_{11}a_{22}a_{33} - a_{11}a_{23}a_{32} + a_{12}a_{23}a_{31}$$

$$- a_{21}a_{12}a_{33} + a_{13}a_{32}a_{21} - a_{31}a_{22}a_{13} \tag{2-14c}$$

Although this pattern can be generalized, it is not a very practical way to evaluate determinants for n larger than 3. Fortunately, another method exists called the Laplace expansion method. This method is most easily explained after the minor and cofactor of a matrix are defined.

The *minor M_{ij}* of a square matrix **A** is the determinant of the $n - 1 \times n - 1$ matrix derived from **A** by eliminating the ith row and the jth column of **A**. Such as the determinant itself, the minor is also scalar-valued and is only defined for square matrices. An $n \times n$ matrix **A** contains n^2 minors.

The *cofactor C_{ij}* of a matrix **A** is easily defined in terms of the minor M_{ij} by

$$C_{ij} = (-1)^{i+j} M_{ij} \qquad \text{(2-15)}$$

Note that when $i + j$ is even, the cofactor and minor are identical; when $i + j$ is odd, the cofactor and minor differ only in the signs. An $n \times n$ matrix **A** has n^2 cofactors.

The determinant of the original $n \times n$ matrix **A** may now be computed utilizing the notion of the cofactor by selecting any arbitrary column j and using

$$\det \mathbf{A} = \sum_{k=1}^{n} a_{kj} C_{kj} \qquad \text{(2-16)}$$

An equivalent expression for evaluating the determinant is given by

$$\det \mathbf{A} = \sum_{m=1}^{n} a_{im} C_{im} \qquad \text{(2-17)}$$

where row i is selected for the expansion. Although these expressions look formidable, they are not difficult to apply in practice, as shown in Example 2-7. Equations (2-16) and (2-17) are used repeatedly until determinants of matrices of size 2×2 need to be evaluated, at which point Eq. (2-14b) is used (see Example 2-7). This method of evaluating the determinant of a matrix is known as the *Laplace expansion method*.

Example 2-7

Using the Laplace expansion method and expanding with respect to the second row, find the determinant of **A** where

$$\mathbf{A} = \begin{bmatrix} 2 & 4 & -6 \\ 3 & 1 & -5 \\ -1 & 2 & -3 \end{bmatrix}$$

Solution

The three minors associated with the second row are

$$M_{21} = \det \begin{bmatrix} 4 & -6 \\ 2 & -3 \end{bmatrix} \qquad M_{22} = \det \begin{bmatrix} 2 & -6 \\ -1 & -3 \end{bmatrix} \qquad M_{23} = \det \begin{bmatrix} 2 & 4 \\ -1 & 2 \end{bmatrix}$$

Each of these 2×2 determinants may be evaluated with the help of Eq. (2-14b) to give

$$M_{21} = 0 \qquad M_{22} = -12 \qquad M_{23} = 8$$

The corresponding cofactors are

$$C_{21} = (-1)^{2+1}(0) = 0 \qquad C_{22} = (-1)^{2+2}(-12) = -12$$

and

$$C_{23} = (-1)^{2+3}(8) = -8$$

Therefore, the determinant of $\mathbf{A}$ is given by

$$\det \mathbf{A} = (3)(0) + (1)(-12) + (-5)(-8) = 28$$

Try to evaluate the determinant of $\mathbf{A}$ by expansion with respect to another row or column. ∎

2-7 ADJOINT AND INVERSE OF A MATRIX [1]

The *adjoint* (adj) of an $n \times n$ matrix $\mathbf{A}$, denoted as adj($\mathbf{A}$), is easily defined in terms of the transpose of the cofactor matrix $\mathbf{C}$. Stated mathematically,

$$\text{adj}(\mathbf{A}) = \mathbf{C}^{T} \qquad \text{(2-18)}$$

The adjoint itself is another $n \times n$ matrix and is particularly useful in providing a concise expression for the inverse of a matrix as described below.

The *inverse* of an $n \times n$ matrix $\mathbf{A}$, denoted as $\mathbf{A}^{-1}$ (and stated A inverse), is another $n \times n$ matrix such that

$$\mathbf{A}\mathbf{A}^{-1} = \mathbf{A}^{-1}\mathbf{A} = \mathbf{I}$$

where $\mathbf{I}$ is the $n \times n$ identity matrix. In other words, the inverse of an $n \times n$ matrix $\mathbf{A}$ is another $n \times n$ matrix such that when it is both premultiplied and postmultiplied by $\mathbf{A}$ itself, the result is the $n \times n$ identity matrix. In terms of the adjoint, the inverse of $\mathbf{A}$ is given by

$$\mathbf{A}^{-1} = \frac{\text{adj}(\mathbf{A})}{\det \mathbf{A}} \qquad \text{(2-19)}$$

If the determinant of $\mathbf{A}$ is zero, then $\mathbf{A}^{-1}$ does not exist and $\mathbf{A}$ is said to be *singular*. When the determinant of $\mathbf{A}$ is nonzero, the inverse of $\mathbf{A}$ is unique and $\mathbf{A}$ is said to be *nonsingular*. The inverse is only defined for square matrices. When Eq. (2-19) is applied to a nonsingular 2×2 matrix $\mathbf{A}$, the inverse is given by

$$\mathbf{A}^{-1} = \frac{1}{\det \mathbf{A}} \begin{bmatrix} a_{22} & -a_{12} \\ -a_{21} & a_{11} \end{bmatrix} \qquad \text{(2-20)}$$

where det **A** is given by Eq. (2-14b). A useful identity is given by

$$(\mathbf{AB})^{-1} = \mathbf{B}^{-1}\mathbf{A}^{-1} \tag{2-21}$$

providing **A** and **B** are both square, of the same size, and nonsingular. Similarly, if in addition **C** is also square, of the same size, and nonsingular, then

$$(\mathbf{ABC})^{-1} = \mathbf{C}^{-1}\mathbf{B}^{-1}\mathbf{A}^{-1} \tag{2-22}$$

The identities given by Eqs. (2-21) and (2-22) may be readily extended to products involving four or more matrices. If $\mathbf{A}^{-1} = \mathbf{A}^T$, then the matrix **A** is said to be *orthogonal*. Orthogonal matrices will arise quite naturally in Chapter 3.

Actually, Eq. (2-19) is not a very practical way to evaluate an inverse. A more practical method is based on Gauss-Jordan elimination (see Chapter 6). In addition, computer programs or subroutines abound that may be used effectively to compute inverses of rather large matrices. One such subroutine is given in Appendix B (subroutine INVDET) [2].

Example 2-8

Evaluate the inverse of **A** where

$$\mathbf{A} = \begin{bmatrix} 2 & 8 \\ -3 & 4 \end{bmatrix}$$

Solution

By Eq. (2-20), $\mathbf{A}^{-1}$ is computed as follows:

$$\mathbf{A}^{-1} = \frac{1}{(2)(4) - (-3)(8)} \begin{bmatrix} 4 & -8 \\ 3 & 2 \end{bmatrix} = \begin{bmatrix} \frac{1}{8} & -\frac{1}{4} \\ \frac{1}{32} & \frac{1}{16} \end{bmatrix} \qquad \blacksquare$$

Example 2-9

Compute the inverse of **A** which is given by

$$\mathbf{A} = \begin{bmatrix} 2 & 8 \\ 1 & 4 \end{bmatrix}$$

Solution

Since det $\mathbf{A} = 0$, $\mathbf{A}^{-1}$ does not exist. The matrix **A** is singular. $\qquad \blacksquare$

2-8   MATRIX PARTITIONING

It is always possible to subdivide or partition matrices into a number of smaller submatrices. If the submatrices are conformable, they may be treated as though they are scalar elements when adding, subtracting, or multiplying two matrices.

The order of the products is not arbitrary, however, as it is if the elements are truly scalars. For example, let us compute the product of **A** and **B** to form **C** in two different ways:

$$C = AB = [A_1 \mid A_2]\begin{bmatrix} B_1 \\ \hline B_2 \end{bmatrix} = A_1B_1 + A_2B_2$$

or

$$C = \begin{bmatrix} A_{11} & A_{12} \\ \hline A_{21} & A_{22} \end{bmatrix}\begin{bmatrix} B_1 \\ \hline B_2 \end{bmatrix} = \begin{bmatrix} A_{11}B_1 + A_{12}B_2 \\ \hline A_{21}B_1 + A_{22}B_2 \end{bmatrix} = \begin{bmatrix} C_1 \\ \hline C_2 \end{bmatrix}$$

The subdivisions of **A** and **B** must be chosen in order to ensure that the products such as A_1B_1 and $A_{11}B_1$ are meaningful (i.e., the submatrices must be conformable). Stated differently, the number of columns in A_1 must equal the number of rows in B_1, etc. Matrix partitioning is not introduced here to provide an alternate means of multiplying two matrices. In Chapter 3, matrix partitioning arises quite naturally when the element stiffness matrix is developed for a two-dimensional truss.

2-9 DEFINITION OF A VECTOR

Let us initially restrict our definition of a vector to that in a three-dimensional space in which a vector is a directed line segment with a unique length (magnitude) and direction as shown in Fig. 2-1. It is convenient to resolve a vector into its three components along three mutually perpendicular axes. These axes establish a co-ordinate system, and since they are mutually perpendicular they are said to be orthogonal. For example, a vector **v** may be written with respect to a cartesian reference frame in terms of its three components v_x, v_y, v_z as

$$v = v_x i + v_y j + v_z k$$

where **i**, **j**, and **k** are the three unit vectors along the x, y, and z axes, respectively. A unit vector has a length of unity (see Sec. 2-11) and is directed along its associated coordinate axis. Figure 2-1 shows a cartesian reference frame and the three unit vectors.

A vector is often written in column matrix form as mentioned in Sec. 2-2. For example, the vector **v** may be written as

$$v = \begin{bmatrix} v_x \\ v_y \\ v_z \end{bmatrix} = \begin{bmatrix} v_1 \\ v_2 \\ v_3 \end{bmatrix}$$

Often numerical subscripts, as shown above, prove to be more convenient than letter subscripts. In this text, a matrix represented by boldfaced lowercase letters (e.g., **v**) will denote a column matrix (or simply a vector). In order to indicate a row matrix, the transpose will be used. In other words, v^T represents a row matrix, whereas **v** represents a column matrix (or a vector). There are two reasons for this

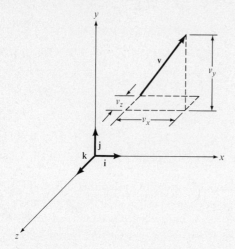

Figure 2-1   Vector **v** in a cartesian reference frame.

convention: the first is that it removes any ambiguity in matrix expressions developed later, and the second is that it conserves space in writing the vector out, as the next example illustrates. It is important to recognize that all vectors may be written in matrix form (as a column matrix), but not all matrices have a vector meaning.

Example 2-10

Write **v** in matrix form where

$$\mathbf{v} = 1\mathbf{i} - 4\mathbf{j} + 3\mathbf{k}$$

Solution

In column matrix or column vector form,

$$\mathbf{v} = \begin{bmatrix} 1 \\ -4 \\ 3 \end{bmatrix}$$

which in turn may be written as

$$\mathbf{v}^T = \begin{bmatrix} 1 & -4 & 3 \end{bmatrix}$$

or as

$$\mathbf{v} = \begin{bmatrix} 1 & -4 & 3 \end{bmatrix}^T$$

Clearly these all convey the same information, but the second two forms require fewer lines.   ∎

Although the notion of a vector was introduced from a geometrical point of view, vectors need not have only three components. In later chapers, vectors with four or more components will naturally arise.

2-10 EQUALITY OF TWO VECTORS

Two vectors **u** and **v** are equal if and only if their respective components are equal and if they are both written with respect to the same coordinate system. If **u** and **v** are given by

$$\mathbf{u} = u_x\mathbf{i} + u_y\mathbf{j} + u_z\mathbf{k}$$

and

$$\mathbf{v} = v_x\mathbf{i} + v_y\mathbf{j} + v_z\mathbf{k}$$

and if **v** = **u**, then

$$u_x = v_x$$

$$u_y = v_y$$

$$u_z = v_z$$

since both are assumed to be written with respect to the same *xyz* coordinate system.

2-11 LENGTH OF A VECTOR

A vector that is directed from the origin of the coordinate system to another point is said to be a *position vector*. In Fig. 2-2, the position vectors **p** and **q**, which are directed from the origin to the points P and Q, respectively, are shown. If point P has coordinates (p_1, p_2, p_3) and Q has coordinates (q_1, q_2, q_3), then **p** and **q** are given by

$$\mathbf{p} = p_1\mathbf{i} + p_2\mathbf{j} + p_3\mathbf{k}$$

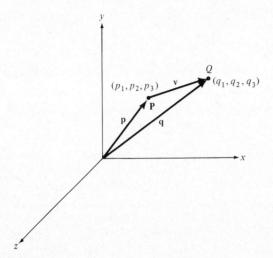

Figure 2-2 Vector **v** defined in terms of the position vectors **p** and **q**.

and

$$\mathbf{q} = q_1\mathbf{i} + q_2\mathbf{j} + q_3\mathbf{k}$$

From Fig. 2-2, it is noted that $\mathbf{p} + \mathbf{v} = \mathbf{q}$ (see Sec. 2-12) and so

$$\mathbf{v} = \mathbf{q} - \mathbf{p} = (q_1 - p_1)\mathbf{i} + (q_2 - p_2)\mathbf{j} + (q_3 - p_3)\mathbf{k}$$

Based on geometrical considerations, the length of the vector $\mathbf{v}$, denoted as $|\mathbf{v}|$, is given by

$$|\mathbf{v}| = |\mathbf{q} - \mathbf{p}| = \sqrt{(q_1 - p_1)^2 + (q_2 - p_2)^2 + (q_3 - p_3)^2} \qquad \textbf{(2-23)}$$

Although Eq. (2-23) is based on geometrical considerations in three dimensions, the notion of vector length may be readily extended to multidimensional space (see Problems 2-41 and 2-42).

Example 2-11

Compute the length of the vector that runs from point P at $(3, -4, 6)$ to point Q at $(2, 5, -7)$.

Solution

The vector $\mathbf{v}$ is easily determined as

$$\mathbf{v} = \mathbf{q} - \mathbf{p} = [2 - 3]\mathbf{i} + [5 - (-4)]\mathbf{j} + [-7 - 6]\mathbf{k}$$

or

$$\mathbf{v} = -1\mathbf{i} + 9\mathbf{j} - 13\mathbf{k}$$

The length of $\mathbf{v}$ is now calculated from Eq. (2-23) as

$$|\mathbf{v}| = \sqrt{(-1)^2 + (9)^2 + (-13)^2} = 15.8 \qquad \blacksquare$$

2-12 VECTOR ALGEBRA

Vector algebra includes the notions of vector addition and subtraction, as well as two types of vector multiplication. In this chapter only one type of vector multiplication, the dot product, is needed and thus reviewed. The other type is reviewed in later chapters.

Two vectors are simply added component by component to obtain the vector sum, provided the vectors are written with respect to the same coordinate system. The rules for vector addition and subtraction are identical to those presented in Sec. 2-4 for matrix addition and subtraction. Addition of two vectors is frequently given a geometrical interpretation by using the parallelogram law. However, simple addition of the respective components is generally more useful in this text.

The dot product of two vectors $\mathbf{p}$ and $\mathbf{q}$ is denoted as $\mathbf{p} \cdot \mathbf{q}$ and is a scalar defined by

$$\mathbf{p \cdot q} = |\mathbf{p}||\mathbf{q}| \cos \alpha \qquad \textbf{(2-24)}$$

where α is the smaller angle included between $\mathbf{p}$ and $\mathbf{q}$. Equation (2-24) implies

$$\mathbf{i \cdot i} = \mathbf{j \cdot j} = \mathbf{k \cdot k} = 1 \qquad \textbf{(2-25)}$$

and

$$\mathbf{i \cdot j} = \mathbf{j \cdot k} = \mathbf{k \cdot i} = 0 \qquad \textbf{(2-26)}$$

With the help of Eqs. (2-25) and (2-26), $\mathbf{p \cdot q}$ may be given in terms of the cartesian components as

$$\mathbf{p \cdot q} = p_1 q_1 + p_2 q_2 + p_3 q_3 \qquad \textbf{(2-27)}$$

which in turn can be written as

$$\mathbf{p \cdot q} = [p_1 \quad p_2 \quad p_3] \begin{bmatrix} q_1 \\ q_2 \\ q_3 \end{bmatrix} = \mathbf{p}^T \mathbf{q} \qquad \textbf{(2-28)}$$

Therefore, the dot product of $\mathbf{p}$ and $\mathbf{q}$ is frequently written in matrix form as $\mathbf{p}^T\mathbf{q}$. The dot product is also frequently referred to as a scalar product and an inner product.

Example 2-12

Given the vectors $\mathbf{p}$ and $\mathbf{q}$ from Example 2-11, determine $\mathbf{p} + \mathbf{q}$, $\mathbf{p} - \mathbf{q}$, and $\mathbf{p \cdot q}$.

Solution

From Example 2-11, we have

$$\mathbf{p} = 3\mathbf{i} - 4\mathbf{j} + 6\mathbf{k}$$

and

$$\mathbf{q} = 2\mathbf{i} + 5\mathbf{j} - 7\mathbf{k}$$

Therefore, the required results are

$$\mathbf{p} + \mathbf{q} = 5\mathbf{i} + 1\mathbf{j} - 1\mathbf{k}$$
$$\mathbf{p} - \mathbf{q} = 1\mathbf{i} - 9\mathbf{j} + 13\mathbf{k}$$

and

$$\mathbf{p \cdot q} = (3)(2) + (-4)(5) + (6)(-7) = -56 \qquad \blacksquare$$

2-13 DIRECTION COSINES

The direction cosines of a vector $\mathbf{v}$ are defined as the cosines of the angles formed by the vector $\mathbf{v}$ and the three coordinate axes. Figure 2-3 shows a vector $\mathbf{v}$ in two

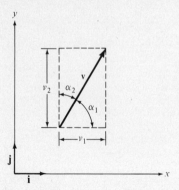

Figure 2-3 The cosines of the angles α_1 and α_2 are the direction cosines.

dimensions. The angles that **v** makes with the x and y axes are α_1 and α_2, respectively. From Fig. 2-3 and the notion of the dot product, it follows that

$$\mathbf{v} \cdot \mathbf{i} = |\mathbf{v}||\mathbf{i}| \cos \alpha_1$$

But $\mathbf{v} \cdot \mathbf{i} = v_1$ and $|\mathbf{i}| = 1$, and solving for $\cos \alpha_1$ gives

$$\cos \alpha_1 = \frac{v_1}{|\mathbf{v}|} \qquad \text{(2-29a)}$$

In a similar fashion, it is readily shown that

$$\cos \alpha_2 = \frac{v_2}{|\mathbf{v}|} \qquad \text{(2-29b)}$$

In three dimensions, if α_3 is the angle between a vector **v** and the z axis, then in addition to the above, we have

$$\cos \alpha_3 = \frac{v_3}{|\mathbf{v}|} \qquad \text{(2-29c)}$$

In general, the direction cosines are not all independent and are related by the following:

$$\cos^2 \alpha_1 + \cos^2 \alpha_2 + \cos^2 \alpha_3 = 1 \qquad \text{(2-30)}$$

The simple proof is left to the reader (see Problem 2-49).

Example 2-13

For the vector **v** in Example 2-11, determine the three direction cosines and verify that Eq. (2-30) is satisfied.

Solution

From Eq. (2-29), we compute

$$\cos \alpha_1 = \frac{-1}{15.8} = -0.0633$$

$$\cos \alpha_2 = \frac{9}{15.8} = +0.570$$

$$\cos \alpha_3 = \frac{-13}{15.8} = -0.823$$

from which we get

$$\alpha_1 = 93.6° \qquad \alpha_2 = 55.3° \qquad \alpha_3 = 145.4°$$

Equation (2-30) is readily verified once α_1, α_2, and α_3 are known.   ■

2-14 SOLUTION TO SYSTEMS OF LINEAR ALGEBRAIC EQUATIONS

In all static stress analysis and other steady-state applications, a system of linear algebraic equations of the form

$$\mathbf{Ka} = \mathbf{f} \qquad\qquad (2\text{-}31)$$

will have to be solved for the vector **a**. The system of equations implied in Eq. (2-31) may be solved by many methods, but only the matrix inversion method is reviewed here. Chapter 6 contains a more practical method based on Gaussian elimination.

The solution to Eq. (2-31) by the matrix inversion method is straightforward. If both sides of Eq. (2-31) are premultiplied by the inverse of the $n \times n$ matrix **K** (assuming the inverse exists) and if it is noted that $\mathbf{K}^{-1}\mathbf{K} = \mathbf{I}$ and $\mathbf{Ia} = \mathbf{a}$, then an explicit expression for **a** results and is given by

$$\mathbf{a} = \mathbf{K}^{-1}\mathbf{f} \qquad\qquad (2\text{-}32)$$

Unfortunately this method is not very practical in large problems because of the time consumed (even by a computer) in obtaining the inverse. Nevertheless, it provides a concise expression for the solution to Eq. (2-31). As mentioned in Sec. 2-7, the inverse of a matrix is most readily performed by a subroutine on a digital computer. Such a subroutine (subroutine INVDET) may be found in Appendix B. Equation (2-32) simply states that to obtain **a** (which may be the nodal displacements, temperatures, velocities, pressures, etc.), we simply invert **K** and post-multiply this result by the vector **f** (which is known).

Example 2-14

Determine the solution to the following system of linear algebraic equations:

$$x_1 - 3x_2 \qquad = 5$$
$$2x_1 + 4x_2 - 8x_3 = 22$$
$$-7x_1 + 8x_2 - x_3 = 16$$

Solution

From the matrix representation, we have

$$\mathbf{K} = \begin{bmatrix} 1 & -3 & 0 \\ 2 & 4 & -8 \\ -7 & 8 & -1 \end{bmatrix} \quad \text{and} \quad \mathbf{f} = \begin{bmatrix} 5 \\ 22 \\ 16 \end{bmatrix}$$

from which $\mathbf{K}^{-1}$ is computed (using subroutine INVDET) to be

$$\mathbf{K}^{-1} = \begin{bmatrix} -0.52632 & 0.02632 & -0.21053 \\ -0.50877 & 0.00877 & -0.07018 \\ -0.38596 & -0.11404 & -0.08772 \end{bmatrix}$$

The solution for x_1, x_2, and x_3 is then easily computed from Eq. (2-32) by a simple matrix multiplication to be

$$x_1 = -5.421$$
$$x_2 = -3.474$$
$$x_3 = -5.842$$

The reader should verify these by direct substitution into the original system of equations. ■

It should be mentioned that $\mathbf{K}$ will always have an inverse for well-posed problems solved by the finite element method. Therefore, the questions as to whether the system of equations represented by Eq. (2-31) has one solution, no solution, or several solutions do not arise. For well-posed problems, Eq. (2-31) has only one solution for $\mathbf{a}$ and therefore the solution is unique. One solution method applicable to nonlinear algebraic equations is described in Chapter 8. The reader may wish to consult references 3 to 12 for additional material relating to matrices and vectors.

REFERENCES

1. Brogan, W. L., *Modern Control Theory*, Quantum, New York, 1974, pp. 45–48.
2. Hornbeck, R. W., *Numerical Methods*, Quantum, New York, 1975.
3. Bellman, R. E., *Introduction to Matrix Analysis*, 2nd ed., McGraw-Hill, New York, 1970.
4. Bronson, R., *Matrix Methods—An Introduction*, Academic, New York, 1969.

5. Campbell, H. G., *Matrices with Applications*, Appleton-Century-Crofts, New York, 1968.
6. Franklin, J. H., *Matrix Theory*, Prentice-Hall, Englewood Cliffs, N.J., 1968.
7. Fuller, L. E., *Basic Matrix Theory*, Prentice-Hall, Englewood Cliffs, N.J., 1962.
8. Hohn, F. E., *Elementary Matrix Algebra*, 3rd ed., Macmillan, New York, 1973.
9. Moore, J. T., *Elements of Linear Algebra and Matrix Theory*, McGraw-Hill, New York, 1968.
10. Pipes, L. A., *Matrix Methods for Engineering*, Prentice-Hall, Englewood Cliffs, N.J., 1963.
11. Schneider, H., and G. P. Barker, *Matrices and Linear Algebra*, Holt, Rinehart and Winston, New York, 1968.
12. Stein, F. M., *Introduction to Matrices and Determinants*, Wadsworth, Belmont, Calif., 1967.

PROBLEMS

2-1 Represent the following systems of equations in the matrix form $\mathbf{Ax} = \mathbf{b}$:

a. $2x_1 + 3x_2 = 4$
$-5x_1 + 6x_2 = -6$

b. $-8x_1 + 3x_2 + 6x_3 = -9$
$4x_2 - 5x_3 = 6$
$x_3 = 5$

c. $4x_1 - 5x_2 + 6x_3 = 5$
$2x_1 + 5x_2 - x_3 = 2$

2-2 Express the following systems of equations in matrix form:

a. $3x_1 + 4x_2 + x_3 = 6$
$2x_2 - 7x_3 = -2$
$-8x_1 + 6x_2 + x_3 = 0$

b. $6x_1 + 5x_2 + 3x_3 + x_4 = 10$
$7x_1 + 3x_2 - 6x_4 = -7$

2-3 What is the 6×6 identity matrix and the 4×4 null matrix?

2-4 For the matrices $\mathbf{A}$ and $\mathbf{B}$ given below, determine if $\mathbf{A}$ and $\mathbf{B}$ are equal:

a. $\mathbf{A} = \begin{bmatrix} 2 & -8 \\ 8 & 6 \end{bmatrix}$ $\mathbf{B} = \begin{bmatrix} 2 & 8 \\ -8 & 6 \end{bmatrix}$

b. $\mathbf{A} = \begin{bmatrix} 6 & 1 \\ -2 & 4 \\ -1 & 8 \end{bmatrix}$ $\mathbf{B} = \begin{bmatrix} 6 & 1 \\ -2 & 4 \end{bmatrix}$

c. $\mathbf{A} = \begin{bmatrix} 8 & 0 \\ 0 & 5 \\ -3 & 2 \end{bmatrix}$ $\mathbf{B} = \begin{bmatrix} 8 & 0 \\ 0 & 5 \\ -3 & 2 \end{bmatrix}$

2-5 Which of the matrices given below are equal (taking two at a time)?

$$A = \begin{bmatrix} 6 & -3 \\ 4 & 8 \end{bmatrix} \qquad B = \begin{bmatrix} 6 & -8 & 2 \\ 5 & 6 & -7 \end{bmatrix}$$

$$C = \begin{bmatrix} 6 & -3 \\ 4 & 8 \end{bmatrix} \qquad D = \begin{bmatrix} 6 & -8 & 2 \\ 5 & 6 & -7 \end{bmatrix}$$

$$E = \begin{bmatrix} 6 & 3 \\ 0 & 8 \end{bmatrix} \qquad F = \begin{bmatrix} 2 & 5 \\ -1 & 8 \\ 0 & -5 \end{bmatrix}$$

2-6 For the matrices in Problem 2-5, determine the sums and differences indicated below. If the matrices are not conformable for addition or subtraction, so state this.

a. A + B b. A − C c. A − D d. A + E
e. B + D f. C + E g. D + F

2-7 Repeat Problem 2-6 for the following sums and differences:

a. B + C b. B − C c. B − D d. C − E
e. C + D f. E − E g. F + F

2-8 For the matrices of Problem 2-5, determine

a. 5A b. −7B c. 8C
d. 0D e. −2E f. 5F

2-9 Repeat Problem 2-8 for

a. −7A b. 0B c. −C
d. 8D e. 4E f. 0F

2-10 For the matrices of Problem 2-5, determine the following products. If the matrices are not conformable for multiplication, so state this.

a. AB b. AC c. DA d. DF e. FD

2-11 Repeat Problem 2-10 for the following products:

a. CB b. BC c. CE d. EF e. FE

2-12 What is the size (or dimensions) of the matrices given in Problem 2-5?

2-13 What is the size of each of the products in Problem 2-10?

2-14 What is the size of each of the products in Problem 2-11?

2-15 For the matrices in Problem 2-5, show that

a. A + E = E + A
b. 6D = D6
c. F(A + E) = FA + FE
d. C − E = C + (−E)

2-16 For the matrices of Problem 2-5, show that

  a. $\mathbf{B} + \mathbf{D} = \mathbf{D} + \mathbf{B}$
  b. $8(\mathbf{C} + \mathbf{E}) = 8\mathbf{C} + 8\mathbf{E}$
  c. $(\mathbf{B} + \mathbf{D})\mathbf{F} = \mathbf{BF} + \mathbf{DF}$
  d. $(\mathbf{A} + \mathbf{C}) + \mathbf{E} = \mathbf{A} + (\mathbf{C} + \mathbf{E})$

2-17 For the matrices of Problem 2-5, show that

  a. $5(\mathbf{A} + \mathbf{C}) = 5\mathbf{A} + 5\mathbf{C}$
  b. $\mathbf{BF} \neq \mathbf{FB}$
  c. $(\mathbf{BF})\mathbf{A} = \mathbf{B}(\mathbf{FA})$
  d. $\mathbf{AE} \neq \mathbf{EA}$

2-18 By writing the matrices on an element basis (e.g., $\mathbf{C}$ is denoted as c_{ij}), prove the commutative laws given by Eqs. (2-4) and (2-5).

2-19 Prove the distributive laws given by Eqs. (2-6) to (2-8) by writing each matrix on an element basis.

2-20 Prove the associative laws given by Eqs. (2-9) to (2-11).

2-21 Consider the matrices given below:

$$\mathbf{A} = \begin{bmatrix} 2 & -6 \\ 3 & 5 \end{bmatrix} \qquad \mathbf{B} = \begin{bmatrix} 2 & -7 & 8 \\ -7 & 3 & 5 \\ 8 & 5 & -6 \end{bmatrix}$$

$$\mathbf{C} = \begin{bmatrix} 0 & 4 & -5 \\ -4 & 0 & 3 \\ 5 & -3 & 0 \end{bmatrix} \qquad \mathbf{D} = \begin{bmatrix} 6 & 7 \\ 7 & -2 \end{bmatrix}$$

$$\mathbf{E} = \begin{bmatrix} 5 & 6 & -3 & 1 \\ 6 & 2 & 5 & 4 \\ -3 & 5 & -8 & -2 \\ 1 & 4 & -2 & 0 \end{bmatrix} \qquad \mathbf{F} = \begin{bmatrix} 0 & 5 & -6 \\ -5 & 0 & -3 \\ 6 & 3 & -2 \end{bmatrix}$$

$$\mathbf{G} = \begin{bmatrix} 2 & -3 & 5 \\ 6 & 4 & 0 \end{bmatrix}$$

  a. Which matrices are symmetric? Why?
  b. Which matrices are skew-symmetric? Why?

2-22 For the matrices of Problem 2-21, determine

  a. $\mathbf{A}^T$ **b.** $\mathbf{B}^T$ **c.** $\mathbf{F}^T$

2-23 For the matrices of Problem 2-21, determine

  a. $\mathbf{C}^T$ **b.** $\mathbf{D}^T$ **c.** $\mathbf{E}^T$ **d.** $\mathbf{G}^T$

2-24 For the matrices of Problem 2-21, show that

  a. $(\mathbf{AD})^T = \mathbf{D}^T\mathbf{A}^T$
  b. $(\mathbf{GBF})^T = \mathbf{F}^T\mathbf{B}^T\mathbf{G}^T$
  c. $\mathbf{D}^T = \mathbf{D}$

2-25   For the matrices of Problem 2-21, show that

   a. $(\mathbf{BC})^T = \mathbf{C}^T\mathbf{B}^T$
   b. $\mathbf{C} = -\mathbf{C}^T$
   c. $(\mathbf{ADG})^T = \mathbf{G}^T\mathbf{D}^T\mathbf{A}^T$

2-26   Prove for all symmetric matrices $\mathbf{M}$ that $\mathbf{P}^T\mathbf{MP}$ is also always symmetric, where $\mathbf{P}$ is not necessarily symmetric (but conformable).

2-27   Using Eqs. (2-14) and the matrices from Problem 2-21, determine

   a. det $\mathbf{A}$
   b. det $\mathbf{B}$

2-28   Repeat Problem 2-27 for

   a. det $\mathbf{D}$
   b. det $\mathbf{E}$

2-29   For the matrix $\mathbf{A}$ of Problem 2-21, determine the following:

   a. The matrix formed by the minors of $\mathbf{A}$
   b. The matrix formed by the cofactors of $\mathbf{A}$
   c. The adjoint of $\mathbf{A}$
   d. The determinant of $\mathbf{A}$
   e. The inverse of $\mathbf{A}$

2-30   Repeat all parts of Problem 2-29 for the matrix $\mathbf{D}$ of Problem 2-21.

2-31   Repeat all parts of Problem 2-29 for the matrix $\mathbf{B}$ of Problem 2-21.

2-32   It may be shown in general that if $\mathbf{A}$ and $\mathbf{B}$ are both square and of the same size, then $\det(\mathbf{AB}) = (\det \mathbf{A})(\det \mathbf{B})$. For the matrices $\mathbf{A}$ and $\mathbf{D}$ of Problem 2-21, show that $\det(\mathbf{AD}) = (\det \mathbf{A})(\det \mathbf{D})$.

2-33   Repeat Problem 2-32 for matrices $\mathbf{B}$ and $\mathbf{C}$ of Problem 2-21.

2-34   Write the following vectors as column vectors:

   a. $\mathbf{p} = 3\mathbf{i} - 2\mathbf{j} + 19\mathbf{k}$
   b. $\mathbf{q} = -8\mathbf{i} + 6\mathbf{j}$
   c. $\mathbf{r} = 6\mathbf{i} + 28\mathbf{k}$

2-35   Write the column vectors implied in Problem 2-34 as row vectors (by using the transpose).

2-36   For the vectors given below, determine which are equal (taking two at a time):

$$\mathbf{p} = 3\mathbf{i} - 6\mathbf{j} + 4\mathbf{k} \qquad \mathbf{q} = 5\mathbf{i} + 7\mathbf{k}$$

$$\mathbf{r} = 5\mathbf{i} + 7\mathbf{j} \qquad \mathbf{s} = 3\mathbf{i} - 6\mathbf{j} + 4\mathbf{k}$$

$$\mathbf{u} = 7\mathbf{i} - 8\mathbf{j} + 15\mathbf{k} \qquad \mathbf{v} = 7\mathbf{i} - 8\mathbf{j} + 15\mathbf{k}$$

2-37   For the vectors $\mathbf{p}$, $\mathbf{q}$, and $\mathbf{u}$ from Problem 2-36, determine

   a. $|\mathbf{p}|$ **b.** $|\mathbf{q}|$ **c.** $|\mathbf{u}|$

2-38 For the vectors **r**, **s**, and **v** from Problem 2-36, determine

    a. $|\mathbf{r}|$ **b.** $|\mathbf{s}|$ **c.** $|\mathbf{v}|$

2-39 A vector **v** is directed from point P at $(3, -2, 5)$ to point Q at $(7, -1, -3)$ in a cartesian frame of reference.

    a. Determine the position vectors for both points.
    b. Determine the vector **v**.
    c. Calculate the length of **v**.

2-40 A vector **u** is directed from point R at $(-5, 4, 2)$ to point S at $(0, 6, 7)$ in a cartesian frame of reference.

    a. Determine the position vectors for both points.
    b. Determine the vector **u**.
    c. Determine the length of **u**.

2-41 By a direct extension of Eq. (2-23), show that the length of **v** may be computed from

$$|\mathbf{v}| = \sqrt{\mathbf{v}^T \mathbf{v}}$$

for three-dimensional vectors. The result is, in fact, valid for n-dimensional vectors.

2-42 By using the result of Problem 2-41, compute the lengths of **u** and **v** where

    a. $\mathbf{u} = [6 \quad -5 \quad 0 \quad 4 \quad 2]^T$
    b. $\mathbf{v} = [-1 \quad 2 \quad 5 \quad -3 \quad 6 \quad 7]^T$

2-43 For the vectors **p**, **r**, and **v** from Problem 2-36, determine

    a. $\mathbf{p} + \mathbf{r}$ **b.** $\mathbf{p} - \mathbf{r}$ **c.** $\mathbf{r} + \mathbf{v}$
    d. $\mathbf{v} - \mathbf{p}$ **e.** $\mathbf{p} \cdot \mathbf{r}$ **f.** $\mathbf{r} \cdot \mathbf{v}$

2-44 For the vectors **q**, **s**, **u**, and **v** of Problem 2-36, determine

    a. $\mathbf{q} + \mathbf{s}$ **b.** $\mathbf{s} - \mathbf{u}$ **c.** $\mathbf{u} + \mathbf{v}$
    d. $\mathbf{u} - \mathbf{v}$ **e.** $\mathbf{u} \cdot \mathbf{v}$ **f.** $\mathbf{q} \cdot \mathbf{u}$

2-45 For the vectors **p**, **r**, and **v** of Problem 2-36, determine the angle between

    a. **p** and **r**   **b.** **r** and **v**

    Hint: Use Eq. (2-24) with the help of Eq. (2-28).

2-46 For the vectors **q**, **u**, and **v** of Problem 2-36, determine the angle between

    a. **u** and **v**   **b.** **q** and **u**

2-47 Determine the direction cosines of the vectors **p**, **q**, and **r** from Problem 2-36 and verify that Eq. (2-30) is satisfied in each case.

2-48 Determine the direction cosines of the vectors **s** and **u** from Problem 2-36 and verify that Eq. (2-30) is satisfied in each case.

2-49 Prove that the sum of the squares of the direction cosines is unity [Eq. (2-30)]. *Hint:* Use Eqs. (2-29) and the definition of the length of a vector.

2-50 Solve the following systems of equations by the matrix inversion method:

    a. $3x_1 + 5x_2 = -6$
        $2x_1 - 4x_2 = 5$
    b. $6x_1 - 3x_2 = 5$
        $-3x_1 + 5x_2 = 3$

2-51 Using the matrix inversion method, solve the following systems of equations:

    a. $5x_1 - 6x_2 = 7$
        $3x_1 + 7x_2 = 8$
    b. $5x_1 + 2x_2 + 3x_3 = 5$
        $2x_1 + 3x_2 + 4x_3 = 7$
        $3x_1 + 4x_2 + 5x_3 = 9$

3

Truss Analysis:
The Direct Approach

3-1 INTRODUCTION

One of the most important practical types of engineering structures is the truss. A
truss may be defined as a structure that is composed of straight members connected
at the joints by smooth pins. Smooth pin joints cannot support a moment. A typical
truss is shown in Fig. 3-1. As shown in this figure, external loads such as **P** and
Q may act only at the joints. A truss must be restrained in some manner so that it
does not undergo free-body motion when loaded. The truss in Fig. 3-1 is restrained
by a pin joint at A and a roller joint at B. The reader may wish to consult a statics
textbook, such as Beer and Johnston [1], for a review of the various types of
supports.

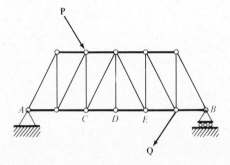

Figure 3-1 A typical two-dimensional truss.

Each member in the truss can only support axial loads, which implies that lateral loads may not act between two joints. All loads, therefore, must act at the joints. Moreover, members in a truss are connected only at their extremities; hence, members may not be continuous through a joint. In Fig. 3-1, for example, there are two distinct members, *CD* and *DE*, and not one member, *CE*. A truss is said to be *rigid* if it does not collapse when loaded. A rigid truss, however, may and will deform slightly under the action of one or more loads. The most common types of trusses are shown in Fig. 3-2.

In this chapter, the finite element formulation of a two-dimensional truss is developed. The direct approach, which does not require the use of variational principles, virtual work principles, or weighted residual methods, is taken. The main purpose for considering this type of problem is to demonstrate how the direct approach is used and to provide a step-by-step illustration of the finite element

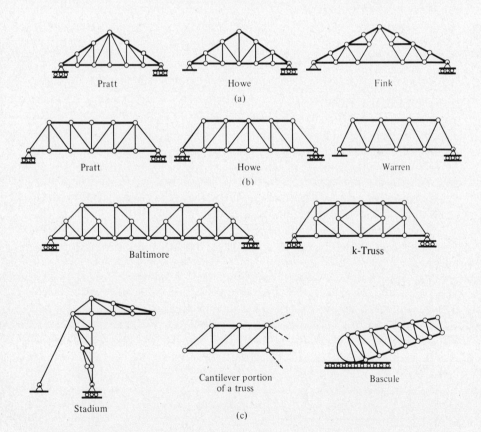

Figure 3-2   Typical trusses. (a) Typical roof trusses. (b) Typical bridge trusses. (c) Other types of trusses. (From *Vector Mechanics for Engineers: Statics and Dynamics* by Bear and Johnston. Copyright © 1962 by McGraw-Hill, New York. Used with permission of McGraw-Hill Book Company.)

method. A specific problem is worked out completely by hand, and the basic steps are then summarized. A simple computer program called TRUSS is described and applied to this problem.

3-2   FINITE ELEMENT FORMULATION: THE DIRECT APPROACH

As mentioned in Sec. 3-1, a truss is assumed to be composed of only uniaxial force members or bar elements. In other words, the joints are all assumed to be pinned and the members cannot withstand bending or torsional moments. This particular formulation is restricted to the two-dimensional truss only and will yield the displacements of each joint as well as the axial elongation, strain, stress, and force for each member. The finite element method is applicable to both statically determinate and indeterminate trusses (and structures in general). The FEM formulation for the three-dimensional truss is given in Sec. 3-5.

Discretization

The first step in any finite element analysis is discretization. Here the truss must be discretized into a number of finite elements. The truss is composed of axial tensile and compressive members, and it seems quite natural to consider each member as an element. A typical truss element e is shown in Fig. 3-3. Note that the element has two nodes, i and j, one at either end. It is at these nodes that the forces are transmitted from one element to the next. These elements are frequently referred to as bar elements, as opposed to beam or frame elements. Bar elements can withstand only axial forces; beam elements allow bending moments; and frame elements include axial forces, bending moments, and torsion [2].

Each joint in the truss is represented by a node that is given a unique number. Nodes are typically numbered consecutively from one to the maximum number present. These node numbers are often referred to as the *global node numbers*, as opposed to the local node "numbers" i and j. The numbering scheme has a strong influence on the computational time to obtain the solution (see Chapter 7). Elements are also numbered consecutively from one to the maximum number of elements. However, the numbering scheme for the elements is totally arbitrary, unlike the nodal numbering scheme. Local node numbers in this book are denoted with lowercase letters i, j, k, etc. For example, element e has only two nodes, and they are denoted as nodes i and j. However, all nodes in a structure have global node numbers and on element 3, for example, the global node numbers may be 10 (node i) and 7 (node j). The corresponding global node numbers are denoted with capital letters

Figure 3-3   Typical finite element in truss model.

I, J, K, etc. The point is that every element in the truss has local node "numbers" i and j with corresponding global node "numbers" I and J. In the example above, I is 10 and J is 7.

An essential part of every finite element analysis and computer program is the data that defines the coordinates of every node present in the truss and the elements. Element data are typically composed of nodal connectivity information and material set flags. Nodal and element data are generally generated within a computer program, but the user must still provide some minimum amount of data to describe the particular problem at hand. This is called *mesh generation*, and one method for accomplishing this is presented in Sec. 3-6.

Example 3-1

Figure 3-4(a) shows a cantilever truss with only five members and four joints. Members A, C, and E are composed of a 0.5-in.-diameter steel rod (with a modulus of elasticity of 30×10^6 psi), and members B and D are composed of a 0.4-in.-diameter aluminum rod (with a modulus of 11×10^6 psi). Discretize the truss showing the nodal coordinate data and the element data in tabular form (hence suitable for use in a computer program). The coordinates (in inches) of the joints are shown in Fig. 3-4(a).

Solution

Since the truss has five members, it is natural to discretize the truss into five elements, each member being an element. Each joint can then be taken as a node. Therefore, four nodes are used. The first step in the discretization process is to

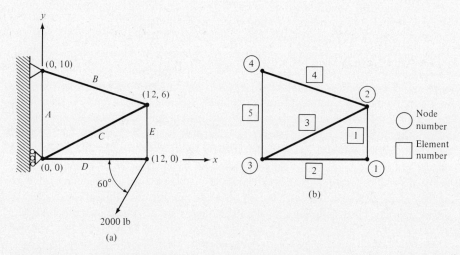

Figure 3-4   (a) A cantilever truss with five members and four joints. (b) Truss discretized into five elements and four nodes.

number each joint or node consecutively from one to the maximum number of nodes present. For the truss under consideration the nodes are numbered 1 to 4, as shown in Fig. 3-4(b). Using the coordinates shown in Fig. 3-4(a) we may summarize the nodal coordinate data as shown below:

Node number	x coordinate, in.	y coordinate, in.
1	12.0	0.0
2	12.0	6.0
3	0.0	0.0
4	0.0	10.0

Let us use material set flags 1 and 2 to denote the 0.5-in. steel and 0.4-in. aluminum, respectively. With the help of Fig. 3-4(b), the element data may be summarized as follows:

Element number	Node i	Node j	Material set flag
1	1	2	1
2	3	1	2
3	3	2	1
4	4	2	2
5	3	4	1

Element Stiffness Relationship in Local Coordinates

It will soon prove to be very convenient if a local coordinate system, $x'y'$, is chosen as shown in Fig. 3-5. This $x'y'$ coordinate system is known as a *local coordinate system*, since it is defined locally on the element. Note that the x' axis is directed

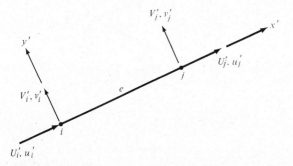

Figure 3-5 Truss element in local coordinate system showing nodal forces and nodal displacements.

along the length of the element from node i to node j. The y' axis must be chosen to be perpendicular to the x' axis, but its sense is arbitrary. Nodal forces in the x' and y' directions at node i are denoted as U_i' and V_i', respectively, and nodal displacements as u_i' and v_i', respectively. Similarly, at node j, the nodal forces are U_j' and V_j' and the nodal displacements are u_j' and v_j' in the x' and y' directions, respectively.

It should be recalled from elementary strength of materials that the axial elongation δ is given by

$$\delta = \frac{PL}{AE} \tag{3-1}$$

where L is the length of the member (or element), P the axial force, A the cross-sectional area, and E the modulus of elasticity (Young's modulus). It is assumed that the elastic range is not exceeded and that A is constant. Solving Eq. (3-1) for P and interpreting it as the axial force at node i (U_i'), we can write

$$U_i' = \frac{AE}{L} (u_i' - u_j') \tag{3-2a}$$

where the effective elongation δ is given by the difference between the nodal displacements along the element. In a similar fashion, we may write

$$U_j' = \frac{AE}{L} (u_j' - u_i') \tag{3-2b}$$

Note that $U_i' = -U_j'$, which must be the case if element e is to be in static equilibrium in the x' direction. Because this particular type of element cannot support transverse forces, we also have

$$V_i' - 0 \tag{3-2c}$$

$$V_j' = 0 \tag{3-2d}$$

Equations (3-2) may be written in matrix form as

$$\frac{AE}{L} \begin{bmatrix} 1 & 0 & -1 & 0 \\ 0 & 0 & 0 & 0 \\ -1 & 0 & 1 & 0 \\ 0 & 0 & 0 & 0 \end{bmatrix} \begin{bmatrix} u_i' \\ v_i' \\ u_j' \\ v_j' \end{bmatrix} = \begin{bmatrix} U_i' \\ V_i' \\ U_j' \\ V_j' \end{bmatrix} \tag{3-3}$$

Since the global nodal coordinates denoted as (x_i, y_i) and (x_j, y_j) are specified for any given truss, the element length may be computed from

$$L = \sqrt{(x_j - x_i)^2 + (y_j - y_i)^2} \tag{3-4}$$

The other two properties, A and E, are also specified for every element, usually with the help of material set flags in a computer program (see Example 3-1 and Secs. 3-3 and 3-6). Equation (3-3) may be written concisely as

$$\mathbf{K}^{e'} \mathbf{a}^{e'} = \mathbf{f}^{e'} \tag{3-5}$$

where $\mathbf{K}^{e'}$ represents the local element stiffness matrix, $\mathbf{a}^{e'}$ the local element nodal displacement vector, and $\mathbf{f}^{e'}$ the local element nodal force vector. The superscript (^{e}) is used throughout this text to denote element and a prime ($'$) to denote the local coordinate system, unless stated otherwise. Before Eq. (3-5) can be applied to the truss on the whole, it must be transformed to the global (xy) coordinate system. This transformation is developed in the next section.

Example 3-2

Determine the local element stiffness matrix for member C (or element 3) for the truss in Example 3-1.

Solution

From Eqs. (3-3) and (3-5), the local element stiffness matrix $\mathbf{K}^{e'}$ for element 3 (or $e = 3$) is given by

$$\mathbf{K}^{(3)'} = \frac{A^{(3)}E^{(3)}}{L^{(3)}} \left[\begin{array}{cc|cc} 1 & 0 & -1 & 0 \\ 0 & 0 & 0 & 0 \\ \hline -1 & 0 & 1 & 0 \\ 0 & 0 & 0 & 0 \end{array} \right]$$

where element 3 has material set flag 1, which corresponds to the 0.50-in. steel or

$$A^{(3)} = \frac{\pi}{4} (0.5)^2 = 0.196 \text{ in.}^2$$

$$E^{(3)} = 30 \times 10^6 \text{ psi}$$

$$L^{(3)} = \sqrt{(x_3 - x_2)^2 + (y_3 - y_2)^2} = \sqrt{(0 - 12)^2 + (0 - 6)^2} = 13.42 \text{ in.}$$

Computing $K^{(3)'}$ gives

$$\mathbf{K}^{(3)'} = 10^3 \left[\begin{array}{cc|cc} 438 & 0 & -438 & 0 \\ 0 & 0 & 0 & 0 \\ \hline -438 & 0 & 438 & 0 \\ 0 & 0 & 0 & 0 \end{array} \right] \text{ lbf/in.}$$

for the local element stiffness matrix for element 3 (member C). The 10^3 scalar multiplier should be noted.   ∎

Transformation from Local to Global Coordinates

A local ($x'y'$) coordinate system was used to derive the stiffness relationship given by Eq. (3-3). The x' axis was purposely taken to be directed from node i to node j. The y' axis, which must be chosen to be perpendicular to the x' axis, is otherwise arbitrary (with respect to its sense). Figure 3-6 shows a global coordinate system in relation to a typical element e. The purpose of this discussion is to find a

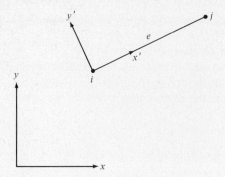

Figure 3-6   Relationship between local and global coordinate systems.

transformation that will enable us to write Eq. (3-5) with respect to a global co-ordinate system.

First, let us set up a global xy system as shown in Fig. 3-7(a) and draw a position vector **r** to an arbitrary point P. The vector **r** may be written in terms of its two cartesian components r_x and r_y as follows:

$$\mathbf{r} = r_x\mathbf{i} + r_y\mathbf{j} \tag{3-6}$$

Next, let us define the same vector **r** in terms of the rotated coordinate system $x'y'$ where, with no loss in generality, the x' axis is purposely taken to be in the same direction as **r** itself, as shown in Fig. 3-7(b). In terms of the cartesian components (r_x' and r_y') in the rotated frame, the vector **r** becomes

$$\mathbf{r} = r_x'\mathbf{i}' + r_y'\mathbf{j}' \tag{3-7}$$

where $\mathbf{i}'$ and $\mathbf{j}'$ are the two unit vectors in the x' and y' directions, respectively. From Eqs. (3-6) and (3-7), we have

$$r_x\mathbf{i} + r_y\mathbf{j} = r_x'\mathbf{i}' + r_y'\mathbf{j}' \tag{3-8}$$

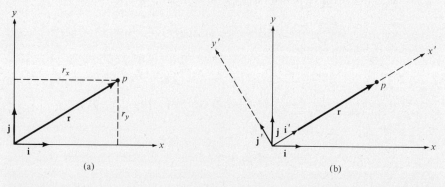

(a)

(b)

Figure 3-7   Position vector **r** (a) in global coordinate system and (b) in local coordinate system.

Taking the dot product of both sides of this equation with respect to $\mathbf{i}$ gives

$$r_x = r'_x \mathbf{i} \cdot \mathbf{i}' + r'_y \mathbf{i} \cdot \mathbf{j}' \tag{3-9}$$

where $\mathbf{i} \cdot \mathbf{i} = 1$ and $\mathbf{i} \cdot \mathbf{j} = 0$ have been used. Note that $\mathbf{i} \cdot \mathbf{i}'$ represents the cosine of the angle between the x and x' axes, and $\mathbf{i} \cdot \mathbf{j}'$ represents the cosine of the angle between the x and y' axes. It should be recalled that these are the direction cosines, which will be denoted as

$$n_{11} = \cos(x,x') = \mathbf{i} \cdot \mathbf{i}' \tag{3-10a}$$

$$n_{12} = \cos(x,y') = \mathbf{i} \cdot \mathbf{j}' \tag{3-10b}$$

where $\cos(x,x')$ represents the cosine of the angle between the x and x' axes, etc. Therefore, with the help of Eqs. (3-10), Eq. (3-9) becomes

$$r_x = n_{11} r'_x + n_{12} r'_y \tag{3-11a}$$

By forming the dot product of Eq. (3-8) with $\mathbf{j}$ in a completely analogous manner we also have

$$r_y = n_{21} r'_x + n_{22} r'_y \tag{3-11b}$$

where n_{21} represents the cosine of the angle between the y and x' axes, and n_{22} represents the cosine of the angle between the y and y' axes. Clearly, in the notation for the direction cosines, the first subscript is associated with the global coordinate system and the second with the local (rotated) system. Equations (3-11) may be written in matrix form as

$$\begin{bmatrix} r_x \\ r_y \end{bmatrix} = \begin{bmatrix} n_{11} & n_{12} \\ n_{21} & n_{22} \end{bmatrix} \begin{bmatrix} r'_x \\ r'_y \end{bmatrix} \tag{3-12a}$$

or more concisely as

$$\mathbf{r} = \mathbf{T}\mathbf{r}' \tag{3-12b}$$

where $\mathbf{T}$ is the so-called transformation matrix that transforms a vector in the local system to one in the global system. It can be shown that the matrix $\mathbf{T}$ is orthogonal (see Problem 3-21) because $\mathbf{T}^{-1} = \mathbf{T}^T$ (see Sec. 2-7). Therefore, it follows from Eq. (3-12b) that

$$\mathbf{r}' = \mathbf{T}^T \mathbf{r} \tag{3-13}$$

This last result may be used to transform a vector in the global system to one in a local system. These transformations are used next to determine the stiffness relationship with respect to the global coordinate system.

Global Element Stiffness Relationship

Before the element stiffness matrices $\mathbf{K}^{e'}$ and nodal force vectors $\mathbf{f}^{e'}$ can be assembled to represent the entire truss, they must be transformed to a global coordinate system. Interpreting $\mathbf{r}'$ as $[u'_i \; v'_i]^T$ and $\mathbf{r}$ as $[u_i \; v_i]^T$ in Eq. (3-13), we have

$$\begin{bmatrix} u_i' \\ v_i' \end{bmatrix} = \begin{bmatrix} n_{11} & n_{21} \\ n_{12} & n_{22} \end{bmatrix} \begin{bmatrix} u_i \\ v_i \end{bmatrix} \tag{3-14}$$

where u_i and v_i are the displacements at node i in the global coordinate directions x and y, respectively. Similarly, for the displacements at node j, it follows from Eq. (3-13) that

$$\begin{bmatrix} u_j' \\ v_j' \end{bmatrix} = \begin{bmatrix} n_{11} & n_{21} \\ n_{12} & n_{22} \end{bmatrix} \begin{bmatrix} u_j \\ v_j \end{bmatrix} \tag{3-15}$$

Figure 3-8(a) shows the relationship between the local and global nodal displacements. Equations (3-14) and (3-15) may be written in one matrix equation as

$$\begin{bmatrix} u_i' \\ v_i' \\ \hline u_j' \\ v_j' \end{bmatrix} = \left[\begin{array}{cc|cc} n_{11} & n_{21} & 0 & 0 \\ n_{12} & n_{22} & 0 & 0 \\ \hline 0 & 0 & n_{11} & n_{21} \\ 0 & 0 & n_{12} & n_{22} \end{array} \right] \begin{bmatrix} u_i \\ v_i \\ \hline u_j \\ v_j \end{bmatrix} \tag{3-16}$$

or more concisely as

$$\mathbf{a}^{e'} = \mathbf{R}\mathbf{a}^e \tag{3-17}$$

where, by definition

$$\mathbf{R} = \left[\begin{array}{c|c} \mathbf{T}^T & \mathbf{0} \\ \hline \mathbf{0} & \mathbf{T}^T \end{array} \right] \tag{3-18}$$

and $\mathbf{a}^e$ represents the vector of nodal displacements for element e referred to the global coordinate system, or

$$\mathbf{a}^e = [u_i \quad v_i \mid u_j \quad v_j]^T \tag{3-19}$$

In a completely analogous manner we may also write

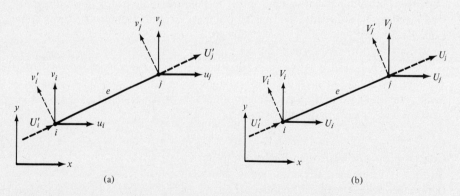

(a)   (b)

Figure 3-8 Typical element e showing global and local (a) nodal displacements and (b) nodal forces. All forces and displacements are shown with a positive sense.

$$\mathbf{f}^{e'} = \mathbf{R}\mathbf{f}^e \tag{3-20}$$

where $\mathbf{f}^e$ is the vector of nodal forces for element e referred to the global coordinate system, or

$$\mathbf{f}^e = [U_i \quad V_i \mid U_j \quad V_j]^T \tag{3-21}$$

Figure 3-8(b) shows the relationship between the local and global nodal forces. With the help of Eqs. (3-17) and (3-20), we may write Eq. (3-5) as

$$\mathbf{K}^{e'}\mathbf{R}\mathbf{a}^e = \mathbf{R}\mathbf{f}^e \tag{3-22}$$

which may then be premultiplied by $\mathbf{R}^T$ to give

$$\mathbf{R}^T\mathbf{K}^{e'}\mathbf{R}\mathbf{a}^e = \mathbf{f}^e \tag{3-23}$$

where $\mathbf{R}^T\mathbf{R} = \mathbf{R}^{-1}\mathbf{R} = \mathbf{I}$ has been used on the right-hand side. Equation (3-23) may be written simply as

$$\mathbf{K}^e\mathbf{a}^e = \mathbf{f}^e \tag{3-24}$$

where $\mathbf{K}^e$, the element stiffness matrix referred to global coordinates, is obtained directly from the element stiffness matrix referred to local coordinates ($\mathbf{K}^{e'}$) by

$$\mathbf{K}^e = \mathbf{R}^T\mathbf{K}^{e'}\mathbf{R} \tag{3-25}$$

The matrix $\mathbf{K}^e$ is often simply referred to as the *global element stiffness matrix* (as opposed to the local element stiffness matrix $\mathbf{K}^{e'}$). If the multiplications indicated in Eq. (3-25) are carried out, we get

$$\mathbf{K}^e = \frac{AE}{L} \begin{bmatrix} n_{11}^2 & n_{21}n_{11} & -n_{11}^2 & -n_{21}n_{11} \\ n_{11}n_{21} & n_{21}^2 & -n_{11}n_{21} & -n_{21}^2 \\ -n_{11}^2 & -n_{21}n_{11} & n_{11}^2 & n_{21}n_{11} \\ -n_{11}n_{21} & -n_{21}^2 & n_{11}n_{21} & n_{21}^2 \end{bmatrix} \tag{3-26}$$

It should be noted that in the expression for $\mathbf{K}^e$ only two of the four possible direction cosines appear: n_{11} and n_{21}. By definition, n_{11} is the cosine of the angle between the x and x' axes, and n_{21} is the cosine of the angle between the y and x' axes. Recall that the x' axis is directed along the element from node i to node j. From Fig. 3-9 it is immediately apparent that

$$n_{11} = \cos \theta_x = \frac{x_j - x_i}{L} \tag{3-27a}$$

and

$$n_{21} = \cos \theta_y = \frac{y_j - y_i}{L} \tag{3-27b}$$

where (x_i,y_i) and (x_j,y_j) for element e are the global coordinates of nodes i and j, respectively, and L is the element length [easily computed from Eq. (3-4)]. It should be recalled that the coordinates of each node are a vital part of the input to every

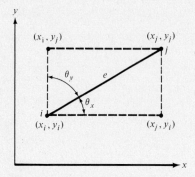

Figure 3-9   Definition of angles θ_x and θ_y used in obtaining the direction cosines.

finite element analysis (and computer program). Consequently, the two needed direction cosines are routinely computed with Eqs. (3-27). It must be emphasized that the important result is Eq. (3-26), which enables us to compute the global element stiffness matrix for every element since A and E are specified, and L, n_{11}, and n_{21} are easily computed (for every element). In the next section, the notion of assemblage is introduced. It is an easy matter to show that $\mathbf{K}^e$ is symmetric because $\mathbf{K}^{e'}$ is itself symmetric (see Problem 2-26).

Example 3-3

For the truss in Example 3-1, determine the global element stiffness matrix for element 3.

Solution

The global element stiffness matrix, denoted as $\mathbf{K}^e$, is given by Eq. (3-26) in terms of the direction cosines n_{11} and n_{21}, which are computed from Eqs. (3-27) for element 3 as follows:

$$n_{11} = \frac{x_2 - x_3}{L^{(3)}} = \frac{12.0 - 0.0}{13.42} = 0.8944$$

$$n_{21} = \frac{y_2 - y_3}{L^{(3)}} = \frac{6.0 - 0.0}{13.42} = 0.4472$$

where the result for the element length from Example 3-2 has been used. The calculation of $\mathbf{K}^{(3)}$ from Eq. (3-26) is now straightforward, and the result is

$$\mathbf{K}^{(3)} = 10^3 \begin{bmatrix} 351 & 176 & -351 & -176 \\ 176 & 88 & -176 & -88 \\ -351 & -176 & 351 & 176 \\ -176 & -88 & 176 & 88 \end{bmatrix} \text{ lbf/in.}$$

where the cross-sectional area $A^{(3)}$ was also taken from Example 3-2. As expected, $\mathbf{K}^{(3)}$ is symmetric.   ∎

Assemblage

The assemblage step is necessary in all finite element analyses. It is in this step that the original truss (or in general, any structure, body, or region being analyzed) is put back together or *assembled* from the individual elements comprising it. Suffice it to say now that the assemblage step is based on the principle of compatibility; that is, the x and y displacements associated with a particular node on any given element must be identical to those associated with the same node on each element that shares this node [3].

Consider a two-dimensional truss with N nodes. Each element e has two nodes i and j with global node numbers I and J. Since each node has two degrees of freedom (i.e., two components of displacement) the 4×4 global element stiffness matrix may be thought of as being partitioned into four submatrices, each of size 2×2, or

$$\mathbf{K}^e = \left[\begin{array}{c|c} \mathbf{K}_{ii}^e & \mathbf{K}_{ij}^e \\ \hline \mathbf{K}_{ji}^e & \mathbf{K}_{jj}^e \end{array} \right] \tag{3-28}$$

These 2×2 submatrices are easily extracted from Eq. (3-26). Since global node number I corresponds to local node i and global node number J to local node j, Eq. (3-28) may be written in terms of the global nodal subscripts as

$$\mathbf{K}^e = \left[\begin{array}{c|c} \mathbf{K}_{I,I}^e & \mathbf{K}_{I,J}^e \\ \hline \mathbf{K}_{J,I}^e & \mathbf{K}_{J,J}^e \end{array} \right] \tag{3-29}$$

The assemblage stiffness matrix, denoted as $\mathbf{K}^a$, is always of the form

$$\mathbf{K}^a = \begin{bmatrix} \mathbf{K}_{1,1}^a & \mathbf{K}_{1,2}^a & \mathbf{K}_{1,3}^a & \cdots & \mathbf{K}_{1,N}^a \\ \mathbf{K}_{2,1}^a & \mathbf{K}_{2,2}^a & \mathbf{K}_{2,3}^a & \cdots & \mathbf{K}_{2,N}^a \\ \mathbf{K}_{3,1}^a & \mathbf{K}_{3,2}^a & \mathbf{K}_{3,3}^a & \cdots & \mathbf{K}_{3,N}^a \\ \vdots & \vdots & \vdots & \ddots & \vdots \\ \mathbf{K}_{N,1}^a & \mathbf{K}_{N,2}^a & \mathbf{K}_{N,3}^a & \cdots & \mathbf{K}_{N,N}^a \end{bmatrix} \tag{3-30}$$

Note that the superscript (^{a}) denotes assemblage. Each of these $\mathbf{K}_{I,J}^a$'s are 2×2 submatrices. To begin the assemblage process, the assemblage stiffness matrix $\mathbf{K}^a$ is zeroed out. Then the individual global element stiffness submatrices (i.e., the $\mathbf{K}_{I,J}^e$'s) are added to the corresponding position in the matrix $\mathbf{K}^a$ (i.e., at $\mathbf{K}_{I,J}^a$) for element 1, then for element 2, and so forth, up to and including the last element. The assemblage of each 2×2 submatrix in Eq. (3-30) may be given by the relationship

$$\mathbf{K}_{I,J}^a = \sum_{e=1}^{M} \mathbf{K}_{I,J}^e \qquad \text{for } I = 1 \text{ to } N \text{ and } J = 1 \text{ to } N \tag{3-31}$$

In Eq. (3-31), M is the maximum number of elements, and $\mathbf{K}_{I,J}^e$ is taken as the 2×2 null matrix if element e does not have nodes I and J. The next example will help to clarify the assemblage process for the truss (see Sec. 3-3 for a numerical example).

Example 3-4

For the truss of Example 3-1, determine the assemblage stiffness matrix in terms of the 2×2 global element stiffness submatrices $\mathbf{K}_{I,J}^e$.

Solution

Element 1 has global node numbers 1 and 2 (i.e., I and J, respectively), and so the global element stiffness matrix for element 1, denoted as $\mathbf{K}^{(1)}$, is given by

$$\mathbf{K}^{(1)} = \left[\begin{array}{c|c} \mathbf{K}_{1,1}^{(1)} & \mathbf{K}_{1,2}^{(1)} \\ \hline \mathbf{K}_{2,1}^{(1)} & \mathbf{K}_{2,2}^{(1)} \end{array} \right]$$

Having first zeroed-out the assemblage stiffness matrix $\mathbf{K}^a$, we may then "add" this result to $\mathbf{K}^a$ to get

$$\mathbf{K}^a = \left[\begin{array}{c|c|c|c} \mathbf{K}_{1,1}^{(1)} & \mathbf{K}_{1,2}^{(1)} & 0 & 0 \\ \hline \mathbf{K}_{2,1}^{(1)} & \mathbf{K}_{2,2}^{(1)} & 0 & 0 \\ \hline 0 & 0 & 0 & 0 \\ \hline 0 & 0 & 0 & 0 \end{array} \right]$$

This is not the complete form of the assemblage stiffness matrix because four more elements have yet to be considered. Element 2 has global node numbers 3 and 1 (i.e., I and J, respectively) and the global element stiffness matrix for element 2 is given by

$$\mathbf{K}^{(2)} = \left[\begin{array}{c|c} \mathbf{K}_{3,3}^{(2)} & \mathbf{K}_{3,1}^{(2)} \\ \hline \mathbf{K}_{1,3}^{(2)} & \mathbf{K}_{1,1}^{(2)} \end{array} \right]$$

After combining this element stiffness matrix with the result from element 1, we get

$$\mathbf{K}^a = \left[\begin{array}{c|c|c|c} \mathbf{K}_{1,1}^{(1)} + \mathbf{K}_{1,1}^{(2)} & \mathbf{K}_{1,2}^{(1)} & \mathbf{K}_{1,3}^{(2)} & 0 \\ \hline \mathbf{K}_{2,1}^{(1)} & \mathbf{K}_{2,2}^{(1)} & 0 & 0 \\ \hline \mathbf{K}_{3,1}^{(2)} & 0 & \mathbf{K}_{3,3}^{(2)} & 0 \\ \hline 0 & 0 & 0 & 0 \end{array} \right]$$

Note the position of each entry in $\mathbf{K}^a$ carefully. After considering element 3 (with nodes 3 and 2), element 4 (with nodes 4 and 2), and element 5 (with nodes 3 and 4), the assemblage stiffness matrix becomes

$$\mathbf{K}^a = \left[\begin{array}{c|c|c|c} \mathbf{K}_{1,1}^{(1)} + \mathbf{K}_{1,1}^{(2)} & \mathbf{K}_{1,2}^{(1)} & \mathbf{K}_{1,3}^{(2)} & 0 \\ \hline \mathbf{K}_{2,1}^{(1)} & \mathbf{K}_{2,2}^{(1)} + \mathbf{K}_{2,2}^{(3)} + \mathbf{K}_{2,2}^{(4)} & \mathbf{K}_{2,3}^{(3)} & \mathbf{K}_{2,4}^{(4)} \\ \hline \mathbf{K}_{3,1}^{(2)} & \mathbf{K}_{3,2}^{(3)} & \mathbf{K}_{3,3}^{(2)} + \mathbf{K}_{3,3}^{(3)} + \mathbf{K}_{3,3}^{(5)} & \mathbf{K}_{3,4}^{(5)} \\ \hline 0 & \mathbf{K}_{4,2}^{(4)} & \mathbf{K}_{4,3}^{(5)} & \mathbf{K}_{4,4}^{(4)} + \mathbf{K}_{4,4}^{(5)} \end{array} \right]$$

Several significant observations should be made. First, each of the individual 2 × 2 element stiffness submatrices is symmetric, as an examination of Eq. (3-26) clearly shows. Mathematically, this may be stated as

$$\mathbf{K}^e_{I,J} = \mathbf{K}^e_{J,I} \tag{3-32}$$

It then follows that $\mathbf{K}^a$ itself is symmetric, which has several advantages as delineated in Chapters 6 and 7. Another property of $\mathbf{K}^a$ is that it is *banded*, which means the nonzero entries in $\mathbf{K}^a$ are gathered along the main diagonal. In the example above, this is not very obvious because the truss has only five elements and four nodes. Because of the symmetric nature of $\mathbf{K}^a$, we usually use the term *half-bandwidth*, denoted as b_w in this text. A general expression for the half-bandwidth is given by

$$b_w = N_{\mathrm{DOF}}(1 + \Delta_{\max}) \tag{3-33}$$

where N_{DOF} is the number of degrees of freedom per node and $\Delta_{\max}$ is the *maximum of the maximum differences* between node numbers on each element considering all elements one at a time. In the truss model there are two degrees of freedom associated with each node (the two components of displacement). From Eq. (3-33) it can be seen that the way in which the nodes are numbered can have a strong influence on the half-bandwidth. In a computer program, the execution time increases with the square of the bandwidth for the usual solution techniques [4]. ■

Example 3-5

Determine the half-bandwidth of the assemblage stiffness matrix from Example 3-4 by direct examination of $\mathbf{K}^a$ and by computation using Eq. (3-33).

Solution

Let us write $\mathbf{K}^a$ in the following form showing each potentially nonzero entry as an x and each zero entry as a 0:

$$
\mathbf{K}^a =
\begin{array}{c}
\overset{\displaystyle |\longleftarrow\ b_w\ \longrightarrow|}{}
\begin{bmatrix}
x & x & x & x & x & x & 0 & 0 \\
x & x & x & x & x & x & 0 & 0 \\
x & x & x & x & x & x & x & x \\
x & x & x & x & x & x & x & x \\
x & x & x & x & x & x & x & x \\
x & x & x & x & x & x & x & x \\
0 & 0 & x & x & x & x & x & x \\
0 & 0 & x & x & x & x & x & x
\end{bmatrix}
\end{array}
$$

A direct count of the number of nonzero entries in the half-bandwidth above yields 6. Equation (3-33), of course, must give the same result for the half-bandwidth, but let us compute it. There are two degrees of freedom per node, hence $N_{\mathrm{DOF}} = 2$. The maximum difference between any two node numbers considering all elements one at a time is 2, or $\Delta_{\max}$ = 2. Therefore, Eq. (3-33) readily gives b_w = 2(1 + 2) = 6 as expected.   ■

Application of Loads

The assemblage nodal force vector $\mathbf{f}^a$ may be obtained by an assemblage process similar to the one used to obtain $\mathbf{K}^a$. It is simpler, however, because each element nodal force vector, $\mathbf{f}^e$, is a column vector (of size 4 × 1), as opposed to a matrix. The assemblage of each element nodal force vector in this manner is illustrated in subsequent chapters. Because the external loads on the truss can only occur at the joints, the effective assemblage nodal force vector $\mathbf{f}^a$ can be obtained by a more direct route as explained below.

The assemblage nodal force vector, with $2N$ entries, is first zeroed-out, where N is the number of nodes present. Let us assume that at global node number K, there is a point load with components F_x and F_y. These components are referred to the global coordinate system as shown in Fig. 3-10. Because the load is applied at node K, we must add F_x and F_y to the $(2K - 1)$st and $2K$th positions in $\mathbf{f}^a$, respectively. Reaction forces are not considered to be loads in finite element analysis. They are handled routinely as restraints (see Sec. 3-3).

Example 3-6

For the loading shown in Fig. 3-4(a) for the truss in Example 3-1, determine the assemblage nodal force vector $\mathbf{f}^a$.

Solution

The 2000-lbf load is the only external load and must be resolved into its two cartesian components, F_x and F_y. Note from Example 3-1 that an xy (global) coordinate system is implied as shown in Fig. 3-4(a). The x and y components of the applied load are given by F_x = $-2000 \cos 60°$ = -1000 lb and F_y = $-2000 \sin 60°$ =

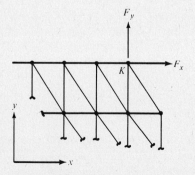

Figure 3-10   Portion of truss showing node K with external load (components F_x and F_y).

-1732 lb. Since this load is applied to node 1, the first [from $2(1) - 1$] and second [from $2(1)$] entries in $\mathbf{f}^a$ must be changed to yield the following result:

$$
\mathbf{f}^a = \begin{bmatrix} -1000 \\ -1732 \\ \hline 0 \\ 0 \\ \hline 0 \\ 0 \\ \hline 0 \\ 0 \end{bmatrix} \quad \begin{array}{l} \left.\begin{array}{l} F_x \\ F_y \end{array}\right\} \text{Node 1} \\ \left.\begin{array}{l} F_x \\ F_y \end{array}\right\} \text{Node 2} \\ \left.\begin{array}{l} F_x \\ F_y \end{array}\right\} \text{Node 3} \\ \left.\begin{array}{l} F_x \\ F_y \end{array}\right\} \text{Node 4} \end{array}
$$

If another force also acts at node 1, the respective components would be added to those above. The restraints on the nodal displacements are considered next. ∎

Application of Restraints on Nodal Displacements and Solution

The assemblage stiffness matrix $\mathbf{K}^a$ and nodal force vector $\mathbf{f}^a$ are obtained as described above. The assemblage system equation is of the form

$$\mathbf{K}^a\mathbf{a} = \mathbf{f}^a \tag{3-34}$$

where $\mathbf{a}$ is the vector of nodal unknowns or, in this case, the vector of unknown nodal displacements. The vector $\mathbf{a}$ is given by

$$\mathbf{a} = [u_1 \quad v_1 \mid u_2 \quad v_2 \mid \cdots \mid u_N \quad v_N]^T \tag{3-35}$$

where N is the maximum number of nodes and u_I, v_I are the x and y components of the displacement at global node number I. In Eq. (3-35), the transpose is used to conserve space—$\mathbf{a}$ itself is a column vector as implied by Eq. (3-34). Equation (3-34) cannot be solved for the vector $\mathbf{a}$ because the restraints on the nodal displacements have not been considered. In fact, *the inverse of $\mathbf{K}^a$ does not exist, in general, until the restraints on displacements are taken into account.* In other words, $\mathbf{K}^a$ is singular. After the restraints are considered, Eq. (3-34) is written as

$$\mathbf{Ka} = \mathbf{f}$$

where $\mathbf{K}$ is no longer singular. Therefore, $\mathbf{K}^{-1}$ exists, which implies a unique solution for the vector of nodal displacements $\mathbf{a}$. Two methods are commonly used to impose the prescribed displacement restraints [5]. Each is discussed below.

Method 1

The first method is most easily explained by an example. Consider the system of equations

$$k_{11}a_1 + k_{12}a_2 + k_{13}a_3 = f_1 \tag{3-36a}$$

$$k_{21}a_1 + k_{22}a_2 + k_{23}a_3 = f_2 \qquad \text{(3-36b)}$$

$$k_{31}a_1 + k_{32}a_2 + k_{33}a_3 = f_3 \qquad \text{(3-36c)}$$

Let us impose a restraint on a_2, in other words, that a_2 is to have a specified value, e.g., a_p (a prescribed displacement in this truss analysis). Let us further assume that the system of equations is symmetric (such as the truss model) and that the symmetry is to be preserved. Let us replace Eq. (3-36b) with

$$a_2 = a_p$$

The system of equations thus becomes

$$k_{11}a_1 + k_{12}a_2 + k_{13}a_3 = f_1$$

$$a_2 \qquad\qquad = a_p$$

$$k_{31}a_1 + k_{32}a_2 + k_{33}a_3 = f_3$$

But the symmetry has been destroyed, and to preserve it we need a zero coefficient on a_2 in the first equation. We can get that by noting that $a_2 = a_p$ (which is known) and writing

$$k_{11}a_1 + k_{13}a_3 = f_1 - k_{12}a_p$$

Note that the term involving a_2 is conveniently transposed to the right-hand side (and a_p replaces a_2). After performing a similar operation on the third equation and using matrix notation, we get

$$\begin{bmatrix} k_{11} & 0 & k_{13} \\ 0 & 1 & 0 \\ k_{31} & 0 & k_{33} \end{bmatrix} \begin{bmatrix} a_1 \\ a_2 \\ a_3 \end{bmatrix} = \begin{bmatrix} f_1 - k_{12}a_p \\ a_p \\ f_3 - k_{32}a_p \end{bmatrix} \qquad \text{(3-37)}$$

Note that f_2 disappears! The implication is that for degrees of freedom in the truss model where an unknown reaction force exists, this unknown force (f_2 above) never really enters the formulation. Instead, a known displacement is imposed. In structural analysis problems, these prescribed displacements are usually zero, and in this case $a_p = 0$ in Eq. (3-37).

Method 2

A much simpler method exists for imposing prescribed displacement restraints. The method is based on the concept of penalty functions, and a proof of why the method works is given in Zienkiewicz [6]. The purpose here is not to prove it but to learn how to apply it.

Let us again consider the system of equations in Eq. (3-36). In this method a large number β is selected, say six to twelve orders of magnitude larger than the largest coefficient k_{ij}, and is added to the coefficient k_{ii} if a_i is to be prescribed. In addition, the right-hand side of the ith equation is changed to β times the prescribed value (of displacement). In the example above, since a_2 is to be prescribed (i.e., $a_2 = a_p$), the second equation becomes

$$k_{21}a_1 + (k_{22} + \beta)a_2 + k_{23}a_3 = \beta a_p$$

Note that β is chosen to be much larger than the coefficients and, for all practical purposes, this last equation is really equivalent to

$$\beta a_2 = \beta a_p$$

or

$$a_2 = a_p$$

which is the desired result. The other equations are unchanged. In matrix form the system of equations becomes

$$\begin{bmatrix} k_{11} & k_{12} & k_{13} \\ k_{21} & k_{22} + \beta & k_{23} \\ k_{31} & k_{32} & k_{33} \end{bmatrix} \begin{bmatrix} a_1 \\ a_2 \\ a_3 \end{bmatrix} = \begin{bmatrix} f_1 \\ \beta a_p \\ f_3 \end{bmatrix} \tag{3-38}$$

Note that this simple method automatically preserves symmetry. In a computer program, Method 2 is extremely easy to apply. Examples that demonstrate Methods 1 and 2 are given in Secs. 4-8 and 3-3, respectively.

Both methods result in the system of equations represented by the matrix equation

$$\mathbf{Ka} = \mathbf{f} \tag{3-39}$$

(without any superscripts). This equation may be solved for the nodal displacements contained in the vector $\mathbf{a}$ by any method applicable to the solution of linear, algebraic equations. In fact, some solution techniques, such as the one in Sec. 6-8, take advantage of the symmetric and banded nature of the stiffness matrix $\mathbf{K}$, but let us use the matrix inversion method here. If $\mathbf{K}^{-1}$ is computed and if Eq. (3-39) is premultiplied by it, we get

$$\mathbf{a} = \mathbf{K}^{-1}\mathbf{f} \tag{3-40}$$

This method is used in Sec. 3-3 in the solution to an example problem. A much more practical method of solution is presented in Sec. 6-8.

Computation of the Element Resultants

By definition, the *element resultants* for the truss include quantities such as the axial elongations, strains, stresses, and forces. Each of these is computed more or less directly from the nodal displacements $\mathbf{a}$ that were computed to be the solution of $\mathbf{Ka} = \mathbf{f}$.

The elongation δ for element e may be computed in terms of the x' components of nodal displacements at nodes i and j by

$$\delta = u'_j - u'_i \tag{3-41}$$

But from Eq. (3-14), we have

$$u'_i = n_{11}u_i + n_{21}v_i \tag{3-42a}$$

and from Eq. (3-15),

$$u'_j = n_{11}u_j + n_{21}v_j \tag{3-42b}$$

Note that in Eqs. (3-42) only two of the direction cosines appear (n_{11} and n_{21}): these are easily computed for every element by Eqs. (3-27). In Eqs. (3-42), u_i, v_i, u_j, and v_j are known once the nodal displacements are determined (from the solution of $\mathbf{Ka} = \mathbf{f}$). Therefore, for any element e, the elongation δ is readily computed.

The axial strain ε for a typical element e is computed from

$$\varepsilon = \frac{\delta}{L} \tag{3-43}$$

where L is the element length computed from Eq. (3-4). The material is assumed to be elastic so that Hooke's law applies. Therefore, the axial stress σ in the element is computed from

$$\sigma = E\varepsilon \tag{3-44}$$

Finally the axial force F in the element is determined by

$$F = \sigma A \tag{3-45}$$

In Eqs. (3-41), (3-43), (3-44), and (3-45), positive values of δ, ε, σ, and F denote tension, whereas negative values denote compression. The formulation of this section is utilized in the next section in an example.

3-3 APPLICATION TO A SPECIFIC EXAMPLE

In this section, a simple problem is executed completely by hand in order to illustrate the basic steps in the finite element method. Let us reconsider the truss in Example 3-1 shown in Fig. 3-4(a). It should be recalled that members A, C, and E are made of a 0.5-in.-diameter steel rod and members B and D of a 0.4-in.-diameter aluminum rod. The modulus of elasticity of steel and aluminum are 30×10^6 and 11×10^6 psi, respectively. The truss is constrained and loaded as shown in Fig. 3-4(a), which also shows the coordinates of each joint.

The first step in any finite element analysis is discretization. Let us use the results of Example 3-1 where we took each member of the truss to be an element. It should be recalled from that example that the discretization step may be divided into two separate but related tasks: (1) specification of the nodal coordinates and (2) specification of the element data, including material set definitions. Because this step is so vitally important, let us summarize the results of the discretization from Example 3-1:

Nodal Coordinate Data

Node number	x coordinate, in.	y coordinate, in.
1	12.0	0.0
2	12.0	6.0
3	0.0	0.0
4	0.0	10.0

Element Data

Element number	Node i	Node j	Material set
1	1	2	1
2	3	1	2
3	3	2	1
4	4	2	2
5	3	4	1

Note that member A is taken as element 5, member B as element 4, and so forth. In addition, material set 1 is the 0.5-in. steel and material set 2, the 0.4-in. aluminum. The nodal numbering scheme used above will result in the smallest possible bandwidth of the assemblage stiffness matrix.

In Sec. 3-2, three properties were needed for each element: the cross-sectional area, the modulus of elasticity, and the length. The length is easily calculated by Eq. (3-4) from the nodal coordinates. However, the remaining two "material properties" must be specified:

Material set	Cross-sectional area, in.^{2}	Modulus of elasticity, psi
1	0.1963	30×10^6
2	0.1257	11×10^6

All of the above information is used frequently throughout the finite element solution process. In summary, the truss model will be composed of four nodes and five elements involving two different types of materials. Not surprisingly, this same information becomes an integral part of the input to the computer program described in Sec. 3-6.

The next step is to determine the local element stiffness matrix for element 1, to transform it to the global element stiffness matrix, and finally to "add" it into the assemblage stiffness matrix (which is always zeroed-out before processing the first element). From Eqs. (3-3) and (3-5), the local element stiffness matrix is given by

$$\mathbf{K}^{e'} = \frac{AE}{L} \left[\begin{array}{cc|cc} 1 & 0 & -1 & 0 \\ 0 & 0 & 0 & 0 \\ \hline -1 & 0 & 1 & 0 \\ 0 & 0 & 0 & 0 \end{array} \right] \tag{3-46}$$

Element 1 is composed of material 1, and so we have

$$\mathbf{K}^{(1)'} = \frac{(0.1963)(30 \times 10^6)}{\sqrt{(12. - 12.)^2 + (6. - 0.)^2}} \left[\begin{array}{cc|cc} 1 & 0 & -1 & 0 \\ 0 & 0 & 0 & 0 \\ \hline -1 & 0 & 1 & 0 \\ 0 & 0 & 0 & 0 \end{array} \right] \tag{3-47a}$$

or

$$\mathbf{K}^{(1)'} = 10^3 \left[\begin{array}{cc|cc} 982 & 0 & -982 & 0 \\ 0 & 0 & 0 & 0 \\ \hline -982 & 0 & 982 & 0 \\ 0 & 0 & 0 & 0 \end{array} \right] \text{ lbf/in.} \tag{3-47b}$$

Although only three or four significant digits are shown here, all calculations were carried out using nine digits and, therefore, numbers may not add exactly as shown during the assemblage. The reader should note the scalar multiplier of 10^3 and verify that each entry in $\mathbf{K}^{(1)'}$ has units of pound-force per inch. The stiffness terms in structural models will always have units of force per unit length—similar to the units of the stiffness for a spring. Although $\mathbf{K}^{(1)'}$ can be transformed to $\mathbf{K}^{(1)}$ by applying Eq. (3-25) to element 1, it is easier to use the result for $\mathbf{K}^e$ given by Eq. (3-26). Recall that only two of the direction cosines are needed, which, for element 1, are readily computed as follows:

$$n_{11} = \frac{x_j - x_i}{L} = \frac{12.0 - 12.0}{6.0} = 0.0 \tag{3-48a}$$

$$n_{21} = \frac{y_j - y_i}{L} = \frac{6.0 - 0.0}{6.0} = 1.0 \tag{3-48b}$$

The reader should examine the orientation of element 1 in Fig. 3-4(b) to verify these direction cosines by inspection. With the help of Eq. (3-26), the global element stiffness matrix for element 1 may be verified to be

$$\mathbf{K}^{(1)} = 10^3 \left[\begin{array}{cc|cc} 0 & 0 & 0 & 0 \\ 0 & 982 & 0 & -982 \\ \hline 0 & 0 & 0 & 0 \\ 0 & -982 & 0 & 982 \end{array} \right] \text{ lbf/in.} \tag{3-49}$$

This result must now be added into the assemblage stiffness matrix (which at this point is simply a null matrix). Since element 1 connects nodes 1 and 2, the assemblage stiffness matrix $\mathbf{K}^a$ becomes

$$\mathbf{K}^a = \begin{bmatrix} 0 & 0 & 0 & 0 & 0 & 0 & 0 & 0 \\ 0 & 982 & 0 & -982 & 0 & 0 & 0 & 0 \\ 0 & 0 & 0 & 0 & 0 & 0 & 0 & 0 \\ 0 & -982 & 0 & 982 & 0 & 0 & 0 & 0 \\ 0 & 0 & 0 & 0 & 0 & 0 & 0 & 0 \\ 0 & 0 & 0 & 0 & 0 & 0 & 0 & 0 \\ 0 & 0 & 0 & 0 & 0 & 0 & 0 & 0 \\ 0 & 0 & 0 & 0 & 0 & 0 & 0 & 0 \end{bmatrix} \text{ lbf/in.} \qquad \textbf{(3-50)}$$

The reader should corroborate this result for the assemblage stiffness matrix (after processing the first element) with Example 3-4.

Although the nodal force vectors $\mathbf{f}^{e'}$, $\mathbf{f}^e$, and $\mathbf{f}^a$ could be considered at this point for element 1, it proves to be more convenient to obtain $\mathbf{f}^a$ directly as indicated by Example 3-6. This same comment applies, of course, to the nodal force vectors for elements 2 through 5.

Element 2 is considered next, and since it is composed of material set 2, $\mathbf{K}^{(2)'}$ is given by

$$\mathbf{K}^{(2)'} = \frac{(0.1257)(11 \times 10^6)}{12.0} \begin{bmatrix} 1 & 0 & -1 & 0 \\ 0 & 0 & 0 & 0 \\ -1 & 0 & 1 & 0 \\ 0 & 0 & 0 & 0 \end{bmatrix} \qquad \textbf{(3-51a)}$$

or

$$\mathbf{K}^{(2)'} = 10^3 \begin{bmatrix} 115 & 0 & -115 & 0 \\ 0 & 0 & 0 & 0 \\ -115 & 0 & 115 & 0 \\ 0 & 0 & 0 & 0 \end{bmatrix} \qquad \textbf{(3-51b)}$$

The reader may verify that for element 2 we have $n_{11} = 1.0$ and $n_{21} = 0.0$, both by inspection and by using Eqs. (3-27). Therefore, we have $\mathbf{K}^{(2)} = \mathbf{K}^{(2)'}$, or the global element stiffness matrix is identical to the local element stiffness matrix [see Eq. (3-26)]. Since element 2 is connected by nodes 3 and 1, the assemblage stiffness matrix becomes (after processing two elements)

$$\mathbf{K}^a = 10^3 \begin{bmatrix} 115 & 0 & 0 & 0 & -115 & 0 & 0 & 0 \\ 0 & 982 & 0 & -982 & 0 & 0 & 0 & 0 \\ 0 & 0 & 0 & 0 & 0 & 0 & 0 & 0 \\ 0 & -982 & 0 & 982 & 0 & 0 & 0 & 0 \\ -115 & 0 & 0 & 0 & 115 & 0 & 0 & 0 \\ 0 & 0 & 0 & 0 & 0 & 0 & 0 & 0 \\ 0 & 0 & 0 & 0 & 0 & 0 & 0 & 0 \\ 0 & 0 & 0 & 0 & 0 & 0 & 0 & 0 \end{bmatrix} \qquad \textbf{(3-52)}$$

Again the reader should examine this result in light of Example 3-4.

Element 3 is a little more interesting than the previous two elements because it is oriented neither horizontally nor vertically. The length of element 3 is easily computed from Eq. (3-4) to be

$$L = \sqrt{(12. - 0.)^2 + (6. - 0.)^2} = 13.42 \text{ in.}$$

and $\mathbf{K}^{(3)'}$ becomes

$$\mathbf{K}^{(3)'} = \frac{(0.1963)(30 \times 10^6)}{13.42} \begin{bmatrix} 1 & 0 & -1 & 0 \\ 0 & 0 & 0 & 0 \\ -1 & 0 & 1 & 0 \\ 0 & 0 & 0 & 0 \end{bmatrix} \qquad \textbf{(3-53a)}$$

or

$$\mathbf{K}^{(3)'} = 10^3 \begin{bmatrix} 439 & 0 & -439 & 0 \\ 0 & 0 & 0 & 0 \\ -439 & 0 & 439 & 0 \\ 0 & 0 & 0 & 0 \end{bmatrix} \qquad \textbf{(3-53b)}$$

In order to transform $\mathbf{K}^{(3)'}$ into $\mathbf{K}^{(3)}$, the direction cosines are needed. Recall from Example 3-3, for element 3 we calculated $n_{11} = 0.894$, $n_{21} = 0.447$. With the help of Eq. (3-26), the global element stiffness matrix for element 3 is computed to be

$$\mathbf{K}^{(3)} = 10^3 \begin{bmatrix} 351 & 176 & -351 & -176 \\ 176 & 88 & -176 & -88 \\ -351 & -176 & 351 & 176 \\ -176 & -88 & 176 & 88 \end{bmatrix} \qquad \textbf{(3-54)}$$

This matrix must now be included in the assemblage matrix $\mathbf{K}^a$, and since element 3 connects nodes 3 and 2, we get the result

$$\mathbf{K}^a = 10^3 \begin{bmatrix} 115 & 0 & 0 & 0 & -115 & 0 & 0 & 0 \\ 0 & 982 & 0 & -982 & 0 & 0 & 0 & 0 \\ 0 & 0 & 351 & 176 & -351 & -176 & 0 & 0 \\ 0 & -982 & 176 & 1070 & -176 & -88 & 0 & 0 \\ -115 & 0 & -351 & -176 & 466 & 176 & 0 & 0 \\ 0 & 0 & -176 & -88 & 176 & 88 & 0 & 0 \\ 0 & 0 & 0 & 0 & 0 & 0 & 0 & 0 \\ 0 & 0 & 0 & 0 & 0 & 0 & 0 & 0 \end{bmatrix} \qquad \textbf{(3-55)}$$

Once again the reader should reexamine Example 3-4. This last result is the assemblage stiffness matrix after processing three elements.

The procedure to be used on element 4 is identical to that used above. The reader should verify that for element 4 we get

$$\mathbf{K}^{(4)'} = 10^3 \begin{bmatrix} 109 & 0 & -109 & 0 \\ 0 & 0 & 0 & 0 \\ -109 & 0 & 109 & 0 \\ 0 & 0 & 0 & 0 \end{bmatrix} \text{ lbf/in.} \qquad \textbf{(3-56)}$$

and with $n_{11} = 0.949$ and $n_{21} = -0.316$,

$$
\mathbf{K}^{(4)} = 10^3 \left[\begin{array}{cc|cc}
98 & -33 & -98 & 33 \\
-33 & 11 & 33 & -11 \\
\hline
-98 & 33 & 98 & -33 \\
33 & -11 & -33 & 11
\end{array}\right] \text{ lbf/in.} \tag{3-57}
$$

After processing four elements and noting that element 4 connects nodes 4 and 2, the assemblage matrix $\mathbf{K}^a$ becomes

$$
\mathbf{K}^a = 10^3 \left[\begin{array}{cc|cc|cc|cc}
115 & 0 & 0 & 0 & -115 & 0 & 0 & 0 \\
0 & 982 & 0 & -981 & 0 & 0 & 0 & 0 \\
\hline
0 & 0 & 450 & 143 & -351 & -176 & -98 & 33 \\
0 & -981 & 143 & 1080 & -176 & -88 & 33 & -11 \\
\hline
-115 & 0 & -351 & -176 & 466 & 176 & 0 & 0 \\
0 & 0 & -176 & -88 & 176 & 88 & 0 & 0 \\
\hline
0 & 0 & -98 & 33 & 0 & 0 & 98 & -33 \\
0 & 0 & 33 & -11 & 0 & 0 & -33 & 11
\end{array}\right] \tag{3-58}
$$

The pertinent results for element 5 are now summarized. The local element stiffness matrix $\mathbf{K}^{(5)'}$ is given by

$$
\mathbf{K}^{(5)'} = 10^3 \left[\begin{array}{cc|cc}
589 & 0 & -589 & 0 \\
0 & 0 & 0 & 0 \\
\hline
-589 & 0 & 589 & 0 \\
0 & 0 & 0 & 0
\end{array}\right] \tag{3-59}
$$

and with $n_{11} = 0.0$ and $n_{21} = 1.0$, the global element stiffness matrix becomes

$$
\mathbf{K}^{(5)} = 10^3 \left[\begin{array}{cc|cc}
0 & 0 & 0 & 0 \\
0 & 589 & 0 & -589 \\
\hline
0 & 0 & 0 & 0 \\
0 & -589 & 0 & 589
\end{array}\right] \tag{3-60}
$$

Since element 5 connects nodes 3 and 4 (in this order), the assemblage stiffness matrix becomes

$$
\mathbf{K}^a = 10^3 \left[\begin{array}{cc|cc|cc|cc}
115 & 0 & 0 & 0 & -115 & 0 & 0 & 0 \\
0 & 982 & 0 & -982 & 0 & 0 & 0 & 0 \\
\hline
0 & 0 & 450 & 143 & -351 & -176 & -98 & 33 \\
0 & -982 & 143 & 1080 & -176 & -88 & 33 & -11 \\
\hline
-115 & 0 & -351 & -176 & 466 & 176 & 0 & 0 \\
0 & 0 & -176 & -88 & 176 & 677 & 0 & -589 \\
\hline
0 & 0 & -98 & 33 & 0 & 0 & 98 & -33 \\
0 & 0 & 33 & -11 & 0 & -589 & -33 & 600
\end{array}\right] \tag{3-61}
$$

All five elements have now been processed, and this last version of the assemblage stiffness matrix represents the complete assemblage stiffness matrix for the entire truss. Note that it is symmetric and slightly banded with a half-bandwidth of 6 as observed from the third row (also see Example 3-5).

The next step is to determine the assemblage nodal force vector. Because the truss and loading in the present example are the same as those in Example 3-6, we may use the result of that example directly. For completeness, $\mathbf{f}^a$ is restated here:

$$\mathbf{f}^a = [-1000 \quad -1732 \mid 0 \quad 0 \mid 0 \quad 0 \mid 0 \quad 0]^T \tag{3-62}$$

Note the use of the transpose to denote $\mathbf{f}^a$, a column vector in row matrix form.

Actually, there are three other loads as a result of the reactions. As mentioned in Sec. 3-2, these reaction forces need not be considered in the construction of $\mathbf{f}^a$. However, let us include these reactions in $\mathbf{f}^a$ to review why this is the case. Let us denote the reaction in the x direction at node 3 as U_3 and denote the reactions in the x and y directions at node 4 as U_4 and V_4, respectively. If these external forces are included in the assemblage nodal force vector, we get

$$\mathbf{f}^a = [-1000 \quad -1732 \mid 0 \quad 0 \mid U_3 \quad 0 \mid U_4 \quad V_4]^T \tag{3-63}$$

The assemblage system equation $\mathbf{K}^a\mathbf{a} = \mathbf{f}^a$ becomes

$$10^3 \begin{bmatrix} 115 & 0 & 0 & 0 & -115 & 0 & 0 & 0 \\ 0 & 982 & 0 & -982 & 0 & 0 & 0 & 0 \\ 0 & 0 & 450 & 143 & -351 & -176 & -98 & 33 \\ 0 & -982 & 143 & 1080 & -176 & -88 & 33 & -11 \\ -115 & 0 & -351 & -176 & 466 & 176 & 0 & 0 \\ 0 & 0 & -176 & -88 & 176 & 677 & 0 & -589 \\ 0 & 0 & -98 & 33 & 0 & 0 & 98 & -33 \\ 0 & 0 & 33 & -11 & 0 & -589 & -33 & 600 \end{bmatrix} \begin{bmatrix} u_1 \\ v_1 \\ u_2 \\ v_2 \\ u_3 \\ v_3 \\ u_4 \\ v_4 \end{bmatrix} = \begin{bmatrix} -1000 \\ -1732 \\ 0 \\ 0 \\ U_3 \\ 0 \\ U_4 \\ V_4 \end{bmatrix}$$

$$\tag{3-64}$$

At this point, the matrix on the left is singular; that is, its inverse does not exist and, therefore, it is impossible to solve for $\mathbf{a}$, the vector of unknown nodal displacements. However, the restraints on some of the nodal displacements must be considered, namely $u_3 = 0$, $u_4 = 0$, and $v_4 = 0$. Let us use the second method presented in Sec. 3-2 to impose these restraints. (The first method is illustrated numerically in Chapter 4.) An arbitrary value of 1.0×10^{18} is taken as β, and Eq. (3-64) becomes

$$10^3 \begin{bmatrix} 115 & 0 & 0 & 0 & -115 & 0 & 0 & 0 \\ 0 & 982 & 0 & -982 & 0 & 0 & 0 & 0 \\ 0 & 0 & 450 & 143 & -351 & -176 & -98 & 33 \\ 0 & -982 & 143 & 1080 & -176 & -88 & 33 & -11 \\ -115 & 0 & -351 & -176 & 10^{15} & 176 & 0 & 0 \\ 0 & 0 & -176 & -88 & 176 & 677 & 0 & -589 \\ 0 & 0 & -98 & 33 & 0 & 0 & 10^{15} & -33 \\ 0 & 0 & 33 & -11 & 0 & -589 & -33 & 10^{15} \end{bmatrix} \begin{bmatrix} u_1 \\ v_1 \\ u_2 \\ v_2 \\ u_3 \\ v_3 \\ u_4 \\ v_4 \end{bmatrix} = \begin{bmatrix} -1000 \\ -1732 \\ 0 \\ 0 \\ 0 \times 10^{18} \\ 0 \\ 0 \times 10^{18} \\ 0 \times 10^{18} \end{bmatrix}$$

$$\tag{3-65}$$

The 10^{15} (and not 10^{18}) is entered on the appropriate diagonal entries because of the 10^3 scalar multiplier. Note that after the restraints are considered, the right-hand side is effectively

$$\mathbf{f} = [-1000 \quad -1732 \mid 0 \quad 0 \mid 0 \quad 0 \mid 0 \quad 0]^T$$

which is precisely the same as Eq. (3-62). Therefore, the loads as a result of the reactions need not be considered—a very fortuitous result.

If Eq. (3-65) is solved by the matrix inversion method presented in Sec. 2-14, the resulting nodal displacements, in inches, are

$$u_1 = -0.00868 \qquad u_2 = +0.00996$$
$$v_1 = -0.03528 \qquad v_2 = -0.03351$$

$$u_3 = 0.00000 \qquad u_4 = 0.00000$$
$$v_3 = -0.00176 \qquad v_4 = 0.00000$$

Note that the restrained degrees of freedom all have zero displacements, as expected.

The only remaining task is to determine the so-called element resultants, which include the axial elongations, strains, stresses, and forces. Element 3 is considered in detail to illustrate the use of Eqs. (3-41) to (3-45). The final results for all elements are subsequently summarized below. Recall that element 3 connects nodes 3 and 2 (in this order). From Eqs. (3-41) and (3-42) with $n_{11} = 0.894$ and $n_{21} = 0.447$, the axial elongation δ for element 3 becomes

$$\delta^{(3)} = (n_{11}u_j + n_{21}v_j) - (n_{11}u_i + n_{21}v_i)$$

or

$$\delta^{(3)} = [(0.894)(0.00996) + (0.447)(-0.03351)]$$
$$\qquad - [(0.894)(0.0) + (0.447)(-0.00176)]$$
$$= -0.00529 \text{ in.}$$

The strain in element 3 is readily computed by Eq. (3-43) as

$$\varepsilon^{(3)} = \frac{\delta}{L} = \frac{-0.00529}{13.42} = -394 \times 10^{-6} \text{ in./in.}$$

the stress from Eq. (3-44) as

$$\sigma^{(3)} = \varepsilon E = (-394 \times 10^{-6})(30 \times 10^6) = -11,800 \text{ psi}$$

and finally the force as

$$F^{(3)} = \sigma A = (-11,800)(0.1963) = 2320 \text{ lbf compression}$$

The reader should determine the axial elongations, strains, stresses, and forces for the remaining elements (see Problems 3-53 to 3-56). The results are summarized below.

Element number	δ, in.	ε, in./in.	σ, psi	F, lbf
1	0.00176	0.000294	8,820	1730
2	−0.00868	−0.000723	−7,960	−1000
3	−0.00529	−0.000395	−11,800	−2320
4	0.02004	0.001585	17,400	2190
5	0.00176	0.000176	5,290	1040

Note that elements 1, 4, and 5 (members E, B, and A) are in tension and elements 2 and 3 (members D and C) are in compression. The maximum force of 2320 (compression) occurs in element 3 (member C) but the maximum stress of 17,400 psi (tension) occurs in element 4 (member B). In Problem 3-57 the reader is asked to do a force balance on each node, which serves as a check on these results and also yields the (unknown) reaction forces U_3, U_4, and V_4. In Problem 3-58, the reader is asked to determine these reaction forces directly with the help of Eq. (3-64).

This completes the two-dimensional truss example. In Sec. 3-4 the basic steps in all finite element analyses are summarized. This is followed by the three-dimensional truss formulation and a description of a two-dimensional truss program, called TRUSS, which automates in a computer program all of the steps executed above. The program may be applied to the problem here, as well as to many other more complicated two-dimensional truss problems.

3-4   SUMMARY OF THE BASIC STEPS

The simple problem in Sec. 3-3 illustrated all the basic steps in the finite element method. In the steps that are summarized below, it is assumed that the problem is well-posed, a global coordinate system is defined, and all pertinent geometrical and material data are known.

Step 1: Discretization

The structure, body, or region being analyzed must be discretized into a suitable number of elements. Each of the elements has several nodes associated with it. For example, each element in the truss model has two nodes—one at either end of the element. Discretization results in the specification of the finite element mesh and involves two distinct but related tasks: nodal definitions and element definitions.

The nodes are always numbered consecutively from one to the total number of nodes present. The nodal numbering pattern has a strong influence on execution time in a computer program (for large problems), as explained in Chapters 7 and 8. The basic idea is to number the nodes in such a way so as to minimize the bandwidth of the assemblage stiffness matrix. Many large, canned finite element programs renumber the nodes automatically to ensure this. Nodal definitions are complete when the coordinates of each of the nodes are also specified.

To define the elements, we first number them consecutively from one to the maximum number of elements present. The element numbering scheme is completely arbitrary. The nodes associated with each element must be specified. In addition, the material property data to be used for each element are specified with the help of material set flags. Using material set flags reduces significantly the amount of material property data to be supplied in any computer program.

It should be mentioned that any practical finite element program must have provisions for automatic or semiautomatic mesh generation. A simple mesh generator is discussed in Sec. 3-6.

Step 2: Determination of the Local Element Characteristics

Expressions for the local element characteristics must be derived or taken from a suitable reference. By *element characteristics* we mean the element stiffness matrices and nodal force vectors. The word "local" refers to the fact that the element characteristics are derived in a local reference system, which usually changes from element to element. The proper expressions for the element characteristics in structural problems may be derived by the direct approach, the variational approach, including the principle of minimum potential energy, the virtual-work approach, or the weighted-residual approach. In the truss problem, the direct approach was used. The other approaches are covered in detail in several of the following chapters. In nonstructural problems, the variational and weighted-residual approaches are used most often. With expressions in hand for the local element characteristics, the local element stiffness matrices (and nodal force vectors) may be determined numerically for each element.

Step 3: Transformation of the Element Characteristics

The element characteristics from Step 2 must be transformed from the local coordinate systems to the global system. This is accomplished with the help of a transformation matrix as explained in Sec. 3-2 for the truss model. The transformation of the local element characteristics needs to be performed only when the unknown parameter function is a vector, such as the (nodal) displacements in the truss problem, and then only when a local coordinate system is used.

Step 4: Assemblage of the Global Element Characteristics

The global element stiffness matrices and global element nodal force vectors must be assembled to form the assemblage stiffness matrix and assemblage nodal force vector. The basic idea behind Step 4 is that the unknown parameter function must have the same value at any given node regardless of the element containing the node. The assemblage procedure for the truss was illustrated symbolically in Example 3-4 and numerically in Sec. 3-3.

Step 5: Application of the Prescribed Displacements

After the assemblage step, the assemblage stiffness matrix and assemblage nodal force vector must be modified according to one of the two methods presented in Sec. 3-2 in order to impose the constraints on the restrained degrees of freedom. In structural and stress analysis problems, these are typically the nodal displacements (or deflections and slopes in models involving bending). In thermal analysis problems, the prescribed nodal temperatures must be imposed in a similar manner. In fluid flow problems, the prescribed nodal velocities and pressures must be similarly imposed. The resulting stiffness matrix after Step 5 is nonsingular for well-posed problems.

Step 6: Solution

The resulting system of equations $\mathbf{Ka} = \mathbf{f}$ must be solved for the vector of nodal unknowns $\mathbf{a}$. In the truss problem, the vector $\mathbf{a}$ contains the nodal displacements as given by Eq. (3-35). The solution may be obtained by any of the methods suitable to a system of linear algebraic equations. In this chapter, only the matrix inversion method of solution is used. In Chapter 6, another method is presented.

In thermal analysis problems the vector $\mathbf{a}$ is generally composed of the nodal temperatures, and in fluid flow problems it is composed of the nodal velocities and nodal pressures.

Step 7: Calculation of the Element Resultants

The element resultants, such as axial elongations, strains, stresses, and internal forces, are computed more or less directly from the nodal displacements from the solution step. In thermal analysis and fluid flow problems, the element resultants are the internal heat-flows and fluid stresses, which are computed from the nodal temperatures and nodal velocities, respectively.

3-5   THREE-DIMENSIONAL TRUSS FORMULATION

Let us now extend the development in Sec. 3-2 to the three-dimensional truss. Consider the global xyz coordinate system shown in Fig. 3-11(a). Note the position vector $\mathbf{r}$ which defines an arbitrary point P. Let us write the vector $\mathbf{r}$ in terms of its three cartesian components as follows

$$\mathbf{r} = r_x\mathbf{i} + r_y\mathbf{j} + r_z\mathbf{k} \qquad \text{(3-66)}$$

where $\mathbf{i}$, $\mathbf{j}$, and $\mathbf{k}$ are the three unit vectors in the x, y, and z directions, respectively, as shown on the figure.

As in the two-dimensional truss formulation, let us write the same vector $\mathbf{r}$ in terms of its components in the rotated coordinate system $x'y'z'$ where, with no loss of generality, the x' axis is purposely taken to be in the same direction as $\mathbf{r}$ itself, as shown in Fig. 3-11(b). Writing the vector $\mathbf{r}$ in terms of its components $(r'_x, r'_y, \text{ and } r'_z)$ in the rotated system, we have

$$\mathbf{r} = r'_x\mathbf{i}' + r'_y\mathbf{j}' + r'_z\mathbf{k}' \qquad \text{(3-67)}$$

where $\mathbf{i}'$, $\mathbf{j}'$, and $\mathbf{k}'$ are the unit vectors in the x', y', and z' directions, respectively. Again the $x'y'z'$ coordinate system is commonly referred to as a local coordinate system. From Eqs. (3-66) and (3-67), we have

$$r_x\mathbf{i} + r_y\mathbf{j} + r_z\mathbf{k} = r'_x\mathbf{i}' + r'_y\mathbf{j}' + r'_z\mathbf{k}' \qquad \text{(3-68)}$$

Forming the dot product of this last equation with i gives

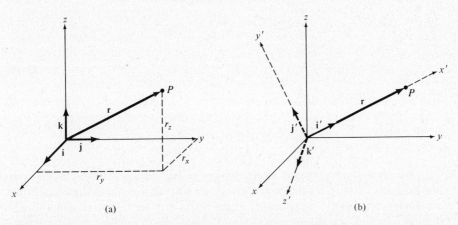

Figure 3-11   Position vector $\mathbf{r}$ in a three-dimensional (a) global coordinate system and (b) local coordinate system.

$$r_x = r_x'\mathbf{i}\cdot\mathbf{i}' + r_y'\mathbf{i}\cdot\mathbf{j}' + r_z'\mathbf{i}\cdot\mathbf{k}' \tag{3-69}$$

But

$$\mathbf{i}\cdot\mathbf{i}' = \cos(x,x') = n_{11} \tag{3-70a}$$

$$\mathbf{i}\cdot\mathbf{j}' = \cos(x,y') = n_{12} \tag{3-70b}$$

and

$$\mathbf{i}\cdot\mathbf{k}' = \cos(x,z') = n_{13} \tag{3-70c}$$

Therefore, Eq. (3-69) becomes

$$r_x = n_{11}r_x' + n_{12}r_y' + n_{13}r_z' \tag{3-71}$$

In a completely analogous manner, we also have

$$r_y = n_{21}r_x' + n_{22}r_y' + n_{23}r_z' \tag{3-72}$$

and

$$r_z = n_{31}r_x' + n_{32}r_y' + n_{33}r_z' \tag{3-73}$$

The last three equations may be written in matrix notation as

$$\begin{bmatrix} r_x \\ r_y \\ r_z \end{bmatrix} = \begin{bmatrix} n_{11} & n_{12} & n_{13} \\ n_{21} & n_{22} & n_{23} \\ n_{31} & n_{32} & n_{33} \end{bmatrix} \begin{bmatrix} r_x' \\ r_y' \\ r_z' \end{bmatrix} \tag{3-74a}$$

or more concisely as

$$\mathbf{r} = \mathbf{T}\mathbf{r}' \tag{3-74b}$$

Equation (3-74b) transforms the components of an arbitrary vector in the local coordinate system to the components in the global system. Since it can be shown that the transformation matrix $\mathbf{T}$ is orthogonal, we have

$$\mathbf{T}^T = \mathbf{T}^{-1}$$

and

$$\mathbf{r}' = \mathbf{T}^{-1}\mathbf{r} = \mathbf{T}^T\mathbf{r} \tag{3-75}$$

where

$$\mathbf{T}^T = \begin{bmatrix} n_{11} & n_{21} & n_{31} \\ n_{12} & n_{22} & n_{32} \\ n_{13} & n_{23} & n_{33} \end{bmatrix}$$

We are now in a position to develop the global element stiffness matrix from the local one for the three-dimensional truss. First of all, we must extend the notation to include the z and z' components of the nodal displacements and nodal forces. Let us represent the three components of displacement as u, v, w and the three components of nodal forces as U, V, W. The subscripts i and j are used as usual to denote the two respective nodes on each element. A prime (') denotes that the variable in question is referenced to the local system (i.e., the $x'y'z'$ coordinate system), while the absence of a prime refers to the global system (i.e., the xyz coordinate system). Figure 3-12 may help to clarify the notation; only nodal displacements are shown, since the figure showing nodal forces would look identical except that uppercase letters would be used to denote the forces.

As in the development for the two-dimensional truss, the two nodal displacements in the x' direction, u'_i and u'_j, are related to the two nodal forces in the same direction, U'_i and U'_j, by

$$U'_i = \frac{AE}{L} (u'_i - u'_j) \tag{3-77a}$$

and

$$U'_j = \frac{AE}{L} (u'_j - u'_i) \tag{3-77b}$$

where A represents the cross-sectional area of the element, E the modulus of elasticity, and L the element length. Now L may be computed from

$$L = \sqrt{(x_j - x_i)^2 + (y_j - y_i)^2 + (z_j - z_i)^2} \tag{3-78}$$

For the truss element only axial loads are allowed, and so we have

$$V'_i = W'_i = 0 \tag{3-77c}$$

and

$$V'_j = W'_j = 0 \tag{3-77d}$$

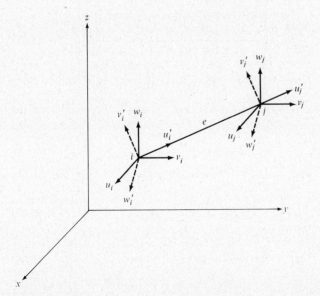

Figure 3-12 Nodal displacements in xyz and $x'y'z'$ coordinate systems.

In other words, all joints are assumed to be pinned. Let us write Eqs. (3-77) in matrix form as

$$\frac{AE}{L}
\begin{bmatrix}
1 & 0 & 0 & -1 & 0 & 0 \\
0 & 0 & 0 & 0 & 0 & 0 \\
0 & 0 & 0 & 0 & 0 & 0 \\
-1 & 0 & 0 & 1 & 0 & 0 \\
0 & 0 & 0 & 0 & 0 & 0 \\
0 & 0 & 0 & 0 & 0 & 0
\end{bmatrix}
\begin{bmatrix}
u_i' \\ v_i' \\ w_i' \\ u_j' \\ v_j' \\ w_j'
\end{bmatrix}
=
\begin{bmatrix}
U_i' \\ V_i' \\ W_i' \\ U_j' \\ V_j' \\ W_j'
\end{bmatrix}
\tag{3-79a}$$

or much more concisely as

$$\mathbf{K}^{e'}\mathbf{a}^{e'} = \mathbf{f}^{e'} \tag{3-79b}$$

where $\mathbf{K}^{e'}$ represents the local element stiffness matrix, $\mathbf{a}^{e'}$ the local element nodal displacement vector, and $\mathbf{f}^{e'}$ the local element nodal force vector.

Let us now determine the global element stiffness matrix. From the development in Sec. 3-2, it should be obvious that

$$\mathbf{a}^{e'} = \mathbf{R}\mathbf{a}^e \tag{3-80}$$

and

$$\mathbf{f}^{e'} = \mathbf{R}\mathbf{f}^e \tag{3-81}$$

where

$$\mathbf{R} = \begin{bmatrix} \mathbf{T}^T & 0 \\ 0 & \mathbf{T}^T \end{bmatrix} =
\begin{bmatrix}
n_{11} & n_{21} & n_{31} & 0 & 0 & 0 \\
n_{12} & n_{22} & n_{32} & 0 & 0 & 0 \\
n_{13} & n_{23} & n_{33} & 0 & 0 & 0 \\
0 & 0 & 0 & n_{11} & n_{21} & n_{31} \\
0 & 0 & 0 & n_{12} & n_{22} & n_{32} \\
0 & 0 & 0 & n_{13} & n_{23} & n_{33}
\end{bmatrix}
\tag{3-82}$$

As usual, $\mathbf{a}^e$ and $\mathbf{f}^e$, respectively, denote the vector of nodal displacements and the vector of nodal forces for element e, referred to the global system. These are given explicitly by

$$\mathbf{a}^e = [\; u_i \quad v_i \quad w_i \;|\; u_j \quad v_j \quad w_j \;]^T \tag{3-83}$$

and

$$\mathbf{f}^e = [\; U_i \quad V_i \quad W_i \;|\; U_j \quad V_j \quad W_j \;]^T \tag{3-84}$$

Equation (3-79b) becomes

$$\mathbf{K}^{e'}\mathbf{R}\mathbf{a}^e = \mathbf{R}\mathbf{f}^e \tag{3-85}$$

which may be premultiplied by $\mathbf{R}^{-1}$ (which equals $\mathbf{R}^T$) to give

$$\mathbf{K}^e\mathbf{a}^e = \mathbf{f}^e \tag{3-86}$$

where $\mathbf{K}^e$ represents the global element stiffness matrix and is obviously defined by

$$\mathbf{K}^e = \mathbf{R}^T \mathbf{K}^{e'} \mathbf{R} \tag{3-87}$$

If $\mathbf{K}^e$ is determined explicitly, one gets

$$\mathbf{K}^e = \frac{AE}{L} \begin{bmatrix} n_{11}^2 & n_{21}n_{11} & n_{31}n_{11} & -n_{11}^2 & -n_{21}n_{11} & -n_{31}n_{11} \\ n_{11}n_{21} & n_{21}^2 & n_{31}n_{21} & -n_{11}n_{21} & -n_{21}^2 & -n_{31}n_{21} \\ n_{11}n_{31} & n_{21}n_{31} & n_{31}^2 & -n_{11}n_{31} & -n_{21}n_{31} & -n_{31}^2 \\ -n_{11}^2 & -n_{21}n_{11} & -n_{31}n_{11} & n_{11}^2 & n_{21}n_{11} & n_{31}n_{11} \\ -n_{11}n_{21} & -n_{21}^2 & -n_{31}n_{21} & n_{11}n_{21} & n_{21}^2 & n_{31}n_{21} \\ -n_{11}n_{31} & -n_{21}n_{31} & -n_{31}^2 & n_{11}n_{31} & n_{21}n_{31} & n_{31}^2 \end{bmatrix} \tag{3-88}$$

Interestingly, the global element stiffness matrix $\mathbf{K}^e$ is a function of only three easily calculated direction cosines, namely, n_{11}, n_{21}, and n_{31}, in addition to A, E, and L. By definition of the direction cosines, we have

$$n_{11} = \cos(x,x') \tag{3-89a}$$

$$n_{21} = \cos(y,x') \tag{3-89b}$$

$$n_{31} = \cos(z,x') \tag{3-89c}$$

Let us define the vector $\mathbf{r}_{ij}$ which runs from node i to j in the global system as

$$\mathbf{r}_{ij} = (x_j - x_i)\mathbf{i} + (y_j - y_i)\mathbf{j} + (z_j - z_i)\mathbf{k} \tag{3-90}$$

But $|\mathbf{r}_{ij}| = L$, and

$$\mathbf{i}' = \frac{\mathbf{r}_{ij}}{|r_{ij}|} = \frac{\mathbf{r}_{ij}}{L} \tag{3-91}$$

so we have

$$\mathbf{i}' = \left(\frac{x_j - x_i}{L}\right)\mathbf{i} + \left(\frac{y_j - y_i}{L}\right)\mathbf{j} + \left(\frac{z_j - z_i}{L}\right)\mathbf{k} \tag{3-92}$$

But dotting both sides of Eq. (3-92) with $\mathbf{i}$ gives

$$\mathbf{i} \cdot \mathbf{i}' = \frac{x_j - x_i}{L}$$

where $\mathbf{i} \cdot \mathbf{i}' = \cos(x,x') = n_{11}$. Similarly dotting Eq. (3-92) with $\mathbf{j}$ and $\mathbf{k}$ gives

$$\mathbf{j} \cdot \mathbf{i}' = \frac{y_j - y_i}{L} = n_{21}$$

and

$$\mathbf{k \cdot i'} = \frac{z_j - z_i}{L} = n_{31}$$

In summary, the three direction cosines n_{11}, n_{21}, and n_{31} are given by

$$n_{11} = \cos(x,x') = \frac{x_j - x_i}{L} \qquad \textbf{(3-93a)}$$

$$n_{21} = \cos(y,x') = \frac{y_j - y_i}{L} \qquad \textbf{(3-93b)}$$

$$n_{31} = \cos(z,x') = \frac{z_j - z_i}{L} \qquad \textbf{(3-93c)}$$

where L, of course, is the length of the element given by Eq. (3-78). For any element, the global nodal coordinates are known since they are an integral part of the input to any finite element program. Therefore, the global element stiffness matrix $\mathbf{K}^e$ is easily computed from Eq. (3-88).

The assemblage process remains the same as described in Sec. 3-2 except that each of the four 2×2 submatrices in $\mathbf{K}^e$ is now a 3×3 submatrix. Prescribed displacement boundary conditions are applied in the usual manner. External nodal forces are also applied in the same manner as described in Sec. 3-2 for the two-dimensional truss. The resulting assemblage system equation

$$\mathbf{Ka} = \mathbf{f} \qquad \textbf{(3-94)}$$

is solved for the nodal displacements, which in turn can be used to solve for the internal axial forces and stresses in each element. The procedure is summarized below. The elongation δ of an element is given by

$$\delta = u'_j - u'_i \qquad \textbf{(3-95)}$$

where from Eq. (3-75), we have

$$u'_i = n_{11}u_i + n_{21}v_i + n_{31}w_i \qquad \textbf{(3-96a)}$$

and

$$u'_j = n_{11}u_j + n_{21}v_j + n_{31}w_j \qquad \textbf{(3-96b)}$$

Since the direction cosines are readily calculated with the help of Eq. (3-93), and since the nodal displacements u_i, v_i, w_i, u_j, v_j, and w_j are now known (i.e., from the solution for $\mathbf{a}$ in $\mathbf{Ka} = \mathbf{f}$), the elongation δ is easily found for each element. The axial strains ε, stresses σ, and forces F within each element can then be determined by using Eqs. (3-43) to (3-45). Negative values of ε, σ, and F denote compression (when $\delta < 0$). This completes the formulation of the three-dimensional truss problem.

3-6   DESCRIPTION OF A SIMPLE COMPUTER PROGRAM: TRUSS

A simple FORTRAN program for the analysis of two-dimensional trusses is presented in this section. A complete listing of the program is provided in Appendix B. The program is capable of analyzing trusses with up to 20 joints and 30 members. No more than 5 different types of "materials" may be present with no more than 15 different nodal forces and 5 different prescribed displacements. Although nearly 20 subroutines are used, the program is quite simple. In fact, the use of subroutines in this manner should facilitate one's understanding of the program. In general, each subroutine performs only one task or a few related tasks. Any consistent set of units may be used.

The program is written in the American National Standards Institute FORTRAN language as described in the ANSI document X3.9-1966 (approved on March 7, 1966). The newer FORTRAN 77 is not used, but the program should run with no significant changes on systems using FORTRAN 77. The program is meant to be run from a console where the variable LCONSL in the main program is the console device number (e.g., 3 on an Apple II microcomputer, 5 on the VAX 11/780). The user should change this number accordingly on his or her particular system.

The input to the program is described in detail in Appendix B. Table 3-1 contains a ready reference for the input to TRUSS and should be used in conjunction with Appendix B. The input is divided into seven sections. Section 1 begins with an *80-column title*, which may be used to document the case being run. This title is followed by the *master control data*, which includes the number of nodes, the number of elements, the number of materials, the number of different prescribed displacements, the number of different point loads, and the output device number. Each of the remaining six input sections is always preceded by a dummy subtitle whose only purpose is to make the input file more readable.

Input Sections 2 and 3 are used to define the mesh, including *nodal coordinate and element data*, as well as the material set flags for each element. As explained in Appendix B, the nodal coordinate and element data may be defined on a one-by-one basis or they may be generated. Input Section 4 contains the *material property data* for each material set present. For the truss, these properties are the cross-sectional area and modulus of elasticity. Input Sections 5, 6, and 7 include the *boundary condition flag data*, the imposed *nodal force data*, and the imposed *nodal displacement data*. Input Sections 2 through 7 must be terminated with a mandatory *blank line*.

Description of the Program

A simplified flowchart of the main program is shown in Figure 3-13. This flowchart is typical of all finite element programs that are based on the so-called stiffness approach. The main program does not do any of the finite element computations—these are relegated to the subroutines.

Table 3-1 Summary of Input to TRUSS Program

The input to TRUSS is summarized to provide a ready reference once the detailed input explanations are understood (see Appendix B). All titles are read under 20A4 formats, all integer input variables are read with I8 formats, and all real input variables are read with F8.0 formats. Integer variables always begin with the letters I to N. Microsoft's FORTRAN-80 allows the user to enter the data in free format style with commas or tabs (control-I) separating the input fields.

Section 1 Input
TITLE (80-column title)
NNODES   NELEM   NMATLS   NPDIS   NPLDS   LOUT

Section 2 Input
SUBT   (e.g., NODAL COORDINATE DATA)
NI   NG   NF
XI   YI   XF   YF
    (blank line)

Section 3 Input
SUBT (e.g., ELEMENT DATA)
LI   MS   LG   LF   NG
NI   NJ
    (blank line)

Section 4 Input
SUBT (e.g., MATERIAL PROPERTY DATA)
MSNO   AREA   ELMOD
    (blank line)

Section 5 Input
SUBT (e.g., BOUNDARY CONDITION FLAG DATA)
NI   IBCX   IBCY   NG   NF
    (blank line)

Section 6 Input
SUBT (e.g., NODAL FORCE DATA)
NFORCE   FORCE
    (blank line)

Section 7 Input
SUBT (e.g., NODAL DISPLACEMENT DATA)
NDISP   DISP
    (blank line)

The main program begins by setting LCONSL to the console device number and by writing on the screen:

INPUT THE NUMBER OF THE INPUT FILE (6-10):

to which the user responds "6" if the input file is FORTO6.DAT, "7" if the input file is FORTO7.DAT, etc. If the user enters "0" or presses the RETURN key, the program stops. This gives the capability to run multiple cases. This is the only portion of the program that is system-dependent and it may need to be modified for use on other systems. The description given here applies to the Microsoft version of FORTRAN called FORTRAN-80 [7], and the program runs without changes on the Apple II Plus microcomputer with a Z-80 card installed, as well as on other microcomputers with the CP/M[1] operating system.

The main program then reads the 80-column title and the master control data. The mesh is generated by calls to subroutines NODGEN and ELEGEN. In subroutine NODGEN, Section 2 of the input file is read and the nodal coordinates are specified either on a node-by-node basis or by generation. The x and y coordinates of node I are stored in XCOOR(I) and YCOOR(I), respectively. In subroutine ELEGEN, Section 3 of the input file, which is used to define the elements and to specify the material set flags for each element, is read. Again, these data may either be given on a element-by-element basis or by generation. The method used to generate the mesh is described later in this section. For element L, the global node numbers I and J are stored in NODI(L) and NODJ(L), respectively, and the material set flag is stored in MATFLG(L). The material property data in Input Section 4 are read next by a call to subroutine MATERL. For material set number MSNO, the cross-sectional area and the modulus of elasticity are read and stored in DATMAT(MSNO,1) and DATMAT(MSNO,2), respectively.

Subroutine BCOND is then called; it reads the boundary condition flags, the imposed nodal forces, and the imposed nodal displacements. The boundary condition flags may be specified either on a node-by-node basis or by generation. For node I, the two boundary condition flags in the x and y direction are stored in NBCX(I) and NBCY(I), respectively. For nodal force number NFORCE, the value of the force is read and stored in FORCE(NFORCE). Similarly, for nodal displacement number NDISP, the value of the displacement is read and stored in DISP(NDISP). The use of the boundary condition flags is explained in detail in Appendix B. Suffice it to say now that negative flags are used to denote forces and positive flags to denote displacements. The input data are then printed in summary form by a call to subroutine SUMMRY.

In the next section of the main program, the assemblage nodal force vector and the assemblage stiffness matrix are zeroed-out. The respective array names are ANFV and ASM. Since 20 joints or nodes are allowed and each node has two degrees of freedom (x and y components of displacement), the dimensions of ANFV and ASM are 40×1 and 40×40, respectively.

In the DO loop labeled "2500" the global element stiffness matrices are generated by a call to subroutine STIFF with the help of prior calls to subroutines COORDS, LENGTH, PROPTY, and TRANSF. Subroutine COORDS simply re-

---

[1]CP/M is a registered trademark of Digital Research.

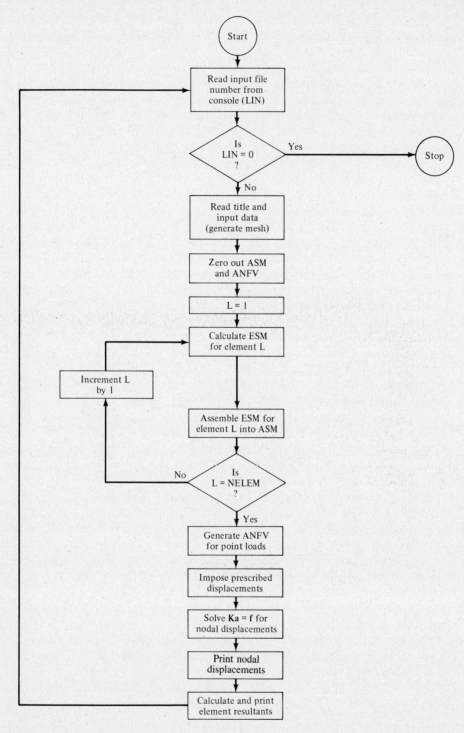

Figure 3-13 Flowchart of program TRUSS.

turns the x and y coordinates of nodes I and J for element L and stores them in XX(1), YY(1) for node I and XX(2), YY(2) for node J. Subroutine LENGTH computes the length of a straight line between two points and, of course, is used to compute the element length. Subroutine PROPTY returns the cross-sectional area (AREA) and the elastic modulus (ELMOD) to be used for element L from the DATMAT array. Subroutine TRANSF returns the two direction cosines, n_{11} and n_{21}, which are needed to obtain the global element stiffness matrix for element L. These direction cosines are stored in DIRCOS(1) and DIRCOS(2), respectively.

Subroutine STIFF then generates the global element stiffness matrix directly by using Eq. (3-26). As in the example problem in Sec. 3-3, it is more expedient to use Eq. (3-26) rather than Eq. (3-25) since the matrix multiplications result in the disappearance of n_{12} and n_{22} as shown in Sec. 3-2. The 4×4 global element stiffness matrix, stored in array ESM, for element L is immediately assembled into the assemblage stiffness matrix ASM by a call to subroutine ASSEMK. In other words, the element stiffness matrix for element 1 is generated and assembled, then for element 2, and so forth. This reduces the memory required since the individual element stiffness matrices are not saved. The reader should examine the listing of subroutine ASSEMK in Appendix B carefully because it gives the algorithm for the assemblage step in a very explicit form.

The assemblage nodal force vector ANFV is generated by subroutine ANFVEC that applies the nodal forces. Then the assemblage stiffness matrix and the assemblage nodal force vector are modified to impose the prescribed displacements by a call to subroutine PDBC. Method 1 of Sec. 3-2 is used to impose these displacements. The system of equations $\mathbf{Ka} = \mathbf{f}$ is then solved by calling subroutine EQSOLV which returns the solution for the nodal displacements in the vector SOLN. Subroutine EQSOLV obtains the solution by calling the matrix inversion subroutine INVDET [8] (which computes $\mathbf{K}^{-1}$), and the matrix-vector multiplication subroutine MATVEC (which computes $\mathbf{a}$ by postmultiplying $\mathbf{K}^{-1}$ by the vector $\mathbf{f}$). The nodal displacements are printed by calling subroutine PRINTN, and the element resultants are computed and printed via a call to the postprocessor subroutine POSTPR. The element resultants include the following for every element: the axial elongation, strain, stress, and force.

The main program then returns to the beginning for a new case. The user may then enter the input file number for a second case. If no more cases are to be run, the user simply presses the RETURN key on the console to stop execution. Cases may also be stacked in one input file.

Mesh Generation

All practical finite element programs must have provisions for some type of node and element generation—referred to collectively as *mesh generation*. The TRUSS program can be used to illustrate one simple mesh generation scheme. Some schemes generate the nodes and elements in two separate steps (such as the one to be presented here), whereas others generate nodes and elements in one integrated step. References

[9–11] give other mesh-generating algorithms. The scheme used here is patterned after that presented by Zienkiewicz [12].

Node Generation

The easiest way to introduce the node generation scheme used in the TRUSS program is to illustrate it with an example. Consider the straight line shown in Fig. 3-14(a) along which six equally spaced nodes are to be placed. Let us assume that the nodes are numbered in such a way that the difference between two successive node numbers is a constant, as shown in the figure. Let NI be the *initial* node, NG the *nodal increment*, and NF the *final* node. In the present example, we have NI = 3, NG = 4, and NF = 23. Let us further assume that the coordinates of nodes NI and NF are also known and are given by (XI,YI) and (XF,YF), respectively. The number of equally spaced divisions, DIV, is given by

$$DIV = \frac{NF - NI}{NG}$$

which is always a whole number (providing the nodes are numbered properly). The increments in the x and y directions, DX and DY, are then given by

$$DX = \frac{XF - XI}{DIV}$$

and

$$DY = \frac{YF - YI}{DIV}$$

The first node in this generation sequence is completely specified since its node number is NI and its coordinates are (XI,YI). The second node is defined from the first with the global nodal number given by NI + NG, the x coordinate by XI + DX, and the y coordinate as YI + DY. The third node is generated from the second in a similar fashion. This process is repeated until the last node (whose global number is NF) is generated. In order to avoid round-off errors, it is prudent to assign the known coordinates (XF,YF) to node NF. In the present example, DIV = (23 − 3)/4 = 5,

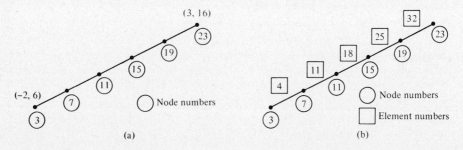

Figure 3-14   Illustration of mesh generation for (a) the nodes and (b) the elements.

DX = [3 − (−2)]/5 = 1, and DY = (16 − 6)/5 = 2. The resulting nodal coordinates are summarized as follows.

Node number	x coordinate	y coordinate
3	−2	6.
7	−1	8.
11	0	10.
15	1.	12.
19	2.	14.
23	3.	16.

In summary, the following variables must be provided as input: NI, NG, NF, XI, YI, XF, YF. The reader should compare these variables with those required in Input Section 2 described in Appendix B and summarized in Table 3-1. In Chapter 7, this method is modified slightly so that the nodes are not necessarily equally spaced. This modification allows the practical use of graded meshes.

Element Generation

Like nodal generation, element generation is explained most easily by an illustration. Let us consider the elements shown in Fig. 3-14(b). It should be noted that not only is there a constant difference between two successive node numbers, but there is also a constant difference between two successive element numbers. Let LI be the *initial* element number, LG the *element number increment*, LF the *final* element number, and NG the *nodal number increment*. Furthermore, let us assume that nodes *i* and *j* for the initial element LI are given by NI and NJ, respectively. This is also a very convenient place to specify the material set to be used for all elements in this generation sequence. Let us denote the material set flag as MS. It should be apparent that the first element is completely defined because the element number is LI, nodes *i* and *j* are NI and NJ, and the material set number is MS. From this element, it is a simple matter to define the second element: the element number is LI + LG, node *i* is NI + NG, node *j* is NJ + NG, and again the material set number is MS. The third element is defined from the second, and so on, until the last element LF is defined. For the example above, LI = 4, LG = 7, LF = 32, and NG = 4. Let us assume arbitrarily that material set 2 is to be used for these elements. The resulting element definitions in this element generation sequence are summarized as follows:

Element number	Node i	Node j	Material set number
4	3	7	2
11	7	11	2
18	11	15	2
25	15	19	2
32	19	23	2

In summary, the following variables must be specified for each element generation sequence: LI, MS, LG, LF, NG, NI, and NJ. These variables should be compared with those required in Input Section 3 (Table 3-1 and Appendix B). The reader should also examine subroutines NODGEN and ELEGEN in Appendix B carefully so that this method of mesh generation is thoroughly understood.

Example 3-7

Using the TRUSS program, resolve the example problem in Sec. 3-3. More specifically, determine the displacements of each joint, and for each member determine the axial elongation, strain, stress, and internal force. The truss is shown in Fig. 3-4(a).

Solution

For convenience, let us use the discretization (or mesh) shown in Fig. 3-4(b). A complete listing of the input file is given in Table 3-2. The reader should study this input with frequent reference to Table 3-1 and Appendix B. A few key points with respect to the input are made below.

First note that following the TITLE are the master control data: four nodes, five elements, two materials, one prescribed displacement, two different point loads, and 3 for the output device number (the Apple II console). This is immediately followed by the NODAL COORDINATE DATA subtitle and the actual input. Note that nodes 1 and 2 are defined on a node-by-node basis (NG = 0), while nodes 3 and 4 are defined via the generation feature (with NG = 1). These input lines are followed by a mandatory blank line.

The ELEMENT DATA subtitle and data follow next. Elements 1, 3, and 5 are defined in the first-generation sequence and elements 2 and 4 in the second-generation sequence. From the first sequence note that element 3 would be defined as having nodes 2 and 3 (in this order). It is desirable to use the discretization from Example 3-1, and for this reason element 3 is redefined in the input file as having nodes 3 and 2. The TRUSS program always uses the latest nodal coordinate and element data specified. Again a mandatory blank line terminates this input section (as well as all the following sections).

The MATERIAL DATA subtitle and material property definitions then follow. Note that the material data can be specified in any order. The BOUNDARY CONDITION FLAGS subtitle and data follow. Note that the flags for node 2 are not defined, so IBCX = 0 and IBCY = 0 are used (by default) for this node. The input file is completed by the NODAL FORCE LOADS and CONSTRAINED NODAL DISPLACEMENTS subtitles and data. The material data, force data, and displacement data may be specified more than once for each condition (like the nodal coordinate, element, and boundary condition flag data). The program simply uses the last values read when conflicting input data are provided.

The output from the program is shown in Table 3-3 and is self-explanatory. The nodal displacements and element resultants agree with the results from Sec. 3-3. ■

Table 3-2   Input Data File for Example 3-7

EXAMPLE 3-7: MODEL OF TRUSS SHOWN IN FIG. 3-4					
4	5	2	1	2	3

NODAL COORDINATE DATA

1			
12.0	0.0		
2			
12.0	6.0		
3	1	4	
0.	0.	0.	10.

ELEMENT DATA

1	1	2	5	1
1	2			
2	2	2	4	1
3	1			
3	1	0	0	0
3	2			

MATERIAL DATA (2 DIFFERENT MATERIALS)

2	0.1257	11.E+06
1	0.1963	30.E+06

BOUNDARY CONDITION FLAGS

1	-1	-2	0	0
3	1	0	0	0
4	1	1	0	0

NODAL FORCE LOADS

1	-1000.
2	-1732.

CONSTRAINED NODAL DISPLACEMENTS

1	0.00

    (blank line)

3-7 REMARKS

It is important to put this rather lengthy chapter into perspective. The two-dimensional truss was defined, and the complete finite element formulation was presented. A simple problem was solved by hand to illustrate the process. The basic steps were then summarized and extended to the three-dimensional truss. A simple program called TRUSS was described; it may be used to solve more complicated two-dimensional truss problems. The program illustrates the basic features of most finite element programs.

Table 3-3 Output from Program TRUSS for Example 3-7

EXAMPLE 3-7: MODEL OF TRUSS SHOWN IN FIG. 3-4

```
        NUMBER OF NODES:         4
        NUMBER OF ELEMENTS:      5
        NUMBER OF MATERIALS:     2
        NUMBER OF PRES DISP:     1
        NUMBER OF PT LOADS:      2
        OUTPUT UNIT NUMBER:      3
```

NODE NO.	IBCX	IBCY	X-COORD	Y-COORD
1	-1	-2	12.000	0.0000
2	0	0	12.000	6.0000
3	1	0	0.0000	0.0000
4	1	1	0.0000	10.0000

ELEMENT NO.	NODE I	NODE J	MAT SET FLAG
1	1	2	1
2	3	1	2
3	3	2	1
4	4	2	2
5	3	4	1

MATERIAL	AREA	ELASTIC MODULUS
1	.1963	.3000E+08
2	.1257	.1100E+08

SUMMARY OF DIFFERENT EXTERNAL LOADS

NUMBER	NODAL FORCE
1	-1000.0
2	-1732.

SUMMARY OF PRESCRIBED NODAL DISPLACEMENTS

NUMBER	NODAL DISP.
1	0.000

SUMMARY OF NODAL DISPLACEMENTS

NODE NO.	X-COMPONENT	Y-COMPONENT
1	-.86787E-02	-.35277E-01
2	.99552E-02	-.33513E-01
3	0.0000	-.17646E-02
4	0.0000	0.0000

SUMMARY OF ELEMENT RESULTANTS

ELEMENT NO.	ELONGATION	STRAIN	STRESS	FORCE
1	.17646E-02	.29411E-03	8823.2	1732.0
2	-.86787E-02	-.72322E-03	-7955.4	-1000.00
3	-.52939E-02	-.39459E-03	-11838.	-2323.7
4	.20042E-01	.15845E-02	17429.	2190.8
5	.17646E-02	.17646E-03	5293.9	1039.2

Although the approach illustrated in this chapter is referred to as *the* finite element method, this is but one type of many finite element methods. The method presented here and throughout this book is referred to as the *stiffness approach*, where the primary unknowns **a** are related to the nodal forces **f** by the relationship **Ka** = **f**. Not surprisingly, **K** is referred to as the *stiffness matrix*. An alternate finite element method exists, called the *flexibility method* or the *force matrix method*, where the primary unknowns are the nodal forces. The form of the global system equation is **Lf** = **a**, where **a**, as usual, represents the (now known) nodal displacements. Obviously, the flexibility matrix **L** is related to the inverse of the stiffness matrix **K**. The stiffness approach is much more powerful and popular than the flexibility approach. For this reason, the stiffness approach is used exclusively in this book.

Although the finite element method is generally regarded as an approximate solution technique, it does give the *exact* solution for the two-dimensional truss (and the three-dimensional truss). The reason for this is that the elongation δ is given *exactly* by PL/AE (providing that A and E are constant in any member). The assumption of smooth pin joints, which is also accommodated *exactly* in the finite element formulation, also contributes to the *exact* solution by the finite element method. Some argue that because the truss model is exact it should not be considered a bona fide application of the finite element method. Needless to say, the author does not agree with this point of view.

Section 3-2 deserves some special attention. It will become evident in later chapters that most of the steps delineated in this section are routine and are readily extended to other applications in structural analysis, as well as to problems in other disciplines. The reader will soon come to appreciate the following steps as being routine in structural analysis: (1) discretization, (2) assemblage of the stiffness matrix and nodal force vector, (3) application of the prescribed displacements, (4) solution for the nodal displacements, and (5) calculation of element resultants. The steps that are different for different models are the ones in which we determine the local and/or global element stiffness matrices and nodal force vectors—conveniently referred to as the *element characteristics*. Any one of several methods may be used to derive the element characteristics: (1) the direct approach, (2) the variational approach, (3) the virtual-work approach, or (4) the weighted-residual approach. In this chapter, only the direct approach is used (although it may not seem to be very direct to the newcomer to the finite element method). Nevertheless, a direct and well-known relationship, namely,

$$P = \frac{AE}{L}\delta$$

was used to obtain the element characteristics as given by Eq. (3-3) in the local coordinate system and by Eq. (3-24) [and Eq. (3-26)] in the global system. The direct approach is not easy to apply when the elements to be used are not obvious. In these cases other, more indirect, approaches are taken. These other approaches are discussed in several of the following chapters.

REFERENCES

1. Beer, F. P., and E. R. Johnston, Jr., *Vector Mechanics for Engineers: Statics and Dynamics*, McGraw-Hill, New York, 1962, p. 111.
2. Cheung, Y. K., and M. F. Yeo, *A Practical Introduction to Finite Element Analysis*, Pitman, London, 1979, p. 1.
3. Desai, C. S., *Elementary Finite Element Method*, Prentice-Hall, Englewood Cliffs, N.J., 1979, pp. 30–31.
4. Zienkiewicz, O. C., *The Finite Element Method*, McGraw-Hill (UK), London, 1977, pp. 135–136.
5. Huebner, K. H., *The Finite Element Method for Engineers*, Wiley, New York, 1975, pp. 51–53.
6. Zienkiewicz, O. C., *The Finite Element Method*, McGraw-Hill (UK), London, 1977, p. 85.
7. *FORTRAN-80 User's Guide, Microsoft FORTRAN-80 for the Apple II Computer*, Microsoft Consumer Products, © 1980, Bellevue, Wash.
8. Hornbeck, R. W., *Numerical Methods*, Quantum, New York, 1975.
9. Zienkiewicz, O. C., *The Finite Element Method*, McGraw-Hill (UK), London, 1977, pp. 204–207.
10. Zienkiewicz, O. C., and D. V. Phillips, "An Automatic Mesh Generation Scheme for Plane and Curved Element Domains," *Int. J. Numer. Methods Eng.*, vol. 3, pp. 519–528, 1971.
11. Gordon, W. J., and C. A. Hall, "Construction of Curvilinear Coordinate Systems and Application to Mesh Generation," *Int. J. Numer. Methods Eng.*, vol. 7, pp. 461–477, 1973.
12. Zienkiewicz, O. C., *The Finite Element Method*, McGraw-Hill (UK), London, 1977, p. 693.

PROBLEMS

3-1 Discretize the simple Warren truss shown in Fig. P3-1 by making tables similar to those in Example 3-1. Use the global coordinate system shown in the figure. The truss is composed of equilateral triangles whose sides are 3 feet long. Members *A*, *C*, *E*, *G*, *I*, and *K* are made of 0.5-in.-diameter steel rods and members *B*, *D*, *F*, *H*, and *J* of 0.25-in.-diameter aluminum rods.

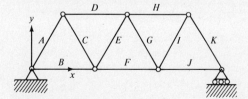

Figure P3-1

3-2 Discretize the Pratt truss shown in Fig. P3-2 by making tables similar to those in Example 3-1. Use the global coordinate system shown in the figure. All members are square in cross section and are made of steel. The horizontal, slanted, and vertical members have dimensions 1 × 1 cm, 0.75 × 0.75 cm, and 0.5 × 0.5 cm, respectively.

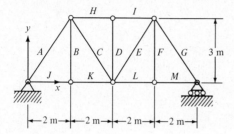

Figure P3-2

3-3 Discretize the cantilever truss shown in Fig. P3-3 by making tables similar to those in Example 3-1. Use the global coordinate system shown in the figure. Members A, B, and C are made of 0.75-in.-diameter steel rods; members J and K are of 0.5-in.-diameter steel rods; and the remaining members are of 0.6-in.-square aluminum bars.

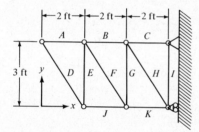

Figure P3-3

3-4 Discretize the K truss shown in Fig. P3-4 by making tables similar to those in Example 3-1. Use the global coordinate system shown in the figure. All horizontal members are 0.75-cm-diameter steel rods, and all other members are 1-cm-square steel bars.

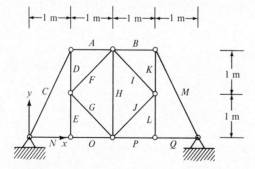

Figure P3-4

3-5   Compute the local element stiffness matrix for element 5 in the discretized Warren truss shown in Fig. P3-5. The truss is composed of equilateral triangles with sides of length L, and the members are steel rods with diameter D, where $L = 2$ m and $D = 1$ cm.

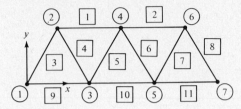

Figure P3-5

3-6   Repeat Problem 3-5 for element 8 with $L = 6$ ft and $D = 0.5$ in.

3-7   For the discretized Pratt truss shown in Fig. P3-7, compute the local element stiffness matrix for element 7 if $L = 6$ ft and $h = 9$ ft. Element 7 has a 0.5×1.0 in. rectangular cross section and is composed of aluminum.

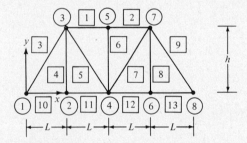

Figure P3-7

3-8   Repeat Problem 3-7 for element 5 if $L = 1$ m and $h = 1.5$ m. Element 5 has a 1×2 cm rectangular cross-section and is composed of steel.

3-9   A cantilever truss is discretized as shown in Fig. P3-9. For element 6, determine the local element stiffness matrix if $L = 5$ m and $h = 7$ m. Element 6 is composed of a 1-cm-diameter steel rod.

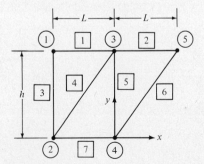

Figure P3-9

3-10   Repeat Problem 3-9 for element 2, which is composed of 0.5-in.-diameter steel. Take $L = 10$ ft and $h = 14$ ft.

3-11   For the discretized K truss shown in Fig. P3-11, compute the local element stiffness matrix for element 9, which is composed of a 0.75-in.-diameter aluminum rod. Take $L = 4$ ft and $h = 2$ ft.

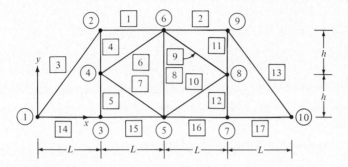

Figure P3-11

3-12   Repeat Problem 3-11 for element 10, which is composed of a 2-cm-diameter aluminum rod. Take $L = 1.5$ m and $h = 0.75$ m.

3-13   For element 5 of the Warren truss in Problem 3-5, determine the global element stiffness matrix if element 5 connects

   a. Nodes 3 and 4 (in this order)
   b. Nodes 4 and 3 (in this order)

3-14   For element 8 of the truss in Problem 3-6, determine the global element stiffness matrix if element 8 connects

   a. Nodes 6 and 7 (in this order)
   b. Nodes 7 and 6 (in this order)

3-15   For element 7 of the Pratt truss in Problem 3-7, determine the global element stiffness matrix if element 7 connects

   a. Nodes 4 and 7 (in this order)
   b. Nodes 7 and 4 (in this order)

3-16   For element 5 of the Pratt truss in Problem 3-8, determine the global element stiffness matrix if element 5 connects

   a. Nodes 3 and 4 (in this order)
   b. Nodes 4 and 3 (in this order)

3-17   For element 6 of the cantilever truss in Problem 3-9, determine the global element stiffness matrix if element 6 connects

   a. Nodes 4 and 5 (in this order)
   b. Nodes 5 and 4 (in this order)

3-18   For element 2 of the cantilever truss in Problem 3-10, determine the global element stiffness matrix if element 2 connects

  a. Nodes 3 and 5 (in this order)
  b. Nodes 5 and 3 (in this order)

3-19   For element 9 of the K truss in Problem 3-11, determine the global element stiffness matrix if element 9 connects

  a. Nodes 6 and 8 (in this order)
  b. Nodes 8 and 6 (in this order)

3-20   For element 10 of the K truss in Problem 3-12, determine the global element stiffness matrix if element 10 connects

  a. Nodes 5 and 8 (in this order)
  b. Nodes 8 and 5 (in this order)

3-21   Show that the matrix $\mathbf{T}$ defined in Sec. 3-2 is orthogonal; i.e., show that $\mathbf{T}^{-1} = \mathbf{T}^{T}$.

3-22   Show that the matrix $\mathbf{R}$ defined by Eq. (3-18) is orthogonal; i.e., show that $\mathbf{R}^{-1} = \mathbf{R}^{T}$.

3-23   Indicate how the global element stiffness matrix is added to the assemblage stiffness matrix for element 5 of the Warren truss in Problem 3-13 if element 5 connects
  a. Nodes 3 and 4 (in this order)
  b. Nodes 4 and 3 (in this order)

3-24   Indicate how the global element stiffness matrix is added to the assemblage stiffness matrix for element 8 of the Warren truss in Problem 3-14 if element 8 connects

  a. Nodes 6 and 7 (in this order)
  b. Nodes 7 and 6 (in this order)

3-25   Indicate how the global element stiffness matrix is added to the assemblage stiffness matrix for element 7 of the Pratt truss in Problem 3-15 if element 7 connects

  a. Nodes 4 and 7 (in this order)
  b. Nodes 7 and 4 (in this order)

3-26   Indicate how the global element stiffness matrix is added to the assemblage stiffness matrix for element 5 of Problem 3-16 if element 5 connects

  a. Nodes 3 and 4 (in this order)
  b. Nodes 4 and 3 (in this order)

3-27   How is the global element stiffness matrix for element 6 of Problem 3-17 added to the assemblage stiffness matrix if element 6 connects

  a. Nodes 4 and 5 (in this order)
  b. Nodes 5 and 4 (in this order)

3-28   Indicate how the global element stiffness matrix for element 2 of Problem 3-18 is added to the assemblage stiffness matrix if element 2 connects

  a. Nodes 3 and 5 (in this order)
  b. Nodes 5 and 3 (in this order)

3-29 How does the global element stiffness matrix for element 9 of Problem 3-19 contribute to the assemblage stiffness matrix if element 9 connects

    a. Nodes 6 and 8 (in this order)
    b. Nodes 8 and 6 (in this order)

3-30 Indicate how the global element stiffness matrix for element 10 of Problem 3-20 is added to the assemblage stiffness matrix if element 10 connects

    a. Nodes 5 and 8 (in this order)
    b. Nodes 8 and 5 (in this order)

3-31 Determine the half-bandwidth of the assemblage stiffness matrix for the discretized Warren truss in Problem 3-5:

    a. By direct examination of $\mathbf{K}^a$
    b. By Eq. (3-33)

3-32 What is the half-bandwidth of the assemblage stiffness matrix for the discretized Pratt truss of Problem 3-7:

    a. By direct examination of $\mathbf{K}^a$
    b. By Eq. (3-33)

3-33 Determine the half-bandwidth of the assemblage stiffness matrix for the discretized cantilever truss of Problem 3-9:

    a. By direct examination of $\mathbf{K}^a$
    b. By Eq. (3-33)

3-34 What is the half-bandwidth of the assemblage stiffness matrix for the discretized K truss of Problem 3-11:

    a. By direct examination of $\mathbf{K}^a$
    b. By Eq. (3-33)

3-35 A stadium truss is loaded as shown in Fig. P3-35. The global node numbers are also shown. What is the assemblage nodal force vector if there are no other loads present?

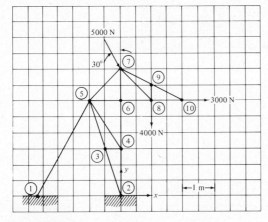

Figure P3-35

3-36   A Fink truss to be used in a roof is loaded as shown in Fig. P3-36. The global node numbers are also shown. Determine the assemblage nodal force vector if there are no other loads present. Note that two loads act on node 8.

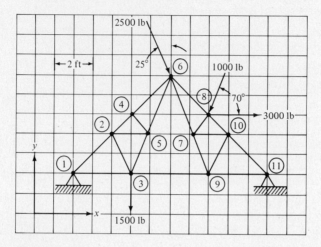

Figure P3-36

3-37   A cantilever truss is loaded as shown in Fig. P3-37. The global node numbers are also shown. Determine the assemblage nodal force vector. Note the two loads acting at node 7.

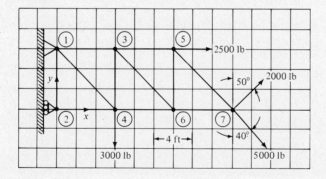

Figure P3-37

3-38 A Howe truss is loaded as shown in Fig. P3-38. The global node numbers are also
shown. Determine the assemblage nodal force vector.

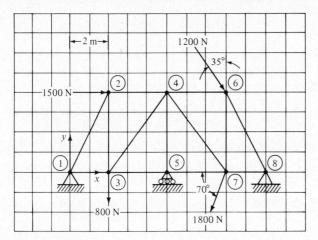

Figure P3-38

3-39 For the stadium truss in Problem 3-35:

  a. Specify which degrees of freedom are restrained.
  b. Indicate how the assemblage system equation $\mathbf{K}^a\mathbf{a} = \mathbf{f}^a$ is modified to impose
  these prescribed displacements.

3-40 For the Fink truss in Problem 3-36:

  a. Specify which degrees of freedom are restrained.
  b. Indicate how the assemblage system equation $\mathbf{K}^a\mathbf{a} = \mathbf{f}^a$ is modified to impose
  these prescribed displacements.

3-41 For the cantilever truss in Problem 3-37:

  a. Specify which degrees of freedom are restrained.
  b. Indicate how the assemblage system equation $\mathbf{K}^a\mathbf{a} = \mathbf{f}^a$ is modified to impose
  these prescribed displacements.

3-42 For the Howe truss in Problem 3-38:

  a. Specify which degrees of freedom are restrained.
  b. Indicate how the assemblage system equation $\mathbf{K}^a\mathbf{a} = \mathbf{f}^a$ is modified to impose
  these prescribed displacements.

3-43 Consider the following system of equations:

$$3x_1 - 4x_2 - 5x_3 = 7$$

$$-4x_1 + 7x_2 - 2x_3 = 2$$

$$-5x_1 - 2x_2 + x_3 = 5$$

Using Method 1 of Sec. 3-2, modify (but do not solve) the system of equations to
impose $x_3 = -2$. Do not destroy the symmetry.

3-44   Repeat Problem 3-43 using Method 2 of Sec. 3-2.

3-45   Consider the following system of four equations and four unknowns:

$$6x_1 + 2x_2 + 4x_3 + 5x_4 = 8$$

$$2x_1 + 5x_2 - 7x_3 + 2x_4 = 7$$

$$4x_1 - 7x_2 + 8x_3 - 7x_4 = 2$$

$$5x_1 + 2x_2 - 7x_3 + 3x_4 = 5$$

Using Method 1 of Sec. 3-2, modify (but do not solve) the system of equations to impose $x_1 = 5$ and $x_3 = 3$ (concurrently). Do not destroy the symmetry.

3-46   Repeat Problem 3-45 using Method 2 of Sec. 3-2.

3-47   Solve the resulting system of equations from Problem 3-43 for the unknowns (after imposing $x_3 = -2$). Use the matrix-inversion method.

3-48   Solve the resulting system of equations from Problem 3-45 for the unknowns (after imposing $x_1 = 5$ and $x_3 = 3$). Use the matrix-inversion method.

3-49   Reconsider the Warren truss in Problem 3-5. If nodes 3 and 4 are found to have displacements

$$u_3 = 0.0100 \text{ cm} \qquad u_4 = -0.0150 \text{ cm}$$

$$v_3 = 0.0200 \text{ cm} \qquad v_4 = 0.0015 \text{ cm}$$

for some loading condition, determine for element 5 the axial

a. Elongation   **b.** Strain   **c.** Stress   **d.** Force

3-50   Reconsider the Pratt truss in Problem 3-7. If for some loading condition, nodes 4 and 7 are found to have displacements

$$u_4 = 0.0100 \text{ in.} \qquad u_7 = -0.0020 \text{ in.}$$

$$v_4 = 0.0300 \text{ in.} \qquad v_7 = 0.0045 \text{ in.}$$

Determine for element 7 the axial

a. Elongation   **b.** Strain   **c.** Stress   **d.** Force

3-51   For some loading condition, nodes 4 and 5 for the cantilever truss in Problem 3-9 are known to have the following displacements:

$$u_4 = 0.0100 \text{ cm} \qquad u_5 = -0.0315 \text{ cm}$$

$$v_4 = 0.0000 \text{ cm} \qquad v_5 = -0.0126 \text{ cm}$$

Determine for element 6 the axial

a. Elongation   **b.** Strain   **c.** Stress   **d.** Force

3-52   For some particular loading condition, nodes 6 and 8 for the K truss in Problem 3-11 are known to have the following displacements:

$$u_6 = 0.0000 \text{ in.} \qquad u_8 = -0.0269 \text{ in.}$$

$$v_6 = -0.0158 \text{ in.} \qquad v_8 = +0.0105 \text{ in.}$$

Determine for element 9 the axial

a. Elongation   **b.** Strain   **c.** Stress   **d.** Force

3-53   For the truss described in Sec. 3-3 and shown in Fig. 3-4, determine the element resultants for element 1 by using the results in Sec. 3-3 (show the calculations).

3-54   For the truss described in Sec. 3-3 and shown in Fig. 3-4, determine the element resultants for element 2 by using the results in Sec. 3-3 (show the calculations).

3-55   For the truss described in Sec. 3-3 and shown in Fig. 3-4, determine the element resultants for element 4 by using the results in Sec. 3-3 (show the calculations).

3-56   For the truss described in Sec. 3-3 and shown in Fig. 3-4, determine the element resultants for element 5 by using the results in Sec. 3-3 (show the calculations).

3-57   Verify the values obtained for the forces in each member of the truss in Sec. 3-3 by doing a force balance on each node. Remember to include the unknown reaction force U_3 at node 3 and forces U_4 and V_4 at node 4. From the force balances, determine the values of U_3, U_4, and V_4 for static equilibrium.

3-58   With the help of Eq. (3-64) and the results for the nodal displacements, determine the unknown reaction forces U_3, U_4, and V_4 for the truss in Sec. 3-3. Are these results consistent with those from static equilibrium considerations?

3-59   Reconsider the stadium truss shown in Fig. P3-35. The nodal coordinates may be obtained from the superimposed grid. All horizontal members are made of 1-cm-diameter steel rods and all remaining members of 0.75-cm-diameter aluminum rods. The truss is loaded and restrained as shown in the figure. Using the TRUSS program, determine the nodal displacements and the element resultants (for all nodes and elements, respectively).

3-60   A Fink truss is loaded and restrained as shown in Fig. P3-36. All horizontal members are made of 0.5-in.-diameter steel rods and all other members of 0.75-in. square aluminum bars. The nodal coordinates may be taken directly from the figure with the help of the superimposed grid. Using the TRUSS program, determine the nodal displacements and element resultants.

3-61   Reconsider the cantilever truss shown in Fig. P3-37. The horizontal, slanted, and vertical members are made of 0.75-in.-, 0.5-in.-, and 1.0-in.-diameter steel rods. The nodal coordinates may be taken directly from the figure with the help of the superimposed grid. The truss is loaded and restrained as shown. Using the TRUSS program, determine the nodal displacements and the element resultants.

3-62   The Howe truss shown in Fig. P3-38 is to be analyzed. The truss is loaded and restrained as shown. All vertical members are made of 1.5-cm square steel bars and all other members of 2.0-cm-diameter aluminum rods. The nodal coordinates may be obtained directly from the figure. With the help of the TRUSS program, determine the nodal displacements and the element resultants.

3-63   Modify the TRUSS program in Appendix B so that it can be used to analyze three-dimensional trusses. Make use of the formulation in Sec. 3-5.

4

Variational and Weighted Residual Formulations

4-1 INTRODUCTION

In Chapter 3 the direct approach was introduced when the two-dimensional truss problem was formulated. It was pointed out that alternate, more indirect approaches exist that may be used to develop finite element models. Some of these methods of formulation are introduced in this chapter. The reader will soon come to appreciate that virtually all problems that are describable by ordinary and partial differential equations can be solved by the finite element method. Moreover, linear and non-linear problems may be solved in nearly the same way, although nonlinear problems generally require an iterative solution.

In this chapter only steady-state problems are studied. By *steady-state* we mean that the field variable is a function of spacial coordinates only and not a function of time. Problems that allow for time-varying field variables are said to be unsteady, transient, dynamic, or time-dependent. In some texts transient problems are referred to as propagation problems [1]. In structural and stress analysis, time-independent problems are referred to as static or equilibrium problems, whereas time-dependent problems are almost exclusively referred to as dynamic.

In all steady-state, static, or equilibrium problems, a system of algebraic equations results that is always in the form

$$\mathbf{Ka} = \mathbf{f}$$

It should be recalled that the truss model resulted in a system of equations in exactly this form [see Eq. (3-39)]. Furthermore, the vector or column matrix $\mathbf{a}$ always contains the nodal unknowns, which are really the values of the field variables at the nodal points. It cannot be emphasized enough that in structural models (without bending), these unknowns are the nodal displacements; in thermal models, they are the nodal temperatures; and in fluid flow problems, they are the nodal velocities and pressures. When a fluid mechanics problem is formulated in terms of velocities and pressures, the formulation is said to involve the primitive variables. This is in contrast to stream function and vorticity formulations that are less direct [2]. The main purpose of a finite element analysis is to determine the values of the field variable(s) at the node points. Other quantities such as the stresses or heat-flows may then be determined in subsequent calculations.

This chapter is further restricted to the study of only one-dimensional problems. This means that the field variable of interest is a function of only one variable, e.g., x. Two- and three-dimensional problems are covered in subsequent chapters. In addition, variational calculus is introduced in this chapter to familiarize the reader with one of the most popular types of finite element formulations.

The basic difference between ordinary differential calculus and variational calculus may be explained as follows. If we want to determine the value of x that maximizes or minimizes some function $y = f(x)$, we simply take the derivative of y with respect to x, set the result equal to zero, and solve for x. The sign of the second derivative at this value of x indicates whether a minimum, maximum, or point of inflection has been found. On the other hand, if we want to know what *function* results in a certain *definite integral* taking on a minimum or maximum value, then the calculus of variations is required. For example, if we want to obtain the equation of the "curve" that results in the shortest distance between two points in a plane, the calculus of variations may be used. This very example is illustrated later in this chapter and gives the expected result: the equation is that of a straight line that connects the two points.

4-2 SOME APPROXIMATE SOLUTION METHODS

Oftentimes exact solutions to differential equations are not easily obtained. In fact, quite frequently, it is not even possible to obtain an exact solution with currently available mathematical techniques. This is particularly true when the geometry is irregular, or the properties vary spacially, or perhaps, the properties are a function of the field variable. Problems of the latter variety are said to be nonlinear. Engineers, mathematicians, and applied scientists find approximate solution methods indispensable in these cases.

Among the approximate solution techniques that are useful to a thorough understanding of how the finite element method works are the following: the Ritz method, the variational, or Rayleigh-Ritz, method, and the weighted-residual method. Each of these is said to be an integral method because we work with an integral form of the problem statement instead of the governing differential equation directly. The Ritz and variational methods are illustrated in this section.

The Ritz method is quite simple, requires no additional mathematics beyond calculus, and is, in fact, a special case of a particular weighted-residual method (see Sec. 4-6). The variational method, on the other hand, requires some knowledge of variational calculus—a definite disadvantage. For this reason, Secs. 4-3 to 4-5 are devoted to the development of some of the more basic concepts in variational calculus. The weighted residual methods are deferred until Sec. 4-6. The reader is referred to the book by Arpaci [3] for a review of the approximate solution methods based on integral formulations to problems in heat conduction.

In some books the Ritz method is referred to as the Rayleigh-Ritz method. Unfortunately, the variational method is also referred to as the Rayleigh-Ritz method. In this text, the Ritz method always refers to the nonvariational integral formulation and the Rayleigh-Ritz method always refers to the variational formulation.

General Concepts

Before actually illustrating the Ritz and Rayleigh-Ritz methods with a specific problem, let us establish some general concepts. So as not to obscure the basic ideas behind the approximate solution methods, let us restrict the present discussion to a single governing differential equation with only one independent variable. Let us represent the governing equation as

$$f[T(x)] = 0 \qquad \text{in } \Omega \qquad \textbf{(4-1)}$$

where T represents the function sought (e.g., temperature) that is a function of x only. The symbol Ω represents the domain of the region governed by Eq. (4-1). In addition, let us specify the boundary conditions symbolically in the form

$$g_1[T(x)] = 0 \qquad \text{on } \Gamma_1 \qquad \textbf{(4-2a)}$$

and

$$g_2[T(x)] = 0 \qquad \text{on } \Gamma_2 \qquad \textbf{(4-2b)}$$

etc., where Γ_1 and Γ_2 include only those parts of the domain Ω that are on the boundary. Figure 4-1 may help to clarify the notation.

Let us approximate the solution to Eqs. (4-1) and (4-2) with the approximate function T' where

$$T' = T'(x; a_1, a_2, \ldots, a_n) = \sum_{i=1}^{n} a_i N_i(x) \qquad \textbf{(4-3)}$$

which has one or more unknown (but constant) parameters $a_1, a_2, \ldots, a_n$ and that satisfies the boundary conditions given by Eqs. (4-2) exactly. Note that the

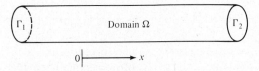

Domain Ω

Γ_1 Γ_2

0 x

Figure 4-1 Schematic of one-dimensional problem domain Ω with two global boundaries Γ_1 and Γ_2.

prime (′) denotes an approximate solution (not a derivative). The functions $N_i(x)$ are referred to as *trial functions*. The problem reduces to having to make somewhat judicious choices for these trial functions and solving for the parameters a_1, a_2, . . . , a_n. In general, if a sequence of approximations could be made, such as

$$T' = T'(x;a_1) = a_1 N_1(x) \qquad \qquad \textbf{(4-4a)}$$

$$T' = T'(x;a_1,a_2) = a_1 N_1(x) + a_2 N_2(x) \qquad \qquad \textbf{(4-4b)}$$

$$T' = T'(x;a_1,a_2,a_3) = a_1 N_1(x) + a_2 N_2(x) + a_3 N_3(x) \qquad \qquad \textbf{(4-4c)}$$

then presumably better accuracy could be obtained with each successive higher-order approximation. Equations (4-4) are referred to as the first-order, second-order, and third-order approximations, respectively. The reader is referred to Becker, Carey, and Oden [4] for a discussion on the necessary properties of the trial functions. Suffice it to say here that they must be continuous and differentiable up to the highest order present in the integral form of the governing equation.

It should not be surprising that if T' given in Eq. (4-3) is substituted for T in Eq. (4-1), the governing equation will not be satisfied exactly. Instead of getting $f(T')$ equal to zero, we get a residual R. Mathematically, we may write

$$f[T'(x; a_1, a_2, \ldots , a_n)] = R(x; a_1, a_2, \ldots , a_n) \qquad \qquad \textbf{(4-5)}$$

where the notation is supposed to indicate that the residual R is a function of x and the parameters a_1, a_2, etc. The exact solution results when the residual R is zero for all points in the domain Ω. For the approximate solution methods, the residual is not in general zero everywhere in Ω, although it may be zero at some selected points.

The Ritz Method

Only the first-order Ritz approximation is considered here. Refer to the book by Arpaci [5] for an illustration of the second-order Ritz approximation. The basic Ritz method for the first-order approximation simply requires that the integral of the residual $R(x;a_1)$ with respect to x be zero over the domain Ω, or

$$\int_\Omega R(x;a_1)\, dx = 0 \qquad \qquad \textbf{(4-6)}$$

Note that since x is essentially a dummy variable in the integral, Eq. (4-6) results in an algebraic equation in the unknown parameter a_1. For well-posed problems and a well-behaved trial function $N_1(x)$, this equation may be solved for a_1. Well-behaved trial functions include those included in the set of polynomials, circular functions, and other continuous and differentiable functions. This simple method is best illustrated by an example.

Example 4-1

Solve the ordinary linear differential equation

$$\frac{d^2T}{dx^2} + 1000x^2 = 0 \qquad 0 \le x \le 1 \tag{4-7}$$

subject to the boundary conditions

$$T(0) = 0 \tag{4-8a}$$

and

$$T(1) = 0 \tag{4-8b}$$

by using the Ritz method with the trial function

$$N_1(x) = x(1 - x^2) \tag{4-9}$$

This particular trial function was chosen because it allows T' to satisfy the boundary conditions exactly, and it does not grossly violate the physics of the problem as explained later.

Solution

The first-order approximation T' from Eq. (4-4a) becomes

$$T' = a_1N_1(x) = a_1x(1 - x^2) \tag{4-10}$$

An expression for the residual R is needed, but first verify that T' given by Eq. (4-10) satisfies the boundary conditions exactly (see Problem 4-1). The residual R by definition may be computed from

$$R = \frac{d^2T'}{dx^2} + 1000x^2$$

And since

$$\frac{d^2T'}{dx^2} = -6a_1x$$

we have

$$R(x;a_1) = -6a_1x + 1000x^2 \tag{4-11}$$

Note that as the notation $R(x;a_1)$ indicates, the residual is indeed a function of x and the parameter a_1. By Eq. (4-6), the first-order Ritz method requires

$$\int_0^1 (-6a_1x + 1000x^2) \, dx = 0$$

which may be readily integrated to give

$$\left[-3a_1x^2 + 1000\frac{x^3}{3} \right]\Bigg|_0^1 = 0$$

Evaluating and solving for a_1 gives

$$a_1 = {}^{1000}\!/_9 \tag{4-12}$$

Therefore, an approximate solution is given by

$$T'(x) = {}^{1000}\!/_9\, x(1 - x^2) \tag{4-13}$$

In Sec. 4-6, this approximate solution is compared with the other approximate solutions as well as with the exact solution. ■

Let us try to give some physical significance to the governing differential equation given by Eq. (4-7). The reader may recognize this as a special form of the heat conduction equation. More specifically, Eq. (4-7) represents one-dimensional heat conduction in a bar of unit length insulated around the periphery with an internal heat source (internal energy generation) that is proportional to the square of x as depicted in Fig. 4-2. Therefore, higher temperatures are expected near the end $x = 1$, although the temperatures at both ends must be zero by Eqs. (4-8). In other words, the maximum temperature T is expected to occur somewhere between $x = 0.5$ and $x = 1.0$. It is for this reason that the trial function given by Eq. (4-9) is most appropriate, for it too shows a maximum temperature between $x = 0.5$ and $x = 1.0$ (in fact, the reader may wish to show that the maximum occurs at $x = 0.577$). In short, the trial function (and hence, the approximate solution) used in Example 4-1 is said to satisfy, albeit approximately, the physics of the problem. This may be stated differently as follows: the assumed temperature distribution does not grossly violate the physics of the problem.

It is emphasized that two requirements must be met by the assumed trial function(s): (1) the physics of the problem must not be grossly violated and (2) the boundary conditions must be satisfied exactly. Later it will be seen that the finite

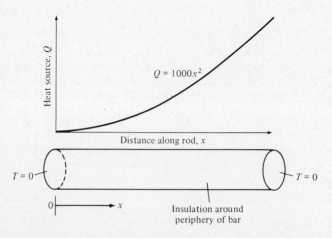

Figure 4-2 Schematic of problem posed in Example 4-1.

element method completely eliminates these rather restrictive conditions, which are very difficult to satisfy in most real-world problems.

The Variational Method (Rayleigh-Ritz)

Although variational calculus is covered in the next several sections, it is illustrative to consider the main idea behind the variational method here in a numerical example. A typical one-dimensional problem in variational calculus is one in which we try to find a function T that minimizes or maximizes integrals of the form

$$I = \int_a^b F[x, T(x), T_x(x)] \, dx \qquad \textbf{(4-14a)}$$

where T_x denotes the first derivative of T with respect to x. The processes of minimization and maximization are frequently referred to collectively as *extremization*, and the integral is said to be extremized or to be made stationary. This integral I is referred to as a *functional*; sometimes the integrand F is also referred to as the functional. This is not terribly surprising because the word "functional" means function of functions. Since F is a function of x, T, and T_x, and T is a function of x itself, clearly I is also a function of functions. It should be emphasized that the notation T_x denotes the derivative of T with respect to x. The prime (') is used to represent an approximation, not the derivative (i.e., T' is the approximation to T, not the derivative of T). The subscript notation for the derivatives proves to be very convenient in two- and three-dimensional variational formulations (see Chapter 8).

It will be shown in Sec. 4-5 that the functional F that corresponds to Eq. (4-7) from Example 4-1 is

$$F = -\frac{1}{2}\left(\frac{dT}{dx}\right)^2 + 1000x^2T \qquad \textbf{(4-15)}$$

Consequently, the variational formulation of the problem in Example (4-1) becomes

$$I = \int_0^1 \left[1000x^2T - \frac{1}{2}\left(\frac{dT}{dx}\right)^2 \right] dx \qquad \textbf{(4-14b)}$$

The idea is to find the function $T(x)$ that extremizes I. Before illustrating how Eq. (4-14b) is used in an approximate solution, several key points must be made.

For a well-posed problem such as that in Example 4-1, the function $T(x)$ that extremizes I in Eq. (4-14b) is exactly the same as that which satisfies the original differential equation and boundary conditions. This implies that the solution $T(x)$ is unique, which is always the case for well-posed problems. A second observation is that the original differential equation [Eq. (4-7)] contains a second-order derivative of T, whereas the variational formulation given by Eq. (4-14b) contains only a first-order derivative! Therefore, the variational form may be used to obtain solutions to problems that are not readily admitted by the differential formulation. An example of this is a body with two different materials and, hence, thermal conductivities. At the point where the two materials meet, the second derivative of the temperature required by the differential formulation may not exist. A variational formulation of the problem would readily yield the correct solution, since the second-derivative in this example is not needed in the formulation. For this reason, the variational formulation of a physical problem is often referred to as the *weak formulation*.

Although the discussion in the previous paragraph alluded to an exact solution of Eq. (4-14b) for the function $T(x)$ that extremizes the functional I, we shall use this equation to obtain an approximate solution for $T(x)$, as shown in the next example. In this text we will only be concerned with the use of variational formulations in approximate solutions, except for a simple illustrative problem in Sec. 4-3.

Example 4-2

Solve the problem posed in Example 4-1 by the variational method by extremizing I given by Eq. (4-14b) with respect to parameter a_1 in the approximation T' (to T) given by

$$T'(x) = a_1 N_1(x) = a_1 x(1 - x^2) \tag{4-16}$$

Note that this same form of the approximate solution was assumed in Example 4-1.

Solution

In terms of the approximation T', Eq. (4-14b) may be written

$$I(a_1) = \int_0^1 \left[1000x^2 T' - \frac{1}{2}\left(\frac{dT'}{dx}\right)^2 \right] dx \tag{4-17}$$

Note that I is a function of a_1 [as indicated by the notation $I(a_1)$] because T' is a function of x and the parameter a_1, and the x is integrated-out. The idea is to determine a_1 such that

$$\frac{dI}{da_1} = 0 \tag{4-18}$$

or, in words, such that the functional I is extremized (or made stationary). To determine whether a minimum or maximum has been found, we may check the second derivative of I with respect to a_1 at the value of a_1 that extremizes I in the first place. If $d^2I/da_1^2 > 0$, a minimum has been found; if $d^2I/da_1^2 < 0$, a maximum has been found. The first derivative of $T'(x)$ is needed in Eq. (4-17), which is readily computed from Eq. (4-16) to be

$$T_x' = \frac{dT'}{dx} = a_1(1 - 3x^2) \tag{4-19}$$

Therefore, Eq. (4-17) becomes

$$I(a_1) = \int_0^1 \{1000x^2[a_1x(1 - x^2)] - \tfrac{1}{2}[a_1(1 - 3x^2)]^2\}\, dx$$

Integration and evaluation readily yields

$$I(a_1) = {}^{1000}\!/_{12}a_1 - \tfrac{2}{5}a_1^2 \tag{4-20}$$

from which we may compute

$$\frac{dI}{da_1} = {}^{1000}\!/_{12} - \tfrac{4}{5}a_1$$

Setting the right-hand side to zero and solving for a_1 gives

$$a_1 = {}^{5000}\!/_{48} \tag{4-21}$$

An approximate solution to Eq. (4-7) by the variational method subject to the boundary conditions given by Eqs. (4-8) is

$$T'(x) = {}^{5000}\!/_{48}\, x(1 - x^2) \tag{4-22}$$

The reader should compare this approximate result with that from the Ritz method of Example 4-1. The approximate solutions from Examples 4-1 and 4-2 are shown for comparison in Fig. 4-3. These approximate solutions are compared to the exact solution later in this chapter.   ■

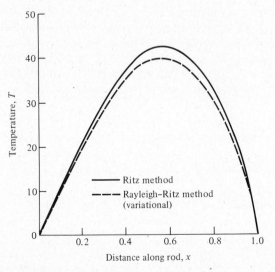

Figure 4-3 Comparison of approximate solutions to problem posed in Example 4-1.

4-3 VARIATIONAL CALCULUS: AN INTRODUCTION

The calculus of variations is introduced in this section for functions that are a function of one independent variable only, e.g., $y = f(x)$. The basic approach that is taken utilizes only ordinary differential and integral calculus. The result is then interpreted in light of the calculus of variations. In what follows, when a derivative is needed the subscript notation from Sec. 4-2 is used; e.g., dT/dx is denoted as T_x (not as T') and dy/dx as y_x.

In general, we usually know the differential equation that describes a phenomenon. However, the variational formulation is usually not obvious. For example, is it obvious that Eq. (4-14b) is the variational formulation that corresponds to the differential formulation given by Eqs. (4-7) and (4-8)? The idea then is to develop a systematic procedure by which we may derive the variational form of a problem from the differential formulation.

It is very important to put the material in this section into proper perspective. The author is of the opinion that although the variational formulation allows us to obtain a firmer grasp of the underlying concepts behind the finite element method, it is by no means absolutely essential to the application of the method to practical problems. It will be learned later that the weighted-residual methods are far easier to apply and can be used even when no variational formulation or principle exists. This last point should be clarified. All variational formulations have a corresponding differential formulation, but, unfortunately, the converse is not true: some differential formulations have no *classical* variational principle. The reader is referred to Zienkiewicz [6] for a procedure that yields a *nonclassical* variational formulation to some of these problems. This does not mean that the finite element method cannot be used in these cases, because we simply resort to the more powerful (and simpler) weighted-residual approach.

Before a systematic procedure is developed that will allow us to obtain the variational form of a problem from the differential formulation, let us examine one of the classical problems in variational calculus, that of finding the equation of the curve that has the shortest length between two points in a plane. Obviously, the curve is a straight line joining the two points, but let us show how the calculus of variations may be used to prove this intuitively obvious result. First, we must set up an integral I that represents the length of the curve. Figure 4-4 shows a family of curves that connects points A and B in the plane of the figure. By minimizing the integral I, we will obtain the equation of the line that results in the shortest possible length.

An elemental arc length ds from the Pythagorean theorem is given by

$$ds = \sqrt{(dx)^2 + (dy)^2} \tag{4-23}$$

which may be written as

$$ds = \sqrt{1 + \left(\frac{dy}{dx}\right)^2}\, dx \tag{4-24}$$

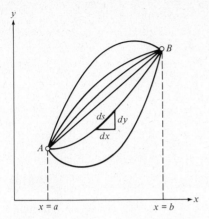

Figure 4-4   Family of curves that connects points A and B in the plane of the paper.

Therefore, the total length of the curve from $x = a$ at point A to $x = b$ at point B is given by

$$I = \int_a^b \sqrt{1 + \left(\frac{dy}{dx}\right)^2}\, dx \tag{4-25}$$

Let us denote this integral in a slightly more general form as

$$I = \int_a^b F(x, y, y_x)\, dx \tag{4-26}$$

where x and y are included for generality, and y_x denotes dy/dx. It is shown in Sec. 4-5 that if $y(x)$ is found such that it minimizes the integral in Eq. (4-26), the functional F must satisfy the following differential equation:

$$\frac{\partial F}{\partial y} - \frac{d}{dx}\left(\frac{\partial F}{\partial y_x}\right) = 0 \tag{4-27}$$

From Eqs. (4-26) and (4-25) we have

$$F = (1 + y_x^2)^{1/2} \tag{4-28}$$

Therefore, application of Eq. (4-27) yields

$$-\frac{d}{dx}\left[\frac{y_x}{(1 + y_x^2)^{1/2}}\right] = 0 \tag{4-29}$$

It follows that the expression in the brackets is a constant, e.g., C_1, or

$$\frac{y_x}{(1 + y_x^2)^{1/2}} = C_1$$

from which it may be concluded that y_x itself is (another) constant, or

$$y_x = C_2$$

But $y_x = dy/dx$ and so $dy/dx = C_2$, which may be integrated to give

$$y = C_2 x + C_3 \qquad \textbf{(4-30)}$$

The integration constants C_2 and C_3 could then be determined for a particular problem by using the boundary conditions

$$y(a) = y_1 \qquad \textbf{(4-31a)}$$

$$y(b) = y_2 \qquad \textbf{(4-31b)}$$

In any event, Eq. (4-30) represents a straight line as expected.

Figure 4-5 shows the function $y(x)$ that actually results in the shortest length between the two points A and B. In addition, the variation of y, denoted as δy, is also shown. If $y(x)$ and its variation δy are added, we get other possible curves $\tilde{y}(x)$ for which the integral I may be evaluated. It is emphasized that in this case only one particular curve, namely that labeled $y(x)$, actually results in the shortest length.

4-4 SOME ADDITIONAL MATHEMATICS

Several concepts from ordinary calculus are reviewed in this section. An extension of these concepts to variational calculus is also provided. Since our goal is to develop an operational understanding of these concepts, no attempt is made to prove them here. The reader may wish to consult the references at the end of this chapter for additional information.

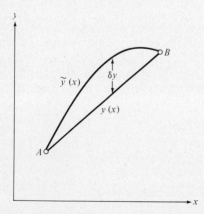

Figure 4-5   The curve that results in the shortest length connecting A and B is shown along with the variation δy and the curve $\tilde{y}(x)$.

Integration by Parts [7]

If du/dx and dv/dx are continuous functions of x over a suitable interval, e.g., from $x = a$ to $x = b$, then the formula for integration by parts applies and is given by

$$\int_a^b u\frac{dv}{dx}dx = uv\bigg|_a^b - \int_a^b v\frac{du}{dx}dx \tag{4-32}$$

The first term on the right-hand side is frequently referred to as the *integrated term*.

Taylor's Series [8]

A function $f(x)$ that is continuous and differentiable over an interval that includes $x = a$ may be written in terms of an infinite series such as

$$f(x) = f(a) + f_x(a)\frac{x - a}{1!} + f_{xx}(a)\frac{(x - a)^2}{2!} + \cdots \tag{4-33}$$

This infinite series is the *Taylor series expansion* for the interval. Note that $f_x(a)$ denotes that the first derivative of f with respect to x is to be evaluated at $x = a$, that $f_{xx}(a)$ denotes that the second derivative of f with respect to x is likewise to be evaluated at $x = a$, and so on. Oftentimes the terms involving the factors $(x - a)^n/n!$ for $n \geq 2$ are small in comparison to the first-order term and the truncated series given by

$$f(x) = f(a) + f_x(a)(x - a) \tag{4-34}$$

is often referred to as a *first-order Taylor expansion* (of the function f about the point $x = a$). It is emphasized that Eq. (4-34) may be used only for values of x very close to a.

The Differential of a Function

Consider a continuous and differentiable function $f(x,y)$ that has two independent variables, x and y. The differential df is then given by

$$df = \frac{\partial f}{\partial x}dx + \frac{\partial f}{\partial y}dy \tag{4-35}$$

For a function $f(x,y,z)$ with three independent variables x, y, and z, we have

$$df = \frac{\partial f}{\partial x}dx + \frac{\partial f}{\partial y}dy + \frac{\partial f}{\partial z}dz \tag{4-36}$$

and so on. Note that we may take the differential of a function but not of a functional, which is a function of functions. Instead we speak of a variation, as discussed next.

The Variation of a Functional

Let us now consider a functional $F(x,y,y_x)$. It should be obvious from the example depicted in Fig. 4-5 that the variation is not taken on the variable x but rather on the function $y(x)$ and possibly some of its derivatives. Therefore, the variation of x, written δx, is identically zero, and the variation of F is given by

$$\delta F = \frac{\partial F}{\partial y}\delta y + \frac{\partial F}{\partial y_x}\delta y_x \tag{4-37}$$

It is hinted here that the usual rules of ordinary calculus apply in the calculus of variations, and in most cases they do. For example, Eq. (4-37) follows from Eq. (4-36) if the differential operator d is replaced by the variational operator δ and, if further, it is recognized that $\delta x \equiv 0$. Of course, f in Eq. (4-36) is a true function, whereas F in Eq. (4-37) is a function of functions; i.e., a functional. One other subtle difference exists in the expressions for the differential: the variables x, y, and z are truly independent, whereas in the expression for the variation, y and y_x are treated as though they are independent (they are actually related by $y_x = dy/dx$). Equation (4-37) may be readily extended to cases in which the functional F is a function of more than two functions (see Problem 4-15).

The Commutative Properties

One of the most important properties of variational calculus is the commutative property that states that the differential of the variation of a function y is identical to the variation of the differential of the same function. The commutative property may be written mathematically as

$$d(\delta y) = \delta(dy) \tag{4-38}$$

A simple proof of this relationship is provided in reference [9]. It is assumed here that the function y is differentiable.

   Another commutative property is the one that states that the variation of the integral of a functional F is the same as the integral of the variation of the same functional, or mathematically as

$$\delta \int_a^b F\,dx = \int_a^b \delta F\,dx \tag{4-39}$$

Note that the two integrals must be evaluated between the same two limits.

Miscellaneous Rules of Variational Calculus

From the definitions of differentiation and variation, it can be shown that all the rules of differentiation have variational counterparts. Let y and z be any continuous and differentiable functions. Then we may write

$$\delta(y + z) = \delta y + \delta z \tag{4-40}$$

$$\delta(yz) = y\,\delta z + z\,\delta y \tag{4-41}$$

$$\delta\left(\frac{y}{z}\right) = \frac{z\,\delta y - y\,\delta z}{z^2} \tag{4-42}$$

$$\delta(y^n) = ny^{n-1}\delta y \tag{4-43}$$

and so forth. The similarity between these and each of their counterparts in differential calculus should be obvious.

4-5   THE EULER-LAGRANGE EQUATION: GEOMETRIC AND NATURAL BOUNDARY CONDITIONS

In this section it will be seen how we may obtain the variational formulation from the differential. As mentioned earlier, a systematic procedure to accomplish this is desirable because, more often than not, the differential equation that describes a particular problem is easily derived and the variational form is not. Obtaining the variational form from the differential allows us to take advantage of some of the approximate solution techniques—including the finite element method itself (see Sec. 4-8).

The necessary condition for the existence of an extremum of the functional

$$I = \int_a^b F(x,y,y_x)\,dx \tag{4-44}$$

is that its first variation δI must be zero, or

$$\delta I = \delta \int_a^b F(x,y,y_x)\,dx = 0 \tag{4-45}$$

provided that

$$\left.\frac{\partial F}{\partial y_x}\delta y\right|_a^b = 0 \tag{4-46}$$

is satisfied [9]. We now want to find the condition(s) that $y(x)$ must satisfy such that the integral I is extremized or made stationary. Let us proceed by noting that by Eq. (4-39) the variation of the integral is the same as the integral of the variation, and so Eq. (4-45) becomes

$$\delta I = \int_a^b \delta F(x,y,y_x)\,dx = 0 \tag{4-47}$$

Now with the help of Eq. (4-37), we may write

$$\delta I = \int_a^b \left(\frac{\partial F}{\partial y}\delta y + \frac{\partial F}{\partial y_x}\delta y_x\right)dx \tag{4-48}$$

Let us rewrite δy_x in a more convenient form by using Eq. (4-38):

$$\delta y_x = \frac{\delta(dy)}{dx} = \frac{d(\delta y)}{dx} \tag{4-49}$$

Therefore, Eq. (4-48) becomes

$$\delta I = \int_a^b \left[\frac{\partial F}{\partial y} \delta y + \frac{\partial F}{\partial y_x} \frac{d(\delta y)}{dx} \right] dx \tag{4-50}$$

Let us now integrate by parts the second term in the integral [see Eq. (4-32)] with

$$u = \frac{\partial F}{\partial y_x} \qquad \text{and} \qquad \frac{dv}{dx} = \frac{d(\delta y)}{dx} \tag{4-51}$$

from which it follows that

$$\frac{du}{dx} = \frac{d}{dx}\left(\frac{\partial F}{\partial y_x} \right) \qquad \text{and} \qquad v = \delta y \tag{4-52}$$

Therefore, Eq. (4-50) becomes

$$\delta I = \frac{\partial F}{\partial y_x} \delta y \bigg|_a^b + \int_a^b \left[\frac{\partial F}{\partial y} - \frac{d}{dx}\left(\frac{\partial F}{\partial y_x} \right) \right] \delta y \, dx = 0 \tag{4-53}$$

One way for Eq. (4-53) to be true is if

$$\frac{\partial F}{\partial y_x} \delta y \bigg|_a^b = 0 \tag{4-54}$$

for then we have

$$\delta I = \int_a^b \left[\frac{\partial F}{\partial y} - \frac{d}{dx}\left(\frac{\partial F}{\partial y_x} \right) \right] \delta y \, dx = 0 \tag{4-55}$$

Note that the condition expressed by Eq. (4-54) is identical to that already stipulated by Eq. (4-46). Since Eq. (4-55) is to hold for all arbitrary variations δy, the only way for Eq. (4-55) to hold in general is if the expression in the brackets itself is zero or

$$\frac{\partial F}{\partial y} - \frac{d}{dx}\left(\frac{\partial F}{\partial y_x} \right) = 0 \tag{4-56}$$

This last result is really a differential equation with y and y_x as the *pseudoindependent variables*. In fact, Eq. (4-56) is precisely the same differential equation that results from the differential formulation in the first place. Actually F is a functional because it is a function of two functions, y and y_x. Equation (4-56) is referred to as the Euler or *Euler-Lagrange equation* for the problem whose functional is given by Eq. (4-44).

Before interpreting these results, let us show that the functional to be extremized for the problem in Example 4-1 is, in fact, given by Eq. (4-14b).

Example 4-3

For the problem posed in Example 4-1, show that the functional to be extremized is given by Eq. (4-14b). Recall that the problem in Example 4-1 was stated as follows:

$$\frac{d^2T}{dx^2} + 1000x^2 = 0 \qquad 0 \leqslant x \leqslant 1 \tag{4-7}$$

$$T(0) = 0 \tag{4-8a}$$

$$T(1) = 0 \tag{4-8b}$$

Solution

First, let us check to see whether or not the condition expressed in Eq. (4-46) [or Eq. (4-54)] is met. In terms of the notation in this problem, Eqs. (4-56) and (4-46) may be written

$$\frac{\partial F}{\partial T} - \frac{d}{dx}\left(\frac{\partial F}{\partial T_x}\right) = 0 \tag{4-57a}$$

and

$$\left. \frac{\partial F}{\partial T_x} \delta T \right|_0^1 = 0 \tag{4-57b}$$

Since the value of T is to be held fixed at either end of the interval, $0 \leqslant x \leqslant 1$, the variation of T or δT must be zero at these two points. Therefore, Eq. (4-57b) is satisfied by virtue of the conditions $T(0) = 0$ and $T(1) = 0$, since $\delta T = 0$ at these points. This type of boundary condition is referred to as a *geometric boundary condition* [10]. When $\partial F/\partial T_x = 0$ on the boundary, we have a *natural boundary condition* (see Example 4-4). Let us now turn to Eq. (4-57a) and make a term-by-term comparison with the governing equation expressed in Eq. (4-7), which yields

$$\frac{\partial F}{\partial T} = 1000x^2 \tag{4-58}$$

and

$$-\frac{d}{dx}\left(\frac{\partial F}{\partial T_x}\right) = -\frac{d}{dx}\left(-\frac{dT}{dx}\right) = -\frac{d(-T_x)}{dx}$$

A further comparison of the latter yields

$$\frac{\partial F}{\partial T_x} = -T_x \tag{4-59}$$

The reader may verify that Eq. (4-58) implies

$$F = 1000x^2 T + f(T_x) \qquad \text{(4-60)}$$

and Eq. (4-59) implies

$$F = -\tfrac{1}{2}T_x^2 + g(T) \qquad \text{(4-61)}$$

by direct substitutions. A further comparison of Eq. (4-60) with Eq. (4-61) yields

$$f(T_x) = -\tfrac{1}{2}T_x^2$$

and

$$g(T) = 1000x^2 T$$

so that

$$F = 1000x^2 T - \tfrac{1}{2}T_x^2 \qquad \text{(4-15)}$$

Actually a constant term may be added to the right-hand side since the resulting expression for F would still satisfy Eq. (4-56); however, this constant in no way affects the extremum of the functional I. From Eqs. (4-44) and (4-15), the functional is given by

$$I = \int_0^1 \left[1000x^2 T - \frac{1}{2}\left(\frac{dT}{dx}\right)^2 \right] dx \qquad \text{(4-14b)}$$

This last result is identical to that presented in Eq. (4-14b) for this same problem. It is emphasized that the function $T(x)$ that extremizes the functional I given in Eq. (4-14b) is the same one that satisfies the governing differential equation and boundary conditions given by Eqs. (4-7) and (4-8), respectively.   ∎

In Example 4-4, Eq. (4-14b) is derived in an alternate but completely equivalent manner.

Example 4-4

By starting with Eq. (4-55) derive Eq. (4-14b) for the problem stated in Example 4-1.

Solution

We now recognize the bracketed term in Eq. (4-55). When it is equated to zero, it becomes the differential equation for the problem. We may write from Eq. (4-55)

$$\delta I = \int_0^1 \left[1000x^2 + \frac{d}{dx}\left(\frac{dT}{dx}\right) \right] \delta T \, dx \qquad \text{(4-62)}$$

This may be broken into two separate integrals to give

$$\delta I = \int_0^1 1000x^2 \, \delta T \, dx + \int_0^1 \frac{d}{dx}\left(\frac{dT}{dx}\right) \delta T \, dx = 0 \qquad \textbf{(4-63)}$$

Integrating the second integral by parts with

$$u = \delta T \qquad \text{and} \qquad dv = \frac{d}{dx}\left(\frac{dT}{dx}\right) dx$$

yields

$$\delta I = \int_0^1 1000x^2 \, \delta T \, dx + \frac{dT}{dx}\delta T \bigg|_0^1 - \int_0^1 \frac{dT}{dx}\frac{d(\delta T)}{dx} \, dx \qquad \textbf{(4-64)}$$

Working with the second integral and using Eq. (4-38), we have

$$\int_0^1 \frac{dT}{dx}\frac{d(\delta T)}{dx} \, dx = \int_0^1 \frac{dT}{dx}\delta\left(\frac{dT}{dx}\right) dx = \int_0^1 \tfrac{1}{2}\delta\left(\frac{dT}{dx}\right)^2 dx \qquad \textbf{(4-65)}$$

If in Eq. (4-64) we require that

$$\frac{dT}{dx}\delta T \bigg|_0^1 = 0 \qquad \textbf{(4-66)}$$

we are left with

$$\delta I = \int_0^1 \left[1000x^2\delta T - \tfrac{1}{2}\delta\left(\frac{dT}{dx}\right)^2 \right] dx \qquad \textbf{(4-67)}$$

Since the variation of x^2 is zero (recall that only T and T_x have a variation), Eq. (4-67) may be written with the help of Eq. (4-39) as

$$\delta I = \delta \int_0^1 \left[1000x^2 T - \frac{1}{2}\left(\frac{dT}{dx}\right)^2 \right] dx \qquad \textbf{(4-68)}$$

from which we conclude that the functional I itself is given by

$$I = \int_0^1 \left[1000x^2 T - \frac{1}{2}\left(\frac{dT}{dx}\right)^2 \right] dx \qquad \textbf{(4-69)}$$

which again is identical to Eq. (4-14b). ■

Let us examine the condition expressed by Eq. (4-66) a little more closely. For the geometric boundary conditions in Example 4-1, we have $\delta T = 0$ at $x = 0$ and $x = 1$. Another way in which Eq. (4-66) could be satisfied is when dT/dx is zero on the boundaries, i.e., at $x = 0$ and $x = 1$ in this problem. This type of boundary condition is the so-called natural boundary condition. If T represents temperature, then dT/dx is zero when insulation is present. In a similar fashion, in Eq. (4-57b) the condition that $\partial F/\partial T_x = 0$ is also referred to as a natural boundary condition.

If in Eq. (4-64) we do not require the integrated term to be zero, the reader may show that Eq. (4-68) becomes

$$\delta I = \delta \left\{ \left. \frac{dT}{dx} T \right|_0^1 + \int_0^1 \left[1000x^2 T - \frac{1}{2} \left(\frac{dT}{dx} \right)^2 \right] dx \right\} \tag{4-70}$$

which introduces a mechanism by which nonzero values of dT/dx on the boundaries may be accommodated. In Eq. (4-70) it is implied that

$$\left. \delta \left(\frac{dT}{dx} T \right) \right|_0^1 = \left. \frac{dT}{dx} \delta T \right|_0^1 \tag{4-71}$$

or that the variation is made on T only and not on dT/dx (in the integrated term only).

In this section, we have learned how to obtain the variational formulation from the differential in two different but equivalent ways. The class of problems has been restricted to a single second-order differential equation in a single independent variable x, i.e., one-dimensional problems. The Euler-Lagrange equation was introduced for the class of problems whose functional F is given by Eq. (4-44), providing we also satisfy

$$\left. \frac{\partial F}{\partial y_x} \delta y \right|_a^b = 0 \tag{4-72}$$

In Problem 4-23, the reader is asked to show that for the functional

$$I = \int_a^b F(x,y,y_x,y_{xx}) \, dx \tag{4-73}$$

the corresponding Euler-Lagrange equation is

$$\frac{\partial F}{\partial y} - \frac{d}{dx}\left(\frac{\partial F}{\partial y_x} \right) + \frac{d^2}{dx^2}\left(\frac{\partial F}{\partial y_{xx}} \right) = 0 \tag{4-74}$$

where y_{xx} denotes the second derivative of $y(x)$ with respect to x. The following conditions are assumed to exist in this case:

$$\left. \left[\frac{\partial F}{\partial y_x} - \frac{d}{dx}\left(\frac{\partial F}{\partial y_{xx}} \right) \right] \delta y \right|_a^b = 0 \tag{4-75a}$$

and

$$\left. \left[\frac{\partial F}{\partial y_{xx}} \right] \delta y_x \right|_a^b = 0 \tag{4-75b}$$

Equations (4-74) and (4-75) are readily derived from Eq. (4-73) by following the first of the two procedures indicated earlier in this section and by integrating by parts twice. From Eq. (4-75a), one set of natural boundary conditions must be given by

$$\left[\frac{\partial F}{\partial y_x} - \frac{d}{dx}\left(\frac{\partial F}{\partial y_{xx}}\right)\right]\Bigg|_a^b = 0 \qquad \textbf{(4-76)}$$

while from Eq. (4-75b), we must have

$$\frac{\partial F}{\partial y_{xx}}\Bigg|_a^b = 0 \qquad \textbf{(4-77)}$$

as another set of natural boundary conditions. The geometric boundary conditions are given by

$$\delta y\Bigg|_a^b = 0 \qquad \textbf{(4-78)}$$

and

$$\delta y_x\Bigg|_a^b = 0 \qquad \textbf{(4-79)}$$

Equation (4-78) is equivalent to the statements that $y(a)$ is prescribed and $y(b)$ is prescribed, i.e., the function $y(x)$ is prescribed on the boundaries. Equation (4-79) implies that dy/dx is prescribed at $x = a$ and $x = b$ as well. What order differential equation does Eq. (4-74) correspond to? What is the highest-order derivative present in the functional? Try to generalize this observation.

In Problems 4-26 and 4-27, the reader will begin to appreciate why classical variational formulations exist for only those problems whose differential formulations do not contain odd-ordered derivatives.

4-6   THE METHOD OF WEIGHTED RESIDUALS

The method of weighted residuals provides an analyst with a very powerful approximate solution procedure that is applicable to a wide variety of problems. It is the existence of the various weighted-residual methods that makes it unnecessary to search for variational formulations to nonstructural problems in order to apply the finite element method to these problems. Before presenting four of the most popular weighted-residual methods, let us present a general framework around which the various methods are developed. In the process of doing this, it is necessary to repeat some of the general concepts given in Sec. 4-2.

General Concepts

Let us restrict the present discussion to a single governing equation with only one independent variable. Let us represent the governing equation as

$$f[T(x)] = 0 \qquad \text{in } \Omega \qquad \textbf{(4-80)}$$

where T represents the function sought (say temperature), which is a function of only x, and where Ω is the domain of the region governed by Eq. (4-80). In addition, let us specify the boundary conditions in the form

$$g_1[T(x)] = 0 \qquad \text{on } \Gamma_1$$

$$g_2[T(x)] = 0 \qquad \text{on } \Gamma_2$$

(4-81)

etc., where Γ_1 and Γ_2 include only those parts of Ω that are on the boundary. Let us again approximate the solution to Eqs. (4-80) and (4-81) with an approximate function T' that is given by

$$T' = T'(x; a_1, a_2, \ldots, a_n) = \sum_{i=1}^{n} a_i N_i(x)$$

(4-82)

which has one or more unknown (but constant) parameters $a_1, a_2, a_3, \ldots, a_n$ and that satisfies the boundary conditions given by Eq. (4-81) exactly. As before, the functions $N_i(x)$ are referred to as the trial functions. If this approximate solution T' is substituted into Eq. (4-80) for $T(x)$, it should not be surprising that it will not necessarily satisfy this equation exactly; that is, some residual error $R(x, a_1, \ldots, a_n)$ results. Therefore, we write

$$f[T'(x, a_1, a_2, \ldots, a_n)] = R(x, a_1, a_2, \ldots, a_n)$$

(4-83)

The method of weighted residuals requires that the parameters $a_1, a_2, \ldots, a_n$ be determined by satisfying

$$\int_{\Omega} w_i(x) R(x; a_1, a_2, \ldots, a_n) \, dx = 0 \qquad i = 1, 2, \ldots, n$$

(4-84)

where the functions $w_i(x)$ are the n arbitrary *weighting functions*. The choice of the weighting functions is left largely up to personal preference, but four particular functions are used most often. The most popular weighted-residual methods are referred to as (1) point collocation, (2) subdomain collocation, (3) least squares, and (4) Galerkin. Each of these is developed in turn, and the problem stated in Example 4-1 is solved in Examples 4-5 to 4-8 with each of the methods, respectively. Of these four methods, the Galerkin method has the widest use in finite element analysis for reasons that will become apparent later. The least-squares method is also used but to a much lesser extent. Some of these methods are used in Chapter 10 when FEM formulations to unsteady problems are considered.

Point Collocation

In the point collocation method, the weighting functions $w_i(x)$ are denoted as $\delta(x - x_i)$ and defined such that

$$\int_a^b \delta(x - x_i) \, dx = 1 \qquad \text{for } x = x_i$$

and

$$\int_a^b \delta(x - x_i) \, dx = 0 \qquad \text{for } x \neq x_i$$

(4-85)

The x_i's are referred to as the collocation points and are selected arbitrarily by the analyst. The δ here has nothing to do with the variational operator but rather is defined by Eq. (4-85). Substitution of $\delta(x - x_i)$ for $w_i(x)$ in Eq. (4-84) gives

$$\int_\Omega \delta(x - x_i) R(x; a_1, a_2, \ldots, a_n) \, dx = 0 \qquad \text{for } i = 1, 2, \ldots, n \quad \textbf{(4-86)}$$

If Eq. (4-86) is evaluated at n collocation points, $x_1, x_2, \ldots, x_n$, then n algebraic equations in n unknowns result:

$$R(x_1; a_1, a_2, \ldots, a_n) = 0$$

$$R(x_2; a_1, a_2, \ldots, a_n) = 0 \qquad\qquad \textbf{(4-87)}$$

$$\vdots$$

$$R(x_n; a_1, a_2, \ldots, a_n) = 0$$

It is emphasized that the n unknowns in Eq. (4-87) are the constant parameters a_1, $a_2, \ldots, a_n$. Once these are determined, the approximate solution is given by Eq. (4-82) and the problem is solved, at least approximately. The constants a_1, a_2, $\ldots, a_n$ are dependent on the choices made for the n trial functions, $N_1(x)$, $N_2(x)$, $\ldots, N_n(x)$, and on the choice made for the n collocation points. For well-posed problems and well-behaved trial functions, the solution of the system of equations implied in Eq. (4-87) will yield the numerical values for the a_i's. Equation (4-87) may be written more concisely as

$$R(x_i; a_1, a_2, \ldots, a_n) = 0 \qquad \text{for } i = 1, 2, \ldots, n \qquad \textbf{(4-88)}$$

where x_i is the ith collocation point. This method is illustrated numerically in Example 4-5.

Example 4-5

Using the trial function assumed in Example 4-1, determine an approximate solution to the problem posed in that example by the point collocation method.

Solution

Note that since only one trial function is used, only one parameter (namely, a_1) may be introduced. This implies that only one collocation point may be selected. It is as good a choice as any to choose the (one and only) collocation point x_1 to be in the middle of the domain, or $x_1 = \frac{1}{2}$. Recall that the residual $R(x; a_1)$ is given by Eq. (4-11) or

$$R(x; a_1) = -6a_1 x + 1000x^2 \qquad \textbf{(4-11)}$$

By Eq. (4-88), we must evaluate this residual at the collocation point or at $x_1 = \frac{1}{2}$ and set the result to zero, or

$$-6a_1(\tfrac{1}{2}) + 1000(\tfrac{1}{2})^2 = 0$$

Solving for a_1 yields

$$a_1 = {}^{1000}\!/_{12} \tag{4-89}$$

Therefore, an approximate solution to the problem by the point collocation method is given by

$$T'_{pc} = {}^{1000}\!/_{12}\, x(1 - x^2) \tag{4-90}$$

The reader should compare this approximate solution with those from Examples 4-1 and 4-2. ∎

Subdomain Collocation

In subdomain collocation, the weighting functions are chosen to be unity over a portion of the domain and zero elsewhere. Mathematically, the $w_i(x)$ are given by

$$w_1(x) = \begin{cases} 1 & \text{for } x \text{ in } \Omega_1 \\ 0 & \text{for } x \text{ not in } \Omega_1 \end{cases}$$

$$w_2(x) = \begin{cases} 1 & \text{for } x \text{ in } \Omega_2 \\ 0 & \text{for } x \text{ not in } \Omega_2 \end{cases}$$

$$\vdots \tag{4-91}$$

$$w_n(x) = \begin{cases} 1 & \text{for } x \text{ in } \Omega_n \\ 0 & \text{for } x \text{ not in } \Omega_n \end{cases}$$

where Ω_1 is a portion of the domain not included in $\Omega_2, \Omega_3, \ldots, \Omega_n$. It is instructive to show these particular weighting functions in graphical form; for example, see Fig. 4-6 for $n = 3$.

Let us return now to our general discussion with n unknown parameters. Substitution of $w_i(x)$ given by Eq. (4-91) into Eq. (4-84) gives n integral equations:

$$\int_{\Omega_1} R(x;\, a_1, a_2, \ldots, a_n)\, dx = 0$$

$$\int_{\Omega_2} R(x;\, a_1, a_2, \ldots, a_n)\, dx = 0$$

$$\vdots \tag{4-92}$$

$$\int_{\Omega_n} R(x;\, a_1, a_2, \ldots, a_n)\, dx = 0$$

The different integration domains should be noted. Since the residual error R is a known function of x, $a_1, a_2, \ldots, a_n$ [see Eq. (4-83)], the integrations in Eq. (4-92) can be carried out, resulting again in n equations in n unknowns. Again it is emphasized that the unknowns are $a_1, a_2, \ldots, a_n$. The values obtained for these

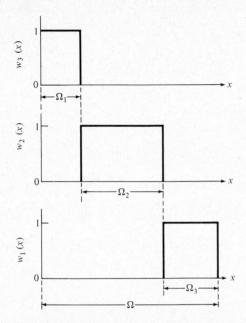

Figure 4-6 Possible subdomains and weighting functions for the subdomain collocation weighted-residual method (for three subdomains).

constants are, in general, dependent on the n subdomains chosen as well as on the choice of the n trial functions.

Example 4-6

Repeat Example 4-1 by using the subdomain collocation method with the same trial function.

Solution

Since there is only one trial function and one unknown parameter a_1, only one integral of the residual (over the entire domain from $x = 0$ to $x = 1$) needs to be evaluated:

$$\int_0^1 R(x;a_1)\, dx = 0 \tag{4-93}$$

Substituting $R(x,a_1)$ given by Eq. (4-11) into Eq. (4-93) gives

$$\int_0^1 (-6a_1x + 1000x^2)\, dx = 0$$

Integrating, evaluating, and solving for a_1 yields

$$a_1 = {}^{1000}\!/_9 \tag{4-94}$$

Therefore, the approximate solution to the problem by the subdomain collocation method is given by

$$T'_{sc} = {}^{1000}\!/\!9\, x(1 - x^2) \tag{4-95}$$

The reader will undoubtedly note that the same value for a_1 was obtained from the Ritz method in Example 4-1. This is not just a coincidence: the first-order Ritz method is identical to the first-order subdomain collocation weighted-residual method. ∎

Least Squares

The method of least squares requires that the integral I of the square of the residual R be minimized. The integral I is given by

$$I = \int_\Omega [R(x_i; a_1, a_2, \ldots, a_n)]^2 \, dx \tag{4-96}$$

The problem is reduced to the determination of the parameters $a_1, a_2, \ldots, a_n$ such that the value of the integral I is minimized. Therefore, we must take the partial derivative of I, first, with respect to a_1 and set the result to zero, then, with respect to a_2 and set that result to zero, and so forth. Mathematically, this may be written as

$$\frac{\partial I}{\partial a_1} = \frac{\partial}{\partial a_1} \int_\Omega [R(x_i, a_1, a_2, \ldots, a_n)]^2 \, dx = 0$$

$$\frac{\partial I}{\partial a_2} = \frac{\partial}{\partial a_2} \int_\Omega [R(x_i, a_1, a_2, \ldots, a_n)]^2 \, dx = 0$$

$$\vdots$$

$$\frac{\partial I}{\partial a_n} = \frac{\partial}{\partial a_n} \int_\Omega [R(x_i, a_1, a_2, \ldots, a_n)]^2 \, dx = 0$$

where Ω includes the entire x domain. Because the limits on the integral are not a function of the a_i's, the order of the integration and differentiation may be interchanged to give

$$\int_\Omega \frac{\partial}{\partial a_1} [R(x_i, a_1, a_2, \ldots, a_n)]^2 \, dx = 0$$

$$\int_\Omega \frac{\partial}{\partial a_2} [R(x_i, a_1, a_2, \ldots, a_n)]^2 \, dx = 0$$

and so forth. Carrying out the differentiations and dividing by the constant factor 2 (and dropping the arguments on R for now) gives

$$\int_\Omega R \frac{\partial R}{\partial a_1} \, dx = 0$$

$$\int_\Omega R \frac{\partial R}{\partial a_2} \, dx = 0$$

$$\vdots$$

$$\int_\Omega R \frac{\partial R}{\partial a_n} \, dx = 0$$

(4-97)

Now it is a simple matter to determine the weighting functions by a direct comparison of Eq. (4-97) with Eq. (4-84) from which we get

$$w_1(x) = \frac{\partial R}{\partial a_1}$$

$$w_2(x) = \frac{\partial R}{\partial a_2}$$

$$\vdots$$

$$w_n(x) = \frac{\partial R}{\partial a_n}$$

(4-98)

These may be summarized concisely as

$$w_i(x) = \frac{\partial R}{\partial a_i} \qquad \text{for } i = 1, 2, \ldots, n$$

(4-99)

Therefore, the least-squares weighted-residual method requires the solution of the n equations given by Eq. (4-84) for the n unknown parameters $a_1, a_2, \ldots, a_n$, where the $w_i(x)$'s are given by Eq. (4-99). Alternately, the system of equations implied in Eq. (4-97) may be solved simultaneously for the a_i's. Again the values obtained for the a_i's are dependent on the n trial functions chosen.

Example 4-7

Resolve (approximately) the problem in Example 4-1 by using the least-squares method and the same trial function.

Solution

The residual error $R(x,a_1)$, given by Eq. (4-11), is repeated here for easy reference:

$$R(x,a_1) = -6a_1x + 1000x^2$$

(4-11)

Since there is only one unknown parameter a_1, only the first of Eq. (4-97) needs to be used, namely,

$$\int_0^1 R \frac{\partial R}{\partial a_1} \, dx = 0 \tag{4-100}$$

Computing $\partial R/\partial a_1$ gives

$$\frac{\partial R}{\partial a_1} = -6x$$

and Eq. (4-100) becomes

$$\int_0^1 (-6x)(-6a_1 x + 1000x^2) \, dx = 0 \tag{4-101}$$

Simplifying the integrand, integrating, evaluating, and solving for a_1 yields

$$a_1 = {}^{1000}\!/\!_8$$

from which we may write the approximate solution from the least-squares method as

$$T'_{ls} = {}^{1000}\!/\!_8 \, x(1 - x^2) \tag{4-102}$$

The reader should note that still another value for a_1 is obtained. ∎

Galerkin

In the Galerkin weighted-residual method, the trial functions $N_i(x)$ themselves are used as the weighting functions, or

$$w_i(x) = N_i(x) \tag{4-103}$$

For the Galerkin method Eq. (4-84) becomes

$$\int_\Omega N_i(x) \, R(x; a_1, a_2, \ldots, a_n) \, dx = 0 \qquad \text{for } i = 1 \text{ to } n \tag{4-104}$$

Because there is one trial function for each unknown parameter, Eq. (4-104) really gives n such equations that, when solved, yield the values of the unknown parameters, $a_1, a_2, \ldots, a_n$. It should be obvious that again the values obtained for the a_i's are dependent on the choice of trial functions. The Galerkin method is seen to be quite simple.

Example 4-8

Resolve the problem in Example 4-1 by using the Galerkin method and the same trial function.

Solution

The trial function $N_1(x)$ is given by Eq. (4-9), or

$$N_1(x) = x(1 - x^2) \tag{4-9}$$

Substituting R from Eq. (4-11) and $N_1(x)$ from Eq. (4-9) into Eq. (4-104) for $i = 1$ only gives

$$\int_0^1 x(1 - x^2)(-6a_1x + 1000x^2)\, dx = 0 \qquad \textbf{(4-105)}$$

Simplifying the integrand, integrating, evaluating, and solving for a_1 yields

$$a_1 = {}^{5000}\!/\!_{48} \qquad \textbf{(4-106)}$$

Therefore, an approximate solution by the Galerkin method is given by

$$T'_G = {}^{5000}\!/\!_{48}\, x(1 - x^2) \qquad \textbf{(4-107)}$$

The reader should note that exactly the same value for a_1 was obtained in Example 4-2. There an approximate solution was obtained by the Rayleigh-Ritz method, which is a variational method. This is not a coincidence for it can be shown that the variational and Galerkin methods must give identical results, providing the problem has a classical variational statement (or principle) in the first place [11]. This profound result carries over into the application of these methods to FEM, as shown in Secs. 4-8 and 4-9 for the same problem. ■

Comparison with the Exact Solution

If the problem posed in Example 4-1 is solved exactly, we get

$$T(x) = {}^{1000}\!/\!_{12}\, x(1 - x^3) \qquad \textbf{(4-108)}$$

Note the exponent of 3 on the x in the parentheses; our approximate solution was assumed to be of the form

$$T'(x) = a_1x(1 - x^2) \qquad \textbf{(4-109)}$$

which is slightly different than the exact solution above. The exact solution from Eq. (4-108) is compared to the four approximate solutions by the various weighted-residual methods in Fig. 4-7. The approximate solution from the Ritz method (Example 4-1) is identical to that obtained from subdomain collocation. Similarly, the Rayleigh-Ritz (variational) and Galerkin solutions are identical.

It is seen that none of the methods seems to do a particularly outstanding job of approximating the exact solution. However, in Secs. 4-8 and 4-9, the variational and Galerkin methods are adapted to a finite element formulation for a solution to the same problem. The fundamental difference between what we have done so far and what we will be doing next lies in the type of trial functions that are assumed. Recall that trial functions were chosen that applied globally to the entire domain being analyzed. From this point onward, trial functions will be chosen that are to be applied locally (over each and every finite element). These trial functions will

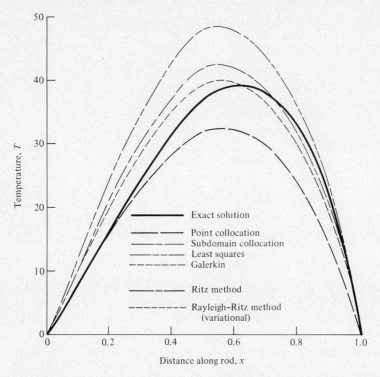

Figure 4-7   Comparison of approximate solutions to problem posed in Example 4-1 with the exact solution.

be said to be *piecewise continuous*. Piecewise continuous trial functions in this text will be referred to as *shape functions*.

This important step is fundamental to the finite element method and has associated with it two important implications: (1) the piecewise continuous trial functions or shape functions need not satisfy the physics of the problem since they are to be applied locally in every element, not globally, and (2) the global boundary conditions need not be satisfied. This is not to say that the shape functions have no restrictions placed on them. These restrictions are delineated in Chapter 6. Before actually illustrating this procedure, we need to review some additional mathematics with which the reader may not be familiar.

4-7   MORE MATHEMATICS

In this section we will review the following: (1) the derivative of a matrix with respect to a scalar, (2) the integral of a matrix with respect to a scalar, (3) the derivative of scalar with respect to a vector, and (4) the integral as a sum of other integrals. Examples of the use of these notions are provided throughout the text in general and in the next two sections in particular.

Derivative of a Matrix with Respect to a Scalar

Consider an $m \times n$ matrix $\mathbf{M}$ whose elements are functions of x. The derivative of $\mathbf{M}$ with respect to x is obtained by taking the derivative of each of the entries (or elements) in $\mathbf{M}$ with respect to x, or

$$\frac{d\mathbf{M}(x)}{dx} = \begin{bmatrix} \dfrac{dM_{11}}{dx} & \dfrac{dM_{12}}{dx} & \cdots & \dfrac{dM_{1n}}{dx} \\[2ex] \dfrac{dM_{21}}{dx} & \dfrac{dM_{22}}{dx} & \cdots & \dfrac{dM_{2n}}{dx} \\[2ex] \vdots & & \ddots & \vdots \\[2ex] \dfrac{dM_{m1}}{dx} & \dfrac{dM_{m2}}{dx} & \cdots & \dfrac{dM_{mn}}{dx} \end{bmatrix} \tag{4-110}$$

The instances when $\mathbf{M}$ is a column matrix ($n = 1$) or when $\mathbf{M}$ is a row matrix ($m = 1$) are seen to be special cases.

Integral of a Matrix with Respect to a Scalar

Let us again consider an $m \times n$ matrix $\mathbf{M}$ whose elements are functions of x. It follows that the integral of $\mathbf{M}$ with respect to x may be found by taking the integral of each element in $\mathbf{M}$, or

$$\int \mathbf{M}(x)\, dx = \begin{bmatrix} \int M_{11}\, dx & \int M_{12}\, dx & \cdots & \int M_{1n}\, dx \\ \int M_{21}\, dx & \int M_{22}\, dx & \cdots & \int M_{2n}\, dx \\ \vdots & \vdots & \ddots & \vdots \\ \int M_{m1}\, dx & \int M_{m2}\, dx & \cdots & \int M_{mn}\, dx \end{bmatrix} \tag{4-111}$$

Although Eq. (4-111) is written with indefinite integrals, the same idea holds for definite integrals (providing the same limits are used on all integrals). Like differentiation, integration of a column matrix and a row matrix are seen to be special cases of Eq. (4-111).

Derivative of a Scalar Function with Respect to a Vector

Let us now consider the following scalar function:

$$y(x) = a_1 N_1(x) + a_2 N_2(x) + \cdots + a_n N_n(x) \tag{4-112}$$

and determine $\partial y / \partial a_i$, or

$$\frac{\partial y}{\partial a_1} = N_1(x) \tag{4-113a}$$

$$\frac{\partial y}{\partial a_2} = N_2(x) \tag{4-113b}$$

$$\vdots$$

$$\frac{\partial y}{\partial a_n} = N_n(x) \tag{4-113c}$$

We may write

$$\frac{\partial y}{\partial a_i} = N_i(x) \qquad \text{for } i = 1, 2, \ldots, n \tag{4-114}$$

If we define **N** and **a** as

$$\mathbf{N} = [N_1(x) \quad N_2(x) \quad \cdots \quad N_n(x)] \tag{4-115}$$

and

$$\mathbf{a} = [a_1 \quad a_2 \quad \cdots \quad a_n]^T \tag{4-116}$$

then Eq. (4-112) may be concisely written as

$$y(x) = \mathbf{Na} \tag{4-117}$$

where it is implied that $\mathbf{N} = \mathbf{N}(x)$. Note the use of the transpose in Eq. (4-116) that is used simply to conserve space since **a** is really a column matrix. From Eqs. (4-113) it follows that

$$\frac{\partial y}{\partial \mathbf{a}} \equiv \begin{bmatrix} \dfrac{\partial y}{\partial a_1} \\[2ex] \dfrac{\partial y}{\partial a_2} \\[2ex] \vdots \\[2ex] \dfrac{\partial y}{\partial a_n} \end{bmatrix} = \begin{bmatrix} N_1(x) \\[1ex] N_2(x) \\[1ex] \vdots \\[1ex] N_n(x) \end{bmatrix} = \mathbf{N}^T \tag{4-118}$$

Therefore, it may be concluded for **N** and **a** defined by Eqs. (4-115) and (4-116), respectively, that

$$\frac{\partial(\mathbf{Na})}{\partial \mathbf{a}} = \mathbf{N}^T \tag{4-119}$$

This result may not be what we would intuitively expect because of the need to transpose **N**. It is particularly useful in the next section where the Rayleigh-Ritz method is adapted to FEM. This result is not needed in the Galerkin method.

The Integral as a Sum of Other Integrals

It follows directly from the definition of integration that an integral may be evaluated by dividing the domain Ω into M nonoverlapping subdomains Ω^e and summing the integrals over each subdomain, or

$$\int_\Omega f(x)\, dx = \sum_{e=1}^{M} \int_{\Omega^e} f(x)\, dx \tag{4-120}$$

Each nonoverlapping subdomain Ω^e will correspond to one finite element in this text. The superscript $(^e)$ will be interpreted to be the element number.

4-8   THE RAYLEIGH-RITZ FINITE ELEMENT METHOD

In Sec. 4-2 the Rayleigh-Ritz method was used to find an approximate solution to the problem posed in Example 4-1. In this section, we wish to obtain another approximate solution but now in the context of the finite element method. As in Example 4-2, the variational statement will be the starting point. However, the fundamental difference in the treatment here compared to that in Sec. 4-2 is the following: piecewise continuous trial functions will be used that apply only to a small portion of the entire domain, i.e., over a finite element. The reader should recall that the trial functions used in Secs. 4-2 to 4-6 were applied globally, i.e., to the entire domain. Now the trial functions will be taken to be *interpolation polynomials* as described below.

Consider the domain that includes the interval $a \le x \le b$ as shown in Fig. 4-8. Let us divide this interval into M subintervals as shown. Actually, this process is called discretization and a typical subinterval from x_j to x_k is really a finite element and is simply denoted as element e. The ends of the subinterval are the locations of the two nodes, nodes j and k. Since the element e has two nodes, a linear interpolating polynomial may be used to describe the behavior of the field variable, e.g., $y(x)$, over the element, or

$$y^e = mx + b \tag{4-121}$$

The superscript $(^e)$ on the field variable serves to remind us that the approximation applies only over element e and not globally. But at $x = x_j$, we must have $y^e = y_j$, and at $x = x_k$, we must have $y^e = y_k$. Therefore, applying these two conditions to Eq. (4-121) yields

$$y_j = mx_j + b \tag{4-122a}$$

$$y_k = mx_k + b \tag{4-122b}$$

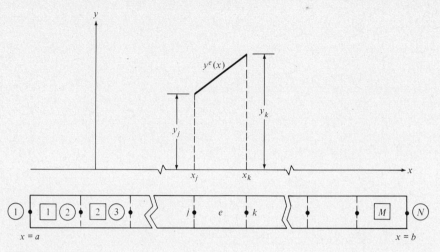

Figure 4-8 Typical one-dimensional problem domain discretized into M elements with N nodes with an assumed linear variation of the field variable $y(x)$ over a typical element e with nodes j and k.

that may be solved for m and b to give

$$m = \frac{y_k - y_j}{x_k - x_j} \tag{4-123a}$$

and

$$b = y_j - \left(\frac{y_k - y_j}{x_k - x_j}\right) x_j \tag{4-123b}$$

Substituting these expressions for m and b in Eq. (4-121) and some rearrangement gives the following:

$$y^e = \left(\frac{x_k - x}{x_k - x_j}\right) y_j + \left(\frac{x - x_j}{x_k - x_j}\right) y_k \tag{4-124}$$

It is observed that Eq. (4-124) is of the form

$$y^e = a_1 N_1(x) + a_2 N_2(x) \tag{4-125a}$$

or

$$y^e = a_j N_j(x) + a_k N_k(x) \tag{4-125b}$$

where

$$a_1 = a_j = y_j \tag{4-126a}$$

$$a_2 = a_k = y_k \tag{4-126b}$$

$$N_1(x) = N_j(x) = \frac{x_k - x}{x_k - x_j} \tag{4-127a}$$

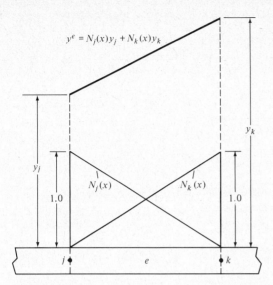

Figure 4-9 Typical element e showing the two shape functions and the resulting variation of the field variable $y^e(x)$ (linear between nodes j and k).

and

$$N_2(x) = N_k(x) = \frac{x - x_j}{x_k - x_j}$$ **(4-127b)**

The functions $N_j(x)$ and $N_k(x)$ are referred to as the *shape functions* for the element.

We see that instead of using higher-order polynomials that apply globally to the entire domain, we will now use two first-order shape functions that apply only to a small subinterval or element. These particular shape functions are shown in Fig. 4-9. Note that they are both linear and at $x = x_j$, we have $N_j = 1$ and $N_k = 0$, and at $x = x_k$, we have $N_j = 0$ and $N_k = 1$. Equation (4-125b) may be written in matrix form as follows:

$$y^e = \mathbf{N}\mathbf{a}^e$$ **(4-128)**

where by definition

$$\mathbf{N} = [N_j(x) \quad N_k(x)]$$ **(4-129a)**

and

$$\mathbf{a}^e = \begin{bmatrix} a_j \\ a_k \end{bmatrix}$$ **(4-129b)**

It should be emphasized that the vector $\mathbf{a}^e$ contains the values of the field variable at the two nodes. The idea is to use Eq. (4-128) to represent $y(x)$ over each element in the discretized domain and to find the values of the a_i's by extremizing the functional I with respect to the a_i's. The set of all a_i's for the entire domain represents the finite element solution to the original problem.

Because the function $y(x)$ itself given by Eq. (4-121) is continuous and the first derivatives are not necessarily continuous (from element to element), the function $y(x)$ is said to be C^0-continuous. A function with continuous first derivatives is said to be C^1-continuous, etc. In a similar fashion, the two shape functions given by Eq. (4-127) are also said to be C^0-continuous or to possess C^0-continuity. Some C^1-continuous shape functions are derived in Chapter 7 for the beam model.

These rather abstract ideas are best illustrated by tackling a numerical example, which again, for comparison purposes, is taken to be the problem posed in Example 4-1.

Example 4-9

Using the Rayleigh-Ritz finite element method, find an approximate solution to the problem posed in Example 4-1 by determining the value of the field variable, T, at six equally spaced node points.

Solution

Before the Rayleigh-Ritz method can be used, the variational statement of the problem is needed. Because of the two prescribed temperature boundary conditions at $x = 0$ and $x = 1$ (i.e., on the global boundary), we could start with the functional given by Eq. (4-69) from Example 4-4. However, let us use the functional implied in Eq. (4-70) so as to show what happens to the *integrated term*. Therefore, the functional to be used here is

$$I = \sum_{e=1}^{M} \left\{ \frac{dT}{dx} T^e \bigg|_{x_j}^{x_k} + \int_{x_j}^{x_k} \left[1000x^2 T^e - \frac{1}{2} \left(\frac{dT^e}{dx} \right)^2 \right] dx \right\} \qquad \textbf{(4-130)}$$

Note that Eq. (4-120) has been used to represent I as a sum of the integrals over each element. The integrated term above will be appreciated later to be completely equivalent to that implied in Eq. (4-70) because the *internal* contributions will cancel during the assemblage step and only the end conditions will survive. The summation sign in Eq. (4-130) unnecessarily clutters the equations; therefore, let us drop the summation and write the functional for a typical element e as

$$I^e = \frac{dT}{dx} T^e \bigg|_{x_j}^{x_k} + \int_{x_j}^{x_k} \left[1000x^2 T^e - \frac{1}{2} \left(\frac{dT^e}{dx} \right)^2 \right] dx \qquad \textbf{(4-131)}$$

where

$$I = \sum_{e=1}^{M} I^e \qquad \textbf{(4-132)}$$

and where M is the number of elements used. This summation is really considered at the assemblage step later. The a_i's in the approximation y^e [to $y(x)$] in element e are to be determined by differentiating I with respect to **a** and setting the result to zero, or

$$\frac{dI}{d\mathbf{a}} = \sum_{e=1}^{M} \frac{dI^e}{d\mathbf{a}^e} \tag{4-133}$$

For now, let us extremize I^e in Eq. (4-131) with respect to the vector $\mathbf{a}^e$ where for T^e we take

$$T^e = \mathbf{N}\mathbf{a}^e \tag{4-134}$$

as implied by Eq. (4-128), or

$$\frac{dI^e}{d\mathbf{a}^e} = \frac{dT}{dx}\frac{dT^e}{d\mathbf{a}^e}\bigg|_{x_j}^{x_k} + \int_{x_j}^{x_k}\left[1000x^2\frac{dT^e}{d\mathbf{a}^e} - \frac{1}{2}\frac{d}{d\mathbf{a}^e}\left(\frac{dT^e}{dx}\right)^2\right]dx = 0 \tag{4-135}$$

But from Eq. (4-119) we have

$$\frac{dT^e}{d\mathbf{a}^e} = \frac{d(\mathbf{N}\mathbf{a}^e)}{d\mathbf{a}^e} = \mathbf{N}^T \tag{4-136}$$

and it also follows that

$$\frac{1}{2}\frac{d}{d\mathbf{a}^e}\left(\frac{dT^e}{dx}\right)^2 = \frac{1}{2}(2)\frac{dT^e}{dx}\frac{d}{d\mathbf{a}^e}\left(\frac{dT^e}{dx}\right) = \frac{d}{d\mathbf{a}^e}\left(\frac{dT^e}{dx}\right)\frac{dT^e}{dx} = \frac{d\mathbf{N}^T}{dx}\frac{d\mathbf{N}}{dx}\mathbf{a}^e \tag{4-137}$$

Using the results from Eqs. (4-136) and (4-137) in Eq. (4-135) yields

$$\frac{dT}{dx}\mathbf{N}^T\bigg|_{x_j}^{x_k} + \int_{x_j}^{x_k}1000x^2\,\mathbf{N}^T\,dx - \int_{x_j}^{x_k}\frac{d\mathbf{N}^T}{dx}\frac{d\mathbf{N}}{dx}\mathbf{a}^e\,dx = 0 \tag{4-138}$$

But the vector $\mathbf{a}^e$ is not a function of x so it can be removed from the right (why not from the left?) to give

$$\left[\int_{x_j}^{x_k}\frac{d\mathbf{N}^T}{dx}\frac{d\mathbf{N}}{dx}\,dx\right]\mathbf{a}^e = \frac{dT}{dx}\mathbf{N}^T\bigg|_{x_j}^{x_k} + \int_{x_j}^{x_k}1000x^2\mathbf{N}^T\,dx \tag{4-139}$$

which is of the form

$$\mathbf{K}^e\mathbf{a}^e = \mathbf{f}^e \tag{4-140}$$

where $\mathbf{K}^e$ may be referred to as the element *stiffness* matrix, $\mathbf{f}^e$ as the element nodal *force* vector, and $\mathbf{a}^e$ as the vector of nodal unknowns (i.e., T_j and T_k). Before Eq. (4-140) can be solved, however, two more important tasks must be completed: (1) assemblage of the $\mathbf{K}^e$'s and $\mathbf{f}^e$'s to form $\mathbf{K}^a$ and $\mathbf{f}^a$ and (2) application of the geometric boundary conditions (prescribed temperatures) at $x = 0$ and $x = 1$. The matrix $\mathbf{K}^a$ is referred to as the *assemblage stiffness matrix* and the vector $\mathbf{f}^a$ is referred to as the *assemblage nodal force vector* (before considering the geometric boundary conditions). After the geometric boundary conditions are applied, we have a system of equations represented by

$$\mathbf{K}\mathbf{a} = \mathbf{f}$$

which may be solved for the values of the field variables at the node points (i.e., $a_1, a_2, \ldots$). From Eqs. (4-139) and (4-140), we define

$$\mathbf{K}^e = \int_{x_j}^{x_k} \frac{d\mathbf{N}^T}{dx} \frac{d\mathbf{N}}{dx} \, dx \tag{4-141}$$

and

$$\mathbf{f}^e = \frac{dT}{dx} \mathbf{N}^T \bigg|_{x_j}^{x_k} + \int_{x_j}^{x_k} 1000x^2 \mathbf{N}^T \, dx \tag{4-142}$$

It should be noted that $\mathbf{K}^e$ is a symmetric 2×2 matrix, whereas $\mathbf{f}^e$ is a 2×1 matrix (or column vector). Obviously, $\mathbf{a}^e$ is also a 2×1 matrix. It is very easy to evaluate $\mathbf{K}^e$ given by Eq. (4-141) if the expressions for $N_j(x)$ and $N_k(x)$ from Eq. (4-127) are used. The final result may be verified to be (see Problem 4-51)

$$\mathbf{K}^e = \frac{1}{x_k - x_j} \begin{bmatrix} 1 & -1 \\ -1 & 1 \end{bmatrix} \tag{4-143}$$

In arriving at Eq. (4-143), the reader should make use of some of the review material in Sec. 4-7, namely, differentiation and integration of a matrix with respect to a scalar.

The *integrated term* in Eq. (4-142) is evaluated as follows:

$$\frac{dT}{dx} \mathbf{N}^T \bigg|_{x_j}^{x_k} = \frac{dT(x_k)}{dx} \mathbf{N}^T(x_k) - \frac{dT(x_j)}{dx} \mathbf{N}^T(x_j)$$

But from Eqs. (4-127) and (4-129a) we have

$$\mathbf{N}^T(x_k) = \begin{bmatrix} 0 \\ 1 \end{bmatrix}$$

and

$$\mathbf{N}^T(x_j) = \begin{bmatrix} 1 \\ 0 \end{bmatrix}$$

Therefore, the integrated term becomes

$$\frac{dT}{dx} \mathbf{N}^T \bigg|_{x_j}^{x_k} = \frac{dT(x_k)}{dx} \begin{bmatrix} 0 \\ 1 \end{bmatrix} + \frac{dT(x_j)}{dx} \begin{bmatrix} -1 \\ 0 \end{bmatrix} \tag{4-144}$$

The evaluation of the remaining part of $\mathbf{f}^e$, namely the integral contribution, is very tedious but routine. The reader should show (see Problem 4-52) that the final result is given by

$$\int_{x_j}^{x_k} 1000x^2 \mathbf{N}^T \, dx = \frac{1000/12}{x_k - x_j} \begin{bmatrix} x_k^4 - 4x_k x_j^3 + 3x_j^4 \\ 3x_k^4 - 4x_j x_k^3 + x_j^4 \end{bmatrix} \tag{4-145}$$

An alternate (and approximate) method of evaluating this integral is outlined in Problem 4-53.

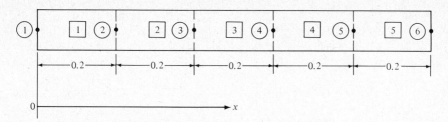

Figure 4-10 Discretized problem domain for problem posed in Example 4-1 and solved by the Rayleigh-Ritz finite element method in Example 4-9.

Before making use of these results in carrying out the numerical computations, we must discretize the global domain into five equally sized elements that require six equally spaced nodes, as shown in Fig. 4-10. It proves to be very convenient to organize the nodal coordinate and element connectivity data as shown in Tables 4-1 and 4-2, respectively. Note that a material set flag is not used because only one material is present in the model.

By using the data in these tables it is easy to compute the element stiffness matrices $\mathbf{K}^{(1)}$, $\mathbf{K}^{(2)}$, etc., and the element nodal force vectors $\mathbf{f}^{(1)}$, $\mathbf{f}^{(2)}$, etc. For example, for $\mathbf{K}^{(1)}$ we have

$$\mathbf{K}^{(1)} = \frac{1}{x_2 - x_1}\begin{bmatrix} 1 & -1 \\ -1 & 1 \end{bmatrix} = \frac{1}{0.2}\begin{bmatrix} 1 & -1 \\ -1 & 1 \end{bmatrix} = \begin{bmatrix} 5 & -5 \\ -5 & 5 \end{bmatrix}$$

Table 4-1 Nodal Coordinate Data for Example 4-9

Node number	x
1	0.0
2	0.2
3	0.4
4	0.6
5	0.8
6	1.0

Table 4-2 Element Data for Example 4-9

Element number	Nodes connected	
	i	j
1	1	2
2	2	3
3	3	4
4	4	5
5	5	6

Obviously, we also have

$$\mathbf{K}^{(2)} = \mathbf{K}^{(3)} = \mathbf{K}^{(4)} = \mathbf{K}^{(5)} = \begin{bmatrix} 5 & -5 \\ -5 & 5 \end{bmatrix}$$

(4-146a)

For $\mathbf{f}^{(1)}$, we have

$$\mathbf{f}^{(1)} = \begin{bmatrix} 0 \\ 1 \end{bmatrix} \frac{dT(x_2)}{dx} + \begin{bmatrix} -1 \\ 0 \end{bmatrix} \frac{dT(x_1)}{dx}$$
$$+ \frac{1000}{(12)(0.2)} \begin{bmatrix} (0.2)^4 - 4(0.2)(0.0)^3 + 3(0.0)^4 \\ 3(0.2)^4 - 4(0.0)(0.2)^3 + (0.0)^4 \end{bmatrix}$$

or

$$\mathbf{f}^{(1)} = \begin{bmatrix} 0 \\ 1 \end{bmatrix} \frac{dT(x_2)}{dx} + \begin{bmatrix} -1 \\ 0 \end{bmatrix} \frac{dT(x_1)}{dx} + \begin{bmatrix} 0.667 \\ 2.00 \end{bmatrix}$$

Similarly, we find

$$\mathbf{f}^{(2)} = \begin{bmatrix} 0 \\ 1 \end{bmatrix} \frac{dT(x_3)}{dx} + \begin{bmatrix} -1 \\ 0 \end{bmatrix} \frac{dT(x_2)}{dx} + \begin{bmatrix} 7.3 \\ 11.3 \end{bmatrix}$$

$$\mathbf{f}^{(3)} = \begin{bmatrix} 0 \\ 1 \end{bmatrix} \frac{dT(x_4)}{dx} + \begin{bmatrix} -1 \\ 0 \end{bmatrix} \frac{dT(x_3)}{dx} + \begin{bmatrix} 22.0 \\ 28.6 \end{bmatrix}$$

(4-146a)

$$\mathbf{f}^{(4)} = \begin{bmatrix} 0 \\ 1 \end{bmatrix} \frac{dT(x_5)}{dx} + \begin{bmatrix} -1 \\ 0 \end{bmatrix} \frac{dT(x_4)}{dx} + \begin{bmatrix} 44.7 \\ 54.0 \end{bmatrix}$$

$$\mathbf{f}^{(5)} = \begin{bmatrix} 0 \\ 1 \end{bmatrix} \frac{dT(x_6)}{dx} + \begin{bmatrix} -1 \\ 0 \end{bmatrix} \frac{dT(x_5)}{dx} + \begin{bmatrix} 75.4 \\ 87.3 \end{bmatrix}$$

The assemblage of the element stiffness matrices to form $\mathbf{K}^a$ follows the same line of reasoning presented in Chapter 3 where the truss was studied. However, the situation here is even simpler because the submatrices in the $\mathbf{K}^e$'s are of size 1×1, or simply scalars. The reader is reminded that the primary factor in deciding where the element stiffness matrix $\mathbf{K}^e$ for element e is added into the assemblage stiffness matrix $\mathbf{K}^a$ is the two global node numbers (I and J) associated with the element. This assemblage step was summarized in Chapter 3 for elements containing only two nodes (such as the present problem) in Eq. (3-31). The details of the assemblage of the stiffness matrices are omitted below, but the reader should show the missing steps if there is still some doubt as to how this important step is accomplished.

The assemblage of the nodal force vectors to form $\mathbf{f}^a$ is done in a completely analogous manner, except that it is even simpler because only a column matrix is involved. If the element nodal force vector is imagined to be partitioned as

$$\mathbf{f}^e = \begin{bmatrix} f_I^e \\ \hline f_J^e \end{bmatrix}$$

(4-147a)

then the assemblage of these to form $\mathbf{f}^a$ may be summarized by the following equation:

$$\mathbf{f}_n^a = \sum_{e=1}^{M} \mathbf{f}_n^e \qquad \text{(4-147b)}$$

where M is the maximum number of elements used in the model and $\mathbf{f}_n^e$ is taken to be zero if element e does not contain node n. With this algorithm, the reader should be able to show that the assemblage nodal force vector in the following matrix equation results:

$$
\begin{bmatrix}
5 & -5 & 0 & 0 & 0 & 0 \\
-5 & 10 & -5 & 0 & 0 & 0 \\
0 & -5 & 10 & -5 & 0 & 0 \\
0 & 0 & -5 & 10 & -5 & 0 \\
0 & 0 & 0 & -5 & 10 & -5 \\
0 & 0 & 0 & 0 & -5 & 5
\end{bmatrix}
\begin{bmatrix}
T_1 \\ T_2 \\ T_3 \\ T_4 \\ T_5 \\ T_6
\end{bmatrix}
=
\begin{bmatrix}
0.667 - \dfrac{dT(x_1)}{dx} \\
9.33 \\
33.3 \\
73.3 \\
129.4 \\
87.3 + \dfrac{dT(x_6)}{dx}
\end{bmatrix}
\qquad \text{(4-148)}
$$

Note how the unknown derivatives (dT/dx) for the interior nodes have canceled. Physically this may be interpreted as follows: although there is a temperature gradient and hence a heat flux at each node and hence between two neighboring elements, these internal heat fluxes cancel. The only heat fluxes that *survive* are those on the global boundary, and then only if there is actually an imposed heat flux there. If the temperature gradients (or heat fluxes) at either or both ends were specified, the value would simply be entered as indicated above. That case would correspond to a prescribed heat flux into or from the ends. In any event, in the problem at hand we have prescribed temperatures at both ends, and by using Method 1 (Method 2 may also be used), Eq. (4-148) is modified to impose $T_1 = 0$ and $T_6 = 0$, which results in

$$
\begin{bmatrix}
1 & 0 & 0 & 0 & 0 & 0 \\
0 & 10 & -5 & 0 & 0 & 0 \\
0 & -5 & 10 & -5 & 0 & 0 \\
0 & 0 & -5 & 10 & -5 & 0 \\
0 & 0 & 0 & -5 & 10 & 0 \\
0 & 0 & 0 & 0 & 0 & 1
\end{bmatrix}
\begin{bmatrix}
T_1 \\ T_2 \\ T_3 \\ T_4 \\ T_5 \\ T_6
\end{bmatrix}
=
\begin{bmatrix}
0.0 \\ 9.3 \\ 33.3 \\ 73.3 \\ 129.4 \\ 0.0
\end{bmatrix}
\qquad \text{(4-149)}
$$

The disappearance of the (unknown) temperature gradients at nodes 1 and 6 should be noted and thoroughly understood: in thermal problems, in general, the boundary nodes must have either a prescribed temperature or a prescribed gradient (heat flux), convective, and/or radiative boundary conditions. It should be noted further that the case of insulation is a special case of a specified gradient since $dT/dx = 0$. This zero-gradient case is the so-called natural boundary condition. In any event, solving Eq. (4-149) for the nodal temperatures yields

$$T_1 = 0$$

$$T_2 = 16.5$$

$$T_3 = 31.2$$

$$T_4 = 39.2$$

$$T_5 = 32.5$$

$$T_6 = 0$$

The reader will note that these values lie right on the curve for the exact solution in Fig. 4-7! The significant point here is that two linear shape functions (or trial functions on a local or piecewise basis) have given results that are extremely close to the exact solution, whereas the use of cubic-order, global trial functions in the Rayleigh-Ritz method resulted in an error as large as 21% (at $x = 0.2$). Except for the Galerkin method, the other global approximate solution methods resulted in even larger errors. ∎

So far it has been seen that the variational (Rayleigh-Ritz) and Galerkin weighted-residual methods perform the best out of all the approximate solution techniques considered here. The variational FEM solution presented in this section resulted in a solution very close to the exact one (with only five elements). The next section will illustrate how the Galerkin method is adapted to FEM.

4-9   THE GALERKIN FINITE ELEMENT METHOD

In Sec. 4-6, four different weighted-residual methods were introduced. In addition, four approximate solutions to the same problem were obtained by each of these methods. Global trial functions were used; these functions applied to the entire domain. In this section we wish to repeat the solution only for the Galerkin method, but this time in the context of the finite element method. The implication is that piecewise continuous trial functions or shape functions will be employed for each element. In fact, as in the Rayleigh-Ritz FEM (see Sec. 4-8), these shape functions will actually be interpolation polynomials.

As in Sec. 4-8, the problem domain is divided or discretized into a number of elements. Figure 4-8 is again applicable to this formulation, and the interval $a \leq x \leq b$ (i.e., the problem domain) is divided into M such elements. A typical element e has two nodes, j and k, with coordinates x_j and x_k, respectively. The field variable y at nodes j and k has values y_j and y_k, respectively. As before, since each element has two nodes and since a unique straight line may be drawn between two points, the interpolation polynomial to be used to describe the field variable, e.g., $y(x)$, must be of the form

$$y^e = mx + b \tag{4-150a}$$

which is identical to Eq. (4-121). Therefore, the results of the first part of Sec. 4-8 are directly applicable here. In other words, we wish to express $y^e(x)$ in the form

$$y^e(x) = y_j N_j(x) + y_k N_k(x) \tag{4-150b}$$

From Eqs. (4-125) to (4-127), we see that at $x = x_j$ we have $y^e(x_j) = y_j$, and at $x = x_k$ we have $y^e(x_k) = y_k$. As in the Rayleigh-Ritz FEM formulation in Sec. 4-8, two piecewise continuous linear trial functions are to be used. The functions $N_j(x)$ and $N_k(x)$ are referred to as shape functions.

Example 4-10

Resolve the problem posed in Example 4-1 (approximately) by the Galerkin finite element method by using six equally spaced nodes.

Solution

Recall that the Galerkin method requires that the trial functions themselves be used as the weighting functions. The integral of the weighted residual R in this case is written as

$$\int_\Omega N_i(x) R(x, a_1, a_2, \ldots, a_n)\, dx = 0 \tag{4-151}$$

But instead of using trial functions $N_i(x)$ that are applied globally to the entire x domain and that satisfy the boundary conditions automatically (and exactly), let us try to represent the solution T^e over a small interval (really a finite element), say from x_j to x_k, as

$$T^e = N_j(x) T_j + N_k(x) T_k \tag{4-152}$$

where T_j and T_k are analogous to the unknown parameters, a_1 and a_2, in the general, weighted-residual method. Moreover, if $N_j(x)$ and $N_k(x)$, in addition to being linear, are such that when $x = x_j$ we have $N_j = 1$ and $N_k = 0$, and when $x = x_k$ we have $N_j = 0$ and $N_k = 1$, then in fact we have appropriate shape functions. The reader should verify that for $N_j(x)$ and $N_k(x)$ given in Eq. (4-127) this is in fact the case. In summary, in one-dimensional C^0-continuous problems with two nodes per element, the shape functions must be linear because there are two nodal points per element and they must satisfy the following conditions:

$$N_j(x_j) = N_k(x_k) = 1 \tag{4-153a}$$

and

$$N_j(x_k) = N_k(x_j) = 0 \tag{4-153b}$$

In the Galerkin method, the two shape functions N_j and N_k are the weighting functions. From Eq. (4-104), we may write the following weighted-residual equations:

$$\sum_{e=1}^{M} \int_{x_j}^{x_k} N_j(x)[R^e(x;T_j,T_k)] \, dx = 0 \tag{4-154a}$$

and

$$\sum_{e=1}^{M} \int_{x_j}^{x_k} N_k(x)[R^e(x;T_j,T_k)] \, dx = 0 \tag{4-154b}$$

where e denotes the element and M is the number of elements (or intervals into which the domain is divided). A typical *finite element* goes from x_j (where the temperature is T_j) to x_k (where the temperature is T_k). If matrix notation is used and $\mathbf{N}^T$ is defined as

$$\mathbf{N}^T = \begin{bmatrix} N_j(x) \\ N_k(x) \end{bmatrix}$$

then Eqs. (4-154) may be written in one equation as

$$\sum_{e=1}^{M} \int_{x_j}^{x_k} \mathbf{N}^T[R^e(x;T_j,T_k)] \, dx = 0 \tag{4-155}$$

But the residual R^e for a typical element e in the problem at hand is, by definition,

$$R^e = \frac{d^2T^e}{dx^2} + 1000x^2 \tag{4-156}$$

where the approximate solution T^e, applicable only to element e, has been substituted for the exact temperature T in the governing equation and the result set equal to the residual R^e for the element. Therefore, Eq. (4-155) becomes

$$\sum_{e=1}^{M} \int_{x_j}^{x_k} \mathbf{N}^T \left(\frac{d^2T^e}{dx^2} + 1000x^2 \right) dx = 0 \tag{4-157}$$

Before breaking this integral into two separate integrals, let us agree to drop the summation sign. The reason for this is simple: it represents the assemblage step that it is hoped is becoming routine. In other words, it is preferred to work with a typical element (element e) and at some later stage we will put the pieces back together. Breaking Eq. (4-157) into two separate integrals gives

$$\int_{x_j}^{x_k} \mathbf{N}^T \frac{d^2T^e}{dx^2} \, dx + \int_{x_j}^{x_k} \mathbf{N}^T (1000x^2) \, dx = 0 \tag{4-158}$$

If the first integral is integrated by parts we get

$$\mathbf{N}^T \frac{dT^e}{dx} \bigg|_{x_j}^{x_k} - \int_{x_j}^{x_k} \frac{d\mathbf{N}^T}{dx} \frac{dT^e}{dx} \, dx + \int_{x_j}^{x_k} \mathbf{N}^T (1000x^2) \, dx = 0 \tag{4-159}$$

But we have already agreed to write T^e in terms of two (linear) shape functions, N_j and N_k, and the two nodal temperatures, T_j and T_k, as

$$T^e = \mathbf{N}\mathbf{a}^e \tag{4-160}$$

where

$$\mathbf{N} = [N_j \quad N_k] \tag{4-161}$$

and

$$\mathbf{a}^e = \begin{bmatrix} T_j \\ T_k \end{bmatrix} \tag{4-162}$$

Using Eq. (4-160) *in the integral terms only* in Eq. (4-159) gives

$$\mathbf{N}^T \frac{dT^e}{dx}\bigg|_{x_j}^{x_k} - \int_{x_j}^{x_k} \frac{d\mathbf{N}^T}{dx} \frac{d\mathbf{N}}{dx} \mathbf{a}^e \, dx + \int_{x_j}^{x_k} \mathbf{N}^T (1000x^2) \, dx = 0 \tag{4-163}$$

The reason for not using Eq. (4-160) in the integrated term is that by not doing so, we have allowed a mechanism for imposing gradient boundary conditions (see Problem 4-56). But $\mathbf{a}^e$ is not a function of x and may be pulled out of the integral from the right to give a rather simple equation of the form

$$\mathbf{K}^e \mathbf{a}^e = \mathbf{f}^e \tag{4-164}$$

where

$$\mathbf{K}^e = \int_{x_j}^{x_k} \frac{d\mathbf{N}^T}{dx} \frac{d\mathbf{N}}{dx} \, dx \tag{4-165}$$

and

$$\mathbf{f}^e = \mathbf{N}^T \frac{dT}{dx}\bigg|_{x_j}^{x_k} + \int_{x_j}^{x_k} \mathbf{N}^T (1000x^2) \, dx \tag{4-166}$$

For all practical purposes, Eqs. (4-165) and (4-166) from the Galerkin method are identical to Eqs. (4-141) and (4-142), respectively, from the variational method! As mentioned before, this is not surprising because when a classical variational principle exists, it must give the same FEM formulation as the Galerkin method. Again the element stiffness matrix is seen to be symmetric.

Equation (4-165) for the element stiffness matrix and Eq. (4-166) for the element nodal force vector are the so-called element characteristics. Note that it is not necessary to distinguish between the local and global element stiffness matrices since a local coordinate system was not used in the development of these equations. Moreover, the function $T(x)$ is a scalar function (such as temperature) and hence no transformation from a local to a global coordinate system would be necessary before the assemblage step. It should be obvious that the remaining part of the

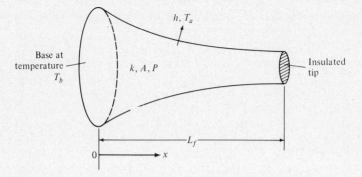

Figure 4-11   Circular pin fin undergoing convection to an ambient fluid.

solution by the Galerkin finite element method is identical to that in Example 4-9. ∎

4-10   APPLICATION: ONE-DIMENSIONAL HEAT TRANSFER IN A PIN FIN

Consider a circular *pin fin* of length L_f and varying cross-sectional area A and perimeter P as shown in Fig. 4-11. Fins such as these are used to increase the heat transfer rate, particularly to gases. Gases generally have low convective heat transfer coefficients. By increasing the surface area, such as in the pin fin, we can increase the heat transfer rate significantly. The heat transfer rate is also referred to as the *heat removal rate* in this text. Depending on the circumstances, fins are also commonly used in liquids.

The temperature distribution in the fin is needed for the case when the base is held at a temperature T_b, the tip is insulated, and the fin itself convects to a fluid at a temperature T_a with a heat transfer coefficient h. The thermal conductivity is denoted as k, and the temperature at any point along the fin is denoted as T. The heat removal rate and fin efficiency (defined below) are to be determined also.

The governing equation may be shown to be (see Problem 4-66)

$$\frac{d}{dx}\left(kA\frac{dT}{dx}\right) - hP(T - T_a) = 0 \qquad \text{for } 0 \leq x \leq L_f \qquad \textbf{(4-167)}$$

subject to the two boundary conditions

$$T(0) = T_b \qquad \textbf{(4-168a)}$$

$$\frac{dT}{dx}(L_f) = 0 \qquad \textbf{(4-168b)}$$

The intent is to obtain an approximate solution using the finite element method. In particular the finite element characteristics will be derived using the Galerkin method.

The Element Characteristics

The derivation begins by forming the integral of the weighted residual and setting the result to zero, or

$$\int_{x_j}^{x_k} \mathbf{N}^T \left[\frac{d}{dx} \left(kA \frac{dT}{dx} \right) - hP(T - T_a) \right] dx = 0 \qquad \textbf{(4-169)}$$

where a typical element e is assumed to connect node j (at $x = x_j$) to node k (at $x = x_k$) as shown in Fig. 4-12. The superscript (^{e}) on the field variable, T in this case, is no longer shown. Note that the weighting function is taken to be the transpose of the shape function matrix, since the Galerkin method is used. At nodes j and k, the temperatures are T_j and T_k, respectively. Note also that each node represents a planar surface and not just a point.

Let us integrate by parts the term with the second-order derivative as follows:

$$\int_{x_j}^{x_k} \mathbf{N}^T \frac{d}{dx} \left(kA \frac{dT}{dx} \right) dx = \mathbf{N}^T kA \frac{dT}{dx} \Big|_{x_j}^{x_k} - \int_{x_j}^{x_k} \frac{d\mathbf{N}^T}{dx} kA \frac{dT}{dx} \, dx \qquad \textbf{(4-170)}$$

As illustrated in Example 4-9, the so-called integrated term cancels for all interior nodes during the assemblage step. Moreover, at the base of the fin, the temperature is prescribed. As shown in Example 4-9, the unknown temperature gradient at the corresponding node is eventually eliminated when the prescribed temperature is imposed on this node. Also, at $x = L_f$, we have $dT(L_f)/dx = 0$. Therefore, this integrated term need no longer be considered here. (See Sec. 8-3 for the case of a noninsulated tip.) Therefore, Eq. (4-169) becomes

$$-\int_{x_j}^{x_k} \frac{d\mathbf{N}^T}{dx} kA \frac{dT}{dx} \, dx - \int_{x_j}^{x_k} \mathbf{N}^T hPT \, dx + \int_{x_j}^{x_k} \mathbf{N}^T hPT_a \, dx = 0 \qquad \textbf{(4-171)}$$

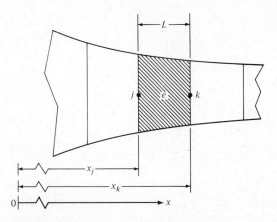

Figure 4-12 One-dimensional element with nodes j and k.

But using $T = \mathbf{N}\mathbf{a}^e$ and noting that only $\mathbf{N}$ is a function of x, we get

$$\mathbf{K}^e\mathbf{a}^e = \mathbf{f}^e \tag{4-172}$$

where

$$\mathbf{K}^e = \mathbf{K}_x^e + \mathbf{K}_{cv}^e \tag{4-173}$$

and

$$\mathbf{K}_x^e = \int_{x_j}^{x_k} \frac{d\mathbf{N}^T}{dx} kA \frac{d\mathbf{N}}{dx} \, dx \tag{4-174}$$

$$\mathbf{K}_{cv}^e = \int_{x_j}^{x_k} \mathbf{N}^T hP\mathbf{N} \, dx \tag{4-175}$$

and

$$\mathbf{f}^e = \int_{x_j}^{x_k} \mathbf{N}^T hPT_a \, dx \tag{4-176}$$

The vector of nodal unknowns $\mathbf{a}^e$ is given by

$$\mathbf{a}^e = [T_j \quad T_k]^T \tag{4-177}$$

and the shape function matrix $\mathbf{N}$ by

$$\mathbf{N} = [N_j(x) \quad N_k(x)] \tag{4-178}$$

where N_j and N_k are, in turn, given by Eqs. (4-127).

In the most general case, $k, h, A, P,$ and T_a may all be functions of x. However, let us perform the integrations in Eqs. (4-174) to (4-176) approximately by taking the following approach. Let us treat these parameters as constants in any given element by using the values at the midpoint of the element where $x = \bar{x}$ and by denoting them as $\bar{k}, \bar{h}, \bar{A}, \bar{P},$ and $\bar{T}_a$. It can be shown that (see Problem 4-69)

$$\mathbf{K}_x^e = \frac{\overline{kA}}{L} \begin{bmatrix} 1 & -1 \\ -1 & 1 \end{bmatrix} \tag{4-179}$$

$$\mathbf{K}_{cv}^e = \frac{\overline{hPL}}{6} \begin{bmatrix} 2 & 1 \\ 1 & 2 \end{bmatrix} \tag{4-180}$$

and

$$\mathbf{f}^e = \frac{\overline{hPL}\bar{T}_a}{2} \begin{bmatrix} 1 \\ 1 \end{bmatrix} \tag{4-181}$$

where L is the element length, or $L = x_k - x_j$.

The element stiffness matrices and nodal force vectors may be determined for every element with the help of Eqs. (4-179) to (4-181). The assemblage of these 2×2 matrices and 2×1 vectors is performed in the usual manner to give a system of N algebraic equations in the N unknown nodal temperatures. This assumes the fin is discretized into M elements with N nodes, where M is equal to $N - 1$. The result is $\mathbf{K}^a\mathbf{a} = \mathbf{f}^a$. The geometric (or prescribed temperature) boundary conditions are applied at this point to yield $\mathbf{Ka} = \mathbf{f}$. This matrix equation may be solved for the nodal temperatures in the vector $\mathbf{a}$. Therefore, it is assumed that the nodal temperatures are known in what follows.

Heat Removal Rate

The heat removal rate can be determined by two different methods, once the nodal temperatures are known. The first method is based on differentiation, and the second method is based on integration. Not surprisingly, the second method is significantly more accurate, as illustrated numerically in Example 4-11 below.

In the first method, the heat removal rate Q_R from the fin is evaluated by computing the heat flux from conduction at the base of the fin (where $x = 0$), and by multiplying this value by the cross-sectional area A. This may be summarized as

$$Q_R = -kA\frac{dT}{dx}\bigg|_{x=0} = -kA\frac{d\mathbf{N}}{dx}\mathbf{a}^{(1)} \tag{4-182}$$

where it is assumed that element 1 is at the base of the fin as shown in Fig. 4-13. It follows (see Problem 4-70) that

$$Q_R = \frac{\bar{k}A}{L}(T_1 - T_2) \tag{4-183}$$

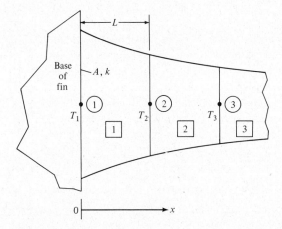

Figure 4-13   Portion of the discretized fin showing some elements near the base.

where it is further assumed that element 1 connects nodes 1 and 2, and where L is the element length. More accurate results are usually obtained if the cross-sectional area at the base is used instead of the area at the element centroid. If Q_R is negative, heat is conducted toward the base of the fin.

In the second method, the heat removal rate is determined by integrating the heat loss from the exposed surfaces of the fin that undergo convection. Let us assume the tip of the fin is insulated, so that only the fin periphery needs to be considered. Therefore, we have

$$Q_R = \int_0^{L_f} hP(T - T_a)\, dx \tag{4-184}$$

However, it is preferred to write this as a sum of M separate integrals over each of the M elements, or

$$Q_R = \sum_{e=1}^{M} \int_{x_j}^{x_k} hP(T - T_a)\, dx \tag{4-185}$$

It can be shown (see Problem 4-70) that if $T = \mathbf{N}\mathbf{a}^e$ is used, and if h, T_a, and P are evaluated at $x = \bar{x}$ (i.e., at the centroid of the element) and denoted as $\bar{h}$, $\bar{T}_a$, and $\bar{P}$, then Eq. (4-185) becomes

$$Q_R = \sum_{e=1}^{M} \bar{h}\bar{P}L\left(\frac{T_j + T_k}{2} - \bar{T}_a\right) \tag{4-186}$$

where T_j and T_k are the temperatures at nodes j and k for element e. These nodal temperatures are known from the solution of $\mathbf{Ka} = \mathbf{f}$. It is emphasized that the heat removal rate from Eq. (4-186) is more accurate than that from Eq. (4-183).

Fin Efficiency

The fin efficiency η_f may be defined in several ways, but only one definition is given here. This definition is based on the hypothetical condition that the entire fin is at the base temperature, because this would result in the maximum possible heat removal rate. This corresponds to an infinite thermal conductivity and, therefore, is not physically realizable. Nonetheless, this definition of fin efficiency does give an indication of the relative effectiveness of the fin in increasing the heat removal rate. Denoting this maximum rate as Q_{max}, we may write

$$\eta_f = \frac{Q_R}{Q_{max}} \tag{4-187}$$

where

$$Q_{max} = \int_0^{L_f} hP(T_b - T_a)\, dx \tag{4-188}$$

If average values of h, P, and T_a are used for each element, Eq. (4-188) becomes

$$Q_{max} = \sum_{e=1}^{M} \overline{hPL}(T_b - \overline{T}_a)$$ **(4-189)**

Again it has been assumed that the tip of the fin is insulated.

Example 4-11

Determine the temperatures, heat removal rate, and efficiency of the circular pin fin shown in Fig. 4-14(a). The fin is made of pure copper and has a thermal conductivity k of 400 W/m-°C. The base is held at $T_b = 85$°C and the ambient fluid temperature T_a is maintained at 25°C. The fin length L_f is 2 cm, and the diameter D is 0.4 cm. The convective heat transfer coefficient h is 150 W/m²-°C and the tip at $x = L_f$ is insulated.

Solution

The first step in the finite element solution is discretization of the fin. Let us arbitrarily assume only two elements (and three nodes) as shown in Fig. 4-14(b). It is convenient to summarize the node and element data in tabular form as shown in Table 4-3. The calculations are summarized below in a form that may be readily implemented in a computer program (see Problem 4-78).

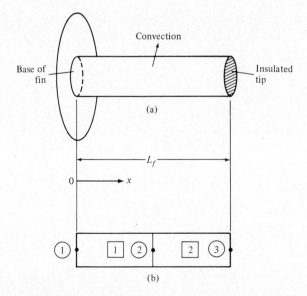

(a)

(b)

Figure 4-14 Circular pin fin (a) analyzed in Example 4-11 and (b) discretized into two elements and three nodes.

Table 4-3   Node and Element Data for Example 4-11

Node number	x, m
1	0.0
2	0.01
3	0.02

Element number	Nodes connected j	k
1	1	2
2	2	3

Element 1:   Node j is 1.   Node k is 2.

$$x_j = 0.0 \text{ m} \qquad \text{and} \qquad x_k = 0.01 \text{ m}$$

$$L = x_k - x_j = 0.01 - 0.00 = 0.01 \text{ m}$$

$$D = 0.004 \text{ m}$$

$$A = \frac{\pi D^2}{4} = \frac{\pi(0.004)^2}{4} = 1.2566 \times 10^{-5} \text{ m}^2$$

$$P = \pi D = \pi(0.004) = 1.2566 \times 10^{-2} \text{ m}$$

$$\mathbf{K}_x^{(1)} = \frac{(400)(1.2566 \times 10^{-5})}{0.01} \begin{bmatrix} 1 & -1 \\ -1 & 1 \end{bmatrix}$$

$$= \begin{bmatrix} 0.50265 & -0.50265 \\ -0.50265 & 0.50265 \end{bmatrix} \text{W/}^\circ\text{C}$$

$$\mathbf{K}_{cv}^{(1)} = \frac{(150)(1.2566 \times 10^{-2})(0.01)}{6} \begin{bmatrix} 2 & 1 \\ 1 & 2 \end{bmatrix}$$

$$= \begin{bmatrix} 0.00628 & 0.00314 \\ 0.00314 & 0.00628 \end{bmatrix} \text{W/}^\circ\text{C}$$

$$\mathbf{K}^{(1)} = \mathbf{K}_x^{(1)} + \mathbf{K}_{cv}^{(1)} = \begin{bmatrix} 0.50893 & -0.49951 \\ -0.49951 & 0.50893 \end{bmatrix} \text{W/}^\circ\text{C}$$

$$\mathbf{K}^a = \begin{bmatrix} 0.50893 & -0.49951 & 0 \\ -0.49951 & 0.50893 & 0 \\ 0 & 0 & 0 \end{bmatrix} \text{W/}^\circ\text{C}$$

$$\mathbf{f}^{(1)} = \frac{(150)(1.2566 \times 10^{-2})(0.01)(25)}{2} \begin{bmatrix} 1 \\ 1 \end{bmatrix} = \begin{bmatrix} 0.23561 \\ 0.23561 \end{bmatrix} \text{W}$$

$$\mathbf{f}^a = \begin{bmatrix} 0.23561 \\ 0.23561 \\ 0 \end{bmatrix} \text{W}$$

Element 2:   Node j is 2.   Node k is 3.

$$x_j = 0.01 \text{ m} \qquad \text{and} \qquad x_k = 0.02 \text{ m}$$

$$L = x_k - x_j = 0.02 - 0.01 = 0.01 \text{ m}$$

$$D = 0.004 \text{ m}$$

$$A = 1.2566 \times 10^{-5} \text{ m}^2$$

$$P = 1.2566 \times 10^{-2} \text{ m}$$

$$\mathbf{K}_x^{(2)} = \mathbf{K}_x^{(1)}$$

$$= \begin{bmatrix} 0.50265 & -0.50265 \\ -0.50265 & 0.50265 \end{bmatrix} \text{W/°C}$$

$$\mathbf{K}_{cv}^{(2)} = \mathbf{K}_{cv}^{(1)}$$

$$= \begin{bmatrix} 0.00628 & 0.00314 \\ 0.00314 & 0.00628 \end{bmatrix} \text{W/°C}$$

$$\mathbf{K}^{(2)} = \mathbf{K}_x^{(2)} + \mathbf{K}_{cv}^{(2)} = \mathbf{K}^{(1)}$$

$$= \begin{bmatrix} 0.50893 & -0.49951 \\ -0.49951 & 0.50893 \end{bmatrix} \text{W/°C}$$

$$\mathbf{K}^a = \begin{bmatrix} 0.50893 & -0.49951 & 0 \\ -0.49951 & 1.01786 & -0.49951 \\ 0 & -0.49951 & 0.50893 \end{bmatrix} \text{W/°C}$$

$$\mathbf{f}^{(2)} = f^{(1)} = \begin{bmatrix} 0.23561 \\ 0.23561 \end{bmatrix} \text{W}$$

$$\mathbf{f}^a = \begin{bmatrix} 0.23561 \\ 0.47122 \\ 0.23561 \end{bmatrix} \text{W}$$

Before application of the prescribed temperatures we have $\mathbf{K}^a\mathbf{a} = \mathbf{f}^a$, or

$$\begin{bmatrix} 0.50893 & -0.49951 & 0 \\ -0.49951 & 1.01786 & -0.49951 \\ 0 & -0.49951 & 0.50893 \end{bmatrix} \begin{bmatrix} T_1 \\ T_2 \\ T_3 \end{bmatrix} = \begin{bmatrix} 0.23561 \\ 0.47122 \\ 0.23561 \end{bmatrix}$$

Note that the assemblage stiffness or *conductance* matrix $\mathbf{K}^a$ is symmetric and banded with a half-bandwidth of 2 [this also follows from Eq. (3-33)]. The prescribed temperature boundary condition at the base of the fin will be imposed by using Method 1 from Sec. 3-2. The first step is to modify the first row (since T_1 is to be imposed) as shown below:

$$\begin{bmatrix} 1 & 0 & 0 \\ -0.49951 & 1.01786 & -0.49951 \\ 0 & -0.49951 & 0.50893 \end{bmatrix} \begin{bmatrix} T_1 \\ T_2 \\ T_3 \end{bmatrix} = \begin{bmatrix} 85 \\ 0.47122 \\ 0.23561 \end{bmatrix}$$

In order to preserve symmetry, the K_{21} entry must be multiplied by the prescribed temperature and transposed to the right-hand side (in the same row). Therefore, f_2 must be replaced by $0.47122 - (-0.49951)(85.)$ or 42.930. This is the only term that needs to be modified because **K** is now symmetric. The system of equations to be solved is given by

$$\begin{bmatrix} 1 & 0 & 0 \\ 0 & 1.01786 & -0.49951 \\ 0 & -0.49951 & 0.50893 \end{bmatrix} \begin{bmatrix} T_1 \\ T_2 \\ T_3 \end{bmatrix} = \begin{bmatrix} 85 \\ 42.930 \\ 0.23561 \end{bmatrix}$$

Solving this system of linear, algebraic equations by the matrix inversion method yields the following nodal temperatures:

$$T_1 = 85.0°C \qquad T_2 = 81.8°C \qquad T_3 = 80.8°C$$

The heat removal rates may now be computed as summarized below:

$$Q_R = \sum_{e=1}^{2} \overline{h} PL \left(\frac{T_j + T_k}{2} - T_a \right)$$

$$= 150\pi(0.004)(0.01) \left(\frac{85 + 81.8}{2} - 25 \right)$$

$$+ \; 150\pi(0.004)(0.01) \left(\frac{81.8 + 80.8}{2} - 25 \right)$$

$$= 1.10 + 1.06 = 2.16 \text{ W}$$

If the alternate (less accurate) method is used, we get

$$Q_R = \frac{\overline{k}A}{L}(T_1 - T_2) = \frac{400\pi(0.004)^2/4}{0.01}(85 - 81.8) = 1.60 \text{ W}$$

This is quite different than the result from the method based on integration. It is shown below that the exact value for the heat removal rate is 2.155 W. Clearly, the method based on integration is significantly more accurate.

Finally, the fin efficiency is readily computed to be

$$\eta_f = \frac{Q_R}{Q_{max}} = \frac{2.16}{150\pi(0.004)(0.02)(85 - 25)}$$

$$= \frac{2.16}{2.26} = 0.956 \quad \text{or} \quad 95.6\%$$

The results from Example 4-11 are compared in Table 4-4 to the exact solution given by [13]

$$T(x) = T_a + (T_b - T_a) \frac{\cosh \lambda(L_f - x)}{\cosh \lambda L_f}$$

and

$$Q_R = \lambda kA(T_b - T_a) \tanh \lambda L_f$$

where

$$\lambda = \sqrt{\frac{hP}{kA}}$$

Note that the results for 4 and 8 elements are also given in this table. These results were obtained by the computer program described in Problem 4-78. The nodal temperatures, heat removal rates, and efficiencies in Table 4-4 converge to the exact solution as the number of elements is increased. Five significant digits are shown for the purpose of comparison only.

A detailed discussion of error predictions and convergence is beyond the scope of this text. Introductory material on these topics may be found in the book by Becker, Carey, and Oden [14].

4-11 REMARKS

This chapter began with an introduction to some of the more important integral methods that lead to the present-day finite element method in nonstructural applications. Only one-dimensional problems were considered, but the basic groundwork

Table 4-4 Summary of Results for Example 4-11

	Temperatures, °C			
x, cm	Exact	2 Elements	4 Elements	8 Elements
0.0	85.000	85.000	85.000	85.000
0.25	83.998			83.998
0.50	83.134		83.133	83.134
0.75	82.407			82.407
1.00	81.814	81.807	81.812	81.814
1.25	81.354			81.354
1.50	81.027		81.024	81.026
1.75	80.830			80.830
2.00	80.765	80.756	80.762	80.765

	Heat removal rate, W		
Case	Eq. (4-183)	Eq. (4-186)	Fin efficiency, %*
2 Elements	1.6050	2.1618	95.6
4 Elements	1.8771	2.1569	95.4
8 Elements	2.0149	2.1557	95.3
Exact	2.1552	2.1552	95.3

*Based on Eq. (4-187) with Q_R from Eq. (4-186).

has been laid for extension to two and three dimensions in later chapters. Among the integral methods introduced here were the Ritz method, the variational or Rayleigh-Ritz method, and the weighted-residual method. All these approximate solution methods may be used to solve differential equations. However, we generally work with an integral form of the governing equation. Each of these methods was introduced on a global basis, which meant that the trial functions were assumed to apply globally to the entire problem domain.

The Ritz method is the simplest and least powerful of the methods. A specific example problem was considered. Following this, the Rayleigh-Ritz method was introduced and applied to the same problem. This method requires a rudimentary knowledge of variational calculus, which is the main disadvantage of this approach. Some of the more important concepts in the calculus of variations were reviewed. Then four of the most popular weighted-residual methods were introduced, which included the point collocation method, the subdomain collocation method, the least-squares method, and the Galerkin method. These weighted-residual methods do not require any advanced mathematics beyond ordinary calculus (i.e., variational calculus is not needed in these methods). Moreover, it was stated that the weighted-residual methods could be used even when a classical variational principle does not exist. Each of these methods was illustrated by solving the same example problem, and the results of each of the approximate solution methods were compared with the exact solution. None of them did a particularly good job of matching the exact solution, but the Galerkin and Rayleigh-Ritz methods were among the best.

After some additional mathematics was reviewed, the Rayleigh-Ritz and Galerkin methods were cast into a form that is directly useful in the finite element method. The same example problem was solved, approximately, by using trial functions that no longer applied globally to the entire problem domain but rather applied locally over each element. These piecewise continuous trial functions gave rise to two other functions that are referred to as shape functions. These particular shape functions allow the field variable to be continuous within each element and at the ends of each element. However, the derivatives at the ends of each element are not necessarily continuous. Therefore, these particular shape functions are said to be C^0-continuous, or to possess C^0-continuity. In Chapter 7, shape functions with continuous first derivatives will be introduced; shape functions with continuous first derivatives are said to be C^1-continuous, or to possess C^1-continuity. It was seen that accurate results could be obtained with the use of the piecewise continuous trial functions, which are really interpolating polynomials. The reader is referred to the book by Myers [12] for additional material relating to the variational finite element method as it applies to one-dimensional problems in heat transfer.

When the Rayleigh-Ritz method is applied to a second-order differential equation with no first-order derivative present, it results in an integral formulation that has only a first-order derivative. Therefore, the integral formulation is said to be a *weak formulation*. In a similar fashion, the Galerkin method also results in a similar result if the term containing the second-order derivative is integrated by parts. This is a fundamental step in the finite element method and one that we will see throughout the book. The implication is that the finite element method can be used to obtain solutions to problems that possess real material discontinuities, such as composite materials with two or more thermal conductivities.

The emphasis in this chapter has been the solution of one-dimensional, second-order differential equations. In the next chapter, as in Chapter 3, we turn to the structural analysis area where the general finite element formulation to problems in stress analysis is presented. Two additional methods of finite element formulations, which are more readily applied to these problems, are introduced.

REFERENCES

1. Hinton, E., and D. R. J. Owen, *An Introduction to Finite Element Computations*, Pineridge Press, Swansea, UK, 1979. pp. 3–5.
2. Huebner, K. H., *The Finite Element Method for Engineers*, Wiley, New York, 1975, pp. 339–357.
3. Arpaci, V. S., *Conduction Heat Transfer*, Addison-Wesley, Reading, Mass., 1966.
4. Becker, E. B., G. F. Carey, and J. T. Oden, *Finite Elements: An Introduction, vol. 1*, Prentice-Hall, Englewood Cliffs, N.J., 1981, pp. 6–9.
5. Arpaci, V. S., *Conduction Heat Transfer*, Addison-Wesley, Reading, Mass., 1966, pp. 163–164.
6. Zienkiewicz, O. C., *The Finite Element Method*, McGraw-Hill (UK), London, 1977, pp. 74–75.
7. Schwartz, A., *Calculus and Analytic Geometry*, Holt, Rinehart and Winston, New York, 1967, p. 357.
8. Schwartz, A., *Calculus and Analytic Geometry*, Holt, Rinehart and Winston, New York, 1967, pp. 798–799.
9. Arpaci, V. S., *Conduction Heat Transfer*, Addison-Wesley, Reading, Mass., 1966, pp. 439–440.
10. Huebner, K. H., *The Finite Element Method for Engineers*, Wiley, New York, 1975, p. 445.
11. Zienkiewicz, O. C., *The Finite Element Method*, McGraw-Hill (UK), London, 1977, pp. 70–72.
12. Myers, G. E., *Analytical Methods in Conduction Heat Transfer*, McGraw-Hill, New York, 1971, pp. 320–362.
13. Incropera, F. P., and D. P. DeWitt, *Fundamentals of Heat Transfer*, Wiley, New York, 1981, pp. 105–117.
14. Becker, E. B., G. F. Carey, and J. T. Oden, *Finite Elements: An Introduction*, vol. 1, Prentice-Hall, Englewood Cliffs, N.J., 1981, pp. 36–38.

PROBLEMS

4-1 For the problem posed in Example 4-1, verify that the approximate solution given by Eq. (4-10) satisfies the boundary conditions given by Eq. (4-8) exactly.

4-2 Reconsider the problem posed in Example 4-1. It is desired to obtain another approximate solution by the first-order Ritz method by using the following trial function:

$$N_1(x) = x(1 - x^4)$$

a. Does the use of this trial function satisfy, at least approximately, the physics of the problem?
b. Does the assumed approximate solution satisfy the boundary conditions exactly?
c. Determine the value of a_1 and hence the approximate solution corresponding to the above trial function.
d. How does the result from part (c) compare to the exact solution? Show the comparison on a graph that includes the approximate solution from Example 4-1.

4-3 For the problem posed in Example 4-1, another first-order Ritz solution is sought by assuming the following trial function:

$$N_1(x) = \sin \pi x$$

a. Does this trial function satisfy, at least approximately, the physics of the problem?
b. Does the assumed approximate solution satisfy the boundary conditions exactly?
c. Determine the value of a_1 and hence the approximate solution corresponding to the above trial function.
d. How does the result from part (c) compare to the exact solution? Show the comparison on a graph that includes the approximate solution from Example 4-1.

4-4 Recall that the exact solution to the problem posed in Example 4-1 is given by

$$T(x) = {}^{1000}\!/_{12}\, x(1 - x^3)$$

a. Obtain the first-order Ritz approximation to the problem by assuming the following trial function:

$$N_1(x) = x(1 - x^3)$$

b. How does the resulting approximate solution compare to the exact solution? Try to generalize this result.

4-5 Consider the following differential equation:

$$\frac{d^2T}{dx^2} + 100 = 0 \qquad 0 \le x \le 10$$

subject to the boundary conditions

$$T(0) = 0$$
$$T(10) = 0$$

a. What phenomenon might these equations represent? *Hint:* see Example 4-1.
b. Consider the following trial function to be used in the first-order Ritz solution:

$$N_1 = x(10 - x)$$

Verify that the approximate solution that utilizes this trial function satisfies the boundary conditions exactly.
c. Does the assumed form of the approximate solution satisfy the physics of the problem, at least approximately?
d. Obtain the first-order Ritz solution to the problem.
e. Determine the exact solution and compare it with the Ritz solution from part (d). Try to generalize this result.

4-6 Reconsider the problem posed in Problem 4-5.

    a. Verify that the first-order approximate solution with the following trial function satisfies the boundary conditions exactly:

$$N_1(x) = \sin \frac{\pi x}{10}$$

    b. Is the physics of the problem satisfied, at least approximately?

    c. Obtain the first-order Ritz solution to the problem and compare it to the exact solution.

4-7 Reconsider the problem posed in Example 4-1 except that now an approximate solution is sought by using the Rayleigh-Ritz method.

    a. For the trial function

$$N_1(x) = x(1 - x^4)$$

verify that the boundary conditions are satisfied exactly and that the physics of the problem is not grossly violated.

    b. Obtain the corresponding approximate solution by the Rayleigh-Ritz (variational) method.

    c. Compare the result from part (b) with the exact solution.

4-8 Solve all parts of Problem 4-3 by using the Rayleigh-Ritz method.

4-9 **a.** Obtain the first-order Rayleigh-Ritz solution to the problem posed in Example 4-1 by assuming the following trial function:

$$N_1(x) = x(1 - x^3)$$

    b. Compare the result from part (a) with the exact solution. Try to generalize this result.

4-10 For the problem posed in Problem 4-5, verify that the functional I is given by

$$I = \int_0^{10} \left[-\frac{1}{2} \left(\frac{dT}{dx} \right)^2 + 100T \right] dx$$

Use the method in Example 4-3.

4-11 Redo Problem 4-10 by using the method in Example 4-4.

4-12 Reconsider the problem posed in Problem 4-5.

    a. Determine the approximate solution by the first-order Rayleigh-Ritz method by using the trial function

$$N_1(x) = \sin \frac{\pi x}{10}$$

The functional I is given in Problem 4-10.

    b. Compare the result from part (a) to the exact solution.

    c. How does the result from part (a) compare to the approximate solution from the Ritz method (see Problem 4-6)?

4-13 Reconsider the problem posed in Problem 4-5.

a. Determine the approximate solution by the second-order Rayleigh-Ritz method by assuming

$$T'(x) = a_1 \sin \frac{\pi x}{10} + a_2 \sin \frac{3\pi x}{10}$$

The functional to be extremized is given in Problem 4-10.

b. Compare the result from part (a) with the approximate solution from Problem 4-12 and with the exact solution.

c. Make an educated guess as to what would happen if a series such as

$$T'(x) = \sum_{k=1}^{n} a_k \sin \frac{(2k - 1)\pi x}{10}$$

were used in the Rayleigh-Ritz method for increasingly larger values of n.

4-14 Reconsider the problem posed in Example 4-1.

a. Obtain the second-order Rayleigh-Ritz solution that corresponds to the following trial functions:

$$N_1(x) = x(1 - x^2)$$
$$N_2(x) = x(1 - x^4)$$

b. Compare the results from part (a) with the approximate solution from Example 4-2 and with the exact solution.

c. Compare the results from part (a) with the approximate solution from Problem 4-7.

4-15 Extend Eq. (4-37) to the case when the functional F is a function of x, $y(x)$, $y_x(x)$, and $y_{xx}(x)$, where the subscripts are used to denote derivatives.

4-16 Consider the following differential equation:

$$\frac{d^2 y}{dx^2} + 6y = 10x \qquad 0 \le x \le 2$$

subject to the boundary conditions

$$y(0) = 1$$
$$y(2) = 0$$

a. By using the method in Example 4-3, verify that the functional to be extremized is given by

$$I = \int_0^2 \left[3y^2 - 10xy - \frac{1}{2}\left(\frac{dy}{dx}\right)^2 \right] dx$$

b. Are the boundary conditions given above geometric or natural?

c. Do the boundary conditions satisfy Eq. (4-46)? Why or why not?

4-17 Reconsider the differential equation given in Problem 4-16. State whether each of the following sets of boundary conditions is geometric, natural, or neither:

a. $y(0) = 5$ **b.** $y_x(0) = 5$ **c.** $y(0) = 6$
     $y(2) = 3$ $y(2) = 3$ $y_x(2) = 0$

4-18 Reconsider the problem posed in Problem 4-16. Repeat part (a) by using the method illustrated in Example 4-4.

4-19 Nonlinear differential equations may also have a variational formulation. For example, consider the differential equation

$$\frac{d^2T}{dx^2} + 9T^2 = 5x^3 \qquad 0 \leq x \leq 5$$

subject to

$$T(0) = 9$$
$$T(5) = 0$$

a. Which term (or terms) makes the differential equation nonlinear?
b. Show that the corresponding variational formulation is given by

$$I = \int_0^5 \left[3T^3 - 5x^3T - \frac{1}{2}\left(\frac{dT}{dx}\right)^2 \right] dx$$

if the method illustrated in Example 4-3 is used.

4-20 Repeat part (b) of Problem 4-19 by using the method in Example 4-4.

4-21 Consider the following nonlinear differential equation:

$$\frac{d}{dx}\left(\frac{dT}{dx}\right)^2 + 5T = 0 \qquad 0 \leq x \leq 4$$

subject to the boundary conditions

$$\frac{dT(0)}{dx} = 0$$

and

$$T(4) = 100$$

a. Classify the boundary conditions: i.e., are they geometric, natural, or neither?
b. Show that the corresponding functional I is given by

$$I = \int_0^4 \left[\tfrac{5}{2}T^2 - \frac{1}{3}\left(\frac{dT}{dx}\right)^3 \right] dx$$

by using the method in Example 4-3.

4-22 Repeat Part b of Problem 4-21 by using the method in Example 4-4.

4-23 The necessary condition for the existence of an extremum of the functional

$$I = \int_a^b F(x,y,y_x,y_{xx})\, dx \qquad\qquad \textbf{(4-73)}$$

is that its first variation δI must be zero provided that

$$\left[\frac{\partial F}{\partial y_x} - \frac{d}{dx}\left(\frac{\partial F}{\partial y_{xx}}\right)\right]\delta y\,\Bigg|_a^b = 0 \qquad \textbf{(4-75a)}$$

and

$$\frac{\partial F}{\partial y_{xx}}\,\delta y_x\,\Bigg|_a^b = 0 \qquad \textbf{(4-75b)}$$

a. Show that the corresponding Euler-Lagrange equation is given by Eq. (4-74).
b. What order differential equation does Eq. (4-74) correspond to? *Hint:* See Problem 4-25.
c. What is the highest-order derivative present in the functional F?
d. Try to generalize the results from parts (b) and (c). Explain why the variational formulation may be called the *weak* formulation.

4-24 In Problem 4-23, identify:

a. The geometric boundary conditions and explain physically what these imply.
b. The natural boundary conditions and explain physically what these imply.

4-25 Consider the following fourth-order differential equation:

$$\frac{d^4y}{dx^4} + 8\frac{d^2y}{dx^2} + 4y = 10 \qquad 0 \le x \le 5$$

subject to the boundary conditions

$$y(0) = 0 \qquad y(5) = 100$$
$$y_x(0) = 1 \qquad y_x(5) = 0$$

a. Indicate whether each of the boundary conditions is geometric, natural, or neither. *Hint:* See Eqs. (4-73) to (4-79) and Problems 4-23 and 4-24.
b. Show that the corresponding functional I is given by

$$I = \int_0^5 \left[\frac{1}{2}\left(\frac{d^2y}{dx^2}\right)^2 - 4\left(\frac{dy}{dx}\right)^2 + 2y^2 - 10y\right] dx$$

c. Do the boundary conditions meet the conditions expressed in Eqs. (4-75)? Which ones?
d. What is the highest-order derivative present in the governing differential equation? In the variational formulation (i.e., in the functional I)?
e. From the results of part (d), explain why the variational formulation may be called the *weak* formulation?

4-26 It is not possible to obtain a classical variational formulation to problems whose governing differential equation contains an odd-ordered derivative. For example, consider

$$\frac{d^2y}{dx^2} + 4\frac{dy}{dx} + 5y = 3$$

Using the method illustrated in Example 4-3, *try* to obtain an expression for the functional I. Describe clearly the difficulties that are encountered.

4-27 Reconsider the dilemma posed in Problem 4-26. Using the method illustrated in Example 4-4, *try* to obtain an expression for the functional *I*. Describe clearly the difficulties that are encountered.

4-28 Reconsider the problem posed in Example 4-1 and obtain additional approximate solutions by using the trial function

$$N_1(x) = x(1 - x^4)$$

and the following weighted-residual methods:

a. Point collocation   **b.** Subdomain collocation
c. Least squares   **d.** Galerkin

4-29 Reconsider the problem posed in Example 4-1 and obtain additional approximate solutions by using the trial function

$$N_1(x) = \sin \pi x$$

and the following weighted-residual methods:

a. Point collocation   **b.** Subdomain collocation
c. Least squares   **d.** Galerkin

4-30 Recall that the exact solution to the problem posed in Example 4-1 is given by

$$T(x) = {}^{1000}\!/_{12}\, x(1 - x^3)$$

Consider the following trial function

$$N_1(x) = x(1 - x^3)$$

a. Obtain the corresponding approximate solution by using each of the four weighted-residual methods.
b. How do the approximate solutions from part (a) compare with the exact solution? Try to generalize this result.

4-31 Solve the problem posed in Example 4-1 by using the point collocation method if the following two trial functions are used in a second-order approximation:

$$N_1(x) = x(1 - x^2)$$
$$N_2(x) = x(1 - x^4)$$

Take the two collocation points at $x = \frac{1}{3}$ and $x = \frac{2}{3}$.

4-32 Repeat Problem 4-31 by using the subdomain collocation method. Use two equally spaced intervals.

4-33 Repeat Problem 4-31 by using the least-squares method.

4-34 Repeat Problem 4-31 by using the Galerkin method.

4-35 Solve the problem posed in Example 4-1 by using the point collocation method if the following two trial functions are used in a second-order approximation:

$$N_1(x) = \sin \pi x$$
$$N_2(x) = \sin 3\pi x$$

4-36   Repeat Problem 4-35 by using the subdomain collocation method. Use two equally spaced intervals.

4-37   Repeat Problem 4-35 by using the least-squares method.

4-38   Repeat Problem 4-35 by using the Galerkin method.

4-39   Reconsider the problem posed in Example 4-1 and obtain the first-order approximate solution for the point collocation method by using the trial function

$$N_1(x) = x(1 - x^2)$$

a. Assume the following collocation points: $x = 0.2, 0.4, 0.6,$ and 0.8.
b. Plot the four approximate solutions from part (a) on a graph that also includes the exact solution. How do these results compare to that from Example 4-5? Which is the best result?

4-40   Consider the differential equation

$$\frac{d^2T}{dx^2} + 100 = 0 \qquad 0 \le x \le 10$$

subject to the boundary conditions

$$T(0) = 0$$
$$T(10) = 0$$

Assume the trial function

$$N_1(x) = \sin \frac{\pi x}{10}$$

Obtain the corresponding approximate solution from the point collocation method and compare it to the exact solution by showing both on a graph. Take the collocation point at $x = 5$.

4-41   Repeat Problem 4-40 by using the subdomain collocation method.

4-42   Repeat Problem 4-40 by using the least-squares method.

4-43   Repeat Problem 4-40 by using the Galerkin method.

4-44   Consider the differential equation

$$\frac{d^2T}{dx^2} - T + 100x = 0 \qquad 0 \le x \le 10$$

subject to the boundary conditions

$$T(0) = 0$$
$$T(10) = 0$$

a. Assume the trial function

$$N_1(x) = x^2(100 - x^2)$$

and obtain the corresponding approximate solution by using the Ritz method.

    b. Obtain the exact solution and compare it with the result from part (a) by showing both on a graph.

4-45 Repeat Problem 4-44 by using the Rayleigh-Ritz method.

4-46 Repeat Problem 4-44 by using the point collocation method. Take the collocation point at $x = 5$.

4-47 Repeat Problem 4-44 by using the subdomain collocation method.

4-48 Repeat Problem 4-44 by using the least-squares method.

4-49 Repeat Problem 4-44 by using the Galerkin method.

4-50 Consider the following differential equation for the temperature T in a bar undergoing convection to ambient at 25°C:

$$\frac{d^2T}{dx^2} - 3(T - 25) = 0 \qquad 0 \le x \le 2$$

subject to the boundary conditions

$$T(0) = 150$$

$$\frac{dT(2)}{dx} = 0$$

    a. Are the boundary conditions geometric, natural, or neither?
    b. What does the second boundary condition physically represent?
    c. Derive the integral that represents the functional for the problem. Use either of the two methods illustrated in the text.
    d. Explain clearly and concisely why the following approximate solution is appropriate.

$$T'(x) = 150 - a_1 x(4 - x)$$

    e. Obtain the approximate solution corresponding to the Rayleigh-Ritz method.
    f. Obtain the approximate solutions corresponding to the point collocation, subdomain collocation, least-squares, and Galerkin methods.

4-51 Verify that the element stiffness matrix for Example 4-9 [which is given by Eq. (4-141)] evaluates to the result given by Eq. (4-143).

4-52 Verify Eq. (4-145) by performing the exact integration.

4-53 In arriving at Eq. (4-145), an exact integration was performed. Because integrals such as these arise in essentially all finite element formulations, very sophisticated numerical integration methods have been developed. These techniques are illustrated later in the text. One simple method can, however, be introduced here. Let $\bar{x}$ denote the value of the x coordinate at the centroid of the element, or $\bar{x} = (x_j + x_k)/2$. The integral is then evaluated approximately by evaluating the integrand at the centroid, pulling the result through the integral and evaluating the trivial integral that remains. Therefore, an alternate (but approximate) way to evaluate

$$\int_{x_j}^{x_k} 1000x^2 \mathbf{N}^T \, dx$$

from Example 4-9 (and Example 4-10) is given by

$$\int_{x_j}^{x_k} 1000x^2 \mathbf{N}^T \, dx = 1000 \left(\frac{x_j + x_k}{2} \right)^2 \begin{bmatrix} \frac{1}{2} \\ \frac{1}{2} \end{bmatrix} (x_k - x_j)$$

a. Verify that $\mathbf{N}^T$ is given by

$$\mathbf{N}^T = \begin{bmatrix} \frac{1}{2} \\ \frac{1}{2} \end{bmatrix}$$

when evaluated at $\bar{x} = (x_j + x_k)/2$.

b. Redo Example 4-9 by using this approximate integration method. Repeat only those parts of the finite element solution that are different from those in Example 4-9.

c. How do the resulting nodal temperatures compare to those obtained from the exact integration?

d. Why is it reasonable to expect this approximate integration to yield more accurate nodal temperatures as the number of elements is increased?

4-54 Consider the differential equation

$$\frac{d^2T}{dx^2} + 1000x^2 = 0 \qquad 0 \le x \le 1$$

subject to the boundary conditions

$$T(0) = 25$$
$$T(1) = 50$$

By making use of as much of Example 4-9 as is possible, determine the values of T at six equally spaced nodal points.

4-55 Repeat Problem 4-54 for the following boundary conditions:

$$T(0) = 25$$
$$\frac{dT(1)}{dx} = 0$$

4-56 Repeat Problem 4-54 for the following boundary conditions:

$$T(0) = 50$$
$$\frac{dT(1)}{dx} = 100$$

4-57 Let us determine the governing equation for the uniaxial stress member as follows. Consider the tapered uniaxial stress member shown in Fig. P4-57(a). The cross-sectional area varies with x and may be denoted as $A(x)$, or simply A. The axial stress is denoted as σ, where σ is positive for tensile stresses and negative otherwise. The axial force σA is assumed to vary according to a first-order Taylor expansion as shown in Fig. P4-57(b). Also shown is the body force $bA \, dx$ as a result of gravity, where b is the body force (or weight) per unit volume.

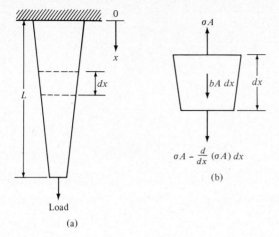

Figure P4-57

a. By considering the elemental volume shown in Fig. P4-57(b) to be in static equilibrium in the x direction, show that

$$\frac{d(\sigma A)}{dx} + bA = 0 \qquad \textbf{(4-190a)}$$

b. Rewrite Eq. (4-190a) in terms of the axial strain ϵ by invoking Hooke's law (which is a constitutive relationship) or $\sigma = E\epsilon$. Note that this implies that we are assuming the material to be linear elastic.

c. Next, eliminate the strain ϵ in the result from part (b) by invoking the strain-displacement relationship, $\epsilon = du/dx$, where u is the axial displacement that is, in general, a function of x. In other words, show that Eq. (4-190a) becomes

$$\frac{d}{dx}\left(AE\frac{du}{dx}\right) + bA = 0 \qquad \textbf{(4-190b)}$$

Equation (4-190b) is now in a suitable form for a Ritz, Rayleigh-Ritz, or weighted-residual solution. It is the governing equation for the problem.

4-58 Reconsider Problem 4-57 in general and Eq. (4-190b) in particular. It is desired to obtain the corresponding functional. Let us proceed as follows: from Eq. (4-55) it follows that

$$\delta I = \int_0^L \left[\frac{d}{dx}\left(AE\frac{du}{dx}\right) + bA\right] \delta u \, dx = 0$$

In the usual manner (by integrating by parts, etc.), show that this may be written as

$$\delta I = AE\frac{du}{dx} \, \delta u \bigg|_0^L - \delta \int_0^L \frac{1}{2} AE\left(\frac{du}{dx}\right)^2 dx + \delta \int_0^L ubA \, dx = 0 \qquad \textbf{(4-191a)}$$

But $E \, du/dx = \sigma$ and $\sigma A = P$, where P is the axial force; therefore, show that I itself is given by

$$I = Pu \Big|_0^L - \frac{1}{2} \int_0^L AE \left(\frac{du}{dx}\right)^2 dx + \int_0^L ubA \, dx \qquad \textbf{(4-191b)}$$

4-59 Reconsider Problems 4-57 and 4-58 in general and Eq. (4-191b) in particular. Let us now obtain the finite element characteristics (i.e., $\mathbf{K}^e$ and $\mathbf{f}^e$) by the Rayleigh-Ritz method. We proceed in the following manner:

a. Rewrite Eq. (4-191b) on an element basis as follows:

$$I^e = Pu^e \Big|_{x_j}^{x_k} - \frac{1}{2} \int_{x_j}^{x_k} AE \left(\frac{du^e}{dx}\right)^2 dx + \int_{x_j}^{x_k} u^e bA \, dx \qquad \textbf{(4-192a)}$$

where a typical element e connects node j at $x = x_j$ and node k at $x = x_k$ as shown in Fig. P4-59. Note that a typical node actually represents a plane and not just a point as shown.

b. Since the shape functions from Secs. 4-8 and 4-9 are applicable here (why?), the axial displacement u^e over a typical element e can be represented as

$$u^e = \mathbf{N}a^e \qquad \textbf{(4-192b)}$$

where $\mathbf{N}$ is given by

$$\mathbf{N} = \left[\frac{x_k - x}{x_k - x_j} \quad \frac{x - x_j}{x_k - x_j} \right] \qquad \textbf{(4-192c)}$$

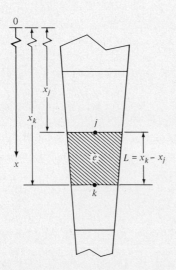

Figure P4-59   Note that L is defined here to be the element length, or $L = x_k - x_j$.

and $\mathbf{a}^e$ by

$$\mathbf{a}^e = \begin{bmatrix} u_j \\ u_k \end{bmatrix} \tag{4-192d}$$

where, in turn, u_j and u_k are the axial displacements at nodes j and k, respectively (positive in the $+x$ direction). Substitute for u^e in Eq. (4-192a) with u^e given by Eq. (4-192b), compute $dI^e/d\mathbf{a}^e$, set the result to zero, and rearrange the terms to show that we get

$$\mathbf{K}^e \mathbf{a}^e = \mathbf{f}^e \tag{4-192e}$$

where

$$\mathbf{K}^e = \int_{x_j}^{x_k} \frac{d\mathbf{N}^T}{dx} AE \frac{d\mathbf{N}}{dx} dx \tag{4-192f}$$

and

$$\mathbf{f}^e = P \mathbf{N}^T \Big|_{x_j}^{x_k} + \int_{x_j}^{x_k} \mathbf{N}^T bA \, dx \tag{4-192g}$$

c. Treating A, b, and E as constants in any given element by using the values at the midpoint of the element where $x = \bar{x}$ and denoting them as $\bar{A}$, $\bar{b}$, and $\bar{E}$, respectively, show that

$$\mathbf{K}^e = \frac{\bar{A}\bar{E}}{L} \begin{bmatrix} 1 & -1 \\ -1 & 1 \end{bmatrix} \tag{4-192h}$$

and

$$\mathbf{f}^e = \begin{bmatrix} -P_j \\ P_k \end{bmatrix} + \frac{\bar{b}\bar{A}L}{2} \begin{bmatrix} 1 \\ 1 \end{bmatrix} \tag{4-192i}$$

where L is the element length, or $L = x_k - x_j$; and P_j and P_k are the axial forces at nodes j and k, respectively (positive in the $+x$ direction).

d. After the assemblage step and the geometric boundary conditions (prescribed displacements) have been imposed, the following matrix equation may be used to represent the entire uniaxial stress member:

$$\mathbf{Ka} = \mathbf{f}$$

This equation is solved in the usual manner for the vector of nodal unknowns $\mathbf{a}$. Therefore, the axial displacement at each nodal point is now known. Show that the average strain $\bar{\epsilon}$ within the element is given by

$$\bar{\epsilon} = \frac{u_k - u_j}{L} \tag{4-192j}$$

the average axial stress $\bar{\sigma}$ by

$$\bar{\sigma} = \frac{\bar{E}(u_k - u_j)}{L} \tag{4-192k}$$

and the average axial force $\overline{F}$ by

$$\overline{F} = \frac{\overline{AE}(u_k - u_j)}{L} \tag{4-192m}$$

These average axial strains, stresses, and forces are generally associated with the centroid of the element, i.e., at $x = \overline{x} = (x_j + x_k)/2$.

4-60 Reconsider Problem 4-57 in general and Eq. (4-190b) in particular. Let us now obtain the finite element characteristics (i.e., $\mathbf{K}^e$ and $\mathbf{f}^e$) by the Galerkin weighted-residual method. We proceed in the following manner.

a. Form the weighted-residual equation for a typical element in the usual manner by writing

$$\int_{x_j}^{x_k} \mathbf{N}^T \left[\frac{d}{dx} \left(AE \frac{du}{dx} \right) + bA \right] dx = 0 \tag{4-193a}$$

where a typical element e connects node j at $x = x_j$ and node k at $x = x_k$ as shown in Fig. P4-59. Note that a typical node actually represents a plane and not a point as shown. Integrate by parts the term in Eq. (4-193a) containing the second derivative and show that

$$\mathbf{N}^T AE \frac{du}{dx} \bigg|_{x_j}^{x_k} - \int_{x_j}^{x_k} \frac{d\mathbf{N}^T}{dx} AE \frac{du}{dx} \, dx + \int_{x_j}^{x_k} \mathbf{N}^T bA \, dx = 0 \tag{4-193b}$$

b. Note that

$$AE \frac{du}{dx} = AE\epsilon = A\sigma = P$$

where ϵ is the axial strain, σ is the axial stress, and P is the axial force. Use this result only in the integrated term in Eq. (4-193b) to give

$$P\mathbf{N}^T \bigg|_{x_j}^{x_k} - \int_{x_j}^{x_k} \frac{d\mathbf{N}^T}{dx} AE \frac{du^e}{dx} \, dx + \int_{x_j}^{x_k} \mathbf{N}^T bA \, dx = 0 \tag{4-193c}$$

Note that superscript (^{e}) has now been added to u to give u^e because the assumed displacement function applies only on an element basis, not globally.

c. Since the shape functions from Secs. 4-8 and 4-9 are applicable here (why?), we can represent the axial displacement over a typical element e as

$$u^e = \mathbf{N}\mathbf{a}^e \tag{4-193d}$$

where N is given by Eqs. (4-129a) and (4-127) and $\mathbf{a}^e$ is given by

$$\mathbf{a}^e = \begin{bmatrix} u_j \\ u_k \end{bmatrix} \tag{4-193e}$$

where u_j and u_k are the axial displacements at nodes j and k, respectively (positive in the $+x$ direction). Show that Eq. (4-193c) can be written in the form

$$\mathbf{K}^e \mathbf{a}^e = \mathbf{f}^e \tag{4-193f}$$

where
$$\mathbf{K}^e = \int_{x_j}^{x_k} \frac{d\mathbf{N}^T}{dx} AE \frac{d\mathbf{N}}{dx}\, dx \tag{4-193g}$$

and
$$\mathbf{f}^e = P\,\mathbf{N}^T\Big|_{x_j}^{x_k} + \int_{x_j}^{x_k} \mathbf{N}^T b A\, dx \tag{4-193h}$$

Note that these expressions for the finite element characteristics are identical to those obtained by the Rayleigh-Ritz method in Problem 4-59.

d. Do part (c) in Problem 4-59.

e. Do part (d) in Problem 4-59.

4-61 Consider the tapered uniaxial stress member shown in Fig. P4-61(a). The top of the member is fixed, and a load P is applied to the bottom end. It is desired to obtain the displacements and the internal strains, stresses, and forces. Although it is possible to obtain the exact solution, a finite element solution is sought as described below. The member has a circular cross section with $D = 1.0$ in., $d = 0.5$ in., $H = 12$ in., and $P = 5000$ lbf. In addition, the material is 0.6% carbon steel with $E = 30 \times 10^6$ psi and $b = 0.283$ lbf/in.^{3}. Using the discretization shown in Fig. P4-61(b) with five equally spaced nodes and making use of the results of Problem 4-59 (or Problem 4-60), do the following.

a. Determine the displacement at each nodal point.

b. Determine the element resultants that are defined here to include the strains, stresses, and forces within each element.

c. Show that the body force contribution to the assemblage nodal force vector may be neglected with very little error.

d. Compare these results with those from the exact solution.

4-62 Solve Problem 4-61 if the member has a circular cross section with $D = 2$ cm, $d = 1$ cm, $H = 25$ cm, and $P = 18$ kN. In addition, the material is 0.6% carbon steel with $E = 21 \times 10^{10}$ N/m^2 and $b = 77$ kN/m^3.

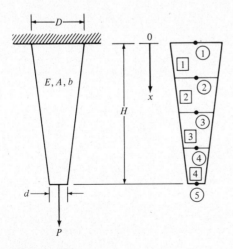

Figure P4-61

4-63   Solve Problem 4-61 if the cross section of the member is square.

4-64   Solve Problem 4-62 if the cross section of the member is square.

4-65   What is the half-bandwidth of the assemblage stiffness matrix for the discretized uniaxial stress member in Problem 4-61? Does Eq. (3-33) give the correct result?

4-66   Let us derive the governing equation for the temperature $T(x)$ in a circular pin fin of varying cross-sectional area A and perimeter P as shown in Fig. P4-66(a). We proceed by taking an infinitesimal slice of the fin with thickness dx and represent the heat transfer rate to the slice from conduction as $q_x A$, where q_x is the heat flux. As shown in Fig. P4-66(b), the corresponding heat transfer rate leaving the slice is obtained by assuming a first-order Taylor expansion. In addition, the convective heat transfer from the exposed surface of the fin is given by $hP(T - T_a)\, dx$, where h is the convective heat transfer coefficient and T_a is the ambient fluid temperature far removed from the fin. The length of the fin is L_f.

a. From a steady-state energy balance on the infinitesimal slice shown in Fig. P4-66(b), show that

$$-\frac{d}{dx}(q_x A) - hP(T - T_a) = 0 \qquad \textbf{(4-194a)}$$

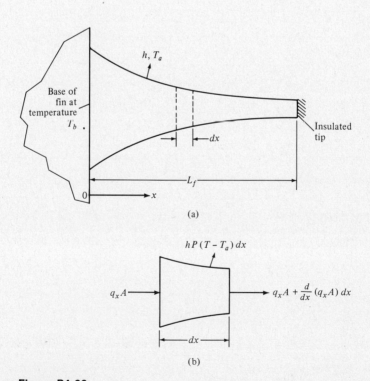

(a)

(b)

Figure P4-66

b. Next, invoke Fourier's law of heat conduction

$$q_x = -k\frac{dT}{dx} \tag{4-194b}$$

where k is the thermal conductivity. Equation (4-194b) is referred to as a *constitutive relationship*. The minus sign is necessary because q_x is assumed to be positive in the direction of decreasing temperature. With the help of Eq. (4-194b), show that Eq. (4-194a) becomes

$$\frac{d}{dx}\left(kA\,\frac{dT}{dx}\right) - hP\,(T - T_a) = 0 \tag{4-194c}$$

Equation (4-194c) is now in a suitable form for a Ritz, Rayleigh-Ritz, or weighted-residual solution and represents the governing equation for the problem.

4-67 Reconsider Problem 4-66 in general and Eq. (4-194c) in particular. It is desired to obtain the corresponding functional. Let us proceed as follows. From Eq. (4-55) it follows that

$$\delta I = \int_0^{L_f} \left[\frac{d}{dx}\left(kA\,\frac{dT}{dx}\right) - hP(T - T_a)\right] \delta T\,dx = 0$$

In the usual manner (by integrating by parts, etc.) show that this may be written as

$$\delta I = kA\,\frac{dT}{dx}\,\delta T\bigg|_0^{L_f} - \delta \int_0^{L_f} \frac{1}{2}\,kA\left(\frac{dT}{dx}\right)^2 dx$$

$$- \delta \int_0^{L_f} \frac{1}{2}\,hPT^2\,dx + \delta \int_0^{L_f} hPT_aT\,dx = 0 \tag{4-195a}$$

But $k\,dT/dx = -q_x$, where q_x is the heat flux from conduction. Therefore, show that I itself is given by

$$I = -q_xAT\bigg|_0^{L_f} - \int_0^{L_f} \frac{1}{2}\,kA\left(\frac{dT}{dx}\right)^2 dx$$

$$- \frac{1}{2}\int_0^{L_f} hPT^2\,dx + \int_0^{L_f} hPT_aT\,dx \tag{4-195b}$$

4-68 Reconsider Problems 4-66 and 4-67 in general and Eq. (4-195b) in particular. Let us now obtain the finite element characteristics (i.e., $\mathbf{K}^e$ and $\mathbf{f}^e$) by the Rayleigh-Ritz method by proceeding in the following manner:

a. Rewrite Eq. (4-195b) on an element basis as

$$I^e = -q_xAT^e\bigg|_{x_j}^{x_k} - \frac{1}{2}\int_{x_j}^{x_k} kA\left(\frac{dT^e}{dx}\right)^2 dx$$

$$- \frac{1}{2}\int_{x_j}^{x_k} hP(T^e)^2\,dx + \int_{x_j}^{x_k} hPT_aT^e\,dx \tag{4-196a}$$

where a typical element e connects node j at $x = x_j$ and node k at $x = x_k$, as shown in Fig. 4-12. Each node actually represents a planar surface, not just a point.

b. Since the shape functions from Secs. 4-8 and 4-9 are applicable here (why?), the temperature T^e over a typical element e can be represented as

$$T^e = \mathbf{N}\mathbf{a}^e \qquad \text{(4-196b)}$$

where $\mathbf{N}$ is given by

$$\mathbf{N} = \begin{bmatrix} \dfrac{x_k - x}{x_k - x_j} & \dfrac{x - x_j}{x_k - x_j} \end{bmatrix} \qquad \text{(4-196c)}$$

and $\mathbf{a}^e$ by

$$\mathbf{a}^e = \begin{bmatrix} T_j \\ T_k \end{bmatrix} \qquad \text{(4-196d)}$$

where T_j and T_k are the temperatures at nodes j and k, respectively. Substitute for T^e in Eq. (4-196a) with T^e given by Eq. (4-196b), compute $dI^e/d\mathbf{a}^e$, set the result to zero, and rearrange the terms to yield

$$\mathbf{K}^e\mathbf{a}^e = \mathbf{f}^e \qquad \text{(4-196e)}$$

where

$$\mathbf{K}^e = \mathbf{K}_x^e + \mathbf{K}_{\text{cv}}^e \qquad \text{(4-196f)}$$

and

$$\mathbf{f}^e = \mathbf{f}_q^e + \mathbf{f}_{\text{cv}}^e \qquad \text{(4-196g)}$$

and, further,

$$\mathbf{K}_x^e = \int_{x_j}^{x_k} \frac{d\mathbf{N}^T}{dx} kA \frac{d\mathbf{N}}{dx} \, dx \qquad \text{(4-196h)}$$

$$\mathbf{K}_{\text{cv}}^e = \int_{x_j}^{x_k} \mathbf{N}^T hP\mathbf{N} \, dx \qquad \text{(4-196i)}$$

$$\mathbf{f}_q^e = -q_x A\mathbf{N}^T \Big|_{x_j}^{x_k} \qquad \text{(4-196j)}$$

and

$$\mathbf{f}_{\text{cv}}^e = \int_{x_j}^{x_k} \mathbf{N}^T hPT_a \, dx \qquad \text{(4-196k)}$$

Note that $\mathbf{K}_x^e$ and $\mathbf{K}_{\text{cv}}^e$ are 2×2 element *stiffness* or conductance matrices, and the nodal *force* vectors $\mathbf{f}_q^e$ and $\mathbf{f}_{\text{cv}}^e$ are both of size 2×1.

c. It should be noted that k, h, A, P, and T_a may all be functions of x. However, let us perform the integrations in Eqs. (4-196h) to (4-196k) approximately by taking the following approach. Let us treat these parameters as constants in any given element by using the values at the midpoint of the element where $x = \bar{x}$ and by denoting them as $\bar{k}$, $\bar{h}$, $\bar{A}$, $\bar{P}$, and $\bar{T}_a$. Show that

$$\mathbf{K}_x^e = \frac{\bar{k}\bar{A}}{L} \begin{bmatrix} 1 & -1 \\ -1 & 1 \end{bmatrix} \qquad \text{(4-196m)}$$

$$\mathbf{K}^e_{cv} = \frac{\overline{h}PL}{6} \begin{bmatrix} 2 & 1 \\ 1 & 2 \end{bmatrix} \qquad \textbf{(4-196n)}$$

$$\mathbf{f}^e_q = \begin{bmatrix} +q_jA_j \\ -q_kA_k \end{bmatrix} \qquad \textbf{(4-196p)}$$

$$\mathbf{f}^e_{cv} = \frac{\overline{h}PT_aL}{2} \begin{bmatrix} 1 \\ 1 \end{bmatrix} \qquad \textbf{(4-196q)}$$

where q_j and q_k represent the heat fluxes from conduction at nodes j and k; A_j and A_k are the values of the cross-sectional areas at nodes j and k; and L is the element length (i.e., $L = x_k - x_j$). In Chapter 8 it will be seen how convection, radiation, and prescribed heat flux boundary conditions may be included. In the fin model, heat fluxes are not usually imposed on the ends of the fin; therefore, $\mathbf{f}^e_q$ is identically zero for all elements.

4-69 For the pin fin problem formulated in Sec. 4-10, begin with the expressions for the element characteristics [given by Eqs. (4-174) to (4-176)] and derive Eqs. (4-179) to (4-181).

4-70 The heat removal rate from a fin may be computed by Eq. (4-183) or by Eq. (4-186) once the nodal temperatures are known.

a. Derive Eq. (4-183) by beginning with Eq. (4-182).
b. Derive Eq. (4-186) by beginning with Eq. (4-185).
c. Which of the two methods gives more accurate results? Why?

4-71 Consider the fin shown in Fig. P4-71(a). The base of the fin (at $x = 0$) is held at a fixed temperature T_b, the surface of the fin (except the tip) is exposed to a fluid at temperature T_a, and the tip of the fin is insulated. It is desired to obtain the steady-state temperatures within the fin and the heat removal rate. Although it may be possible to obtain an exact solution, a finite element solution is sought. The fin has a circular cross section with $D = 0.5$ cm, $d = 0.2$ cm, and $L_f = 3$ cm. It undergoes convection to a fluid at $T_a = 25°C$ with $h = 75$ W/m²-°C. The fin is made of cast aluminum (4.5% copper) with $k = 168$ W/m-°C and has a base temperature T_b of $90°C$. Using five equally spaced nodes as shown in Fig. P4-71(b) and making use of the results of Sec. 4-10 or Problem 4-68:

a. Determine the temperature at each nodal point.
b. Determine the heat removal rate from the fin using the two methods given in Sec. 4-10. Which of these is more accurate? Determine the fin efficiency.

4-72 Solve Problem 4-71 for a fin of circular cross section with $D = 0.20$ in., $d = 0.10$ in., and $L_f = 1.5$ in. The fin is made of mild steel (1% carbon) that has a thermal conductivity of 26 Btu/hr-ft-°F and undergoes convection to a fluid at 70°F with a heat transfer coefficient of 50 Btu/hr-ft²-°F. The base temperature is 250°F.

4-73 Consider the straight fin of a rectangular cross section shown in Fig. P4-73. The fin is made of mild steel (1% carbon) that has a thermal conductivity of 26 Btu/hr-ft-°F. Both the upper and lower surfaces undergo convection to a fluid at 35°F with a heat transfer coefficient of 30 Btu/hr-ft²-°F. The base temperature is 120°F. Take $L_f = 1.625$ in. and $t = 0.375$ in. The fin width W is much greater than L_f and so a one-dimensional temperature distribution is sought.

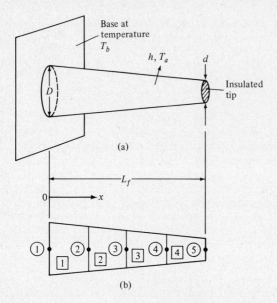

(a)

(b)

Figure P4-71

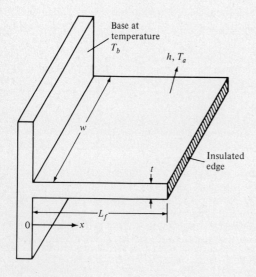

Figure P4-73

a. Discretize the fin into four equal-length elements with five nodes.

b. Using the results of Sec. 4-10 or Problem 4-68, determine the temperature at each nodal point.

c. Determine the heat removal rate in two different ways. Which of these two results is more accurate? What is the fin efficiency?

d. Compare the results from parts (b) and (c) with those from the exact solution.

4-74 Solve Problem 4-73 if the fin is made of cast aluminum (4.5% copper) with a thermal conductivity of 168 W/m-°C. The convective heat transfer coefficient is 85 W/m²-°C on both the upper and lower surfaces to a fluid at 5°C. The base temperature is at 95°C. Take $L_f = 3$ cm and $t = 0.75$ cm.

4-75 Consider the straight fin of triangular cross section shown in Fig. P4-75. The fin material is aluminum alloy 2024-T6, and has a thermal conductivity of 177 W/m-°C. Both the upper and lower surfaces convect to a fluid at 35°C through a heat transfer coefficient of 150 W/m²-°C. Assume a base temperature of 100°C and take $L_f = 1.0$ cm and $H = 0.4$ cm. Since the fin width W is much greater than L_f, a one-dimensional temperature distribution is expected.

a. Discretize the fin into four equal-length elements with five nodes.

b. Use the results of Sec. 4-10 or Problem 4-68, and determine the temperature at each nodal point.

c. Determine the heat removal rate in two different ways. Which of these two results is more accurate? What is the fin efficiency?

4-76 Solve all parts of Problem 4-75 if the fin is made of mild steel with $k = 26$ Btu/hr-ft-°F, $h = 150$ Btu/hr-ft²-°F, $T_b = 200$°F, $T_a = 75$°F, $H = 0.25$ in., and $L_f = 1.0$ in.

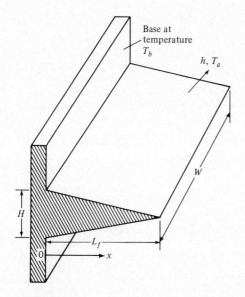

Figure P4-75

4-77   Convert the TRUSS program in Appendix B into one that could be used to find the displacements, strains, stresses, and forces within a uniaxial stress member. Allow up to 30 elements (and 31 nodes) with up to five different materials. The program should be able to handle prescribed displacements as well as imposed nodal forces. Take advantage of the developments in Problem 4-59 (or 4-60). A uniform taper may be assumed.

4-78   Convert the TRUSS program in Appendix B into one that could be used to find the temperature distribution within, the heat removal rate from, and the efficiency of an arbitrary fin. Allow up to 30 elements (and 31 nodes) with up to five different materials. The program needs to handle only prescribed temperature boundary conditions (and insulation, which, being a natural boundary condition, is automatically included). Take advantage of the development in Sec. 4-10. Allow for spacially dependent thermal conductivities, cross-sectional areas, heat transfer coefficients, perimeters, and ambient fluid temperatures by taking the following approach. Change SUBROUTINE PROPTY to the following:

```
            SUBROUTINE PROPTY (L, MATFLG, DATMAT,
     1             THERCN, AREA, HTC, PERIM, TAMB)

C....       DEFINES THE THERMAL CONDUCTIVITY, CROSS-SECTIONAL
C           AREA, HEAT TRANSFER COEFFICIENT, PERIMETER, AND
C           AMBIENT TEMPERATURE FOR ELEMENT L

            DIMENSION MATFLG(30), DATMAT(5,5)
            NFLAG    = MATFLG(L)

            THERCN   = DATMAT(NFLAG, 1)
            IF (THERCN .LT. 0.)   CALL VPROP (THERCN)

            AREA     = DATMAT(NFLAG, 2)
            IF (AREA .LT. 0.)   CALL VPROP (AREA)

            HTC      = DATMAT(NFLAG, 3)
            IF (HTC .LT. 0.)   CALL VPROP (HTC)

            PERIM    = DATMAT(NFLAG, 4)
            IF (PERIM .LT. 0.)   CALL VPROP (PERIM)

            TAMB     = DATMAT(NFLAG, 5)
            IF (TAMB .LT. 0.)   CALL VPROP (TAMB)

            RETURN
            END
```

Note that if a negative value is read for any property, then SUBROUTINE VPROP is called, which is given by

```
            SUBROUTINE VPROP (PROP)
            COMMON /CONST/ C(10)
            COMMON /VPROPS/ X
```

```
C....                    MATHEMATICAL CONSTANTS:
                         PI = 3.141593

                         IPROP = -PROP
                         GO TO (1, 2, 3, 4, 5), IPROP

1                        PROP  =  definition of first variable property
                                  as a function of x and the C(i)'s
                         RETURN

2                        PROP  =  definition of second variable
                                  property and so forth.
                         RETURN

3                        CONTINUE
                         RETURN

4                        CONTINUE
                         RETURN

5                        CONTINUE
                         RETURN

                         END
```

Note that if PROP is negative, it is converted to a positive integer IPROP. The *computed go to* then transfers control to the line whose label is numerically equal to IPROP. Thus by simply including the proper FORTRAN statements in SUBROUTINE VPROP, variable properties are easily accommodated. The C(i)'s are user-defined constants that should be read in Section 1 of the input file (after NNODES, NELEM, etc.) and may be used in SUBROUTINE VPROP. By way of example, let us say that we want to use this technique to allow for the varying cross-sectional area for the fin in Problem 4-75. Let us define $C(1) = 0.004$ (for H in meters) and $C(2) = 0.01$ (for L_f in meters). If we use a "-3" as the input for the area A in Section 4 of the input file (see Appendix B for a description of the input to the TRUSS program), the statement with label "3" should read

$$3 \quad PROP = C(1) * (1. -X/C(2))$$

since this in effect gives

$$A = WH(1. - x/L_f)$$

which represents the correct area variation with x (W is unity since a unit width is assumed). Note that the variable X represents the x coordinate and is passed to VPROP via the *labeled common* (namely, VPROPS).

4-79 Use the computer program from Problem 4-77 (or one furnished by the instructor) to solve Problem 4-61.

a. Run the program for 5, 9, and 17 equally spaced nodes.
b. What is concluded from the results of part (a)?

4-80 Use the computer program from Problem 4-77 (or one furnished by the instructor) to solve Problem 4-63.

      a. Run the program for 5, 9, and 17 equally spaced nodes.
      b. What is concluded from the results of part (a)?

4-81 Use the computer program from Problem 4-78 (or one furnished by the instructor) to solve Problem 4-71. Use SUBROUTINE VPROP to include the spacially varying cross-sectional area and perimeter.

      a. Run the program for 5, 9, and 17 equally spaced nodes.
      b. What is concluded from the results of part (a)?

4-82 Solve Problem 4-72 by using the computer program from Problem 4-78 (or one furnished by the instructor) to solve Problem 4-72. Use SUBROUTINE VPROP to include the spacially varying cross-sectional area and perimeter.

      a. Run the program for 5, 9, and 17 equally spaced nodes.
      b. Is there a significant difference in the results from part (a)? Explain.

4-83 Solve Problem 4-73 by using the computer program from Problem 4-78 (or one furnished by the instructor). Use 5, 9, and 17 nodes to demonstrate increased accuracy as the element size decreases.

4-84 Solve Problem 4-74 by using the computer program from Problem 4-78 (or one furnished by the instructor). Use 5, 9, and 17 nodes to demonstrate increased accuracy as the element size decreases.

4-85 Solve Problem 4-75 by using the computer program from Problem 4-78 (or one furnished by the instructor). Use 5, 9, and 17 nodes and SUBROUTINE VPROP to model the cross-sectional area variation with x.

4-86 Solve Problem 4-76 by using the computer program from Problem 4-78 (or one furnished by the instructor). Assume 5, 9, and 17 nodes, and use SUBROUTINE VPROP to model the cross-sectional area variation with x.

4-87 Consider the annular fin of constant thickness as shown in Fig. P4-87. The outer edge at radius R_o is insulated. The inner edge at radius R_i is held at the base temperature T_b. Both top and bottom surfaces convect to a fluid at temperature T_a through a heat transfer coefficient h.

      a. Show that the governing equation is given by

$$\frac{d}{dr}\left(ktr\,\frac{dT}{dr}\right) - 2hr(T - T_a) = 0$$

      and the boundary conditions by

$$T(R_i) = T_b$$

$$\frac{dT}{dr}(R_o) = 0$$

      b. Explain why the finite element formulation from Sec. 4-10 or Problem 4-68 is applicable providing that we use

$$A = 2\pi rt$$

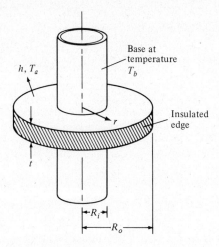

Figure P4-87

and

$$P = 4\pi r$$

4-88 Use the program from Problem 4-78 (or one furnished by the instructor) to determine the temperature distribution within, the heat removal rate from, and the efficiency of the brass annular fin shown in Fig. P4-87. Take $k = 60$ Btu/hr-ft-°F, $t = 0.125$ in., $R_i = 0.625$ in., $R_o = 1.5$ in., $T_b = 300$°F, $h = 20$ Btu/hr-ft²-°F, and $T_a = 70$°F. Use 30 nodes.

Hint: See Problems 4-78 and 4-87 and make use of SUBROUTINE VPROP to take the variable cross-sectional area A and perimeter P into account. The variable X in SUBROUTINE VPROP should be interpreted to be the radial coordinate r.

4-89 Do Problem 4-88 for an annular fin made of AISI 1010 carbon steel with $k = 64$ W/m-°C, $t = 5$ mm, $R_i = 1.0$ cm, $R_o = 2.0$ cm, $T_b = 160$°C, $h = 15$ W/m²-°C, and $T_a = 40$°C. Use 30 nodes. (See Problem 4-88 for a hint.)

4-90 Consider the differential equation and boundary conditions in Problem 4-5.

   a. Derive the integral expressions for the element characteristics. Use the variational method.
   b. Evaluate the integrals from part (a) by using the shape functions given by Eqs. (4-127) for the two-node one-dimensional element.
   c. Use four equally spaced nodes (three elements) and determine the value of T at these nodes.

4-91 Solve all parts of Problem 4-90 by using the Galerkin method.

4-92 Consider the differential equation and boundary conditions in Problem 4-16.

   a. Derive the integral expressions for the element characteristics. Use the variational method.

b. Evaluate the integrals from part (a) by using the shape functions given by Eqs. (4-127) for the two-node one-dimensional element.

c. Use four equally spaced nodes (three elements) and determine the value of y at these nodes.

4-93 Solve all parts of Problem 4-92 by using the Galerkin method.

4-94 Consider the differential equation and boundary conditions in Problem 4-25. Try to derive the integral expressions for the element characteristics. Use the variational method. Can the shape functions for the two-node one-dimensional element be used to evaluate the integrals? Why or why not? *Hint:* Integration by parts must be used twice.

4-95 Do Problem 4-94 by using the Galerkin method.

4-96 The differential equation given in Problem 4-26 does *not* have a classical variational formulation.

a. Use the Galerkin method and derive the integral expressions for the finite element characteristics. Allow for only natural and geometric boundary conditions.

b. Is each of the stiffness matrices symmetric?

c. What can be concluded when the differential equation contains an odd-order derivative?

4-97 Consider the differential equation and boundary conditions in Problem 4-44.

a. Derive the integral expressions for the element characteristics. Use the variational method.

b. Evaluate the integrals from part (a) by using the shape functions given by Eqs. (4-127) for the two-node one-dimensional element.

c. Use four equally spaced nodes (three elements) and determine the value of T at these nodes.

4-98 Solve all parts of Problem 4-97 by using the Galerkin method.

4-99 Consider the differential equation and boundary conditions in Problem 4-50.

a. Derive the integral expressions for the element characteristics. Use the variational method.

b. Evaluate the integrals from part (a) by using the shape functions given by Eqs. (4-127) for the two-node one-dimensional element.

c. Use four equally spaced nodes (three elements) and determine the value of T at these nodes.

4-100 Solve all parts of Problem 4-99 by using the Galerkin method.

4-101 Consider the differential equation and boundary conditions in Problem 4-87.

a. Derive the integral expressions for the element characteristics. Use the variational method. Note that the elements in this case are concentric annuli with nodes j and k at radii r_j and r_k, respectively.

b. Evaluate the integrals from part (a) by using the shape functions given by Eqs. (4-127) for the two-node one-dimensional element. Note that x must be replaced by r, x_j by r_j, and x_k by r_k. Do not evaluate the integrals exactly. Instead, evaluate the integrand at the element centroid, and proceed.

4-102 Solve all parts of Problem 4-101 by using the Galerkin method.

4-103 Solve Problem 4-88 by hand using two elements and the formulation from Problem 4-101 or Problem 4-102.

4-104 Solve Problem 4-89 by hand using two elements and the formulation from Problem 4-101 or Problem 4-102.

4-105 Use the variational approach to determine the integral expressions for the element characteristics for the differential equation and boundary conditions given in Problem 4-19. Note that the differential equation is nonlinear. Comment on how this problem might be solved numerically with the finite element method.

4-106 Solve Problem 4-105 by using the Galerkin method to derive the integral expressions for the element characteristics.

5

General Approach
to Structural Analysis

5-1 INTRODUCTION

One of the most important and challenging steps in all finite element analyses is the derivation of the finite element characteristics. It should be recalled that these are defined to be the expressions for the element stiffness matrix $\mathbf{K}^e$ and nodal force vector $\mathbf{f}^e$. The superscript (^{e}) serves as a reminder that the expressions are derived on an element basis. In the last chapter these finite element characteristics were derived based on the Rayleigh-Ritz and Galerkin methods. The former is a variational method, whereas the latter is based on the weighted-residual method. These methods, particularly the Galerkin method, are easily applied to nonstructural two- and three-dimensional analyses, as will be seen in later chapters. However, in stress and structural analysis applications, the finite element characteristics are more easily derived by invoking one of the following: (1) the principle of minimum potential energy, (2) the principle of virtual displacements, (3) the minimum complementary energy principle, (4) the principle of virtual stress, (5) Reissner's principle, or (6) Hamilton's principle [1].

The principles of minimum potential energy and of virtual displacements are used in this chapter. Both approaches result in the so-called stiffness (or displacement) method, wherein the primary unknowns are the nodal displacements (as opposed to the nodal forces). The finite element formulation based on these two principles is presented in Sec. 5-5 after each is introduced. The four remaining methods are outside the scope of this book. This chapter is further restricted to static stress analysis.

The chapter begins with a review of the basic concepts of elasticity, which include the notions of stress and strain at a point, principal stresses, and constitutive relationships (generalized Hooke's law); the equations of static equilibrium and compatibility; and the strain-displacement relationships. This is followed by the development of the principles of minimum potential energy and virtual displacements for direct use in the finite element method. The piecewise continuous trial functions that are used to represent the displacements within an element will be referred to as displacement functions. The criteria that these functions must satisfy to ensure convergence are delineated in Chapter 6. Although the approach is illustrated with a simple one-dimensional example, it should be emphasized that the expressions to be derived for the finite element characteristics are applicable to one-, two-, and three-dimensional stress analysis problems, as well as to those with axisymmetry. In fact, these results can even be used in the analysis of plates and shells.

5-2   BASIC CONCEPTS IN ELASTICITY

The terms stress and strain as used below are defined quantitatively in subsequent sections. For now let us define *stress* as force per unit area and *strain* as elongation per unit length. In this book only *linear elastic* materials are considered. This particular class of materials behaves in such a way that when loaded and subsequently unloaded, the stress-strain relationship is linear, as shown in Fig. 5-1(a), and the material returns to its original undeformed state. A material that returns to its original shape with a stress-strain relationship, as depicted in Fig. 5-1(b), is said to be *nonlinear elastic*. The elastic limit of the material is the stress at which permanent deformation occurs. The proportional limit is the same as the elastic limit for linear elastic materials. The analysis of structures outside the linear elastic range is beyond the scope of this book.

For a more formal development of the material in this section, the reader is referred to references [2–4] for several excellent books on the subject. In keeping

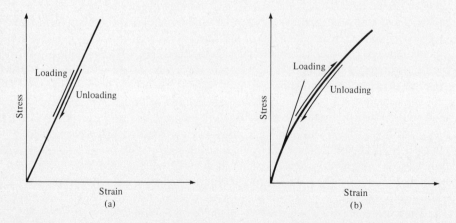

Figure 5-1   Stress-strain curves for (a) linear elastic and (b) nonlinear elastic materials.

with the introductory nature of this book, indicial notation (cartesian tension no-tation) is not used.

It should be pointed out that the equations of static equilibrium and compatibility that are reviewed here are not explicitly needed in the finite element method. There are, however, two reasons for including them here. The first is that reference will be made to these equations, and the reader should have no doubt as to which set of equations the reference is being made. The second reason is that a review of these equations will help to put the finite element method in the proper context in this chapter as well as in Chapter 7.

Stress at a Point

The review begins with the definition of *stress*. Consider a body in static equilibrium, loaded as shown in Fig. 5-2(a). The external forces F_1, F_2, etc., are transmitted through the body in a complex manner. Based on static equilibrium considerations,

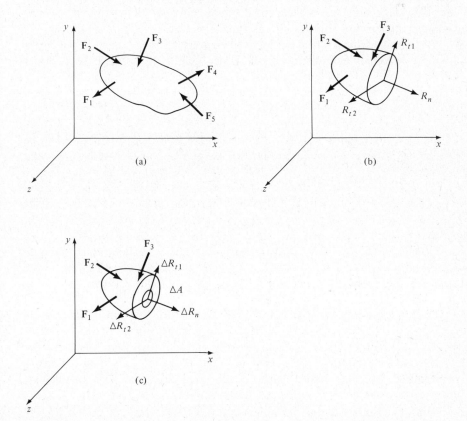

Figure 5-2   (a) External forces acting on body; (b) forces acting on cut surface; (c) forces acting on an infinitesimally small surface area ΔA on cut surface.

when the body is cut along some plane in general, a force **R** is required to maintain static equilibrium. Such a plane is shown in Fig. 5-2(b), where the force **R** has been resolved into a normal component R_n and two tangential and orthogonal (perpendicular) components R_{t1} and R_{t2}. The latter two components are in the plane of the cut. Instead of working with the entire area A, let us work with a much smaller area ΔA such that the respective forces are ΔR_n, ΔR_{t1}, and ΔR_{t2} as shown in Fig. 5-2(c). The *normal stress* σ_n is then defined as

$$\sigma_n = \lim_{\Delta A \to 0} \frac{\Delta R_n}{\Delta A} \tag{5-1}$$

and the two shear (or tangential) stresses as

$$\sigma_{t1} = \lim_{\Delta A \to 0} \frac{\Delta R_{t1}}{\Delta A} \quad \text{and} \quad \sigma_{t2} = \lim_{\Delta A \to 0} \frac{\Delta R_{t2}}{\Delta A} \tag{5-2}$$

In other words, as the area over which the force acts is reduced to zero at a point, three different stresses arise: one normal to the plane containing the point and the other two tangential. The two tangential stresses are referred to as *shear stresses*. The shear stresses are orthogonal to each other as well as to the normal stress.

In the United States where the English engineering system of units is still being used, the customary unit of stress is pounds per square inch (psi) or kilopounds per square inch (ksi). In the International System of Units (SI), the unit of stress is newtons per square meter (N/m^2). For conversion purposes, we may use

$$1 \text{ psi} = 6895 \text{ N/m}^2 \tag{5-3}$$

The trend in the United States in engineering has been and will continue to be toward the SI system of units. It is for this reason that both sets of units are used in this text.

Let us now consider an infinitesimal volume element of a body positioned at a point in a global *xyz*-coordinate system as shown in Fig. 5-3(a). There are six faces to the cube, and on each face one normal and two shear stresses act as shown. Note that the two subscripts are identical for normal stresses, whereas they are different for shear stresses. Thus, no ambiguity should result in using the same symbol to represent both types of stresses. This nomenclature makes use of the following convention. The first subscript is associated with the outward normal to the plane over which the stress acts, and the second subscript is associated with the direction of the stress. *Positive faces* are defined to have outward normals in the same direction as the coordinate axes; *negative faces* have outward normals opposite the coordinate axes. The three visible faces in Fig. 5-3(a) happen to be positive faces, while the three hidden faces are negative. An infinitesimal two-dimensional element is shown in Fig. 5-3(b), where the top and right sides are positive faces and the bottom and left sides are negative faces.

With the notions of positive and negative faces in hand, the following sign convention is adopted. Normal and shear stresses are considered to be positive if directed in the positive coordinate direction on a positive face. Normal and shear stresses are also considered to be positive if directed in the negative coordinate

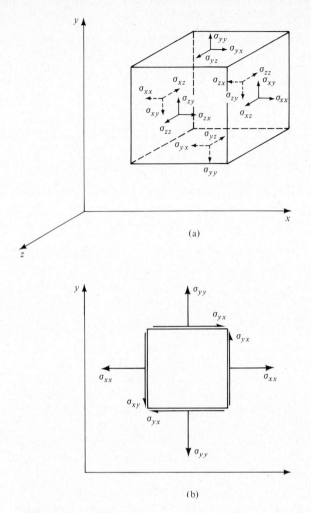

(a)

(b)

Figure 5-3 Definition of positive normal and shear stresses in (a) three dimensions and (b) two dimensions.

direction on a negative face. Otherwise, the stress is considered to be negative. All stresses shown in Fig. 5-3 have a positive sense. This convention ensures that positive normal stresses are tensile, whereas negative normal stresses are compressive.

The Equations of Static Equilibrium

Let us now consider an infinitesimal two-dimensional volume element (with thickness t) as shown in Fig. 5-4. The normal and shear stresses are shown in Fig. 5-4, and it is assumed that these stresses vary from point to point in the body in

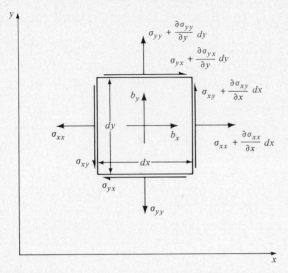

Figure 5-4   Infinitesimal two-dimensional element with stresses and body forces shown.

some continuous manner. Consequently, a first-order Taylor expansion is used to represent the stresses. Each of the stresses in Fig. 5-4 is shown with a positive sense. The two components of the body force per unit volume (b_x and b_y) are also shown. The reader may show that the following equations result if static equilibrium is to be assured (see Problem 5-1):

$$\frac{\partial \sigma_{xx}}{\partial x} + \frac{\partial \sigma_{yx}}{\partial y} + b_x = 0 \tag{5-4a}$$

$$\frac{\partial \sigma_{xy}}{\partial x} + \frac{\partial \sigma_{yy}}{\partial y} + b_y = 0 \tag{5-4b}$$

and

$$\sigma_{xy} = \sigma_{yx} \tag{5-5}$$

These equations can be extended to the three-dimensional case by taking an elemental volume $dx\, dy\, dz$ and applying the equations of static equilibrium in a similar manner (see Problem 5-2). The final result is given by

$$\frac{\partial \sigma_{xx}}{\partial x} + \frac{\partial \sigma_{yx}}{\partial y} + \frac{\partial \sigma_{zx}}{\partial z} + b_x = 0 \tag{5-6a}$$

$$\frac{\partial \sigma_{xy}}{\partial x} + \frac{\partial \sigma_{yy}}{\partial y} + \frac{\partial \sigma_{zy}}{\partial z} + b_y = 0 \tag{5-6b}$$

$$\frac{\partial \sigma_{xz}}{\partial x} + \frac{\partial \sigma_{yz}}{\partial y} + \frac{\partial \sigma_{zz}}{\partial z} + b_z = 0 \tag{5-6c}$$

with

$$\sigma_{xy} = \sigma_{yx} \qquad \sigma_{yz} = \sigma_{zy} \qquad \text{and} \qquad \sigma_{zx} = \sigma_{xz} \qquad \textbf{(5-7)}$$

Obviously, b_z is the z component of the body force per unit volume. Equations (5-4) and (5-6) are known as the equilibrium equations in two and three dimensions, respectively.

Principal Stresses

The reader will recall that Mohr's circle may be used to obtain the two-dimensional stresses at a point on a plane whose normal makes an angle θ with the x axis. Such a plane is shown in Fig. 5-5. Because a computer program implementation is our ultimate goal, the transformation equations from σ_{xx}, σ_{yy}, and σ_{xy} to $\sigma_{x'x'}$, $\sigma_{y'y'}$, and $\sigma_{x'y'}$ are given here:

$$\sigma_{x'x'} = \tfrac{1}{2}(\sigma_{xx} + \sigma_{yy}) + \tfrac{1}{2}(\sigma_{xx} - \sigma_{yy}) \cos 2\theta + \sigma_{xy} \sin 2\theta \qquad \textbf{(5-8a)}$$

$$\sigma_{y'y'} = \tfrac{1}{2}(\sigma_{xx} + \sigma_{yy}) - \tfrac{1}{2}(\sigma_{xx} - \sigma_{yy}) \cos 2\theta - \sigma_{xy} \sin 2\theta \qquad \textbf{(5-8b)}$$

$$\sigma_{x'y'} = -\tfrac{1}{2}(\sigma_{xx} - \sigma_{yy}) \sin 2\theta + \sigma_{xy} \cos 2\theta \qquad \textbf{(5-8c)}$$

Note that $\sigma_{y'y'}$ is not shown in Fig. 5-5 and is the normal stress on the plane with a normal in the y' direction.

Now let us summarize the equation for the principal stresses in two dimensions. Recall that the principal stresses occur on the so-called principal planes whose

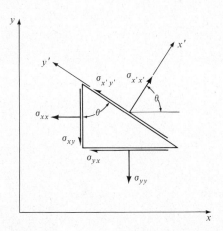

Figure 5-5   Element for transformation of stress at a point.

outward normals are rotated through an angle θ_p from the x and y axis, respectively, where

$$\tan 2\theta_p = \frac{2\sigma_{xy}}{\sigma_{xx} - \sigma_{yy}} \tag{5-9}$$

The two principle stresses, σ_1 and σ_2, are given by

$$\sigma_{1,2} = \frac{\sigma_{xx} + \sigma_{yy}}{2} \pm \sqrt{\left(\frac{\sigma_{xx} - \sigma_{yy}}{2}\right)^2 + \sigma_{xy}^2} \tag{5-10}$$

By definition σ_1 is taken to be the algebraically larger principal stress, as shown on the Mohr's circle in Fig. 5-6. The reader should recall from solid mechanics that Mohr's circle of stress is a circle on the σ-τ plane, where σ and τ denote general normal and shear stresses, respectively. Therefore, the maximum principal stress is σ_1 and the minimum principal stress is σ_2. In terms of these principal stresses, the maximum shear stress $\tau_{\max}$ is given by

$$\tau_{\max} = \frac{\sigma_1 - \sigma_2}{2} \tag{5-11}$$

and occurs on a plane oriented 45° from the principal planes. For the purpose of plotting Mohr's circle or using Eqs. (5-8), a shear stress is positive if it would cause a counterclockwise rotation about a point within the element and is negative otherwise.

The principal stresses for the general three-dimensional state of stress are given by the solution of the following cubic equation:

$$\sigma^3 - I_1\sigma^2 + I_2\sigma - I_3 = 0 \tag{5-12a}$$

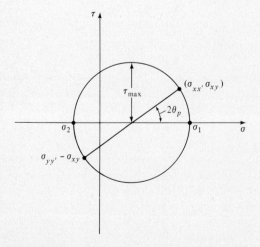

Figure 5-6 Mohr's circle for the two-dimensional state of stress.

where I_1, I_2, and I_3 are known as the stress tensor invariants that are defined by

$$I_1 = \sigma_{xx} + \sigma_{yy} + \sigma_{zz} \tag{5-12b}$$

$$I_2 = \sigma_{xx}\sigma_{yy} + \sigma_{yy}\sigma_{zz} + \sigma_{zz}\sigma_{xx} - \sigma_{xy}^2 - \sigma_{yz}^2 - \sigma_{zx}^2 \tag{5-12c}$$

$$I_3 = \det \begin{bmatrix} \sigma_{xx} & \sigma_{xy} & \sigma_{xz} \\ \sigma_{yx} & \sigma_{yy} & \sigma_{yz} \\ \sigma_{zx} & \sigma_{zy} & \sigma_{zz} \end{bmatrix} \tag{5-12d}$$

Since Eq. (5-12a) is third order in σ, three roots may be obtained. These roots represent the three principal stresses.

Example 5-1

For the two-dimensional state of stress shown in Fig. 5-7:

a. Determine the principle stresses.
b. Show the orientation of the element that results in the stresses from part (a).
c. Determine the maximum shear stress and the associated normal stresses.
d. Show these results on Mohr's circle of stress.

Solution

   a. The principal stresses are readily computed from Eq. (5-10) to be

$$\sigma_{1,2} = \frac{1000 + (-3000)}{2} \pm \sqrt{\left[\frac{1000 - (-3000)}{2}\right]^2 + (-2000)^2}$$

or

$$\sigma_{1,2} = -1000 \pm 2828$$

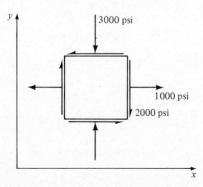

Figure 5-7 State of stress for Example 5-1.

or

$$\sigma_1 = 1828 \text{ psi} \quad \text{and} \quad \sigma_2 = -3828 \text{ psi}$$

Note that σ_1 is taken to be the algebraically larger stress [see the Mohr's circle in part (d)].

b. The angle θ_p is easily computed from Eq. (5-9) to be

$$\theta_p = \frac{1}{2} \tan^{-1}\left[\frac{2(-2000)}{1000 - (-3000)}\right]$$

or

$$\theta_p = -22.5°$$

Therefore, the principal stresses from part (a) result if the element is rotated $-22.5°$ (or 22.5° clockwise) as shown in Fig. 5-8(a). Note that the shear stress is zero in this state of stress.

c. The maximum shear stress is given by

$$\tau_{max} = \frac{\sigma_1 - \sigma_2}{2} = \frac{1828 - (-3828)}{2} = 2828 \text{ psi}$$

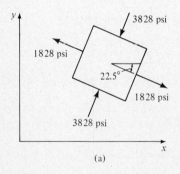

(a)

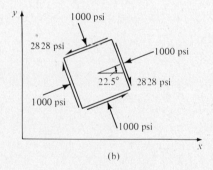

(b)

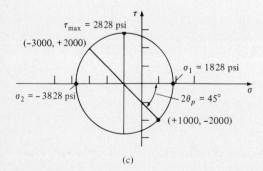

(c)

Figure 5-8   Element orientations for (a) principal stress state and (b) maximum shear stress state. (c) Mohr's circle of stress.

and occurs on an element rotated 45° counterclockwise from the principal stress orientation, or at $\theta = 45° + (-22.5°) = 22.5°$ as shown in Fig. 5-8(b). From Eqs. (5-8),

$$\sigma_{x'x'} = \frac{1000 + (-3000)}{2} + \frac{1000 - (-3000)}{2} \cos [2(22.5°)]$$

$$+ (-2000) \sin [2(22.5°)] = -1000 \text{ psi}$$

$$\sigma_{y'y'} = \sigma_{x'x'} = -1000 \text{ psi}$$

$$\sigma_{x'y'} = -\frac{1000 - (-3000)}{2} \sin [2(22.5°)]$$

$$+ (-2000) \cos [2(22.5°)] = -2828 \text{ psi}$$

The negative sign of $\sigma_{x'y'}$ means that the shear stress on the $x'x'$ face is directed as shown in Fig. 5-8(b), since this would cause the element to rotate clockwise about some internal point.

d. Mohr's circle is shown in Fig. 5-8(c) and clearly indicates that the shear stress is zero on the principal planes and that the maximum shear stress results when the element is rotated 45° from the principal stress orientation. Recall that a counterclockwise element rotation θ is represented on Mohr's circle with a counterclockwise rotation 2θ. ∎

Strain at a Point

Strain like stress generally varies from one point to another in a body under load. Consider a square element whose sides are of unit length as shown in Fig. 5-9(a). Under the action of some external loading, the element will deform such that the sides of the square are no longer perpendicular. In Fig. 5-9(b) the element is shown after such a deformation. The normal strains ϵ_{xx} and ϵ_{yy} are defined with the help of Fig. 5-9(b) as

$$\epsilon_{xx} = x'_A - x_A \tag{5-13a}$$

and

$$\epsilon_{yy} = y'_B - y_B \tag{5-13b}$$

Recall that these definitions are based on the fact that OA and OB are both of unit length before the deformation. Also, from this figure it is seen that the shear strain ϵ_{xy} is defined as

$$\epsilon_{xy} = \gamma_1 + \gamma_2 \tag{5-13c}$$

These normal and shear strains are all shown with a positive sense. Hence, the shear strain ϵ_{xy} is positive when the angle BOA becomes smaller than $\pi/2$. Note that two subscripts are again used: for normal strains the two subscripts are identical. It is seen, therefore, that positive normal and shear stresses will give positive normal

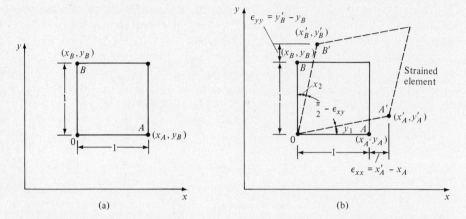

Figure 5-9 Definition of normal and shear strains. (a) Unstrained element with sides of unit length and (b) distorted element.

and shear strains, retaining the notion that positive normal stresses place the element in tension. Note also that positive shear stresses will cause the element to deform as shown in Fig. 5-9(b). All of the concepts of stress transformation and principal stresses applies here for the strain provided that σ_{xy} is replaced with $\epsilon_{xy}/2$ [5]. Obviously, strain is dimensionless, but it is customary to report strain in units of inch per inch (in./in.), meter per meter (m/m), or as microstrain (μs) which is 10^{-6} in./in.

Strain-Displacement Relations

In stress analysis, the Lagrangian point of view is usually adopted in that we follow a point in the body as it moves by virtue of the loading and subsequent deformation. An arbitrary point P therefore is said to undergo displacements in the x, y, and z directions, which are denoted as u, v, and w. This is in contrast to the Eulerian point of view, which is much more convenient in problems involving fluid motion, whereby we concentrate on a fixed point and observe the fluid velocities and pressure at this point. In any event, the strains and displacements are not independent but rather are related for small deformations in two-dimensional problems as

$$\epsilon_{xx} = \frac{\partial u}{\partial x} \qquad \epsilon_{yy} = \frac{\partial v}{\partial y} \qquad \epsilon_{xy} = \frac{\partial u}{\partial y} + \frac{\partial v}{\partial x} \qquad \textbf{(5-14)}$$

and in three-dimensional problems as

$$\epsilon_{xx} = \frac{\partial u}{\partial x} \qquad \epsilon_{yy} = \frac{\partial v}{\partial y} \qquad \epsilon_{zz} = \frac{\partial w}{\partial z}$$

$$\epsilon_{xy} = \frac{\partial u}{\partial y} + \frac{\partial v}{\partial x} \qquad \epsilon_{yz} = \frac{\partial w}{\partial y} + \frac{\partial v}{\partial z} \qquad \epsilon_{zx} = \frac{\partial u}{\partial z} + \frac{\partial w}{\partial x} \qquad \textbf{(5-15)}$$

Equations (5-14) and (5-15) are known as the strain-displacement relations or the kinematic relations. Relations similar to those given in Eq. (5-14) hold for problems in axisymmetric stress analysis, although it will be seen in Chapter 7 that an additional relation must be satisfied (see Problem 5-8). It should be emphasized that these equations hold only for small deformations. Stress analysis involving large deformations is beyond the scope of this text.

It will prove to be convenient in Sec. 5-7 and in most of Chapter 7 to write Eq. (5-14) [and Eq. (5-15)] in matrix form as

$$\boldsymbol{\epsilon} = \mathbf{L}\mathbf{u} \tag{5-16}$$

where in two dimensions, we define the strain vector $\boldsymbol{\epsilon}$ as

$$\boldsymbol{\epsilon} = [\epsilon_{xx} \quad \epsilon_{yy} \quad \epsilon_{xy}]^T \tag{5-17}$$

the linear-operator matrix $\mathbf{L}$ as

$$\mathbf{L} = \begin{bmatrix} \dfrac{\partial}{\partial x} & 0 \\[2mm] 0 & \dfrac{\partial}{\partial y} \\[2mm] \dfrac{\partial}{\partial y} & \dfrac{\partial}{\partial x} \end{bmatrix} \tag{5-18}$$

and the displacement vector $\mathbf{u}$ as

$$\mathbf{u} = [u \quad v]^T \tag{5-19}$$

The reader should determine $\mathbf{L}$ in three dimensions (see Problem 5-9) given that $\boldsymbol{\epsilon}$ and $\mathbf{u}$ are defined as

$$\boldsymbol{\epsilon} = [\epsilon_{xx} \quad \epsilon_{yy} \quad \epsilon_{zz} \quad \epsilon_{xy} \quad \epsilon_{yz} \quad \epsilon_{zx}]^T \tag{5-20}$$

and

$$\mathbf{u} = [u \quad v \quad w]^T \tag{5-21}$$

Note that $\boldsymbol{\epsilon}$ is written in Eq. (5-20) in column vector form instead of its symmetrical tensor form because the former proves to be very convenient in finite element analysis as shown later in this chapter and throughout Chapter 7. The matrix $\mathbf{L}$ is seen to be a *linear-operator matrix*; it may also be referred to as the strain-displacement matrix.

The Compatibility Equation

The strain components themselves are not independent but rather are related in two-dimensional problems undergoing small deformations by

$$\frac{\partial^2 \epsilon_{xx}}{\partial y^2} + \frac{\partial^2 \epsilon_{yy}}{\partial x^2} = \frac{\partial^2 \epsilon_{xy}}{\partial x \, \partial y} \tag{5-22}$$

The relationship stems from the fact that there are two displacements (in two dimensions) and three strains. Therefore, the strains must be related to the displacements with the relationship given by Eq. (5-22).

In Problem 5-10, the reader is asked to show that Eq. (5-22) holds and in Problem 5-11 to state the six compatibility equations that must hold in three dimensions. Since these equations are automatically satisfied in the stiffness approach to FEM, the three-dimensional form of these equations is not explicitly stated here.

A Constitutive Relationship—Hooke's Law

As mentioned, only linear elastic materials that undergo small deformations are considered in this text. It is assumed that the deformations do not result in strains that would cause the material to exceed the proportional limit; that is, the material is assumed to remain within the elastic limits of Hooke's law. It should be recalled that in the most simple form, Hooke's law states that the normal stress σ is proportional to the normal strain ϵ in a uniaxial state of stress, or

$$\sigma = E\epsilon \qquad \qquad \textbf{(5-23)}$$

where E, the proportionality constant, is known as Young's modulus, the elastic modulus, or the modulus of elasticity. The reader will recall that this important material constant was used in the two-dimensional truss model developed in Chapter 3. Note that since the strain ϵ is dimensionless, E has units of newtons per square meter (N/m^2) or pounds per square inch (psi), which are the same as the units of stress.

Equation (5-23) is more appropriately referred to as a *constitutive relationship* as opposed to a law. A *law* in engineering typically holds for all conditions and under all circumstances (barring relativistic effects). Examples of laws of nature are the conservation of mass, the first and second laws of thermodynamics, and Newton's laws of motion. Examples of constitutive relationships are Fourier's law of heat conduction, Newton's viscosity law, Fick's law of diffusion, and Hooke's law. In some texts constitutive relationships are referred to as particular laws. Note that these relationships only hold for a certain class of materials operating in some limited range. However, without these relationships it would not be possible to ascertain stress distributions, temperature distributions, velocity and pressure distributions, and so forth. The reason for this is simple: constitutive relationships are used to bring the total number of equations used in a given model up to the total number of unknowns.

For example, in three-dimensional stress analysis, there are fifteen unknowns: three displacements, six stresses, and six strains. With the constitutive relationship, there are a total of fifteen equations: three equilibrium equations, six strain-displacement equations, and six constitutive (stress-strain) relationships. The compatibility equations are derived from the strain-displacement relations and, therefore, are automatically satisfied if the other fifteen equations and boundary conditions are satisfied. In other words, statically indeterminate problems become solvable largely because of the use of some relevant constitutive relationship that describes the material behavior. The importance of constitutive relationships cannot be over-emphasized.

Hooke's law in the form of Eq. (5-23) is of rather limited use. A much more general form is given by [6]

$$\boldsymbol{\sigma} = \mathbf{D}(\boldsymbol{\epsilon} - \boldsymbol{\epsilon}_0) + \boldsymbol{\sigma}_0 \tag{5-24}$$

where $\mathbf{D}$ is referred to as the material property matrix, $\boldsymbol{\sigma}$ the stress vector, $\boldsymbol{\epsilon}$ the strain vector, $\boldsymbol{\epsilon}_0$ the self-strain vector, and $\boldsymbol{\sigma}_0$ the initial (or residual) stress vector. In three dimensions, $\mathbf{D}$ is a 6×6 matrix and each of the remaining column vectors is of size 6×1. Specific $\mathbf{D}$ matrices are given in Chapter 7 for plane stress, plane strain, axisymmetric stress, and three-dimensional stress analyses. In three-dimensional analysis, we define $\boldsymbol{\sigma}$, $\boldsymbol{\epsilon}$, $\boldsymbol{\sigma}_0$, and $\boldsymbol{\epsilon}_0$ as follows:

$$\boldsymbol{\sigma} = [\sigma_{xx} \quad \sigma_{yy} \quad \sigma_{zz} \quad \sigma_{xy} \quad \sigma_{yz} \quad \sigma_{zx}]^T \tag{5-25}$$

$$\boldsymbol{\epsilon} = [\epsilon_{xx} \quad \epsilon_{yy} \quad \epsilon_{zz} \quad \epsilon_{xy} \quad \epsilon_{yz} \quad \epsilon_{zx}]^T \tag{5-26}$$

$$\boldsymbol{\sigma}_0 = [\sigma_{xx_0} \quad \sigma_{yy_0} \quad \sigma_{zz_0} \quad \sigma_{xy_0} \quad \sigma_{yz_0} \quad \sigma_{zx_0}]^T \tag{5-27}$$

$$\boldsymbol{\epsilon}_0 = [\epsilon_{xx_0} \quad \epsilon_{yy_0} \quad \epsilon_{zz_0} \quad \epsilon_{xy_0} \quad \epsilon_{yz_0} \quad \epsilon_{zx_0}]^T \tag{5-28}$$

We have already reviewed the stress and strain components contained in $\boldsymbol{\sigma}$ and $\boldsymbol{\epsilon}$. However, $\boldsymbol{\sigma}_0$ and $\boldsymbol{\epsilon}_0$ deserve comment.

The initial stress vector $\boldsymbol{\sigma}_0$ represents prestresses that are known to exist in a material before it is loaded. The finite element method (or any method) cannot predict these. They must be specified by the analyst.

The self-strain vector $\boldsymbol{\epsilon}_0$ may be a result of crystal growth, shrinkage, or, most commonly, temperature changes. Various forms of $\boldsymbol{\epsilon}_0$ are presented in Chapter 7 for the different types of stress analysis. The material property matrix $\mathbf{D}$ is symmetric for both isotropic and anisotropic materials. In the isotropic case, the material property matrix will be seen to contain only two independent material properties (such as Young's modulus E and Poisson's ratio μ).

Surface Tractions

A *surface traction*, or simply traction, is defined to be a distributed external or surface loading per unit area. In this text, the surface traction vector is denoted by $\mathbf{s}$, which may be written in vector form in three dimensions as

$$\mathbf{s} = [s_x \quad s_y \quad s_z]^T \tag{5-29}$$

where s_x, s_y, and s_z are the forces per unit area in the x, y, and z directions, respectively. In two-dimensional applications, we simply take $s_z = 0$. An example of a surface traction is the hydrostatic pressure (a force per unit area) on the water-filled side of a dam. Note that in Eq. (5-29) only a single subscript is used to denote the component of the traction.

For the two-dimensional situation depicted in Fig. 5-10(a), note that $s_x = \sigma_{xx}$ and $s_y = \sigma_{xy}$ because the plane over which the surface traction acts happens to be parallel to the y axis. In a similar fashion, in Fig. 5-10(b), we have $s_x = \sigma_{yx}$ and $s_y = \sigma_{yy}$. On nonplanar surfaces or surfaces that are not parallel to one of the coordinate axes, no such simple relationships exist between the surface tractions and the stresses acting on the surface.

5-3   PRINCIPLE OF MINIMUM POTENTIAL ENERGY

The principle of minimum total potential energy [7] may be stated as follows: out of all the possible displacement fields that satisfy the geometric boundary conditions (i.e., the prescribed displacements), the one that also satisfies the equations of static equilibrium results in the minimum total potential energy of the structure (or body). This has been proved rigorously by Sokolnikoff [8]. The *total potential energy* Π is defined here to be the sum of two different types of potential energy: that associated with the internal potential energy U_i (i.e., the so-called strain energy) and that associated with the external potential energy U_e from the external forces that act on the system, or

$$\Pi = U_i + U_e \tag{5-30}$$

But for conservative force systems the loss in the external potential energy during the loading process must be equal to the work done, W_e, on the system by the external forces, or $-U_e = +W_e$, and Eq. (5-30) becomes

$$\Pi = U_i - W_e \tag{5-31}$$

The principle of minimum total potential energy requires that Π be a minimum for stable equilibrium. Furthermore, since Π is a function of functions (the strain and displacement functions), it is a functional. The minimization of the total potential

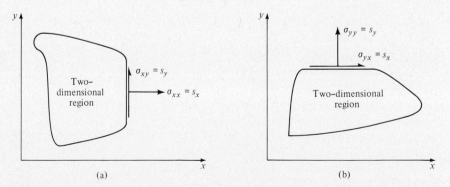

Figure 5-10   Relationship between surface tractions and stresses on surface of two-dimensional body for two special cases: (a) surface parallel to y axis; (b) surface parallel to x axis.

energy is a problem in variational calculus. In order to minimize Π, we insist that the first variation of the total potential energy be zero, that is,

$$\delta \Pi = \delta U_i - \delta W_e = 0 \tag{5-32}$$

or

$$\delta U_i = \delta W_e \tag{5-33}$$

For convenience, a cartesian reference frame is assumed with x, y, and z axes. The term δU_i represents the first variation in the strain energy (as a result of a variation in the strain) or

$$\delta U_i = \int_V (\sigma_{xx} \delta \epsilon_{xx} + \sigma_{yy} \delta \epsilon_{yy} + \sigma_{zz} \delta \epsilon_{zz}$$
$$+ \sigma_{xy} \delta \epsilon_{xy} + \sigma_{yz} \delta \epsilon_{yz} + \sigma_{zx} \delta \epsilon_{zx}) \, dV \tag{5-34a}$$

Equation (5-34a) can be written much more concisely using the matrix notation from Sec. 5-2:

$$\delta U_i = \int_V (\delta \boldsymbol{\epsilon})^T \boldsymbol{\sigma} \, dV \tag{5-34b}$$

where $\delta \boldsymbol{\epsilon}$ represents the first variation of the strain vector. The integral in Eqs. (5-34) is said to be a volume integral because the integration is to be performed over the volume of the body or structure being analyzed.

In considering the variation in the work done, δW_e, by the external forces, we must allow for the following possibilities: a body force $\mathbf{b}$ (per unit volume), a surface traction $\mathbf{s}$, and up to N point loads $\mathbf{f}_p$. The first variation of the work done by these external forces (as a result of a variation in the displacements) is given by

$$\delta W_e = \int_V (b_x \, \delta u + b_y \, \delta v + b_z \, \delta w) \, dV + \int_S (s_x \, \delta u + s_y \, \delta v$$
$$+ s_z \, \delta w) \, dS + \sum_{p=1}^{N} (f_{p_x} \, \delta u + f_{p_y} \, \delta v + f_{p_z} \, \delta w) \tag{5-35a}$$

where the first term represents a volume integral and the second term a surface integral. As the name suggests, a surface integral must be evaluated only on the surface S of the body or structure being analyzed. With the help of matrix notation, Eq. (5-35a) may be written as

$$\delta W_e = \int_V (\delta \mathbf{u})^T \mathbf{b} \, dV + \int_S (\delta \mathbf{u})^T \mathbf{s} \, dS + \sum_{p=1}^{N} (\delta \mathbf{u})^T \mathbf{f}_p \tag{5-35b}$$

where the body force vector $\mathbf{b}$ is given by

$$\mathbf{b} = [b_x \quad b_y \quad b_z]^T \tag{5-36}$$

the surface traction vector $\mathbf{s}$ by

$$\mathbf{s} = [s_x \quad s_y \quad s_z]^T \tag{5-37}$$

and the point load vector $\mathbf{f}_p$ by

$$\mathbf{f}_p = [f_{p_x} \quad f_{p_y} \quad f_{p_z}]^T \tag{5-38}$$

Obviously the vector $(\delta\mathbf{u})^T$ represents the frist variation in the displacement vector, or

$$(\delta\mathbf{u})^T = [\delta u \quad \delta v \quad \delta w] \tag{5-39}$$

where u, v, and w are the displacements in the x, y, and z directions, respectively. With the help of Eqs. (5-34) and (5-35), we may write Eq. (5-33) as

$$\int_V (\delta\boldsymbol{\epsilon})^T \boldsymbol{\sigma} \, dV = \int_V (\delta\mathbf{u}^T)\mathbf{b} \, dV + \int_S (\delta\mathbf{u})^T \mathbf{s} \, dS + \sum_{p=1}^{N} (\delta\mathbf{u})^T \mathbf{f}_p \tag{5-40}$$

Let us now assume a linear elastic material and eliminate $\boldsymbol{\sigma}$ by invoking the constitutive (or stress-strain) relationship given by Eq. (5-24) and rearranging the result (see Problem 5-18) to get

$$\int_V (\delta\boldsymbol{\epsilon})^T \mathbf{D}\boldsymbol{\epsilon} \, dV = \int_V (\delta\boldsymbol{\epsilon})^T \mathbf{D}\boldsymbol{\epsilon}_0 \, dV - \int_V (\delta\boldsymbol{\epsilon})^T \boldsymbol{\sigma}_0 \, dV$$

$$+ \int_V (\delta\mathbf{u})^T \mathbf{b} \, dV + \int_S (\delta\mathbf{u})^T \mathbf{s} \, dS + \sum_{p=1}^{N} (\delta\mathbf{u})^T \mathbf{f}_p \tag{5-41}$$

This equation looks formidable but the reader need not despair. This result will be the starting point in Sec. 5-7, where it is used to obtain the finite element characteristics for essentially all problems in static, linear elasticity.

In Problem 5-19 the reader is asked to show that the total potential energy Π must be given by

$$\Pi = \tfrac{1}{2}\int_V \boldsymbol{\epsilon}^T \mathbf{D}\boldsymbol{\epsilon} \, dV - \int_V \boldsymbol{\epsilon}^T \mathbf{D}\boldsymbol{\epsilon}_0 \, dV + \int_V \boldsymbol{\epsilon}^T \boldsymbol{\sigma}_0 \, dV$$

$$- \int_V \mathbf{u}^T \mathbf{b} \, dV - \int_S \mathbf{u}^T \mathbf{s} \, dS - \sum_{p=1}^{N} \mathbf{u}^T \mathbf{f}_p \tag{5-42}$$

The first three terms represent the strain energy as a result of the strains, the self-strains, and the prestresses, respectively. The last three terms represent the work done on the structure by the body forces, the surface tractions, and the point loads, respectively.

Example 5-2

Use the principle of minimum total potential energy to determine the relationship between the elongation Δ and an axial force P in a uniaxial stress member of uniform cross-sectional area A, length L, and modulus of elasticity E. One end of the bar is restrained and the load P is applied to the free end, as shown in Fig. 5-11.

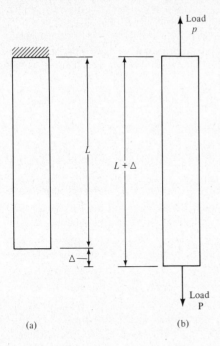

Figure 5-11   Uniaxial stress member of length L in (a) unloaded condition and (b) loaded with axial force P.

Solution

Although it is possible to start with Eq. (5-41), it is easier in this example to use Eq. (5-42) directly by taking

$$\epsilon_0 = 0 \qquad \sigma_0 = 0 \qquad b = 0 \qquad s = 0$$

or

$$\Pi = \tfrac{1}{2} E\epsilon^2 AL - P\Delta$$

since D is simply E for the state of uniaxial stress. The first term in the expression for Π represents the strain energy and the second, the work done by the point load P in moving a distance Δ. Noting that $\epsilon = \Delta/L$ gives

$$\Pi = \tfrac{1}{2} E\left(\frac{\Delta}{L}\right)^2 AL - P\Delta$$

Since we want to determine the elongation Δ for equilibrium and hence the Δ that minimizes Π, we compute $d\Pi/d\Delta$, set the result to zero, solve for Δ, and get

$$\Delta = \frac{PL}{AE}$$

which is a very well-known result. In fact, this result was used in the truss for-mulation in Chapter 3 where the direct approach was illustrated. In Problem 5-20, the reader is asked to show that the potential energy has in fact been minimized by examining the sign on $d^2\Pi/d\Delta^2$.   ∎

5-4   PRINCIPLE OF VIRTUAL DISPLACEMENTS

The principle of virtual displacements is actually a special case of the more general principle of virtual work. Recall from elementary mechanics that *work* is the product of a displacement and the component of the force in the direction of the displacement. *Virtual work* is imagined to occur in one of two ways: (1) when the forces are real and the displacements are virtual (imagined) or (2) when the displacements are real and the forces are virtual. The *principle of virtual displacements* applies to the former, and the *principle of virtual forces* to the latter. One statement of the principle of virtual displacements is the following. If the work done by the external forces on a structural system is equal to the increase in strain energy of the system for any set of admissible virtual displacements, then the system is in equilibrium. By admissable virtual displacements we mean those that satisfy the prescribed dis-placement boundary conditions but are otherwise arbitrary. Before actually devel-oping this principle, let us illustrate it in a simple example.

Example 5-3

Use the principle of virtual displacements to derive an expression for the elongation Δ in a uniaxial stress member of constant cross-sectional area A, length L, and elastic modulus E if a load P is applied to the free end while the other end is completely restrained as shown in Fig. 5-11.

Solution

From Fig. 5-11 it is seen that the work done by the force P is $P\Delta$. Note that this is not given by $P\Delta/2$ because the load P is assumed to be at its full value during the loading (and elongation) process. Note further that the virtual displacement is entirely arbitrary and is purposely (and conveniently) taken to be equal to the actual elongation Δ! Similarly, the increase in the strain energy is given by $\sigma\epsilon V$, or

$$\sigma\epsilon V = (E\epsilon)\epsilon AL = AEL\left(\frac{\Delta}{L}\right)^2$$

where $\epsilon = \Delta/L$ has been used. The principle of virtual displacements states that

$$AEL\left(\frac{\Delta}{L}\right)^2 = P\Delta$$

whereupon solving for Δ, we get

$$\Delta = \frac{PL}{AE}$$

which again is the desired and well-known result. As mentioned in Example 5-2, this result was the basis for the direct approach illustrated by the two-dimensional truss model in Chapter 3. ∎

Let us now develop the appropriate mathematical form of the principle of virtual displacements. For reasons that will become evident at the end of this section, let us denote the virtual displacements in the x, y, and z directions as δu, δv, and δw, respectively. It should be emphasized that δu does not represent the variation in u but instead is the virtual displacement as defined above. We must recognize that the virtual displacements will cause virtual straining, with the virtual strains to be denoted as $\delta \epsilon_{xx}$, $\delta \epsilon_{yy}$, etc. Let us further consider the external loads as a result of body force **b**, surface tractions **s**, and N point loads $\mathbf{f}_p$ defined as

$$\mathbf{b} = [b_x \quad b_y \quad b_z]^T \tag{5-36}$$

$$\mathbf{s} = [s_x \quad s_y \quad s_z]^T \tag{5-37}$$

and

$$\mathbf{f}_p = [f_{p_x} \quad f_{p_y} \quad f_{p_z}]^T \tag{5-38}$$

The principle of virtual displacements may now be stated mathematically [9] as

$$\int_V (\sigma_{xx} \, \delta\epsilon_{xx} + \sigma_{yy} \, \delta\epsilon_{yy} + \sigma_{zz} \, \delta\epsilon_{zz} + \sigma_{xy} \, \delta\epsilon_{xy} + \sigma_{yz} \, \delta\epsilon_{yz} + \sigma_{zx} \, \delta\epsilon_{zx}) \, dV$$
$$= \int_V (b_x \, \delta u + b_y \, \delta v + b_z \, \delta w) \, dV + \int_S (s_x \, \delta u + s_y \, \delta v + s_z \, \delta w) \, dS$$
$$+ \sum_{p=1}^{N} (f_{p_x} \, \delta u + f_{p_y} \, \delta v + f_{p_z} \, \delta w)$$

or, more concisely, in matrix form as

$$\int_V (\delta\boldsymbol{\epsilon})^T \boldsymbol{\sigma} \, dV = \int_V (\delta\mathbf{u})^T \mathbf{b} \, dV + \int_S (\delta\mathbf{u})^T \mathbf{s} \, dS + \sum_{p=1}^{N} (\delta\mathbf{u})^T \mathbf{f}_p \tag{5-43}$$

It is seen that this is identical to the result given by Eq. (5-40), which was obtained by the principle of minimum potential energy. Equation (5-43) [or (5-40)] is known as the *weak form* of the equilibrium equations, because the equilibrium equations are seen to contain second derivatives of the displacements (after Hooke's law is invoked), whereas Eq. (5-43) [or (5-40)] contains only first derivatives. Furthermore, these two equations are valid for nonlinear as well as linear stress-strain (constitutive) relationships. Obviously, if Eq. (5-43) is applied to a linear elastic material with the constitutive relationship given by Eq. (5-24), then Eq. (5-41) results.

It is seen that the principles of minimum potential energy and of virtual displacements give identical results if the virtual displacements (and strains) are associated with the variations in the displacements (and strains), and vice versa. The reason for representing the first variation of each displacement in the same manner as a virtual displacement should now be obvious. In effect, we have proven the equivalence of the principles of minimum potential energy and of virtual displacements.

5-5   THE FINITE ELEMENT BASIS

In Secs. 5-3 and 5-4 the principles of minimum total potential energy and of virtual displacements were used to develop a very important relationship among the internal stresses (and strains) and the external loads. Both approaches gave virtually the same result. In any event, volume and surface integrals arose that looked rather formidable. However, by discretizing the region to be analyzed into a number of suitable finite elements, as shown in Fig. 5-12, the integrations may be performed with relative ease; for example,

$$\int_V (\delta\boldsymbol{\epsilon})^T \boldsymbol{\sigma} \, dV = \sum_{e=1}^{M} \int_{V^e} (\delta\boldsymbol{\epsilon})^T \boldsymbol{\sigma} \, dV \qquad \textbf{(5-44)}$$

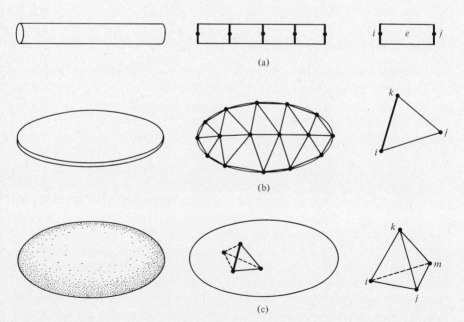

(a)

(b)

(c)

Figure 5-12   Discretization in one, two, and three dimensions with (a) the two-node lineal element, (b) the three-node triangular element, and (c) the four-node tetrahedral element.

and

$$\int_S (\delta\mathbf{u})^T \mathbf{s} \, dS = \sum_{e=1}^{M} \int_{S^e} (\delta\mathbf{u})^T \mathbf{s} \, dS \tag{5-45}$$

where V^e and S^e serve as a reminder that the integrations are to be performed over the element volume and element boundary, respectively. Note that the results from the M nonoverlapping elements are, in effect, added together to obtain the integrals over the original body or region. As we have seen several times before, the summation is not usually carried in the mathematical manipulations because we simply consider it to represent the assemblage step. If Eqs. (5-40) and (5-43) hold over the entire body or region and the region is arbitrarily selected by the analyst (such as the selection of a free body), then these equations must also hold on an element basis, or

$$\int_{V^e} (\delta\boldsymbol{\epsilon})^T \boldsymbol{\sigma} \, dV = \int_{V^e} (\delta\mathbf{u})^T \mathbf{b} \, dV + \int_{S^e} (\delta\mathbf{u})^T \mathbf{s} \, dS + \sum (\delta\mathbf{u})^T \mathbf{f}_p \tag{5-46}$$

where $\mathbf{f}_p$ must be interpreted to be the point loads acting at each node of the element in addition to any other imposed loads acting within the element. The notations V^e and S^e represent the element volume and element boundary or surface, respectively. On internal nodes where no external loading is to be applied, $\mathbf{f}_p$ from one element will cancel (during the assemblage step) that from the adjacent element(s) that shares the same nodes. These same comments hold for the surface traction term (see Sec. 5-8). In other words, although the integral over S^e is nonzero in general on the element boundaries for element e, the contribution to the assemblage nodal force vector will be seen to be effectively zero. The reason for this is that the contribution from the neighboring element will always be of the same magnitude but of the opposite sign (by Newton's third law) unless an external load is explicitly applied there.

Returning to Fig. 5-12, it is seen that the two-node lineal element may be used effectively in one-dimensional problems, the three-node triangular element in two-dimensional (and axisymmetric) problems, and the four-node tetrahedral element in three-dimensional problems. In the next section, the shape function matrix $\mathbf{N}$ is developed in terms of the shape functions N_i, N_j, etc., associated with each of the nodes.

5-6 THE SHAPE FUNCTION MATRIX

Recall from Chapter 4 that the field variable in the finite element method is routinely represented by an interpolating polynomial on a piecewise continuous basis. Let us restrict the present discussion to applications that require only C^0-continuity such as plane stress, plane strain, axisymmetric stress, and three-dimensional stress (but not beam or plate bending or shell analysis). The field variables in the stiffness-based finite element method are always the displacements u, v, and w.

Two-Node Lineal Element

It is instructive to review what happened in Chapter 4 where a scalar field variable was represented in one-dimension with the lineal element. If we now interpret this scalar field variable to be the displacement u in the case of one-dimensional stress analysis, we may take the interpolating polynomial

$$u = c_1 + c_2 x \tag{5-47}$$

to represent this displacement within a typical element. In fact, Eq. (5-47) is referred to as a *displacement function* in particular and as a *parameter function* in general. If we insist that at node i we have $u(x_i) = u_i$ and at node j, $u(x_j) = u_j$, then we can find $N_i(x)$ and $N_j(x)$ such that

$$u = N_i(x)u_i + N_j(x)u_j \tag{5-48}$$

where u_i and u_j are the displacements at nodes i and j, respectively. In Chapter 4 we found it convenient to represent this last result in the matrix form

$$u = \mathbf{N}\mathbf{a}^e \tag{5-49}$$

where $\mathbf{N}$, the shape function matrix, is defined as

$$\mathbf{N} = [N_i(x) \quad N_j(x)] \tag{5-50}$$

and $\mathbf{a}^e$, the vector of nodal unknowns, as

$$\mathbf{a}^e = [u_i \quad u_j]^T \tag{5-51}$$

Because the assumed displacement function is linear in x, the shape functions are also linear in x. Moreover, they must satisfy the following conditions:

$$
\begin{array}{ll}
N_i(x_i) = 1 & N_i(x_j) = 0 \\
N_j(x_i) = 0 & N_j(x_j) = 1
\end{array} \tag{5-52}
$$

Three-Node Triangular Element

In a similar fashion, it seems reasonable to represent the x component of displacement in two-dimensional (and axisymmetric) problems as

$$u = c_1 + c_2 x + c_3 y \tag{5-53}$$

if the three-node triangular element from Fig. 5-12 is used. Note that the triangle has three nodes and, not accidentally, the parameter function for the displacement (i.e., the displacement function) has three constants. Therefore, three shape functions should be expected for this element, or

$$u = N_i(x,y)u_i + N_j(x,y)u_j + N_k(x,y)u_k \tag{5-54}$$

where u_i, u_j, and u_k are the x components of the nodal displacements, and N_i, N_j, and N_k are the three shape functions. Because we insist that $u(x_i,y_i) = u_i$, $u(x_j,y_j) = u_j$, and $u(x_k,y_k) = u_k$, it should not come as any surprise that

$$
\begin{aligned}
N_i(x_i,y_i) &= 1 & N_j(x_i,y_i) &= 0 & N_k(x_i,y_i) &= 0 \\
N_i(x_j,y_j) &= 0 & N_j(x_j,y_j) &= 1 & N_k(x_j,y_j) &= 0 \\
N_i(x_k,y_k) &= 0 & N_j(x_k,y_k) &= 0 & N_k(x_k,y_k) &= 1
\end{aligned}
\tag{5-55}
$$

Moreover, because the assumed function for u given by Eq. (5-53) is linear in x and y, the shape functions themselves must also be linear in x and y. In two-dimensional problems, the y component of displacement v must also be considered, or

$$
v = c_4 + c_5 x + c_6 y
\tag{5-56}
$$

where three new constants are introduced. However, because the conditions expressed by Eq. (5-55) must also hold for the y component of displacement (why?), we may write

$$
v = N_i(x,y)v_i + N_j(x,y)v_j + N_k(x,y)v_k
\tag{5-57}
$$

where v_i, v_j, and v_k are the y components of nodal displacements. Let us now define **u** and $\mathbf{a}^e$ as

$$
\mathbf{u} = [u \quad v]^T
\tag{5-58}
$$

and

$$
\mathbf{a}^e = [u_i \quad v_i \ \vdots \ u_j \quad v_j \ \vdots \ u_k \quad v_k]^T
\tag{5-59}
$$

so that we may write

$$
\mathbf{u} = \mathbf{N}\mathbf{a}^e
\tag{5-60}
$$

where

$$
\mathbf{N} = \begin{bmatrix} N_i & 0 & \vdots & N_j & 0 & \vdots & N_k & 0 \\ 0 & N_i & \vdots & 0 & N_j & \vdots & 0 & N_k \end{bmatrix}
\tag{5-61}
$$

The shape function matrix **N** is seen to be of size 2×6, since each node has two degrees of freedom (the x and y components of displacement) and the element has three nodes. A typical shape function (that for node i) is plotted in the x-y plane in Fig. 5-13. Note how $N_i(x,y)$ varies from unity at node i to zero at nodes j and k in a linear fashion. The reader should make an attempt to plot $N_j(x,y)$ and $N_k(x,y)$ in a similar fashion (see Problem 5-22).

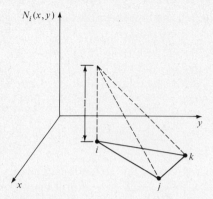

Figure 5-13   Plot of shape function for node i for the three-node triangular element (with C^0-continuity).

Four-Node Tetrahedral Element

Finally, it seems reasonable to represent the x component of displacement in three-dimensional problems by

$$u = c_1 + c_2 x + c_3 y + c_4 z \qquad \textbf{(5-62)}$$

if the four-node tetrahedral element from Fig. 5-12 is used. Note now that the tetrahedral element has four nodes and the assumed displacement function has four constants. Therefore, four shape functions should be expected for this element, and the displacement function for the x component may be written as

$$u = N_i(x,y,z)u_i + N_j(x,y,z)u_j + N_k(x,y,z)u_k + N_m(x,y,z)u_m \qquad \textbf{(5-63)}$$

where u_i, u_j, u_k, and u_m are the x components of the nodal displacements, and N_i, N_j, N_k, and N_m are the four shape functions. Again we insist that $u(x_i,y_i,z_i) = u_i$, $u(x_j,y_j,z_j) = u_j$, and so forth. It should be obvious by now that we must have

$$\begin{aligned}
N_i(x_i,y_i,z_i) &= 1 & N_j(x_i,y_i,z_i) &= 0 & N_k(x_i,y_i,z_i) &= 0 & N_m(x_i,y_i,z_i) &= 0 \\
N_i(x_j,y_j,z_j) &= 0 & N_j(x_j,y_j,z_j) &= 1 & N_k(x_j,y_j,z_j) &= 0 & N_m(x_j,y_j,z_j) &= 0 \\
N_i(x_k,y_k,z_k) &= 0 & N_j(x_k,y_k,z_k) &= 0 & N_k(x_k,y_k,z_k) &= 1 & N_m(x_k,y_k,z_k) &= 0 \\
N_i(x_m,y_m,z_m) &= 0 & N_j(x_m,y_m,z_m) &= 0 & N_k(x_m,y_m,z_m) &= 0 & N_m(x_m,y_m,z_m) &= 1
\end{aligned} \qquad \textbf{(5-64)}$$

In a similar fashion, if we assume the displacement functions for the y and z components of displacement to be

$$v = c_5 + c_6 x + c_7 y + c_8 z \qquad \textbf{(5-65)}$$

and

$$w = c_9 + c_{10} x + c_{11} y + c_{12} z \qquad \textbf{(5-66)}$$

we can in turn use the same shape functions to write these as

$$v = N_i(x,y,z)v_i + N_j(x,y,z)v_j + N_k(x,y,z)v_k + N_m(x,y,z)v_m \quad \textbf{(5-67)}$$

and

$$w = N_i(x,y,z)w_i + N_j(x,y,z)w_j + N_k(x,y,z)w_k + N_m(x,y,z)w_m \quad \textbf{(5-68)}$$

where v_i, v_j, v_k, and v_m are the y components of the nodal displacements, etc. If we now define **u** and $\mathbf{a}^e$ as

$$\mathbf{u} = [u \quad v \quad w]^T \quad \textbf{(5-69)}$$

and

$$\mathbf{a}^e = [u_i \quad v_i \quad w_i \mid u_j \quad v_j \quad w_j \mid u_k \quad v_k \quad w_k \mid u_m \quad v_m \quad w_m]^T \quad \textbf{(5-70)}$$

we may write

$$\mathbf{u} = \mathbf{N}\mathbf{a}^e \quad \textbf{(5-71)}$$

where

$$\mathbf{N} = \begin{bmatrix} N_i & 0 & 0 & N_j & 0 & 0 & N_k & 0 & 0 & N_m & 0 & 0 \\ 0 & N_i & 0 & 0 & N_j & 0 & 0 & N_k & 0 & 0 & N_m & 0 \\ 0 & 0 & N_i & 0 & 0 & N_j & 0 & 0 & N_k & 0 & 0 & N_m \end{bmatrix} \quad \textbf{(5-72)}$$

Now the shape function matrix is seen to be of size 3×12, because each node has three degrees of freedom and the element has four nodes. A typical shape function, e.g., N_i will be seen in Chapter 6 to vary linearly from unity at the node in question (node i) to zero at each of the remaining nodes.

Properties of C^0-Continuous Shape Functions

For the one-, two-, and three-dimensional elements in Fig. 5-12, the shape functions satisfy the following three properties:

1. A shape function associated with a particular node (e.g., node i) must evaluate to unity if evaluated at the coordinates of this node (node i); all other shape functions must evaluate to zero at this same node. Actually, this property is contained in the next property.
2. A shape function associated with a particular node (e.g., node i) varies linearly from unity at the node in question (node i) to zero at each of the remaining nodes.
3. The sum of all the shape functions at any point within the element or on the element boundaries must sum to unity, or

$$\sum N_\beta = 1 \quad \textbf{(5-73)}$$

where the element is assumed to have n nodes and β takes on i, j, k, etc., as appropriate. In this text, whenever a shape function for an arbitrary node is referenced, the subscript β is used and the shape function is denoted as N_β.

The shape functions for the lineal element were derived in Chapter 4; in terms of the nomenclature in this chapter they are

$$N_i(x) = \frac{x_j - x}{x_j - x_i} \tag{5-74a}$$

and

$$N_j(x) = \frac{x - x_i}{x_j - x_i} \tag{5-74b}$$

The shape functions for the three-node triangular and four-node tetrahedral elements are derived in Chapter 6.

Example 5-4

Verify that the shape functions for the one-dimensional, two-node lineal element given by Eqs. (5-74) have the three properties given above.

Solution

First, note that we have

$$N_i(x_i) = \frac{x_j - x_i}{x_j - x_i} = 1 \qquad N_j(x_i) = \frac{x_i - x_i}{x_j - x_i} = 0$$

$$N_j(x_j) = \frac{x_j - x_i}{x_j - x_i} = 1 \qquad N_i(x_j) = \frac{x_j - x_j}{x_j - x_i} = 0$$

which verifies the first property. In fact, these conditions were summarized earlier in Eq. (5-52).

Next, note that $N_i(x)$ is unity at $x = x_i$ (i.e., at node i) and decreases linearly to zero at $x = x_j$ (i.e., at node j). Similarly, note that $N_j(x)$ is unity at $x = x_j$ (i.e., at node j) and decreases linearly to zero at $x = x_i$ (i.e., at node i). Thus the second property is verified.

Finally, note that

$$N_i + N_j = \frac{x_j - x}{x_j - x_i} + \frac{x - x_i}{x_j - x_i} = \frac{x_j - x + x - x_i}{x_j - x_i} = 1$$

for all x. Actually, because the shape functions apply to an element that connects node i at $x = x_i$ to node j at $x = x_j$, we should say that the above condition strictly holds for $x_i \leq x \leq x_j$ (assuming x_i is less than x_j). Thus the third and final property is verified.   ■

Throughout this text, the shape function matrix is referred to quite often. By gaining an appreciation of where the shape functions themselves come from, it is hoped that the development in the next section will be more readily understood. The next section picks up where we left off with Eq. (5-46), which will now be the starting point for the derivation of the finite element characteristics for all problems in static stress analysis. The reader may find it helpful to reread Sec. 5-2, with special attention given to the definitions of $\boldsymbol{\sigma}$, $\boldsymbol{\epsilon}$, $\boldsymbol{\sigma}_0$, $\boldsymbol{\epsilon}_0$, and $\mathbf{L}$.

5-7 THE FINITE ELEMENT CHARACTERISTICS

We are now in a position to formulate the finite element characteristics for all problems in static, linear stress analysis. Equation (5-46) provides the starting point and is restated here for convenience:

$$\int_{V^e} (\delta\boldsymbol{\epsilon})^T \boldsymbol{\sigma} \, dV = \int_{V^e} (\delta\mathbf{u})^T \mathbf{b} \, dV + \int_{S^e} (\delta\mathbf{u})^T \mathbf{s} \, dS + \sum (\delta\mathbf{u})^T \mathbf{f}_p \quad \textbf{(5-46)}$$

Recall that this equation was derived in two seemingly different, but completely equivalent, ways: first from the principle of minimum total potential energy and then from the principle of virtual displacements. In what follows, the symbol $\delta\mathbf{u}$ may represent either the variation on the displacements or the virtual displacements, depending on the reader's preference. Similar comments apply to $\delta\boldsymbol{\epsilon}$ and $\delta\mathbf{a}^e$.

Let us begin by relating the strain vector $\boldsymbol{\epsilon}$ to the vector of nodal unknowns $\mathbf{a}^e$ (i.e., the nodal displacements) by using Eqs. (5-16) and (5-60) [or (5-71)], or

$$\boldsymbol{\epsilon} = \mathbf{L}\mathbf{u} = \mathbf{L}\mathbf{N}\mathbf{a}^e \quad \textbf{(5-75)}$$

Let us refer to the matrix $\mathbf{L}\mathbf{N}$ as the *strain-nodal displacement matrix* and denote it by $\mathbf{B}$, or

$$\mathbf{B} = \mathbf{L}\mathbf{N} \quad \textbf{(5-76)}$$

Therefore, Eq. (5-75) may be written as

$$\boldsymbol{\epsilon} = \mathbf{B}\mathbf{a}^e \quad \textbf{(5-77)}$$

Recall from Sec. 5-2 that $\mathbf{L}$ is a linear operator matrix that contains first derivatives with respect to only x and y in problems requiring only C^0-continuity. If the shape function matrix $\mathbf{N}$ contains only linear functions (in x and y) such as those in Sec. 5-6, then the matrix $\mathbf{B}$ contains only constants. In any event, $\mathbf{B}$ does not contain any of the nodal displacements, which implies that

$$\delta\boldsymbol{\epsilon} = \delta(\mathbf{B}\mathbf{a}^e) = \mathbf{B} \, \delta\mathbf{a}^e \quad \textbf{(5-78)}$$

and

$$(\delta\boldsymbol{\epsilon})^T = (\mathbf{B} \, \delta\mathbf{a}^e)^T = (\delta\mathbf{a}^e)^T \mathbf{B}^T \quad \textbf{(5-79)}$$

Again depending on the reader's preference, $\delta\mathbf{a}^e$ may denote either the variation in the nodal displacements or the virtual nodal displacements. Similarly, we have

$$\delta\mathbf{u} = \delta(\mathbf{N}\mathbf{a}^e) = \mathbf{N} \, \delta\mathbf{a}^e \quad \textbf{(5-80)}$$

and

$$(\delta\mathbf{u})^T = (\mathbf{N} \, \delta\mathbf{a}^e)^T = (\delta\mathbf{a}^e)^T \mathbf{N}^T \quad \textbf{(5-81)}$$

Substituting the results from Eqs. (5-79) and (5-81) into Eq. (5-46), after transposing all the terms to the left-hand side and pulling the $(\delta \mathbf{a}^e)^T$ through the integrals and the summation, yields

$$(\delta \mathbf{a}^e)^T \left\{ \int_{V^e} \mathbf{B}^T \boldsymbol{\sigma} \, dV - \int_{V^e} \mathbf{N}^T \mathbf{b} \, dV - \int_{S^e} \mathbf{N}^T \mathbf{s} \, dS - \sum \mathbf{N}^T \mathbf{f}_p \right\} = 0 \quad \textbf{(5-82)}$$

But the term in braces must be zero because $(\delta \mathbf{a}^e)^T$ is not necessarily zero and is quite arbitrary. If the principle of minimum potential energy path is followed, then $\delta \mathbf{a}^e$ represents the first variation of nodal displacements. This variation is quite arbitrary, and so the term in braces must be zero. If the principle of virtual displacements is the basis for Eq. (5-46), then $\delta \mathbf{a}^e$ is arbitrary because it represents arbitrary virtual displacements. Actually, if the statements of these principles from Secs. 5-3 and 5-4 are read carefully, the variations in the displacements and the virtual displacements are not completely arbitrary because they must satisfy the prescribed displacement boundary conditions. In any event, $\delta \mathbf{a}^e$ is nonzero, in general, and again it is concluded that the term in braces is zero, or

$$\int_{V^e} \mathbf{B}^T \boldsymbol{\sigma} \, dV - \int_{V^e} \mathbf{N}^T \mathbf{b} \, dV - \int_{S^e} \mathbf{N}^T \mathbf{s} \, dS - \sum \mathbf{N}^T \mathbf{f}_p = 0 \quad \textbf{(5-83)}$$

Equation (5-83) is still quite general because a constitutive relationship has not yet been invoked. However, let us now consider only linear elastic materials for which the constitutive (or stress-strain) relationship from Sec. 5-2 applies, or

$$\boldsymbol{\sigma} = \mathbf{D}(\boldsymbol{\epsilon} - \boldsymbol{\epsilon}_0) + \boldsymbol{\sigma}_0 \quad \textbf{(5-24)}$$

where the material property matrix $\mathbf{D}$ contains pertinent material properties such as the elastic modulus and Poisson's ratio. Specific examples of this matrix as well as the self-strain vector $\boldsymbol{\epsilon}_0$ and the prestress vector $\boldsymbol{\sigma}_0$ are presented in Sec. 5-8 and Chapter 7. With the help of Eqs. (5-77) and (5-24), we may write Eq. (5-83) as

$$\left\{ \int_{V^e} \mathbf{B}^T \mathbf{D} \mathbf{B} \, dV \right\} \mathbf{a}^e - \int_{V^e} \mathbf{B}^T \mathbf{D} \boldsymbol{\epsilon}_0 \, dV + \int_{V^e} \mathbf{B}^T \boldsymbol{\sigma}_0 \, dV$$
$$- \int_{V^e} \mathbf{N}^T \mathbf{b} \, dV - \int_{S^e} \mathbf{N}^T \mathbf{s} \, dS - \sum \mathbf{N}^T \mathbf{f}_p = 0 \quad \textbf{(5-84)}$$

Note how the vector $\mathbf{a}^e$ was pulled out of the integral (to the right) because it is not a function of the spacial coordinates. Finally we get an equation of the form

$$\mathbf{K}^e \mathbf{a}^e = \mathbf{f}^e \quad \textbf{(5-85)}$$

where

$$\mathbf{f}^e = \mathbf{f}^e_{\epsilon_0} - \mathbf{f}^e_{\sigma_0} + \mathbf{f}^e_b + \mathbf{f}^e_s + \mathbf{f}^e_{\text{PL}} \quad \textbf{(5-86)}$$

and where we define the element stiffness matrix $\mathbf{K}^e$ by

$$\mathbf{K}^e = \int_{V^e} \mathbf{B}^T \mathbf{D} \mathbf{B} \, dV \quad \textbf{(5-87)}$$

and the five element nodal force vectors by

$$\mathbf{f}^e_{\epsilon_0} = \int_{V^e} \mathbf{B}^T \mathbf{D} \boldsymbol{\epsilon}_0 \, dV \quad \textbf{(5-88)}$$

$$\mathbf{f}^e_{\sigma_0} = \int_{V^e} \mathbf{B}^T \boldsymbol{\sigma}_0 \, dV \quad \textbf{(5-89)}$$

$$\mathbf{f}_b^e = \int_{V^e} \mathbf{N}^T \mathbf{b} \, dV \qquad \text{(5-90)}$$

$$\mathbf{f}_s^e = \int_{S^e} \mathbf{N}^T \mathbf{s} \, dS \qquad \text{(5-91)}$$

and

$$\mathbf{f}_{PL}^e = \Sigma \, \mathbf{N}^T \mathbf{f}_p \qquad \text{(5-92)}$$

For the uniaxial stress member, discretized with the two-node lineal element, the matrix $\mathbf{B}$ is of size 1×2 because $\mathbf{L}$ is a scalar and $\mathbf{N}$ is 1×2 (recall that $\mathbf{B} = \mathbf{LN}$). The matrix $\mathbf{D}$ is actually a scalar (or a 1×1 matrix). The element stiffness matrix $\mathbf{K}^e$ is of size 2×2 as expected, since there are only two nodes on each element and each node has only one degree of freedom. In a similar fashion, it may be shown that each of the five nodal force vectors is of size 2×1 (see Problem 5-24).

Similarly, for problems in two-dimensional stress analysis, analyzed with the three-node triangular element, the matrix $\mathbf{B}$ is of size 3×6 because $\mathbf{L}$ is 3×2 and $\mathbf{N}$ is 2×6. The matrix $\mathbf{D}$ is of size 3×3. The element stiffness matrix $\mathbf{K}^e$ is 6×6 as expected, because there are three nodes per element and each node has two degrees of freedom. Not surprisingly, each of the nodal force vectors must be of size 6×1 (see Problem 5-25).

Finally, for three-dimensional problems in stress analysis analyzed with the four-node tetrahedral element, it can be determined (see Problem 5-26) that $\mathbf{K}^e$ and $\mathbf{f}^e$ are of sizes 12×12 and 12×1, respectively.

In the next section these rather abstract notions are illustrated with a problem in one-dimensional stress analysis.

5-8 APPLICATION: TAPERED UNIAXIAL STRESS MEMBER

In this section the general expressions for the finite element characteristics given by Eqs. (5-87) to (5-92) are applied to a simple stress analysis problem—a tapered uniaxial stress member analyzed with the two-node lineal element. First the formulation is given for a general problem in this class and then it is illustrated numerically in an example for a specific problem.

Consider the tapered uniaxial stress member shown in Fig. 5-14(a). Note that the conditions on the ends of the member are quite arbitrary. In other words, the ends of the member may both be restrained (prescribed displacements), both loaded (via surface tractions and/or point loads), or a combination of both.

The member is discretized into a number of two-node lineal elements as shown in Fig. 5-14(b) with node i (at $x = x_i$) and node j (at $x = x_j$). Note that a node is associated with a planar surface and not a single point. The nodal displacement u_i occurs at $x = x_i$, while the nodal displacement u_j occurs at $x = x_j$.

A typical element e is shown in Fig. 5-14(c) with point loads f_i and f_j and/or surface tractions s_i and s_j acting at the nodes and with a body force $b = \gamma$ as a result of gravity, where γ is the weight density of the material. In addition, this

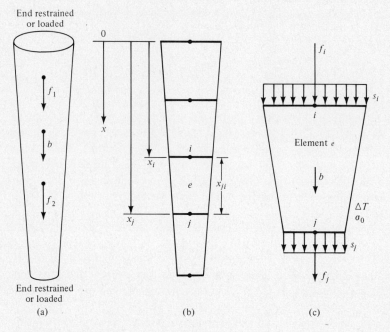

Figure 5-14   General one-dimensional stress problem: (a) uniaxial stress member, (b) discretized into several two-node lineal elements, and (c) typical element with positive point loads and tractions shown on nodal planes.

element is assumed to undergo a temperature change ΔT and to be prestressed to a stress of σ_0.

The Element Characteristics

From Eq. (5-50), the shape function matrix N is of size 1×2 where the shape functions $N_i(x)$ and $N_j(x)$ are given by Eqs. (5-74) and $\mathbf{N}$ itself by

$$\mathbf{N} = \begin{bmatrix} \dfrac{x_j - x}{x_j - x_i} & \dfrac{x - x_i}{x_j - x_i} \end{bmatrix} \tag{5-93}$$

But the strain-nodal displacement matrix $\mathbf{B}$ is related to $\mathbf{N}$ through the linear operator matrix $\mathbf{L}$ by $\mathbf{B} = \mathbf{LN}$, where $\mathbf{L} = d/dx$. This follows from the fact that the strain-displacement relationship for this class of problems is given by $\epsilon = du/dx$ and from the definition of $\mathbf{L}$, namely, $\epsilon = \mathbf{L}u$. Therefore,

$$\mathbf{B} = \frac{d\mathbf{N}}{dx} = \begin{bmatrix} \dfrac{-1}{x_j - x_i} & \dfrac{1}{x_j - x_i} \end{bmatrix} \tag{5-94}$$

The material property matrix **D** is really a scalar here as explained below. The stress-strain relationship for the state of uniaxial stress is given by

$$\sigma = E(\epsilon - \epsilon_0) + \sigma_0 \tag{5-95}$$

Comparing Eqs. (5-95) and (5-24) yields $D = E$. Note that each term in Eq. (5-95) is a scalar. This constitutive relationship is a somewhat more general form of Hooke's law ($\sigma = E\epsilon$) because it includes the possibilities of both self-strains ϵ_0 (e.g., as a result of a temperature change) and prestresses σ_0. The self-strain ϵ_0 is related to the temperature change ΔT by

$$\epsilon_0 = \alpha_t \Delta T \tag{5-96}$$

where α_t is the coefficient of thermal expansion, in units of inches per inch per degree Fahrenheit (in./in.-°F) or meters per meter per degree Celsius (m/m-°C).

We are now in a position to evaluate the element stiffness matrix $\mathbf{K}^e$ by beginning with Eq. (5-87) and writing

$$
\begin{aligned}
\mathbf{K}^e &= \int_{V^e} \mathbf{B}^T \mathbf{D} \mathbf{B} \, dV \\
&= \frac{1}{x_{ji}^2} \int_{x_i}^{x_j} \begin{bmatrix} -1 \\ 1 \end{bmatrix} E [-1 \quad 1] A(x) \, dx
\end{aligned} \tag{5-97}
$$

where $dV = A(x) \, dx$ has been used and A is the cross-sectional area of the element at a specified value of x. The notation x_{ji} is used to represent $x_j - x_i$ in a more concise manner and really represents the element length. If the constants in the integral are pulled through the integration, we get

$$\mathbf{K}^e = \frac{E}{x_{ji}^2} \begin{bmatrix} 1 & -1 \\ -1 & 1 \end{bmatrix} \int_{x_i}^{x_j} A(x) \, dx \tag{5-98}$$

One of the simplest ways to evaluate the remaining integral is to evaluate the area at $x = \bar{x}$, where

$$\bar{x} = \frac{x_i + x_j}{2}$$

denote it as $\bar{A}$, and treat $\bar{A}$ as a constant to get

$$\mathbf{K}^e = \frac{\bar{A}E}{x_{ji}} \begin{bmatrix} 1 & -1 \\ -1 & 1 \end{bmatrix} = \frac{AE}{x_{ji}} \begin{bmatrix} 1 & -1 \\ -1 & 1 \end{bmatrix} \tag{5-99}$$

The bar ($^-$) on the area A is dropped with the understanding that if the cross-sectional area varies with x, a constant value may be used that corresponds to the value at $x = \bar{x}$. Equation (5-99) provides an expression for the element stiffness matrix for the uniaxial stress member. Next we turn to the nodal force vectors.

Let us first determine the element nodal force vector as a result of the self-strain that happens to be caused by a temperature change ΔT. With the help of Eqs. (5-94), (5-96), and $D = E$, it can be shown from Eq. (5-88) that

$$
\mathbf{f}_{\epsilon 0}^e = \int_{V^e} \mathbf{B}^T \mathbf{D} \boldsymbol{\epsilon}_0 \, dV
$$

$$
= \frac{1}{x_{ji}} \int_{x_i}^{x_j} \begin{bmatrix} -1 \\ 1 \end{bmatrix} E \alpha_t \, \Delta T \, A \, dx \tag{5-100}
$$

$$
= AE\alpha_t \, \Delta T \begin{bmatrix} -1 \\ 1 \end{bmatrix}
$$

In Eq. (5-100) it is assumed that α_t and ΔT are constant in any one element. If they vary as a function of x, they may simply be evaluated at $x = \bar{x}$ and treated as constants in any one element. In a similar fashion, it can be shown that (see Problems 5-27 and 5-28)

$$
\mathbf{f}_{\sigma 0}^e = \int_{V^e} \mathbf{B}^T \boldsymbol{\sigma}_0 \, dV = \sigma_0 A \begin{bmatrix} -1 \\ 1 \end{bmatrix} \tag{5-101}
$$

and

$$
\mathbf{f}_b^e = \int_{V^e} \mathbf{N}^T \mathbf{b} \, dV = \frac{\gamma A x_{ji}}{2} \begin{bmatrix} 1 \\ 1 \end{bmatrix} \tag{5-102}
$$

where again σ_0 and γ are assumed to be constant in the element, or suitable average values (i.e., evaluated at $x = \bar{x}$) are used. The nodal force vector as a result of the surface tractions s_i and s_j may be determined from Eq. (5-91) as follows:

$$
\mathbf{f}_s^e = \int_{S^e} \mathbf{N}^T \mathbf{s} \, dS
$$

$$
= \mathbf{N}^T(x_i) \int_{A_i} s_i \, dA + \mathbf{N}^T(x_j) \int_{A_j} s_j \, dA
$$

$$
= s_i A_i \begin{bmatrix} 1 \\ 0 \end{bmatrix} + s_j A_j \begin{bmatrix} 0 \\ 1 \end{bmatrix} = \begin{bmatrix} s_i A_i \\ s_j A_j \end{bmatrix} \tag{5-103}
$$

where A_i and A_j are the cross-sectional areas at nodes i and j, respectively. Finally, the nodal force vector $\mathbf{f}_{PL}^e$ as a result of the point loads at nodes i and j is determined from Eq. (5-92) to be

$$
\mathbf{f}_{PL}^e = \Sigma \mathbf{N}^T \mathbf{f}_p = \mathbf{N}^T(x_i) f_i + \mathbf{N}^T(x_j) f_j
$$

$$
= f_i \begin{bmatrix} 1 \\ 0 \end{bmatrix} + f_j \begin{bmatrix} 0 \\ 1 \end{bmatrix} = \begin{bmatrix} f_i \\ f_j \end{bmatrix} \tag{5-104}
$$

Note that if a point load f_p is imposed within the element, e.g., at $x = x_p$, then $\mathbf{f}_{PL}^e$ is evaluated as follows:

$$
\mathbf{f}_{PL}^e = \mathbf{N}^T(x_p) f_p + \begin{bmatrix} f_i \\ f_j \end{bmatrix}
$$

For example, if $x_p = \bar{x}$, then half of f_p is allocated to node i and half to node j (see Problem 5-29).

This completes the evaluation of the element characteristics for the two-node lineal element in one-dimensional stress analysis. It should be emphasized that s_i, s_j, f_i, and f_j, are positive if directed in the positive direction.

The Element Resultants

The element stiffness matrices and nodal force vectors may be determined for every element with the help of Eqs. (5-99) to (5-104). The assemblage of each of these 2×2 matrices and 2×1 vectors is performed in the usual manner to give a system of N linear algebraic equations in N unknown nodal displacements assuming the member is discretized into $N - 1$ elements with N nodes. The result is $\mathbf{K}^a \mathbf{a} = \mathbf{f}^a$. The geometric (or prescribed displacement) boundary conditions are applied at this point to yield $\mathbf{Ka} = \mathbf{f}$, which may be solved for the nodal displacements contained in the vector $\mathbf{a}$. Therefore, in what follows it is assumed that all the nodal displacements are now known. Each element nodal displacement vector $\mathbf{a}^e$ is also assumed to be known.

The element strains and stresses may be computed from Eq. (5-77) as

$$\bar{\epsilon} = \mathbf{Ba}^e \qquad (5\text{-}105)$$

and from Eq. (5-24) [or Eq. (5-95)] as

$$\bar{\sigma} = E[\mathbf{Ba}^e - \alpha_t \, \Delta T] + \sigma_0 \qquad (5\text{-}106)$$

Note that the axial strain and stress within the element are denoted as $\bar{\epsilon}$ and $\bar{\sigma}$, respectively, where the bars $(\bar{})$ imply *average strain* and *average stress*. The strains and stresses within an element are not necessarily constant. However, Eqs. (5-105) and (5-106) can only result in one strain and one stress for each element (both constant and not a function of x) because $\mathbf{B}$ is a constant matrix in this case. Therefore, the strain $\bar{\epsilon}$ and stress $\bar{\sigma}$ are generally assumed to be the local values of ϵ and σ at $x = \bar{x}$. It follows that the average force $\bar{F}$ within an element is given by $\bar{F} = \bar{\sigma}A$.

The use of the equations developed in this section is illustrated numerically in the next example.

Example 5-5

Determine the displacements, strains, and stresses within the tapered circular, uniaxial stress member loaded as shown in Fig. 5-15(a), which also shows the dimensions of the rod. The rod is fabricated from carbon steel with a modulus of elasticity of 30×10^6 psi (21×10^{10} N/m^2).

Solution

The first step in any finite element analysis is discretization of the region under consideration into a suitable number of finite elements. For the purpose of illustrating the basic approach, let us take only two elements and three nodes as shown in Fig. 5-15(b). It is convenient to summarize the node and element data in a tabular form

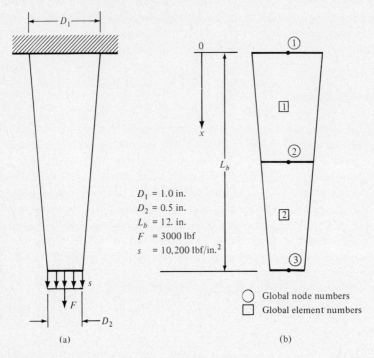

$D_1 = 1.0$ in.
$D_2 = 0.5$ in.
$L_b = 12.$ in.
$F = 3000$ lbf
$s = 10,200$ lbf/in.^{2}

Global node numbers
Global element numbers

(a)   (b)

Figure 5-15   Uniaxial stress member in Example 5-5. (a) Dimensions and loads and (b) discretization into two elements and three nodes.

as shown in Table 5-1. Note that the radius of the rod is supplied as part of the node data. In a computer program, it is somewhat more expedient to include the cross-sectional area in the material property data. This was in fact done in the TRUSS program. The variable area could be taken into account with the help of a variable property routine (see Problem 5-39).

Table 5-1   Node and Element Data for Example 5-5

Node number	x, in.	r, in.
1	0.0	0.500
2	6.0	0.375
3	12.0	0.250

Element number	Node i	Node j
1	1	2
2	2	3

The radius of the bar at $x = \bar{x}$ is denoted as $\bar{r}$ and is easily computed from

$$\bar{r} = \frac{1}{2}\left[\frac{D_1 L_b - (D_1 - D_2)\bar{x}}{L_b}\right]$$

where D_1 is the diameter of the rod at the base, D_2 is the diameter at the tip, and L_b is the length of the bar.

The calculations are summarized below in a form that may be readily implemented in a computer program.

Element 1

Node i is 1. Node j is 2.

$$x_i = 0.0 \text{ in.} \qquad x_j = 6.0 \text{ in.}$$

$$x_{ji} = x_j - x_i = 6 - 0 = 6.0 \text{ in.}$$

$$E = 30 \times 10^6 \text{ psi}$$

$$\bar{x} = \frac{x_i + x_j}{2} = \frac{0 + 6}{2} = 3.0 \text{ in.}$$

$$\bar{r} = \frac{1}{2}\left[1.0 - \frac{(1.0 - 0.5)(3.0)}{12.0}\right] = 0.4375 \text{ in.}$$

$$\bar{A} = \pi\bar{r}^2 = (3.14)(0.4375)^2 = 0.601 \text{ in.}^2$$

$$\mathbf{K}^{(1)} = \frac{(0.601)(30 \times 10^6)}{6.0}\begin{bmatrix} 1 & -1 \\ -1 & 1 \end{bmatrix}$$

$$= \begin{bmatrix} 3007 & -3007 \\ -3007 & 3007 \end{bmatrix} \times 10^3 \text{ lbf/in.}$$

$$\mathbf{K}^a = \begin{bmatrix} 3007 & -3007 & 0 \\ -3007 & 3007 & 0 \\ 0 & 0 & 0 \end{bmatrix} \times 10^3 \text{ lbf/in.}$$

$$\mathbf{f}^{(1)} = \begin{bmatrix} 0 \\ 0 \end{bmatrix} \text{ lbf}$$

$$\mathbf{f}^a = \begin{bmatrix} 0 \\ 0 \\ 0 \end{bmatrix} \text{ lbf}$$

Element 2

Node i is 2. Node j is 3.

$$x_i = 6.0 \text{ in.} \qquad x_j = 12.0 \text{ in.}$$

$$x_{ji} = 12.0 - 6.0 = 6.0 \text{ in.}$$

$$E = 30 \times 10^6 \text{ psi}$$

$$\bar{x} = \frac{6 + 12}{2} = 9.0 \text{ in.}$$

$$\bar{r} = \frac{1}{2}\left[1.0 - \frac{(1.0 - 0.5)(9.0)}{12.0}\right] = 0.3125 \text{ in.}$$

$$\bar{A} = (3.14)(0.3125)^2 = 0.307 \text{ in.}^2$$

$$A_3 = \pi(0.25)^2 = 0.196 \text{ in.}^2$$

$$\mathbf{K}^{(2)} = \frac{(0.307)(30 \times 10^6)}{6.0}\begin{bmatrix} 1 & -1 \\ -1 & 1 \end{bmatrix}$$

$$= \begin{bmatrix} 1534 & -1534 \\ -1534 & 1534 \end{bmatrix} \times 10^3 \text{ lbf/in.}$$

$$\mathbf{K}^a = \begin{bmatrix} 3007 & -3007 & 0 \\ -3007 & 4541 & -1534 \\ 0 & -1534 & 1534 \end{bmatrix} \times 10^3 \text{ lbf/in.}$$

$$\mathbf{f}^{(2)} = f_{\text{PL}}^{(2)} + f_s^{(2)}$$

$$= \begin{bmatrix} 0 \\ 3000 \end{bmatrix} + \begin{bmatrix} 0 \\ (0.196)(10,200) \end{bmatrix} = \begin{bmatrix} 0 \\ 5000 \end{bmatrix}$$

$$\mathbf{f}^a = \begin{bmatrix} 0 \\ 0 \\ 5000 \end{bmatrix} \text{ lbf}$$

Before the application of the prescribed displacement we have $\mathbf{K}^a\mathbf{a} = \mathbf{f}^a$, or

$$\begin{bmatrix} 3007 & -3007 & 0 \\ -3007 & 4541 & -1534 \\ 0 & -1534 & 1534 \end{bmatrix}\begin{bmatrix} u_1 \\ u_2 \\ u_3 \end{bmatrix} = \begin{bmatrix} 0 \\ 0 \\ 5 \end{bmatrix}$$

where both sides have been divided by 10^3. Note that the assemblage stiffness matrix is symmetric and banded, with a half-bandwidth of two [this also follows from Eq. (3-33)]. If the prescribed displacement boundary condition is applied by using Method 1 from Sec. 3-2, we get

$$\begin{bmatrix} 1 & 0 & 0 \\ 0 & 4541 & -1534 \\ 0 & -1534 & 1534 \end{bmatrix}\begin{bmatrix} u_1 \\ u_2 \\ u_3 \end{bmatrix} = \begin{bmatrix} 0 \\ 0 \\ 5 \end{bmatrix}$$

Solving this system of linear, algebraic equations (e.g., by the matrix inversion method) yields the following nodal displacements:

$$u_1 = 0.0 \text{ in.} \qquad u_2 = 0.00166 \text{ in.} \qquad u_3 = 0.00492 \text{ in.}$$

The element resultants are now easily determined as summarized below:

Element 1

$$\mathbf{B} = \begin{bmatrix} \dfrac{-1}{6-0} & \dfrac{1}{6-0} \end{bmatrix} = \frac{1}{6}[-1 \quad 1] \text{ in.}^{-1}$$

$$\bar{\epsilon} = \mathbf{Ba}^{(1)} = \frac{1}{6}[-1 \quad 1]\begin{bmatrix} 0 \\ 0.00166 \end{bmatrix}$$

$$= 0.0002772 \text{ in./in.}$$

$$\bar{\sigma} = E(\bar{\epsilon} - \epsilon_0) + \sigma_0 = (30 \times 10^6)(0.0002772) = 8320 \text{ psi}$$

$$\overline{F} = \bar{\sigma}\overline{A} = (8320)(0.601) = 5000 \text{ lbf}$$

Element 2

$$\mathbf{B} = \begin{bmatrix} \dfrac{-1}{12-6} & \dfrac{1}{12-6} \end{bmatrix} = \frac{1}{6}[-1 \quad 1] \text{ in.}^{-1}$$

$$\bar{\epsilon} = \mathbf{Ba}^{(2)} = \frac{1}{6}[-1 \quad 1]\begin{bmatrix} 0.00166 \\ 0.00492 \end{bmatrix}$$

$$= 0.0005433 \text{ in./in.}$$

$$\bar{\sigma} = E\bar{\epsilon} = (30 \times 10^6)(0.0005433) = 16{,}300 \text{ psi}$$

$$\overline{F} = \bar{\sigma}\overline{A} = (16{,}300)(0.307) = 5000 \text{ lbf} \qquad \blacksquare$$

A few subtle points regarding the calculation of the element nodal force vectors in Example 5-5 need to be discussed. The reader may observe that on an individual element there are surface tractions (or point loads) acting at the nodes as a result of the forces exerted by the adjacent elements or by the restrained end. For example, let us consider a bar discretized into three elements as shown in Fig. 5-16. It should be apparent that the internal surface tractions and/or point loads cancel during the assemblage step. Moreover, if a node is restrained (e.g., node 1), then there is in effect an unknown force at this node. The reader should be convinced, however, that this unknown force never really enters into the formulation because of the application of the prescribed displacement at this node. This same observation was made in Chapter 3 where the two-dimensional truss model was developed (and in Chapter 4 where the pin fin was modeled). The second point to be made is that the stress distribution near the point of application of point loads (and nonuniform surface tractions) is never attained in a one-dimensional formulation, no matter how many elements are used. Predicting such stress distributions requires at least a two-dimensional analysis (see Chapter 7).

The results from Example 5-5 are compared to the exact solution (see Problem 5-37) in Table 5-2, where the finite element solutions for four and eight elements are also given. These FEM results were obtained by the computer program described in Problem 5-39. Stresses are reported with five significant digits in order to show

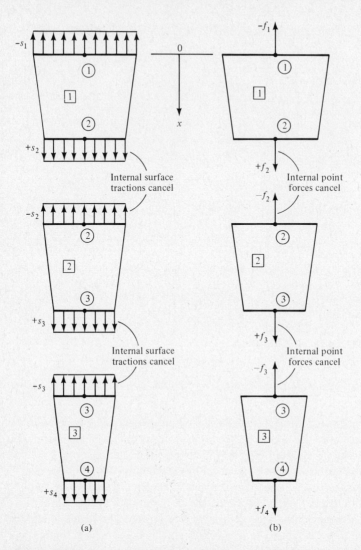

Figure 5-16 Graphical explanation of the cancellation of (a) internal surface tractions and (b) internal point forces that occurs during the assemblage step.

the remarkable accuracy of the results. The nodal displacements in Table 5-2 demonstrate convergence to the exact solution as the number of elements is increased—here the maximum error decreases from -3.3% for two elements to less than -0.3% for eight elements.

A detailed discussion of error predictions and convergence is beyond the scope of this text. The interested reader may find introductory material on this important subject in the book by Becker, Carey, and Oden [10].

Table 5-2 Summary of Results for Example 5-5

	Displacements, in.			
x	Exact	Two Elements	Four Elements	Eight Elements
0.0	0.0	0.0	0.0	0.0
1.5	0.000340			0.000339
3.0	0.000728		0.000724	0.000727
4.5	0.00118			0.00117
6.0	0.00170	0.00166	0.00169	0.00170
7.5	0.00231			0.00231
9.0	0.00306		0.00304	0.00305
10.0	0.00396			0.00395
12.0	0.00509	0.00492	0.00505	0.00508
	Stresses, psi			
x	Exact	Two Elements	Four Elements	Eight Elements
0.75	6,784			6,784
1.50	7,243		7,243	
2.25	7,752			7,752
3.00	8,315	8,315		
3.75	8,942			8,942
4.50	9,644		9,644	
5.25	10,430			10,430
6.00	11,318			
6.75	12,323			12,323
7.50	13,469		13,469	
8.25	14,782			14,782
9.00	16,298	16,297		
9.75	18,058			18,058
10.50	20,120		20,120	
11.25	22,557			22,557

5-9 REMARKS

This chapter began with a review of some of the more important topics from the theory of elasticity and solid mechanics. Following this, the principles of minimum potential energy and virtual displacements were developed into forms that were directly useful in finite element formulations of stress or structural analysis problems. These two principles were seen to be equivalent, and both yielded the stiffness-based finite element method (wherein the primary unknowns are the nodal displacements and not the nodal forces). Moreover, both resulted in a weak form of the equilibrium equations in that at most only first derivatives of the displacements appear in the equations (after the generalized form of Hooke's law is applied). The implication is that lower-order displacement functions may be used in the finite element solution. For example, first-order displacement functions were used throughout this chapter, whereas the equilibrium equations would require at least second-order functions. In the stiffness-based finite element method, the compatibility equations are satisfied exactly by these displacement functions (see Problems 5-49 and 5-50).

The shape function matrix was introduced for the three-node triangular and four-node tetrahedral elements. Although the shape functions themselves were not derived in this chapter, the basic properties that these functions must satisfy for problems requiring only C^0-continuity were presented. In the next chapter, explicit expressions for these shape functions will be derived.

Following this, the finite element characteristics were derived for all problems in linear, static stress or structural analysis. It should be recalled that by finite element characteristics we mean expressions for the element stiffness matrix $\mathbf{K}^e$ and the element nodal force vectors $\mathbf{f}^e$. These expressions are given by Eq. (5-87) and Eqs. (5-88) to (5-92), respectively. Their importance cannot be overemphasized because they may be used to formulate all problems in linear, static stress analysis.

The general approach was illustrated with the simple application of a uniaxial stress member. Admittedly, this example is of little practical importance, but it should help to clarify some of the concepts developed in this chapter. When applied to a numerical example, convergence to the exact solution was demonstrated as the number of elements was increased. The stresses from the finite element solution also compared very favorably with those from the exact solution.

Chapter 6 is devoted to the derivations of the shape functions that we will need in our study of two- and three-dimensional stress analysis in Chapter 7. It will be seen in Chapter 8 that these same shape functions may be used in one-, two-, and three-dimensional thermal analysis and fluid mechanics. Indeed, they also may be used in the analysis of problems in lubrication, electromagnetics, etc. [11].

REFERENCES

1. Huebner, K. H., *The Finite Element Method for Engineers*, Wiley, New York, 1975, pp. 457–462.
2. Fitzgerald, R. W., *Strength of Materials*, Addison-Wesley, Reading, Mass., 1967.
3. Timoshenko, S., and D. H. Young, *Elements of Strength of Materials*, 5th ed., Van Nostrand, New York, 1968.
4. Popov, E. P., *Introduction to Mechanics of Solids*, Prentice-Hall, Englewood Cliffs, N.J., 1968.
5. Popov, E. P., *Introduction to Mechanics of Solids*, Prentice-Hall, Englewood Cliffs, N.J., 1968, pp. 308–309.
6. Zienkiewicz, O. C., *The Finite Element Method*, McGraw-Hill (UK), London, 1977, p. 28.
7. Love, A. E. H., *A Treatise on the Mathematical Theory of Elasticity*, Dover, New York, 1944.
8. Sokolnikoff, I. S., *Mathematical Theory of Elasticity*, 2nd ed., McGraw-Hill, New York, 1956.

9. Ugural, A. C., and S. K. Fenster, *Advanced Strength and Applied Elasticity*, American Elsevier, New York, 1975, pp. 315–316.

10. Becker, E. B., G. F. Carey, and J. T. Oden, *Finite Elements: An Introduction*, vol. 1, Prentice-Hall, Englewood Cliffs, N.J., 1981, pp. 36–38.

11. Zienkiewicz, O. C., *The Finite Element Method*, McGraw-Hill (UK), London, 1977, pp. 423–449.

PROBLEMS

Note: The properties in Appendix A should be used unless stated otherwise in the problem statement.

5-1 With the help of Fig. 5-4, show that the equations of static equilibrium in two dimensions are given by Eq. (5-4) by doing force balances in the x and y directions. Show that Eq. (5-5) holds by doing a moment balance about the center of the infinitesimal element.

5-2 Derive the three-dimensional form of the equilibrium equations given by Eqs. (5-6). Show that Eq. (5-7) holds by doing a moment balance about the center of an infinitesimal cubic element.

5-3 Consider the two-dimensional state of stress shown in Fig. P5-3.

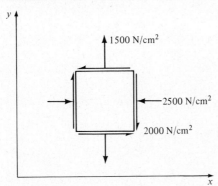

Figure P5-3

a. Plot Mohr's circle of stress.
b. Determine the principal stresses and show the corresponding element orientation.
c. Determine the maximum shear stress and the associated normal stress and element orientation.
d. Use Eqs. (5-9) to (5-11) to corroborate the results from Mohr's circle.

5-4 Consider the two-dimensional state of stress shown in Fig. P5-4.

a. Plot Mohr's circle of stress.
b. Determine the principal stresses and show the corresponding element orientation.
c. Determine the maximum shear stress and the associated normal stress and element orientation.
d. Use Eqs. (5-9) to (5-11) to corroborate the results from Mohr's circle.

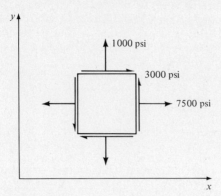

Figure P5-4

5-5 Consider the two-dimensional state of stress shown in Fig. P5-5.

a. Plot Mohr's circle of stress.

b. Determine the principal stresses and show the corresponding element orientation.

c. Determine the maximum shear stress and the associated normal stress and element orientation.

d. With help of Eqs. (5-12), verify the principal stresses from part (b). Assume that σ_{xz}, σ_{yz}, and σ_{zz} are zero. What is the value of the third principal stress in this case?

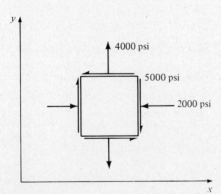

Figure P5-5

5-6 Consider the two-dimensional state of stress shown in Fig. P5-6.

a. Plot Mohr's circle of stress.

b. Determine the principal stresses and show the corresponding element orientation.

c. Determine the maximum shear stress and the associated normal stress and element orientation.

d. With help of Eqs. (5-12), verify the principal stresses from part (b). Assume that σ_{xz}, σ_{yz}, and σ_{zz} are zero. What is the value of the third principal stress in this case?

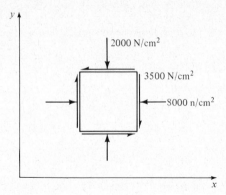

Figure P5-6

5-7 In Chapter 7, the state of plane stress is defined as that state of stress such that the stresses σ_{xz}, σ_{yz}, and σ_{zz} are identically zero (see Sec. 7-2).

  a. What can you conclude about one of the principal stresses? *Hint:* Equations (5-12) must be used.

  b. Derive Eq. (5-10) by beginning with Eqs. (5-12) and using the fact that σ_{xz}, σ_{yz}, and σ_{zz} are zero.

5-8 What are the four strain-displacement or kinematic relationships for small deformations in axisymmetric problems? It is not necessary to derive them. Instead, use a suitable reference.

5-9 What is the linear operator matrix **L** in three-dimensional stress analysis? The matrix **L** is defined by $\epsilon = \mathbf{L}\mathbf{u}$, where ϵ and **u** are given by Eqs. (5-20) and (5-21), respectively.

5-10 Show that the compatibility equation in two dimensions is given by Eq. (5-22).

5-11 By using a suitable reference, *state* the six compatibility equations that must hold in three dimensions. Why are there six relationships?

5-12 A portion of the boundary of a two-dimensional body to be analyzed is shown in Fig. P5-12. On this boundary, a uniform surface traction acts as shown.

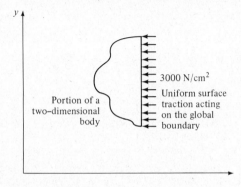

Figure P5-12

a. Determine s_x and s_y on this surface.

b. Suggest an application that could result in this type of surface traction.

5-13 Part of the boundary of a two-dimensional body to be analyzed is shown in Fig. P5-13. On this boundary, a uniform surface traction acts as shown.

a. Determine s_x and s_y on this surface.

b. Suggest an application that could result in this type of surface traction.

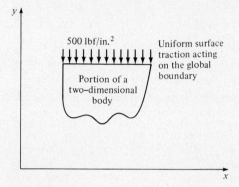

Figure P5-13

5-14 Part of the boundary of a two-dimensional body to be analyzed is shown in Fig. P5-14. On this boundary, a uniform surface traction acts as shown.

a. Determine s_x and s_y on this surface.

b. Suggest an application that could result in this type of surface traction.

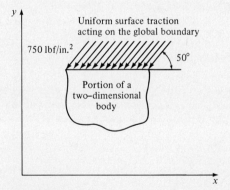

Figure P5-14

5-15 A portion of the boundary of a two-dimensional body to be analyzed is shown in Fig. P5-15. On this boundary, a uniform surface traction acts as shown.

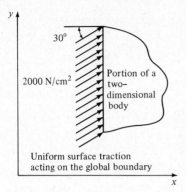

Figure P5-15

a. Determine s_x and s_y on this surface.
b. Suggest an application that could result in this type of surface traction.

5-16 A portion of the boundary of a two-dimensional body to be analyzed is shown in Fig. P5-16. On this boundary, a uniform surface traction acts as shown.

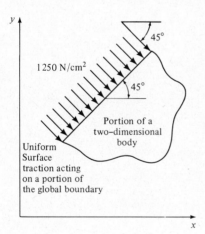

Figure P5-16

a. Determine s_x and s_y on this surface.
b. Suggest an application that could result in this type of surface traction.

5-17 Part of the boundary of a two-dimensional body to be analyzed is shown in Fig. P5-17. On this boundary, a uniform surface traction acts as shown.

a. Determine s_x and s_y on this surface.
b. Suggest an application that could result in this type of surface traction.

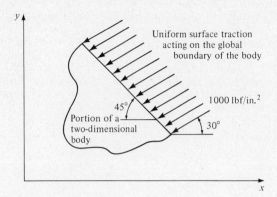

Figure P5-17

5-18   Starting with Eq. (5-40) and using the constitutive relationship for a linear elastic material, show that Eq. (5-41) holds.

5-19   Show that the total potential energy of a structural system in static equilibrium is given by Eq. (5-42), given that the first variation $\delta\Pi$ is given by Eq. (5-32) and that Eqs. (5-34b) and (5-35b) hold.

5-20   Reconsider Example 5-2 and show that potential energy Π has indeed been minimized by computing $d^2\Pi/d\Delta^2$. *Hint:* What is the significance of the sign of this result?

5-21   Show the two shape functions graphically for the two-node lineal element for C^0-continuity given that they are linear and must satisfy Eq. (5-52). Clearly label each of the shape functions.

5-22   For the three-node triangular element, show the shape functions $N_j(x,y)$ and $N_k(x,y)$ in a manner similar to that in Fig. 5-13.

5-23   Derive Eq. (5-84) from Eq. (5-83) with the help of Eqs. (5-77) and (5-24).

5-24   Clearly show why each of the element nodal force vectors given by Eqs. (5-88) to (5-92) is of size 2×1 for the two-node lineal element.

5-25   Show why each of the element nodal force vectors given by Eqs. (5-88) to (5-92) is of size 6×1 for the three-node triangular element. In two-dimensional problems in stress analysis, how many degrees of freedom are there per node? What do the degrees of freedom represent?

5-26   For three-dimensional problems in stress analysis analyzed with the four-node tetrahedral element, show that the element stiffness matrix and nodal force vectors given by Eq. (5-87) and Eqs. (5-88) to (5-92) are of sizes 12×12 and 12×1, respectively. In three-dimensional problems in stress analysis, how many degrees of freedom are there per node? What do the degrees of freedom represent?

5-27   Show that the element nodal force vector from a prestress in a uniaxial stress member is given by the result in Eq. (5-101) if the two-node lineal element is used. What assumptions have to be made in arriving at this result?

5-28 Show that the element nodal force vector from a body force in a uniaxial stress member is given by the result in Eq. (5-102) if the two-node lineal element is used. What assumptions have to be made in arriving at this result?

5-29 Consider the one-dimensional element shown in Fig. P5-29. This element has been extracted from a discretized brass bar in a state of uniaxial stress. The bar has a circular cross section with a constant diameter D and undergoes a temperature change ΔT. A load P is applied as shown at point p. A surface traction s also acts as shown. If $x_i = 2$ cm, $x_j = 5$ cm, $x_p = 3.5$ cm, $D = 1$ cm, $\Delta T = +15°C$, $P = 1300$ N, and $s = 3000$ N/cm^2:

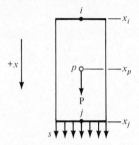

Figure P5-29

a. Determine the element stiffness matrix.
b. Determine the relevant element nodal force vectors. The weight may be neglected.
c. Determine the composite element nodal force vector.

5-30 Solve all parts of Problem 5-29 if $x_i = 2$ in., $x_j = 5$ in., $x_p = 3.5$ in., $D = 0.5$ in., $\Delta T = +10°F$, $P = 1500$ lbf, and $s = 1800$ lbf/in.^{2}.

5-31 Consider the one-dimensional element shown in Fig. P5-31. This element has been extracted from a discretized bar and is in a state of uniaxial stress. The bar is made of hard drawn copper and has a rectangular cross section with dimensions w and h. Two point loads P_1 and P_2 are applied as shown. A surface traction s also acts as shown. The bar undergoes a temperature change ΔT and is initially under a prestress σ_0. If $x_i = 3$ in., $x_j = 5$ in., $x_p = 4.5$ in., $w = 0.75$ in., $h = 0.60$ in., $\Delta T = -20°F$, $P_1 = 1000$ lbf, $P_2 = 200$ lbf, $s = 600$ lbf/in.^{2}, and $\sigma_0 = -400$ lbf/in.^{2}:

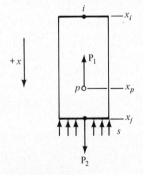

Figure P5-31

a. Determine the element stiffness matrix.
b. Determine the relevant element nodal force vectors. The weight may be neglected.
c. Determine the composite element nodal force vector.

5-32 Solve all parts of Problem 5-31 if $x_i = 1$ cm, $x_j = 3$ cm, $x_p = 2.5$ cm, $w = 1.5$ cm, $h = 1.0$ cm, $\Delta T = -25°C$, $P_1 = 500$ N, $P_2 = 100$ N, $s = 600$ N/cm², and $\sigma_0 = -350$ N/cm².

5-33 Consider the one-dimensional element shown in Fig. P5-33. This element has been extracted from a discretized circular bar and is in a state of uniaxial stress. The bar is made of aluminum alloy 6061 and is tapered such that at nodes i and j the diameters of the bar are D_i and D_j, respectively. Two point loads P_1 and P_2 are applied as shown. A surface traction s also acts as shown. The bar undergoes a temperature change ΔT and is initially under a prestress σ_0. If $x_i = 23$ cm, $x_j = 20$ cm, $x_p = 22$ cm, $D_i = 2.0$ cm, $D_j = 1.50$ cm, $\Delta T = -30°C$, $P_1 = 400$ N, $P_2 = 150$ N, $s = 200$ N/cm², and $\sigma_0 = +400$ N/cm²,

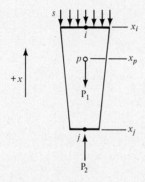

Figure P5-33

a. Determine the element stiffness matrix.
b. Determine the relevant element nodal force vectors. The weight may be neglected.
c. Determine the composite element nodal force vector.

5-34 Solve all parts of Problem 5-33 if $x_i = 10$ in., $x_j = 7$ in., $x_p = 9$ in., $D_i = 0.75$ in., $D_j = 0.50$ in., $\Delta T = -40°F$, $P_1 = 500$ lbf, $P_2 = 200$ lbf, $s = 800$ lbf/in.², and $\sigma_0 = +700$ lbf/in.².

5-35 Consider the uniaxial stress member shown in Fig. P5-35.

a. Solve for the displacements and the element resultants for $D_1 = 1$ cm, $D_2 = 0.75$ cm, $L_b = 6$ cm, $F = 1500$ N, and $s = 5000$ N/cm². Take $E = 21 \times 10^{10}$ N/m². Use only two elements.
b. Compare the results from part (a) with the exact solution given in Problem 5-37.

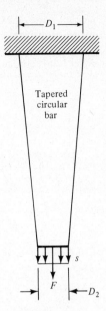

Figure P5-35

5-36 Solve both parts of Problem 5-35 if $D_1 = 0.75$ in., $D_2 = 0.25$ in., $L_b = 5$ in., $s = -8000$ lbf/in.^{2}, $E = 11 \times 10^6$ lbf/in.^{2}, and $F = 1500$ lbf.

5-37 Show that the exact solution for the displacement to the problem posed in Example 5-5 is given by

$$u(x) = \frac{4L_b P}{\pi E(D_1 - D_2)} \left\{ \frac{1}{D_1 - [(D_1 - D_2)/L_b]x} - \frac{1}{D_1} \right\}$$

where P is the sum of the point load and the force from the surface traction acting on the tip of the rod.

5-38 The exact solution for the displacement at the tip of an *untapered* uniaxial stress member is given by $u = FL_b/AE$, where F is the axial force, L_b is the length of the member, A is the cross-sectional area (assumed to be constant), and E is the elastic modulus. Show that the exact solution for the displacements in a *tapered* uniaxial stress member gives this same result if D_1 approaches D_2. The exact solution is given in Problem 5-37.

5-39 Convert the TRUSS program in Appendix B into one that may be used to solve problems in one-dimensional stress analysis. The program should allow for self-strains (from a temperature change), prestresses, body forces, surface tractions, and point loads. Assume no more than 30 elements (and 31 nodes). The following parameters should be taken as the material properties: E, A, α_t, ΔT, σ_0, and b. Allow up to five different types of materials and make use of material set flags. Allow for two different prescribed displacements (using positive boundary condition flags) and up to five different tractions (using negative boundary condition flags), such as the scheme used in the TRUSS program. Allow for point loads by specifying the node number that corresponds to the location of the load. Also specify on the same input line the magnitude of the point load (positive in the positive *x* direction). Use the variable property subroutine VPROP and subroutine PROPTY, both given below, to allow for spacially varying properties.

```
      SUBROUTINE PROPTY (L, MATFLG, DATMAT, ELMOD,
     1      AREA, ALPHA, DELTAT, SIGMAO, BODYF)
C
C   DEFINES THE ELASTIC MODULUS, CROSS-SECTIONAL AREA,
C   COEFFICIENT OF THERMAL EXPANSION, TEMPERATURE CHANGE,
C   PRESTRESS, AND BODY FORCE (PER UNIT VOLUME).
      DIMENSION MATFLG(30), DATMAT(5,6)
      NFLAG    = MATFLG(L)
C
      ELMOD    = DATMAT(NFLAG, 1)
      IF (ELMOD .LT. 0.)  CALL VPROP (ELMOD)
C
      AREA     = DATMAT(NFLAG, 2)
      IF (AREA  .LT. 0.)  CALL VPROP (AREA)
C
      ALPHA    = DATMAT(NFLAG, 3)
      IF (ALPHA .LT. 0.)  CALL VPROP (ALPHA)
C
      DELTAT   = DATMAT(NFLAG, 4)
      IF (DELTAT .LT. 0.) CALL VPROP (DELTAT)
C
      SIGMAO   = DATMAT(NFLAG, 5)
      IF (SIGMAO .LT. 0.)  CALL VPROP (SIGMAO)
C
      BODYF    = DATMAT(NFLAG, 6)
      IF (BODYF .LT. 0.)  CALL VPROP (BODYF)
C
      RETURN
      END


      SUBROUTINE VPROP (PROP)
      COMMON /CONST/ C(10)
      COMMON /VPROPS/ X
C
C     THE C(I)'S ARE TEN USER-DEFINED CONSTANTS READ IN
C     THE FIRST INPUT SECTION (SEE THE TRUSS PROGRAM).
C     NOTE THAT OTHER CONSTANTS, SUCH AS PI, MUST BE
C     DEFINED BEFORE THE COMPUTED GO TO. THE 'X' IS
C     THE VALUE OF THE X COORDINATE WHEN THIS SUBROUTINE
C     IS CALLED.
C
      PI = 3.14159
      IPROP = -PROP
      GO TO (1, 2, 3, 4, 5), IPROP
C
    1      PROP = some function of X, C(1), C(2), etc.
      RETURN
C
    2      PROP = some other function of X, C(1), etc.
      RETURN
```

```
C
    3              CONTINUE
                   RETURN
C
    4              CONTINUE
                   RETURN
C
    5              CONTINUE
                   RETURN
C
                   END
```

Note that if PROP is negative, it is converted to a positive integer IPROP. The *computed go to* then transfers control to the line whose label is numerically equal to IPROP. Thus by simply including the proper FORTRAN statements in SUBROUTINE VPROP, variable properties are easily accommodated. The $C(i)$'s are user-defined constants that should be read in Section 1 of the input file (after NNODES, NELEM, etc.) and may be used in SUBROUTINE VPROP. By way of example, let us say that we want to use this technique to allow for the varying cross-sectional area in the problem posed in Example 5-5. Let us define $C(1) = 1.0$ (for D_1 in inches), $C(2) = 0.5$ (for D_2 in inches), and $C(3) = 12.0$ (for L_b in inches). If we use a "-4" as the input for the area A in Section 4 of the input file (see Appendix B for a description of the input to the TRUSS program), the statements beginning with label "4" should read

```
4       CONTINUE
        DIAM = (C(1)*(C(3)-X) + C(2)*X) / C(3)
        PROP = PI * (DIAM**2) / 4.
        RETURN
```

since this in effect gives

$$D(x) = D_1 \left(\frac{L_b - x}{L_b} \right) + D_2 \left(\frac{x}{L_b} \right)$$

$$A(x) = \frac{\pi D(x)^2}{4}$$

where the expressions for $D(x)$ and $A(x)$ represent the correct diameter and area variations with x. Note that the variable X in the subroutine represents the x coordinate and is passed to VPROP via the *labeled common* (namely, VPROPS). Use the program to verify the results in Example 5-5 and Table 5-2.

5-40   Use the program developed in Problem 5-39 (or one furnished by the instructor) to solve Problem 5-35 with two, four, and eight elements. What happens to the results as the number of elements is increased?

5-41   Use the program developed in Problem 5-39 (or one furnished by the instructor) to solve Problem 5-36 with two, four, and eight elements. What happens to the results as the number of elements is increased?

5-42   Reconsider Example 5-5. Instead of the point load F and traction s acting on the rod, the free end is now also completely restrained. In other words, both ends of the bar are restrained and the bar is assumed to be initially in a stress-free state. The bar then undergoes a temperature *decrease* of 150°F. Assuming $\alpha_t = 6.5 \times 10^{-6}$ in./in.-°F, determine the nodal displacements and element resultants if only two elements are used. Do *not* use a computer program.

5-43   Determine an expression for the exact solution for Problem 5-42. Give the result in terms of the pertinent variables and then apply it to the situation in Problem 5-42.

5-44   Solve Problem 5-42 with the help of the computer program from Problem 5-39 (or one furnished by the instructor) for two, four, and eight elements. What happens to the results as the number of elements is increased?

5-45   Consider the problem posed in Problem 5-35. Instead of the point load F and traction s acting, the free end is now also completely restrained. In other words, both ends of the bar are restrained and the bar is assumed to be initially in a stress-free state. The bar then undergoes a temperature *increase* of 15°C. Assuming $\alpha_t = 11.7 \times 10^{-6}$ m/m-°C, determine the nodal displacements and element resultants if only two elements are used.

5-46   Solve Problem 5-45 with the help of the computer program from Problem 5-39 (or one furnished by the instructor) for two, four, and eight elements. What happens to the results as the number of elements is increased?

5-47   Consider the problem posed in Problem 5-36. Instead of the point load F and traction s acting, the free end is now also completely restrained. In other words, both ends of the bar are restrained and the bar is assumed to be initially in a stress-free state. The bar then undergoes a temperature *decrease* of 120°F. Assuming $\alpha_t = 13 \times 10^{-6}$ in./in.-°F, determine the nodal displacements and element resultants if only two elements are used.

5-48   Solve Problem 5-47 with the help of the computer program from Problem 5-39 (or one furnished by the instructor) for two, four, and eight elements. What happens to the results as the number of elements is increased?

5-49   Show that the trial functions given by Eqs. (5-53) and (5-56) for the three-node triangular element satisfy *exactly* the compatibility equation given by Eq. (5-22).

5-50   Show that the trial functions given by Eqs. (5-62), (5-65), and (5-66) for the four-node tetrahedral element satisfy *exactly* the six compatibility equations.

6

Parameter Functions;
C^0-Continuous Shape Functions;
Simple Integration Formulas; Active
Zone Equation Solvers

6-1 INTRODUCTION

In this chapter, the concept of the parameter function is formalized. This is followed by a review of what is meant by C^0- and C^1-continuous problems. The conditions that the assumed parameter functions must satisfy are given and are discussed in some detail. The shape functions are then derived for some of the more popular one-, two-, and three-dimensional elements. In addition, normalized coordinates, which facilitate the use of these shape functions, are introduced. Axisymmetric elements are also considered. Three special integration formulas, which may be used to evaluate the integrals when the lineal, triangular, and tetrahedral elements are used, are presented. The chapter concludes with an alternate equation solver. Up to now the rather inefficient matrix inversion method has been used to solve the resulting systems of algebraic equations. The equation solver presented at the end of this chapter takes advantage of the banded and symmetric nature of the assemblage stiffness matrix.

6-2 PARAMETER FUNCTIONS

In this section, parameter functions are defined in the context of problems in stress analysis, thermal analysis, and fluid flow analysis. Following this, the requirements that the assumed parameter functions must satisfy are given and discussed.

Definition of a Parameter Function

By now it should be evident that in stress analysis problems to be formulated using the stiffness approach to FEM, the field variables of interest are the displacements. From a knowledge of the displacements, the element resultants such as the stresses and strains may be easily computed. The functions to be used to represent the displacements on an element basis are referred to as *displacement functions* in particular and as *parameter functions* in general. For example, a typical displacement function in one-dimensional stress analysis is given by Eq. (5-47) and in two-dimensional stress analysis by Eqs. (5-53) and (5-56).

   In thermal analysis problems, the field variable of interest is the temperature. On an element basis, the temperature within a typical element is approximated by an assumed temperature function. Again, in general terms, this function is referred to as a parameter function. This situation is slightly simpler than the corresponding stress analysis problem because the temperature field is a scalar, whereas the displacement field is a vector. Not surprisingly, for the three-node triangular element, a suitable parameter function for the temperature is of the same form as that given by Eq. (5-53) for a typical displacement.

   In fluid flow problems, the field variables of interest are the fluid velocities and pressure. If a finite element analysis is to be performed with these so-called *primitive variables*, the parameter functions to be used may be referred to more specifically as *velocity* and *pressure functions*. Because the velocities are vectors, this situation is more closely related to the situation in stress analysis. However, we usually take the Eulerian approach whereby we observe the fluid velocities and pressure at fixed points. This is in contrast to the Lagrangian point of view whereby an individual particle is followed throughout its motion. The reader will recall that in stress analysis, the Lagrangian point of view is adopted.

   Let us now summarize. By parameter function we mean any suitable function that is used to represent the field variable in a typical element. Parameter functions are most often taken to be polynomials because of the ease with which they can be manipulated, as well as for the reasons delineated below. In stress analysis, the parameter functions are really the displacement functions; in thermal analysis, the temperature function; and in fluid flow analysis, the velocity and pressure functions (if the formulation is based on the primitive variables). The reader should be able to extend these notions to other disciplines, such as electromagnetics and mass transfer.

Restrictions on the Parameter Functions

As explained below, the assumed parameter functions must meet two primary requirements: compatibility and completeness. The primary reason for satisfying these requirements is to ensure convergence as the element size is reduced or, more specifically, as the mesh is refined in a regular fashion. Before actually discussing the meaning of these requirements, it may be instructive to review what is meant by a weak formulation.

It should be recalled that a *weak formulation* to a problem is an integral formulation that contains in its integrand derivatives of the field variable that are of a lower order than those in the original governing differential equation. For an example, the reader may wish to compare the differential equation given by Eq. (4-7) with its corresponding weak formulations: Eq. (4-130) from the variational approach and Eq. (4-159) from the Galerkin method.

Furthermore, it should be recalled that a C^0-continuous problem is one whose weak formulation contains at most only first-order derivatives. In a similar fashion, a C^1-continuous problem is one whose weak formulation contains at most only second-order derivatives. This may be generalized as follows: a C^{n-1}-continuous problem is one whose weak formulation contains at most only nth order derivatives. We are now in a position to present the compatibility requirement.

Compatibility

The compatibility requirement may be stated as follows. For C^0-continuous problems, the parameter function itself (not its derivatives) must be continuous along the boundaries of the element. For C^1-continuous problems, the parameter function and its first derivative must be continuous, and not necessarily zero, along the boundaries of the element. This may be generalized as follows. In C^n-continuous problems, the parameter function and its first n derivatives are continuous, and not necessarily zero, along the boundaries of the element.

It is instructive to discuss this requirement as it relates to one-dimensional problems, as shown in Fig. 6-1. In particular, consider Fig. 6-1(a). Note that for the C^0-continuous problem, such as the uniaxial stress member or one-dimensional heat transfer in a fin, the field variable itself is continuous at the interface between any two elements (in this case, at each node). It should be recalled that the weak formulation to these problems involves at most only first-order derivatives.

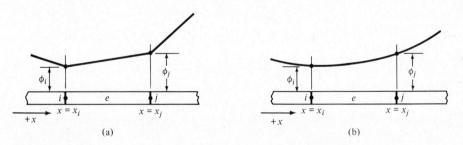

Figure 6-1   One-dimensional element with (a) C^0-continuous parameter functions and (b) C^1-continuous parameter functions. In (a) only the parameter function itself is continuous at each element interface; in (b) the parameter function and its slope are continuous at each element interface.

On the other hand, in C^1-continuous problems, the parameter function itself and its first derivative must be continuous on the interface between any two elements. This situation arises in the case of the beam element shown in Fig. 6-1(b) where both deflection continuity and slope continuity must be guaranteed so that the beam does not develop kinks. Recall from elementary solid mechanics that the slope of the beam at any point is simply the derivative of the deflection at this point (see Sec. 7-5).

The reader should be thoroughly convinced that problems in two-dimensional stress (plane stress and plane strain), axisymmetric stress, and three-dimensional stress are all C^0-continuous problems. The reason for this is that the weak formulation to these problems involves at most first-order derivatives of the displacement(s). Therefore, only the displacement function itself (not any of its derivatives) needs to be continuous along the element boundaries. In Example 6-1 (in Sec. 6-3), the two-node lineal element is shown to satisfy the compatibility requirement for C^0-continuous problems. The reader should be further convinced that the parameter functions assumed in Sec. 5-6 for the triangular and tetrahedral elements do in fact satisfy the compatibility requirement (see Problems 6-3 and 6-4).

On the other hand, in problems involving the bending of plates and shells, the deflections and slopes should be continuous along the element boundaries. This ensures that the plate or shell does not kink. Problems of this type are C^1-continuous because the weak formulation contains at most only second-order derivatives of the deflections.

Elements that obey the compatibility requirement are said to be *conforming*, and those that do not obey this requirement are said to be *nonconforming*. Each of the elements presented in Sec. 5-6 is conforming for problems with C^0-continuity. In the analysis of plate bending with the three-node triangular element, slope continuity along the element boundaries cannot be guaranteed (the slopes at the nodes are continuous). Therefore, this element is said to be nonconforming. However, this element is used in such analyses because it gives results of acceptable accuracy (particularly if special precautions are taken), even though it violates the compatibility requirement [1].

Let us now turn to the second requirement, that of completeness.

Completeness

For C^0-continuous problems, the completeness requirement may be stated as follows. The parameter function must be capable of representing both a constant value of the field variable and constant first partial derivatives as the element size decreases to a point. For C^1-continuous problems, the parameter function must be capable of representing both a constant value of the field variable and constant first and second partial derivatives as the element size decreases to a point. This may be generalized as follows. For C^n-continuous problems, the parameter function must be capable of yielding a constant value of the field variable as well as constant partial derivatives of up to order $n + 1$ as the element size decreases to a point.

Let us examine what this requirement implies for C^0-continuous problems in stress analysis. Consider, for example, the state of uniaxial stress. Constant values of both the displacement itself and the derivative of the displacement (i.e., the strain) must be possible in the assumed form of the displacement function. For the two-node lineal element, we have been assuming a function of the form

$$u = c_1 + c_2 x \tag{6-1}$$

Note that if c_2 is zero, we have $u = c_1$, which is a constant as required. This is referred to as the rigid body mode since the body should be able to undergo a rigid body displacement without straining (when c_2 takes on a value of zero). In addition, $du/dx = c_2$, which is also a constant as required (regardless of the value of c_1). Since the derivative of the displacement is the strain, this second condition requires the displacement function to allow a constant strain in the element. This notion is readily extended to two- and three-dimensional analyses.

Thermal analysis problems involving heat conduction require only C^0-continuous parameter functions because the weak formulations contain at most only first-order derivatives of the temperature. The completeness requirement is equivalent to allowing for constant temperature and constant derivative of the temperature. Since the derivatives of the temperature [with respect to the spatial coordinate(s)] are proportional to the heat fluxes as a result of conduction, the second condition is equivalent to the requirement that constant heat fluxes be possible within the element. Not surprisingly, the assumed forms of the parameter functions in Sec. 5-6 for the displacements are applicable to problems in thermal analysis (see Chapter 8).

Problems in viscous fluid flow analysis that are formulated in terms of the primitive variables are C^0-continuous because the weak formulation is seen to contain derivatives of the velocities of no order greater than one. The completeness requirement requires that it be possible to have both constant velocities and stresses within the element. Potential flow problems formulated in terms of velocity potential function (or stream function) will also be seen to be C^0-continuous. In this case, the completeness requirement is equivalent to the requirements that it be possible to have constant values of the velocity potential function (or stream function) and of the velocity components within the element.

This may be generalized in the case of all C^0-continuous problems as follows. The assumed parameter function must contain a pure constant term in addition to terms that are first order in the spatial coordinates. The reader should review the assumed form of the parameter functions in Sec. 5-6 in order to be thoroughly convinced that these assumed parameter functions satisfy the completeness requirement (see Problems 6-5 and 6-6). Elements that satisfy the completeness requirement are said to be *complete*.

Further discussion of the compatibility and completeness requirements may be found in several other books on the finite element method. The reader may wish to consult the books by Huebner [2], Desai [3], and Martin and Carey [4], as well as other general books on this subject (see references 38 to 46 at the end of Chapter 1).

6-3 ONE-DIMENSIONAL ELEMENT

In this section the familiar two-node lineal element is reexamined. It should be recalled from Chapters 4 and 5 that this element has a node at each end, as shown for a typical element e in Fig. 6-1(a): node i at $x = x_i$ and node j at $x = x_j$. Each node is usually associated with a planar surface, not just a single point.

Let us represent the parameter function in general as ϕ such that it has a value ϕ_i at node i and ϕ_j at node j. The variable ϕ may represent the displacement, temperature, velocity, and so forth. In what follows, the problem is assumed to be C^0-continuous.

The shape functions for the two-node lineal element can be given in terms of several different coordinate systems. However, we will limit the present development to three of the most popular and convenient types of coordinates: (1) global coordinates, (2) serendipity coordinates, and (3) length coordinates. Each of these is discussed. The main reason for introducing the latter two coordinates is to simplify the element integrations that arise.

Global Coordinates

Although the shape functions have already been derived for the two-node lineal element in terms of the global coordinate x in Chapter 4, it is instructive to review the procedure in terms of the present nomenclature. The method to be presented here may be extended easily to two and three dimensions.

We begin by assuming some type of parameter function. In light of the discussion in Sec. 6-2, the form

$$\phi = c_1 + c_2 x \tag{6-2a}$$

or

$$\phi = [1 \quad x]\begin{bmatrix} c_1 \\ c_2 \end{bmatrix} \tag{6-2b}$$

is appropriate, because for C^0-continuous problems it meets both the compatibility and completeness requirements (see Example 6-1 below). At node i, where $x = x_i$, we require $\phi = \phi_i$; and at node j, where $x = x_j$, we require $\phi = \phi_j$. Therefore, we may write

$$\phi_i = c_1 + c_2 x_i \tag{6-3a}$$

$$\phi_j = c_1 + c_2 x_j \tag{6-3b}$$

which may be written in matrix form as

$$\begin{bmatrix} \phi_i \\ \phi_j \end{bmatrix} = \begin{bmatrix} 1 & x_i \\ 1 & x_j \end{bmatrix}\begin{bmatrix} c_1 \\ c_2 \end{bmatrix} \tag{6-4}$$

Solving for the vector of constants yields

$$\begin{bmatrix} c_1 \\ c_2 \end{bmatrix} = \begin{bmatrix} 1 & x_i \\ 1 & x_j \end{bmatrix}^{-1} \begin{bmatrix} \phi_i \\ \phi_j \end{bmatrix} \tag{6-5}$$

Substituting this into Eq. (6-2b) yields

$$\phi = \begin{bmatrix} 1 & x \end{bmatrix} \begin{bmatrix} 1 & x_i \\ 1 & x_j \end{bmatrix}^{-1} \begin{bmatrix} \phi_i \\ \phi_j \end{bmatrix} \tag{6-6}$$

If the 2×2 matrix is inverted and if the matrix multiplications are carried out, we get

$$\phi = \left(\frac{x_j - x}{x_j - x_i} \right) \phi_i + \left(\frac{x - x_i}{x_j - x_i} \right) \phi_j \tag{6-7}$$

which is of the form

$$\phi = N_i(x)\phi_i + N_j(x)\phi_j \tag{6-8}$$

The shape functions are given by

$$N_i(x) = \frac{x_j - x}{x_j - x_i} \tag{6-9a}$$

$$N_j(x) = \frac{x - x_i}{x_j - x_i} \tag{6-9b}$$

which the reader will recall. In Fig. 6-2 these shape functions are shown super-imposed on a typical element. It should be recalled further that these shape functions satisfy the three properties stated in Sec. 5-6, as shown in Example 5-4.

Example 6-1

Show that the following parameter function $\phi = c_1 + c_2 x$ for the two-node lineal element satisfies the compatibility requirement for C^0-continuous problems.

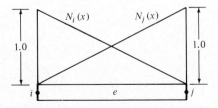

Figure 6-2 Two-node lineal element with shape functions shown for C^0-continuous problems.

Solution

Consider two adjacent elements with global node numbers 1, 2, and 3 as shown in Fig. 6-3. The compatibility requirement requires that, in effect, the value of the parameter function at node 2 (i.e., ϕ_2) must be the same regardless of whether element 1 or element 2 is considered. The assumed form of the parameter function results in the equivalent form given by Eq. (6-7), or

$$\phi(x) = \left(\frac{x_j - x}{x_j - x_i}\right)\phi_i + \left(\frac{x - x_i}{x_j - x_i}\right)\phi_j$$

In terms of the global coordinates, for element 1 at $x = x_2$ we have

$$\phi(x_2) = \left(\frac{x_2 - x_2}{x_2 - x_1}\right)\phi_1 + \left(\frac{x_2 - x_1}{x_2 - x_1}\right)\phi_2 = \phi_2$$

and for element 2 at $x = x_2$ we have

$$\phi(x_2) = \left(\frac{x_3 - x_2}{x_3 - x_2}\right)\phi_2 + \left(\frac{x_2 - x_2}{x_3 - x_2}\right)\phi_3 = \phi_2$$

Since $\phi(x_2)$ from element 1 is the same as $\phi(x_2)$ from element 2 [i.e., both are equal to ϕ_2], the compatibility requirement is satisfied. Obviously, every other shared node in problems with more than two one-dimensional elements can be shown to satisfy the compatibility requirement in a similar fashion. ∎

Note that the element has two nodes and the assumed parameter function has two constants, c_1 and c_2. Therefore, two shape functions should be expected, which was, in fact, what resulted: one for each node. All of this is predicated on the assumption, however, that we are dealing only with C^0-continuous problems. In any event, Eqs. (6-9) provide us with the shape functions for this simple element in terms of the global coordinate (i.e., x).

Next a new coordinate system is introduced that allows us to perform the required integrations more easily, especially when numerical integrations are performed (see Chapter 9).

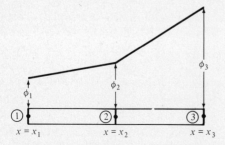

Figure 6-3   Two adjacent two-node lineal elements used in Example 6-1 to show element compatibility.

Serendipity Coordinate

The so-called serendipity coordinate r is a local, normalized coordinate defined in Fig. 6-4 relative to the global coordinate system. The reason for the name serendipity is given in Sec. 6-4. Note that $r = 0$ at $x = \bar{x}$, where $\bar{x} = (x_i + x_j)/2$. Note also that r varies from -1 at $x = x_i$ to $+1$ at $x = x_j$, that is, $-1 \leqslant r \leqslant 1$. It should be obvious that r and x are related by

$$r = \frac{2(x - \bar{x})}{x_j - x_i} \tag{6-10}$$

In terms of this new coordinate, the shape functions become

$$N_i = \tfrac{1}{2}(1 - r) \tag{6-11a}$$

and

$$N_j = \tfrac{1}{2}(1 + r) \tag{6-11b}$$

as shown in Example 6-2 and Problem 6-9.

Example 6-2

Show that the shape function for node i is given by Eq. (6-11a) for the two-node lineal element.

Solution

$$N_i = \frac{1}{2}(1 - r) = \frac{1}{2}\left(1 - \frac{2(x - \bar{x})}{x_j - x_i}\right)$$

$$= \frac{1}{2}\left[\frac{x_j - x_i - 2x + 2[(x_i + x_j)/2]}{x_j - x_i}\right]$$

$$= \frac{1}{2}\left(\frac{2x_j - 2x}{x_j - x_i}\right) = \frac{x_j - x}{x_j - x_i}$$

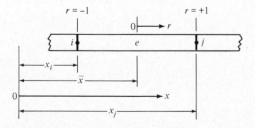

Figure 6-4 Two-node lineal element showing global coordinate x and serendipity coordinate r.

which is the desired result [see Eq. (6-9a)]. The reader should be convinced that the shape functions given by Eqs. (6-11) have the three properties delineated in Sec. 5-6 and that, when plotted on a typical element, the result is the same as Fig. 6-2. ∎

Length Coordinates

The main reason for introducing length coordinates is to take advantage of the simple integration formulas given in Sec. 6-7. Unlike the single global coordinate x and the single serendipity coordinate r, two length coordinates are associated with the lineal element. However, these two length coordinates are not independent, as will be seen shortly. Consider the element shown in Fig. 6-5(a). An internal point p is also shown (point p is *not* a node). Let us now define the so-called length coordinates L_i and L_j as follows:

$$L_i = \frac{\text{length } pj}{\text{length } ij} \tag{6-12a}$$

and

$$L_j = \frac{\text{length } ip}{\text{length } ij} \tag{6-12b}$$

Obviously, we have $0 \le L_i \le 1$ and $0 \le L_j \le 1$. Moreover, we must also have

$$L_i + L_j = 1 \tag{6-13}$$

because the sum of the fractional lengths must equal unity. The two length coordinates are shown on the element in Fig. 6-5(b). If the point p is located at the general global coordinate x, then Eqs. (6-12) are equivalent to

$$L_i = \frac{x_j - x}{x_j - x_i} \tag{6-14a}$$

and

$$L_j = \frac{x - x_i}{x_j - x_i} \tag{6-14b}$$

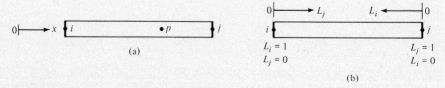

(a)

(b)

Figure 6-5 Two-node lineal element showing (a) interior point p and (b) the two length coordinates L_i and L_j. *Note:* Point p is not a node.

The reader should recognize the expressions on the right-hand sides as N_i and N_j, and we have the fortuitous result that the shape functions, in terms of the length coordinates, are given by

$$N_i = L_i \qquad \text{(6-15a)}$$

and

$$N_j = L_j \qquad \text{(6-15b)}$$

Once again, these shape functions are seen to have all three properties stated in Sec. 5-6 and can be plotted as shown in Fig. 6-2. Like the serendipity coordinate, the length coordinates are also referred to as the local, normalized coordinates (or simply normalized coordinates) because they are defined locally in each element and are normalized to unity.

6-4 TWO-DIMENSIONAL ELEMENTS

In this section two of the most popular two-dimensional elements are considered: the three-node triangular element and the four-node rectangular element. In both cases, the appropriate shape functions are derived in terms of the global coordinates and then cast into forms that utilize suitable normalized coordinates. In both cases, only the shape functions for C^0-continuous problems are considered.

Three-Node Triangular Element

A typical triangular element is shown in Fig. 6-6(a) with nodes i, j, and k. By convention, the nodes associated with a particular element are always given in a *counterclockwise* order. For example, the element shown in Fig. 6-6(b) with global

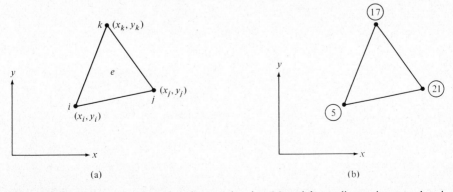

Figure 6-6 Three-node triangular element showing (a) nodal coordinates in general and (b) global node numbers for a typical element.

node numbers 5, 17, and 21 may be defined as having nodes 5, 21, 17 *or* 21, 17, 5 *or* 17, 5, 21 because each of these designations is in accordance with the convention just given.

It is convenient to set up a cartesian reference frame and to define the coordinates of the nodes in terms of these coordinates. In other words, the coordinates of nodes i, j, and k are (x_i, y_i), (x_j, y_j), and (x_k, y_k), respectively. Let us again represent the parameter function as ϕ such that at nodes i, j, and k we have ϕ_i, ϕ_j, and ϕ_k. In other words, the parameter function ϕ must evaluate to ϕ_i at (x_i, y_i), etc. Once again ϕ may be the x or y component of displacement, the temperature, etc.

We are now in a position to derive the shape functions for the three-node triangular element in terms of the global (x, y) coordinates.

Global Coordinates

We begin by assuming some type of parameter function. From Secs. 5-6 and 6-2 it would appear that the form

$$\phi = c_1 + c_2 x + c_3 y \tag{6-16a}$$

or

$$\phi = \begin{bmatrix} 1 & x & y \end{bmatrix} \begin{bmatrix} c_1 \\ c_2 \\ c_3 \end{bmatrix} \tag{6-16b}$$

is appropriate because it meets both the compatibility and the completeness requirements for C^0-continuous problems. At node i where $x = x_i$ and $y = y_i$, we require $\phi = \phi_i$; at node j where $x = x_j$ and $y = y_j$, we require $\phi = \phi_j$; and finally at node k where $x = x_k$ and $y = y_k$, we require $\phi = \phi_k$. From this and Eq. (6-16a), it follows that

$$\phi_i = c_1 + c_2 x_i + c_3 y_i \tag{6-17a}$$

$$\phi_j = c_1 + c_2 x_j + c_3 y_j \tag{6-17b}$$

$$\phi_k = c_1 + c_2 x_k + c_3 y_k \tag{6-17c}$$

Writing these in matrix form gives

$$\begin{bmatrix} \phi_i \\ \phi_j \\ \phi_k \end{bmatrix} = \begin{bmatrix} 1 & x_i & y_i \\ 1 & x_j & y_j \\ 1 & x_k & y_k \end{bmatrix} \begin{bmatrix} c_1 \\ c_2 \\ c_3 \end{bmatrix} \tag{6-18}$$

Solving for the vector of constants and substituting the result into Eq. (6-16b) yields

$$\phi = \begin{bmatrix} 1 & x & y \end{bmatrix} \begin{bmatrix} 1 & x_i & y_i \\ 1 & x_j & y_j \\ 1 & x_k & y_k \end{bmatrix}^{-1} \begin{bmatrix} \phi_i \\ \phi_j \\ \phi_k \end{bmatrix} \tag{6-19}$$

Equation (6-19) is of the form

$$\phi = N_i(x, y)\phi_i + N_j(x, y)\phi_j + N_k(x, y)\phi_k \tag{6-20}$$

where it can be shown (see Problem 6-10) that the shape functions are given by

$$N_i(x, y) = m_{11} + m_{21}x + m_{31}y \tag{6-21a}$$

$$N_j(x, y) = m_{12} + m_{22}x + m_{32}y \tag{6-21b}$$

$$N_k(x, y) = m_{13} + m_{23}x + m_{33}y \tag{6-21c}$$

and in turn

$$
\begin{array}{lll}
m_{11} = (x_j y_k - x_k y_j)/2A & m_{21} = (y_j - y_k)/2A & m_{31} = (x_k - x_j)/2A \\
m_{12} = (x_k y_i - x_i y_k)/2A & m_{22} = (y_k - y_i)/2A & m_{32} = (x_i - x_k)/2A \\
m_{13} = (x_i y_j - x_j y_i)/2A & m_{23} = (y_i - y_j)/2A & m_{33} = (x_j - x_i)/2A
\end{array}
\tag{6-21d}
$$

and

$$A = \tfrac{1}{2} \det \begin{bmatrix} 1 & x_i & y_i \\ 1 & x_j & y_j \\ 1 & x_k & y_k \end{bmatrix} = \text{area of triangle } ijk \tag{6-21e}$$

It should be noted that these m_{ij}'s are strictly a function of the nodal coordinates. Since the area A of the triangular element is never zero, the inverse indicated in Eq. (6-19) will always exist. This in turn implies unique values of the m_{ij}'s for every element. The reader should show that these shape functions satisfy the three properties given in Sec. 5-6 (see Problem 6-11).

Note that the element has three nodes and the assumed parameter function has three constants, c_1, c_2, and c_3. Therefore, three shape functions are to be expected and, indeed, three such functions were found, as given by Eqs. (6-21). Next a new type of normalized coordinate is introduced that allows us to evaluate easily the various integrals that arise.

Area Coordinates

Consider the three-node triangular element shown in Fig. 6-7. The point p is an arbitrary internal point, not a node. The area coordinates L_i, L_j, and L_k are defined as follows:

$$L_i = \frac{\text{area } pjk}{\text{area } ijk} \tag{6-22a}$$

$$L_j = \frac{\text{area } pki}{\text{area } ijk} \tag{6-22b}$$

$$L_k = \frac{\text{area } pij}{\text{area } ijk} \tag{6-22c}$$

From these definitions it follows that the area coordinates are not all independent but are related by

$$L_i + L_j + L_k = 1 \tag{6-23}$$

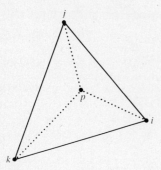

Figure 6-7 Three-node triangular element with interior point p. *Note:* Point p is not a node.

because the sum of the fractional areas must equal unity. It should also be obvious that when point p coincides with node i, $L_i = 1$ and $L_j = L_k = 0$. This last observation is generalized and summarized below:

$$
\begin{array}{lll}
L_i(x_i,y_i) = 1 & L_i(x_j,y_j) = 0 & L_i(x_k,y_k) = 0 \\
L_j(x_i,y_i) = 0 & L_j(x_j,y_j) = 1 & L_j(x_k,y_k) = 0 \\
L_k(x_i,y_i) = 0 & L_k(x_j,y_j) = 0 & L_k(x_k,y_k) = 1
\end{array}
\tag{6-24}
$$

Now consider points p_1 and p_2, which happen to be on a straight line that is parallel to leg jk as shown in Fig. 6-8(a). Because the area of a triangle is given by one-half the product of the base and the altitude, it should be clear that lines of constant area coordinate, e.g., L_i, are parallel to the opposite leg, in this case leg jk. Moreover, as the line containing points p_1 and p_2 is moved parallel to leg jk, the area pjk varies linearly. Therefore, the area coordinates L_i must vary linearly

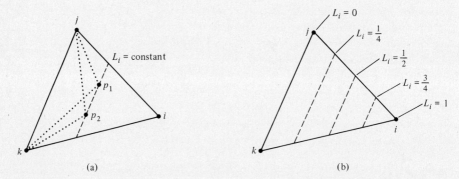

Figure 6-8 Three-node triangular element showing lines of constant area coordinates. In (a) triangles p_1jk and p_2jk have the same areas and hence the dashed line represents a line of constant L_i. In (b) specific lines of constant L_i are shown.

from zero on leg jk to unity at node i. These observations are summarized in Fig. 6-8(b), which shows several lines of constant L_i. Similar conclusions may be drawn about the behavior of L_j and L_k.

The reader may have noted by now that the area coordinates possess the three properties of C^0-continuous shape functions given in Sec. 5-6 for the triangular element. Consequently, the shape functions for nodes i, j, and k may be given in terms of the area coordinates as

$$N_i = L_i \qquad N_j = L_j \qquad \text{and} \qquad N_k = L_k \qquad \text{(6-25)}$$

The implication of this result will be appreciated when integrals must be evaluated that contain the shape functions, as shown in Example 6-8 in Sec. 6-7. Example 6-3 below shows a little more rigorously why $N_i = L_i$, and the reader may show that $N_j = L_j$ and $N_k = L_k$ in a similar fashion (see Problems 6-13 and 6-14).

Example 6-3

From the definition of L_i given by Eq. (6-22a), show that L_i is the same as the shape function N_i given in Eqs. (6-21), and hence $N_i = L_i$.

Solution

From the definition of L_i we have

$$
\begin{aligned}
L_i &= \frac{\text{area } pjk}{\text{area } ijk} = \frac{\frac{1}{2} \det \begin{bmatrix} 1 & x & y \\ 1 & x_j & y_j \\ 1 & x_k & y_k \end{bmatrix}}{A} \\
&= \frac{(x_j y_k - x_k y_j) + (y_j - y_k)x + (x_k - x_j)y}{2A} \\
&= m_{11} + m_{21}x + m_{31}y = N_i
\end{aligned}
$$

where m_{11}, m_{21}, and m_{31} are defined in Eq. (6-21d). ∎

Four-Node Rectangular Element

Figure 6-9 shows a typical rectangular element with nodes i, j, k, and m in a cartesian reference frame. The element must be oriented such that sides ij and km are parallel to the y axis, and sides jk and mi parallel to the x axis. This element is useful in situations in which the geometry is regular. In this text, the nodes associated with a given element must be given by starting at node i and proceeding counterclockwise to node m. For example, the element shown in Fig. 6-9(b) must be defined as having nodes 18, 5, 10, 7, in this order, if the shape functions given below are used. In Chapter 9, these restrictions are lifted at the expense of a little more mathematics.

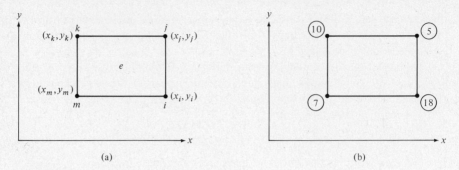

Figure 6-9 Four-node rectangular element showing (a) nodal coordinates in general and (b) global node numbers for a typical element.

Each node is defined by a unique global node number and by its coordinates. In other words, the coordinates of node i are (x_i, y_i); those of node j are (x_j, y_j); etc. Again let us represent the parameter function as ϕ such that at nodes i, j, k, and m, we have ϕ_i, ϕ_j, ϕ_k, and ϕ_m. Stated differently, the parameter function ϕ evaluates to ϕ_i at (x_i, y_i), and so forth. Of course, ϕ may represent the x or y component of displacement, the temperature, and so forth.

The shape functions for the four-node rectangular element may now be derived in terms of the global (x, y) coordinate system.

Global Coordinates

As usual we start by assuming some type of parameter function. The following function meets the compatibility and completeness requirements for C^0-continuous problems from Sec. 6-2:

$$\phi = c_1 + c_2 x + c_3 y + c_4 xy \tag{6-26a}$$

or

$$\phi = \begin{bmatrix} 1 & x & y & xy \end{bmatrix} \begin{bmatrix} c_1 \\ c_2 \\ c_3 \\ c_4 \end{bmatrix} \tag{6-26b}$$

In the usual manner, we require ϕ to be equal to ϕ_i at node i where $x = x_i$ and $y = y_i$, or

$$\phi_i = c_1 + c_2 x_i + c_3 y_i + c_4 x_i y_i \tag{6-27a}$$

Similarly, we have

$$\phi_j = c_1 + c_2 x_j + c_3 y_j + c_4 x_j y_j \tag{6-27b}$$

$$\phi_k = c_1 + c_2 x_k + c_3 y_k + c_4 x_k y_k \tag{6-27c}$$

$$\phi_m = c_1 + c_2 x_m + c_3 y_m + c_4 x_m y_m \tag{6-27d}$$

Writing Eqs. (6-27) in matrix form yields

$$
\begin{bmatrix} \phi_i \\ \phi_j \\ \phi_k \\ \phi_m \end{bmatrix} = \begin{bmatrix} 1 & x_i & y_i & x_i y_i \\ 1 & x_j & y_j & x_j y_j \\ 1 & x_k & y_k & x_k y_k \\ 1 & x_m & y_m & x_m y_m \end{bmatrix} \begin{bmatrix} c_1 \\ c_2 \\ c_3 \\ c_4 \end{bmatrix}
$$

(6-28)

Solving Eq. (6-28) for the vector of constants and substituting into Eq. (6-26b) gives

$$
\phi = \begin{bmatrix} 1 & x & y & xy \end{bmatrix} \begin{bmatrix} 1 & x_i & y_i & x_i y_i \\ 1 & x_j & y_j & x_j y_j \\ 1 & x_k & y_k & x_k y_k \\ 1 & x_m & y_m & x_m y_m \end{bmatrix}^{-1} \begin{bmatrix} \phi_i \\ \phi_j \\ \phi_k \\ \phi_m \end{bmatrix}
$$

(6-29)

which is of the form

$$
\phi = N_i(x,y)\phi_i + N_j(x,y)\phi_j + N_k(x,y)\phi_k + N_m(x,y)\phi_m
$$

(6-30)

The explicit form of the shape functions in terms of the nodal coordinates and the global (x,y) coordinates is left as an exercise (see Problems 6-15 to 6-18). The fact of the matter is that the shape functions for this element are rarely given in terms of the global coordinates. Instead, they are given in terms of the serendipity coordinates as explained next.

Serendipity Coordinates

The word "serendipity" was invented by Horace Walpole in the eighteenth century on the inspiration of a Persian fairy tale, "The Three Princes of Serendip," the heroes of which often made fortuitous discoveries by chance [5]. Evidently, the shape functions for the four-node rectangular element in terms of the normalized coordinates r and s, defined in Fig. 6-10, were also originally discovered by chance;

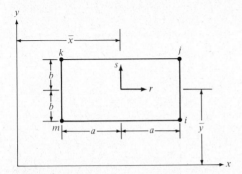

Figure 6-10   Four-node rectangular element showing serendipity coordinates (r,s) relative to the global coordinates (x,y). *Note:* $-1 \le r \le +1$ and $-1 \le s \le +1$. *Also note:* $r = (x - \bar{x})/a$ and $s = (y - \bar{y})/b$.

hence the name *serendipity coordinates*. Note that $r = 0$ and $s = 0$ at $x = \bar{x}$ and $y = \bar{y}$, respectively, where

$$\bar{x} = \frac{x_j + x_k}{2} = \frac{x_i + x_m}{2} \tag{6-31a}$$

and

$$\bar{y} = \frac{y_i + y_j}{2} = \frac{y_k + y_m}{2} \tag{6-31b}$$

The serendipity coordinates r and s are normalized by

$$r = \frac{x - \bar{x}}{a} \tag{6-32a}$$

and

$$s = \frac{y - \bar{y}}{b} \tag{6-32b}$$

where a and b are the element half-lengths as shown in Fig. 6-10. It also follows that $-1 \leqslant r \leqslant 1$ and $-1 \leqslant s \leqslant 1$.

Since the element has four nodes and each node has one shape function associated with it, four shape functions must be found. Although these functions can be derived based on the results of Eqs. (6-29), (6-31), and (6-32) after much algebraic manipulation, let us simply present the shape functions as

$$\begin{aligned} N_i &= \tfrac{1}{4}(1 + r)(1 - s) & N_k &= \tfrac{1}{4}(1 - r)(1 + s) \\ N_j &= \tfrac{1}{4}(1 + r)(1 + s) & N_m &= \tfrac{1}{4}(1 - r)(1 - s) \end{aligned} \tag{6-33}$$

It should be noted that each of these shape functions has the same form as the assumed parameter function [see Eq. (6-26a)]. Moreover, the shape functions given in Eq. (6-33) have the following properties:

1. The shape function associated with a particular node evaluates to unity at this node; all other shape functions evaluate to zero at this node.
2. The shape function associated with a given node varies from unity at the node in question to zero at each of the remaining nodes.
3. The sum of the shape functions is identically equal to unity.

The parameter function may be given in terms of these shape functions and the nodal values ϕ_i, ϕ_j, etc., as

$$\phi = \tfrac{1}{4}(1 + r)(1 - s)\phi_i + \tfrac{1}{4}(1 + r)(1 + s)\phi_j$$
$$+ \tfrac{1}{4}(1 - r)(1 + s)\phi_k + \tfrac{1}{4}(1 - r)(1 - s)\phi_m \tag{6-34}$$

This form of the parameter function meets the compatibility and completeness requirements stated in Sec. 6-2.

Example 6-4

Integrals arise routinely in the finite element method. In Chapter 8, for example, the integral for the *stiffness* matrix as a result of convection from a surface will be seen to contain the shape functions and the convective heat transfer coefficient h. A typical entry in the matrix is given by the integral $\int_A N_i N_j h \, dx \, dy$. With the help of the shape functions given by Eq. (6-33), evaluate this integral for the four-node rectangular element. Assume h to be constant.

Solution

First, from Eqs. (6-32) we note that

$$dx = a \, dr$$

and

$$dy = b \, ds$$

The integral is then evaluated as follows:

$$\int_A N_i N_j h \, dx \, dy = \int_{-1}^{1} \int_{-1}^{1} [\tfrac{1}{4}(1 + r)(1 - s)][\tfrac{1}{4}(1 + r)(1 + s)] h \, ab \, dr \, ds$$

$$= \frac{abh}{16} \int_{-1}^{1} \int_{-1}^{1} (1 + 2r + r^2)(1 - s^2) \, dr \, ds$$

$$= \frac{abh}{16} \left[\int_{-1}^{1} (1 + 2r + r^2) \, dr \right] \left[\int_{-1}^{1} (1 - s^2) \, ds \right]$$

$$= \frac{abh}{16} \left[\left(r + r^2 + \frac{r^3}{3} \right) \Big|_{-1}^{+1} \right] \left[\left(s - \frac{s^3}{3} \right) \Big|_{-1}^{+1} \right]$$

$$= \frac{abh}{16} \left(\frac{8}{3} \right) \left(\frac{4}{3} \right) = \frac{2}{9} abh = \frac{hA}{18}$$

Note that the area A of the element is given by $(2a)(2b)$ and, in fact, has been used above. It will be shown in Chapter 9 how integrals such as this may be evaluated using Gauss-Legendre quadrature (a particular type of numerical integration). ∎

6-5   THREE-DIMENSIONAL ELEMENTS

Two of the most popular three-dimensional elements are now considered: the four-node tetrahedral element and the eight-node brick element. In the case of the former, the appropriate shape functions are derived in terms of the global coordinates and then given in terms of suitable normalized coordinates. In the case of the brick element, the shape functions are simply given in terms of the serendipity coordinates, now r, s, and t. In both cases, only the shape functions for C^0-continuous problems are considered.

Four-Node Tetrahedral Element

A typical tetrahedral element with nodes i, j, k, and m is shown in Fig. 6-11. The nodes associated with a particular element must be given in the order specified by the following convention:

1. Any of the nodes may be specified to be the *first* node.
2. The next three nodes must be taken in a counterclockwise direction as viewed from the first node.

In effect, this gives twelve different ways in which an element may be defined, as summarized in Table 6-1. Zienkiewicz [6] gives a systematic way of splitting an eight-cornered brick into six tetrahedra. Such a process could be relegated to a subroutine and obviously makes the tetrahedral element much more practical because it is easier to picture a three-dimensional body discretized into bricks.

Table 6-1   Proper Element Definitions for the Four-Node Tetrahedral Elements

Nodes			
1	2	3	4
i	j	k	m
i	k	m	j
i	m	j	k
j	k	i	m
j	i	m	k
j	m	k	i
k	i	j	m
k	j	m	i
k	m	i	j
m	i	k	j
m	k	j	i
m	j	i	k

Example 6-5

Consider the element shown in Fig. 6-12. Which of the following element definitions satisfies the ordering convention?

a. 5, 9, 18, 21
b. 9, 5, 18, 21

Solution

The order 5, 9, 18, 21 does satisfy the convention because nodes 9, 18, and 21 occur in a counterclockwise order when viewed from node 5. The order 9, 5, 18,

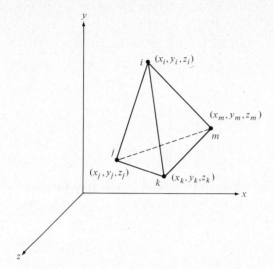

Figure 6-11   Four-node tetrahedral element showing nodal coordinates in general.

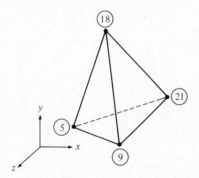

Figure 6-12   Tetrahedral element with global node numbers shown for a typical element.

21 does not satisfy the convention because nodes 5, 18, and 21 do not occur in a counterclockwise direction when viewed from node 9.   ■

Let us set up a cartesian reference frame and define the coordinates of the nodes in terms of these global coordinates. In other words, the coordinates of nodes i, j, k, and m are (x_i, y_i, z_i), (x_j, y_j, z_j), etc. Let us again represent the parameter function as ϕ such that at nodes i, j, k, and m, we have ϕ_i, ϕ_j, ϕ_k, and ϕ_m. Stated differently, the parameter function ϕ must evaluate to ϕ_i at (x_i, y_i, z_i), etc. Of course, ϕ may be the x, y, or z components of displacements, the temperature, etc.

The shape functions for the four-node tetrahedral element are now derived in terms of the global (x, y, z) coordinates by using the procedure illustrated in Secs. 6-2 and 6-3.

Global Coordinates

From Secs. 5-6 and 6-2, it may be concluded that the parameter function

$$\phi = c_1 + c_2 x + c_3 y + c_4 z \qquad \textbf{(6-35a)}$$

or

$$\phi = [1 \quad x \quad y \quad z] \begin{bmatrix} c_1 \\ c_2 \\ c_3 \\ c_4 \end{bmatrix} \qquad \textbf{(6-35b)}$$

is appropriate because it meets both the compatibility and the completeness requirements for C^0-continuous problems. At node i, where $x = x_i$, $y = y_i$, and $z = z_i$, we insist that $\phi = \phi_i$, or

$$\phi_i = c_1 + c_2 x_i + c_3 y_i + c_4 z_i \qquad \textbf{(6-36a)}$$

In a completely analogous manner, we have

$$\phi_j = c_1 + c_2 x_j + c_3 y_j + c_4 z_j \qquad \textbf{(6-36b)}$$

$$\phi_k = c_1 + c_2 x_k + c_3 y_k + c_4 z_k \qquad \textbf{(6-36c)}$$

and

$$\phi_m = c_1 + c_2 x_m + c_3 y_m + c_4 z_m \qquad \textbf{(6-36d)}$$

Writing Eqs. (6-36) in matrix form yields

$$\begin{bmatrix} \phi_i \\ \phi_j \\ \phi_k \\ \phi_m \end{bmatrix} = \begin{bmatrix} 1 & x_i & y_i & z_i \\ 1 & x_j & y_j & z_j \\ 1 & x_k & y_k & z_k \\ 1 & x_m & y_m & z_m \end{bmatrix} \begin{bmatrix} c_1 \\ c_2 \\ c_3 \\ c_4 \end{bmatrix} \qquad \textbf{(6-37)}$$

Solving for the vector of constants and substituting the result into Eq. (6-35b) gives

$$\phi = [1 \quad x \quad y \quad z] \begin{bmatrix} 1 & x_i & y_i & z_i \\ 1 & x_j & y_j & z_j \\ 1 & x_k & y_k & z_k \\ 1 & x_m & y_m & z_m \end{bmatrix}^{-1} \begin{bmatrix} \phi_i \\ \phi_j \\ \phi_k \\ \phi_m \end{bmatrix} \qquad \textbf{(6-38)}$$

which is of the form

$$\phi = N_i(x,y,z)\phi_i + N_j(x,y,z)\phi_j + N_k(x,y,z)\phi_k + N_m(x,y,z)\phi_m \qquad \textbf{(6-39)}$$

It can be shown that the shape functions are given by

$$N_i(x,y,z) = m_{11} + m_{21}x + m_{31}y + m_{41}z \qquad \textbf{(6-40a)}$$

$$N_j(x,y,z) = m_{12} + m_{22}x + m_{32}y + m_{42}z \qquad \textbf{(6-40b)}$$

$$N_k(x,y,z) = m_{13} + m_{23}x + m_{33}y + m_{43}z \qquad \textbf{(6-40c)}$$

$$N_m(x,y,z) = m_{14} + m_{24}x + m_{34}y + m_{44}z \qquad \textbf{(6-40d)}$$

where

$$m_{11} = \frac{1}{6V} \det \begin{bmatrix} x_j & y_j & z_j \\ x_k & y_k & z_k \\ x_m & y_m & z_m \end{bmatrix} \qquad m_{21} = -\frac{1}{6V} \det \begin{bmatrix} 1 & y_j & z_j \\ 1 & y_k & z_k \\ 1 & y_m & z_m \end{bmatrix}$$

$$m_{31} = \frac{1}{6V} \det \begin{bmatrix} 1 & x_j & z_j \\ 1 & x_k & z_k \\ 1 & x_m & z_m \end{bmatrix} \qquad m_{41} = -\frac{1}{6V} \det \begin{bmatrix} 1 & x_j & y_j \\ 1 & x_k & y_k \\ 1 & x_m & y_m \end{bmatrix}$$

(6-40e)

and so forth. In these expressions, V is the volume of the tetrahedron and is given in terms of the nodal coordinates by

$$V = \frac{1}{6} \det \begin{bmatrix} 1 & x_i & y_i & z_i \\ 1 & x_j & y_j & z_j \\ 1 & x_k & y_k & z_k \\ 1 & x_m & y_m & z_m \end{bmatrix} = \text{volume } ijkm \qquad \text{(6-40f)}$$

The reader may want to review Sec. 2-7 in order to see more clearly how these results were obtained. Since the volume of the tetrahedral element is never zero, the indicated inverse in Eq. (6-38) always exists and, therefore, each of the m_{ij}'s may be computed for each element as indicated above. It can be shown that these shape functions satisfy the three properties given in Sec. 5-6 (see Problems 6-30 to 6-32).

Finally, it is noted that the element has four nodes and the assumed parameter function has four constants, which in turn implies four shape functions. This is, in fact, seen to be the case. In effect, this assures us that the four-node tetrahedral element meets the compatibility requirement for C^0-continuous problems.

Next another normalized coordinate is introduced—the volume coordinate. In Sec. 6-7 (and Chapter 9), the reader will come to appreciate these coordinates.

Volume Coordinates

Consider the four-node tetrahedral element shown in Fig. 6-13, where the internal point p is not to be confused with a node. In effect, point p is used to create four

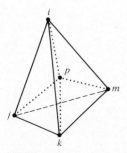

Figure 6-13 Tetrahedral element with interior point p. *Note:* Point p is not a node.

other tetrahedra, each contained in the original tetrahedron. The volume coordinates L_i, L_j, L_k, and L_m are defined as follows:

$$L_i = \frac{\text{volume } pjkm}{\text{volume } ijkm} \qquad L_k = \frac{\text{volume } pijm}{\text{volume } ijkm}$$

$$L_j = \frac{\text{volume } pkmi}{\text{volume } ijkm} \qquad L_m = \frac{\text{volume } pijk}{\text{volume } ijkm} \qquad \textbf{(6-41)}$$

Because the sum of these fractional volumes must equal unity, the four volume coordinates are not independent but rather are related by

$$L_i + L_j + L_k + L_m = 1 \qquad \textbf{(6-42)}$$

Obviously, if the point p coincides with one of the nodes, the volume coordinate associated with that node has a value of unity and all other volume coordinates are zero. Since the volume of a tetrahedron is one-third the product of the area of any face and the height normal to the face, it follows that the volume coordinates vary linearly throughout the tetrahedron. It also follows that planes of constant volume coordinate for a given node are parallel to the face of the tetrahedron that is opposite the node.

Like the length and area coordinates, the volume coordinates possess the same three properties that the shape functions have for C^0-continuous problems. Therefore, the shape function for a particular node is equal to the corresponding volume coordinate, or

$$N_i = L_i \qquad N_j = L_j \qquad N_k = L_k \qquad \text{and} \qquad N_m = L_m \qquad \textbf{(6-43)}$$

Example 6-6 below shows more rigorously why $N_i = L_i$ from which the reader may show in a similar fashion the validity of the remaining equalities in Eq. (6-43) (see Problems 6-33 to 6-35).

Example 6-6

With the help of the definition of L_i given in Eq. (6-41), show that L_i is the same as the shape function N_i given in Eqs. (6-40), and hence $N_i = L_i$.

Solution

Beginning with the definition of L_i, we write

$$L_i = \frac{\text{volume } pjkm}{\text{volume } ijkm} = \frac{\dfrac{1}{6}\det \begin{bmatrix} 1 & x & y & z \\ 1 & x_j & y_j & z_j \\ 1 & x_k & y_k & z_k \\ 1 & x_m & y_m & z_m \end{bmatrix}}{V}$$

or

$$L_i = \frac{1}{6V} \det \begin{bmatrix} x_j & y_j & z_j \\ x_k & y_k & z_k \\ x_m & y_m & z_m \end{bmatrix} - \frac{x}{6V} \det \begin{bmatrix} 1 & y_j & z_j \\ 1 & y_k & z_k \\ 1 & y_m & z_m \end{bmatrix}$$

$$+ \frac{y}{6V} \det \begin{bmatrix} 1 & x_j & z_j \\ 1 & x_k & z_k \\ 1 & x_m & z_m \end{bmatrix} - \frac{z}{6V} \det \begin{bmatrix} 1 & x_j & y_j \\ 1 & x_k & y_k \\ 1 & x_m & y_m \end{bmatrix}$$

or

$$L_i = m_{11} + m_{21}x + m_{31}y + m_{41}z = N_i$$

where m_{11}, m_{21}, m_{31}, and m_{41} are defined in Eqs. (6-40). ■

Eight-Node Brick Element

The eight-node brick element, shown in Fig. 6-14(a) is also known as the *rectangular prismatic element*. Note that the nodes are conveniently numbered 1, 2, . . . , 8. The letter designation (i, j, etc.) for this element is cumbersome and, therefore, is abandoned.

Each element must be defined by its eight global node numbers given in the order 1, 2, . . . , 8. For example, the brick element in Fig. 6-14(b) must be defined as having nodes 15, 18, 14, 10, 21, 62, 81, and 45 (in this order) if the shape functions given below are used. The element must be oriented such that its 1-2-3-4 face is parallel to the x-y plane, the 1-5-6-2 face is parallel to the y-z plane, and so forth. In Chapter 9 these overly restrictive requirements are relaxed.

Each node is typically given a unique global node number and is defined by its nodal coordinates. For example, the coordinates of node 1 are (x_1, y_1, z_1). Let us represent the parameter function as ϕ. It follows that at nodes 1, 2, . . . , 8,

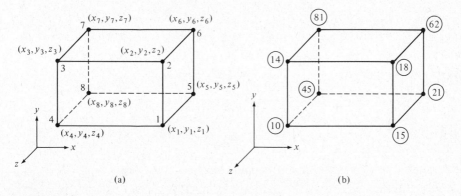

Figure 6-14 (a) Eight-node brick element showing nodal coordinates in general. (b) Eight-node brick element showing global node numbers for a typical element.

the parameter function evaluates to $\phi_1, \phi_2, \ldots, \phi_8$. As usual, ϕ may represent one of the components of the displacement, the temperature, and so forth.

Global Coordinates

The shape functions for the brick element may be derived in terms of the global coordinates by following the standard procedure that has been illustrated in this chapter for each of the other elements. As a starting point, the following parameter function may be assumed:

$$\phi = c_1 + c_2 x + c_3 y + c_4 z + c_5 xy + c_6 yz + c_7 zx + c_8 xyz \quad \textbf{(6-44)}$$

The reader is assured that this form of the parameter function satisfies both the compatibility and completeness requirements for C^0-continuous problems. Although the standard procedure would yield the shape functions (after much algebra), it is far more convenient to use the shape functions for this element if they are given in terms of the serendipity coordinates as given below.

Serendipity Coordinates

Figure 6-15 shows the four-node brick element and a local, normalized coordinate system (r,s,t) with its origin at the centroid $(\bar{x},\bar{y},\bar{z})$ of the brick. The coordinates of the centroid are given by

$$\bar{x} = \frac{x_1 + x_4}{2} = \frac{x_2 + x_3}{2}, \text{ etc.} \quad \textbf{(6-45a)}$$

$$\bar{y} = \frac{y_1 + y_2}{2} = \frac{y_3 + y_4}{2}, \text{ etc.} \quad \textbf{(6-45b)}$$

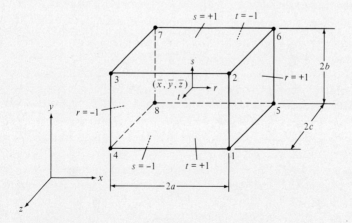

Figure 6-15 Brick element showing the serendipity coordinates r, s, and t. *Note:* $-1 \leq r \leq +1$, $-1 \leq s \leq +1$, and $-1 \leq t \leq +1$. *Also note:* $r = (x - \bar{x})/a$, $s = (y - \bar{y})/b$, and $t = (z - \bar{z})/c$.

$$\bar{z} = \frac{z_1 + z_5}{2} = \frac{z_4 + z_8}{2}, \text{ etc.} \qquad \textbf{(6-45c)}$$

The so-called serendipity coordinates are defined in terms of global coordinates by

$$r = \frac{x - \bar{x}}{a} \qquad s = \frac{y - \bar{y}}{b} \qquad t = \frac{z - \bar{z}}{c} \qquad \textbf{(6-46)}$$

where a, b, and c are the element half-lengths in the x, y, and z directions, respectively. It follows that each serendipity coordinate has a value between -1 and $+1$, or $-1 \leq r \leq +1$, $-1 \leq s \leq +1$, and $-1 \leq t \leq +1$. The shape functions for the eight-node brick element are given below (for C^0-continuous problems):

$$N_1 = \tfrac{1}{8}(1 + r)(1 - s)(1 + t) \qquad N_5 = \tfrac{1}{8}(1 + r)(1 - s)(1 - t)$$

$$N_2 = \tfrac{1}{8}(1 + r)(1 + s)(1 + t) \qquad N_6 = \tfrac{1}{8}(1 + r)(1 + s)(1 - t) \qquad \textbf{(6-47)}$$

$$N_3 = \tfrac{1}{8}(1 - r)(1 + s)(1 + t) \qquad N_7 = \tfrac{1}{8}(1 - r)(1 + s)(1 - t)$$

$$N_4 = \tfrac{1}{8}(1 - r)(1 - s)(1 + t) \qquad N_8 = \tfrac{1}{8}(1 - r)(1 - s)(1 - t)$$

Each of these shape functions has the same form as the assumed parameter function [i.e., Eq. (6-44)]. In addition, the shape functions given by Eqs. (6-47) have the same three properties that the two-dimensional serendipity shape functions possess. Also, the compatibility requirement is met by these shape functions on each face of the element (i.e., on the element boundaries).

Special numerical integration methods make this element quite practical as the reader will come to appreciate in Chapter 9 where Gauss-Legendre quadrature is introduced.

6-6 AXISYMMETRIC ELEMENTS

Many practical problems are axisymmetric in nature. Such problems must meet two conditions to be classified as axisymmetric. The first is that the body must be a body of revolution as shown in Fig. 6-16(a). Consequently, a cylindrical coordinate system is usually adopted with global coordinates r, θ, and z. However, because of the geometric axisymmetry, the geometry is invariant with respect to the θ coordinate. This means that, in effect, only the two remaining coordinates (r,z) need to be considered. The second requirement is that the material properties, the external effects, boundary conditions, and so forth, may not be a function of θ. External effects include surface tractions, body forces, prestresses, and heat sources. Despite these restrictions, many practical problems fall into this class of problems.

The axisymmetry, in effect, allows us to analyze a three-dimensional problem as though it were a two-dimensional problem. This implies that the three-node triangular element and the four-node rectangular element may be used provided that they are rotated through 2π radians as shown in Fig. 6-16. The elemental volume is then given by $dV = 2\pi r \, dr \, dz$. Note that these elements are actually

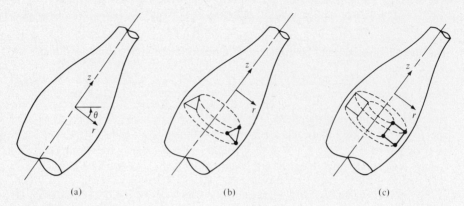

Figure 6-16   (a) Body of revolution discretized into (b) triangular and (c) rectangular donut-shaped elements.

triangular and rectangular donuts or toroids. However, because of the axisymmetry, only half of the planar domain needs to be analyzed, as shown for the axisymmetric body in Fig. 6-17. The *planar domain* is any plane that goes through and is parallel to the centerline of the body. The analysis for all practical purposes is thus *two dimensional*.

The shape functions for the three-node triangular element are given by Eqs. (6-21) [or Eq. (6-25)] provided that x is replaced by r (or z) and y is replaced by z (or r). Of course, each of the nodal coordinates must be similarly reinterpreted. The shape functions for the four-node rectangular element are also readily adapted to this situation.

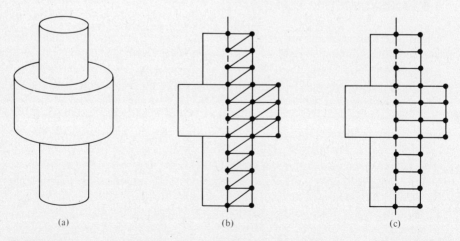

Figure 6-17   (a) Axisymmetric body with its half-plane discretized into (b) triangular elements and (c) rectangular elements.

6-7 SOME SIMPLE INTEGRATION FORMULAS

When the length, area, and volume coordinates were introduced, it was stated that the main reason for doing so was to allow the use of simple integration formulas in evaluating the integrals that routinely arise. These integration formulas are presented in this section and are illustrated in several examples. In each of these formulas the factorial operator is used, where $n! = (n)(n - 1)(n - 2) \cdots (2)(1)$ and $0! = 1$.

Length Coordinates

The appropriate integration formula in terms of the length coordinates L_i and L_j is given by

$$\int_l L_i^\alpha L_j^\beta \, dl = \frac{\alpha! \beta!}{(\alpha + \beta + 1)!} \, l \tag{6-48}$$

where dl is an elemental length between nodes i and j and l is the length of line between nodes i and j, as shown in Fig. 6-18. The exponents α and β must be positive integers.

Example 6-7

With the help of Eq. (6-48) evaluate the element nodal force vector for the body force (that is, $\mathbf{f}_b^e$) as given by Eq. (5-102) for the uniaxial stress member.

Solution

The nodal force vector $\mathbf{f}_b^e$ is given by

$$\mathbf{f}_b^e = \int_{x_i}^{x_j} \mathbf{N}^T \mathbf{b} A \, dx = \int_{x_i}^{x_j} \begin{bmatrix} N_i \\ N_j \end{bmatrix} \gamma A \, dx$$

Let us assume that γ, the weight per unit volume, and A, the cross-sectional area, are both constant, so that the product γA may be removed from the integral.

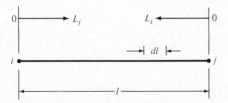

Figure 6-18 One-dimensional element showing infinitesimal length dl as well as the actual element length l.

Moreover, if the shape functions N_i and N_j are given in terms of the length coordinates [see Eqs. (6-15)], we have

$$\mathbf{f}_b^e = \gamma A \int_l \begin{bmatrix} L_i \\ L_j \end{bmatrix} dl = \gamma A \begin{bmatrix} \dfrac{1!0!}{(1 + 0 + 1)!} \, l \\ \dfrac{0!1!}{(0 + 1 + 1)!} \, l \end{bmatrix}$$

Simplifying and noting that l is the element length in this case or $l = x_j - x_i \equiv x_{ji}$, we get

$$\mathbf{f}_b^e = \frac{\gamma A x_{ji}}{2} \begin{bmatrix} 1 \\ 1 \end{bmatrix}$$

which is the expected result [see Eq. (5-102)]. ∎

Area Coordinates

The appropriate integration formula in terms of the area coordinates is given by

$$\int_A L_i^\alpha L_j^\beta L_k^\gamma \, dA = \frac{\alpha!\beta!\gamma!}{(\alpha + \beta + \gamma + 2)!} 2A \tag{6-49}$$

where dA is an elemental area of the element (usually $dx\,dy$ or $dr\,dz$) and A is the area of the triangle formed by nodes i, j, and k. The area A is easily computed from Eq. (6-21e) since the nodal coordinates are always known. The exponents α, β, and γ must be positive integers.

Example 6-8

A typical entry in the stiffness matrix as a result of convection from a thin two-dimensional body is given by $\int_A N_i N_j h \, dx \, dy$ (as shown in Chapter 8). Evaluate this integral if the three-node triangular element is used.

Solution

With the help of Eqs. (6-25) and (6-49), we evaluate the integral as follows:

$$\int_A N_i N_j h \, dx \, dy = \int_A L_i L_j h \, dA$$

$$= h \frac{1!1!0!}{(1 + 1 + 0 + 2)!} 2A = \frac{hA}{12}$$

Volume Coordinates

The appropriate integration formula in terms of the volume coordinates L_i, L_j, L_k, and L_m is given by

$$\int_V L_i^\alpha L_j^\beta L_k^\gamma L_m^\delta \, dV = \frac{\alpha!\beta!\gamma!\delta!}{(\alpha + \beta + \gamma + \delta + 3)!} 6V \qquad (6\text{-}50)$$

where dV is the elemental volume (usually $dx \, dy \, dz$) and V is the volume of the tetrahedron formed by nodes i, j, k, and m. This volume is readily computed from Eq. (6-40f). The exponents α, β, γ, and δ must be positive integers.

Example 6-9

A typical entry in the element nodal force vector as a result of a constant volumetric heat generation rate Q (per unit volume) is seen in Chapter 8 to be given by $\int_V N_j Q \, dx \, dy \, dz$. Assuming a four-node tetrahedral element, evaluate this integral.

Solution

With the help of the second equation in Eq. (6-43), we get

$$\int_V N_j Q \, dx \, dy \, dz = \int_V L_j Q \, dV$$

$$= Q\frac{0!1!0!0!}{(0 + 1 + 0 + 0 + 3)!} 6V = \tfrac{1}{4}QV$$

assuming Q is constant over the element. From this result it is concluded that one-fourth of the heat generation is allocated to node j. In a similar fashion, it may be shown that one-fourth of QV is allocated to each of the remaining three nodes. The total amount is, therefore, given by QV. The validity of this result should be intuitively obvious. ∎

Example 6-10

Redo Example 6-9 with the following spacially dependent volumetric heat generation rate:

$$Q = Q_0 x$$

where Q_0 is a constant.

Solution

The global coordinate x can be represented in the terms of the x coordinates of each of the nodes by writing

$$x = L_i x_i + L_j x_j + L_k x_k + L_m x_m$$

where the volume coordinates have been used. The integral may now be evaluated in the following manner:

$$
\begin{aligned}
\int_V N_j Q \, dx \, dy \, dz &= Q_0 \int_V L_j (L_i x_i + L_j x_j + L_k x_k + L_m x_m) \, dV \\
&= Q_0 \left(\int_V L_j L_i x_i \, dV + \int_V L_j^2 x_j \, dV + \int_V L_j L_k x_k \, dV + \int_V L_j L_m x_m \, dV \right) \\
&= \left[Q_0 \left(x_i \frac{1!1!0!0!}{(1+1+0+0+3)!} 6V + x_j \frac{0!2!0!0!}{(0+2+0+0+3)!} 6V \right. \right. \\
&\qquad \left. \left. + x_k \frac{0!1!1!0!}{(0+1+1+0+3)!} 6V + x_m \frac{0!1!0!1!}{(0+1+0+1+3)!} 6V \right) \right] \\
&= \left(\tfrac{1}{20} x_i + \tfrac{1}{10} x_j + \tfrac{1}{20} x_k + \tfrac{1}{20} x_m \right) Q_0 V \\
&= (x_i + 2x_j + x_k + x_m) \frac{Q_0 V}{20}
\end{aligned}
$$

Unlike the previous example, this result is not at all intuitively obvious.　■

6-8   ACTIVE ZONE EQUATION SOLVER

Up to now we have used only the matrix inversion method to solve the system of algebraic equations implied in $\mathbf{Ka} = \mathbf{f}$. This approach is not very practical in large finite element models because the computation of the inverse of a large matrix can result in excessive computer execution time. A much more efficient solution method is discussed in this section. The basic idea is to decompose the matrix $\mathbf{K}$ into lower and upper triangular matrices $\mathbf{L}$ and $\mathbf{U}$ such that $\mathbf{K} = \mathbf{LU}$. It is shown below how this results in a straightforward solution for the vector $\mathbf{a}$, once this triangular decomposition (or $\mathbf{LU}$ decomposition) has been performed. *Note:* The matrix $\mathbf{L}$ in this section is in no way related to the linear operator matrix (also $\mathbf{L}$) from Chapter 5.

Triangular Decomposition

The triangular decomposition of $\mathbf{K}$ into $\mathbf{L}$ and $\mathbf{U}$ such that $\mathbf{K} = \mathbf{LU}$ is readily summarized in the form of a short algorithm. Let us first illustrate the basis for the algorithm by working on a 3×3 matrix $\mathbf{K}$. It follows from $\mathbf{K} = \mathbf{LU}$ that

$$
\begin{bmatrix} K_{11} & K_{12} & K_{13} \\ K_{21} & K_{22} & K_{23} \\ K_{31} & K_{32} & K_{33} \end{bmatrix} = \begin{bmatrix} L_{11} & 0 & 0 \\ L_{21} & L_{22} & 0 \\ L_{31} & L_{32} & L_{33} \end{bmatrix} \begin{bmatrix} U_{11} & U_{12} & U_{13} \\ 0 & U_{22} & U_{23} \\ 0 & 0 & U_{33} \end{bmatrix}
\tag{6-51}
$$

The number of unknowns on the right-hand side of Eq. (6-51) is seen to be greater than the number of implied equations, that is, twelve unknowns and nine equations. The number of unknowns may always be reduced to the number of implied equations by requiring each diagonal entry in **L** to be unity, or $L_{ii} = 1$. Therefore, Eq. (6-51) becomes

$$
\begin{array}{ccc}
\textcircled{1} & \textcircled{2} & \textcircled{3}
\end{array}
$$

$$
\begin{bmatrix} K_{11} & K_{12} & K_{13} \\ K_{21} & K_{22} & K_{23} \\ K_{31} & K_{32} & K_{33} \end{bmatrix} = \begin{bmatrix} 1 & 0 & 0 \\ L_{21} & 1 & 0 \\ L_{31} & L_{32} & 1 \end{bmatrix}\begin{bmatrix} U_{11} & U_{12} & U_{13} \\ 0 & U_{22} & U_{23} \\ 0 & 0 & U_{33} \end{bmatrix} \qquad \textbf{(6-52)}
$$

Note that the matrix **K** has been divided into (three) zones. In general, an $n \times n$ matrix **K** would be divided into n such zones. If the matrices on the right-hand side are multiplied out and equated with the left-hand side, entry by entry, the following nine equations in nine unknowns (L_{21}, L_{31}, L_{32}, U_{11}, U_{12}, etc.) are obtained:

Zone 1: $\qquad\qquad\qquad K_{11} = U_{11}$

Zone 2: $\qquad\qquad\qquad K_{12} = U_{12}$
$$K_{21} = L_{21}U_{11}$$
$$K_{22} = L_{21}U_{12} + U_{22}$$

Zone 3: $\qquad\qquad\qquad K_{13} = U_{13}$
$$K_{31} = L_{31}U_{11}$$
$$K_{32} = L_{31}U_{12} + L_{32}U_{22}$$
$$K_{23} = L_{21}U_{13} + U_{23}$$
$$K_{33} = L_{31}U_{13} + L_{32}U_{23} + U_{33}$$

These equations may in turn be solved for the nine unknowns to give the following:

Zone 1: $\qquad\qquad\qquad U_{11} = K_{11}$

Zone 2: $\qquad\qquad\qquad U_{12} = K_{12}$

$$L_{21} = \frac{K_{21}}{U_{11}}$$

$$U_{22} = K_{22} - L_{21}U_{12}$$

Zone 3: $\qquad\qquad\qquad U_{13} = K_{13}$

$$L_{31} = \frac{K_{31}}{U_{11}}$$

$$L_{32} = \frac{K_{32} - L_{31}U_{12}}{U_{22}}$$

$$U_{23} = K_{23} - L_{21}U_{13}$$

$$U_{33} = K_{33} - L_{31}U_{13} - L_{32}U_{23}$$

A few subtle points should be made here. Note how the calculations in each zone use only the entries in the **K** matrix that are in this zone and only those entries from the **L** and **U** matrices from previous and present zones (but not from subsequent zones). The *present zone* here is more commonly referred to as the *active zone*; hence, the name active zone equation solver.

The above equations may be generalized for an $n \times n$ matrix $\mathbf{K}$ as follows [7]. For the first active zone, we have

$$U_{11} = K_{11} \tag{6-53}$$

$$L_{11} = 1 \tag{6-54}$$

and for each subsequent active zone j from 2 to n, we have

$$L_{j1} = \frac{K_{j1}}{U_{11}} \tag{6-55}$$

$$U_{1j} = K_{1j} \tag{6-56}$$

and

$$L_{ji} = \frac{K_{ji} - \sum_{m=1}^{i-1} L_{jm}U_{mi}}{U_{ii}} \qquad \text{for } i = 2, 3, \ldots, j-1 \tag{6-57}$$

$$U_{ij} = K_{ij} - \sum_{m=1}^{i-1} L_{im}U_{mj} \qquad \text{for } i = 2, 3, \ldots, j-1 \tag{6-58}$$

and finally

$$L_{jj} = 1 \tag{6-59}$$

$$U_{jj} = K_{jj} - \sum_{m=1}^{j-1} L_{jm}U_{mj} \tag{6-60}$$

For a symmetric $\mathbf{K}$ matrix, the entries in the lower triangular matrix $\mathbf{L}$ can be obtained directly from the upper triangular matrix $\mathbf{U}$ from

$$L_{ji} = \frac{U_{ij}}{U_{ii}}, \qquad \text{for } j \geq i \tag{6-61}$$

In Problem 6-44, the reader is asked to show why Eq. (6-61) holds when $\mathbf{K}$ is symmetric.

The use of these equations is best illustrated by an example.

Example 6-11

Perform the triangular decomposition of the following symmetric matrix:

$$\mathbf{K} = \begin{bmatrix} 8 & 4 & 2 \\ 4 & 8 & 4 \\ 2 & 4 & 8 \end{bmatrix}$$

Solution

The calculations are summarized below for each of the three zones. Although $\mathbf{K}$ is symmetric, Eq. (6-61) is not used in order to illustrate the more general equations.

Zone 1:

$$\begin{array}{ccc} \textcircled{1} & \textcircled{1} & \textcircled{1} \end{array}$$

$$\begin{bmatrix} 8 & 4 & 2 \\ 4 & 8 & 4 \\ 2 & 4 & 8 \end{bmatrix} = \begin{bmatrix} 1 \\ \end{bmatrix} \begin{bmatrix} 8 \\ \end{bmatrix}$$

Note: $L_{11} = 1$ and $U_{11} = 8$

Zone 2:

$$\begin{array}{ccc} \textcircled{2} & \textcircled{2} & \textcircled{2} \end{array}$$

$$\begin{bmatrix} & 4 & 2 \\ 4 & 8 & 4 \\ 2 & 4 & 8 \end{bmatrix} = \begin{bmatrix} 1 \\ 0.5 & 1 \end{bmatrix} \begin{bmatrix} 8 & 4 \\ & 6 \end{bmatrix}$$

Note: $L_{21} = \frac{4}{8} = 0.5$, $U_{12} = 4$, $L_{22} = 1$, and $U_{22} = 8 - (0.5)(4) = 6$.

Zone 3:

$$\begin{array}{ccc} \textcircled{3} & \textcircled{3} & \textcircled{3} \end{array}$$

$$\begin{bmatrix} & & 2 \\ & & 4 \\ 2 & 4 & 8 \end{bmatrix} = \begin{bmatrix} 1 \\ 0.5 & 1 \\ 0.25 & 0.5 & 1 \end{bmatrix} \begin{bmatrix} 8 & 4 & 2 \\ & 6 & 3 \\ & & 6 \end{bmatrix}$$

Note: $L_{31} = \frac{2}{8} = 0.25$, $U_{13} = 2$, $L_{32} = [4 - (0.25)(4)]/6 = 0.5$, $U_{23} = 4 - (0.5)(2) = 3$, $L_{33} = 1$, and $U_{33} = 8 - (0.25)(2) - (0.5)(3) = 6$.

It is concluded that

$$\mathbf{L} = \begin{bmatrix} 1 & 0 & 0 \\ 0.5 & 1 & 0 \\ 0.25 & 0.5 & 1 \end{bmatrix} \quad \text{and} \quad \mathbf{U} = \begin{bmatrix} 8 & 4 & 2 \\ 0 & 6 & 3 \\ 0 & 0 & 6 \end{bmatrix}$$

The reader should verify that the product of $\mathbf{L}$ and $\mathbf{U}$ (in this order) does indeed give the original $\mathbf{K}$ matrix. The $\mathbf{L}$ matrix could also be obtained with Eq. (6-61) since $\mathbf{K}$ is symmetric. ∎

Forward Elimination and Backward Substitution

Let us now reconsider the system of algebraic equations $\mathbf{Ka} = \mathbf{f}$ and try to obtain the solution for the vector $\mathbf{a}$ by taking advantage of the material in the previous section. Beginning with $\mathbf{Ka} = \mathbf{f}$ and using $\mathbf{K} = \mathbf{LU}$, we may write

$$\mathbf{LUa} = \mathbf{f} \tag{6-62}$$

Denoting **Ua** as the vector **z**, we have, in effect, two systems of equations:

$$\mathbf{Lz} = \mathbf{f} \tag{6-63}$$

and

$$\mathbf{Ua} = \mathbf{z} \tag{6-64}$$

Let us write these out explicitly for the case of a 3×3 **K** matrix in order to gain some insight into the general algorithm to be presented. Equation (6-63) implies

$$
\begin{aligned}
z_1 &= f_1 \\
L_{21}z_1 + z_2 &= f_2 \\
L_{31}z_1 + L_{32}z_2 + z_3 &= f_3
\end{aligned}
\tag{6-65}
$$

and Eq. (6-64) implies

$$
\begin{aligned}
U_{11}a_1 + U_{12}a_2 + U_{13}a_3 &= z_1 \\
U_{22}a_2 + U_{23}a_3 &= z_2 \\
U_{33}a_3 &= z_3
\end{aligned}
\tag{6-66}
$$

From Eq. (6-65) it follows that the z_i's may be obtained by a forward sweep as follows:

$$
\begin{aligned}
z_1 &= f_1 \\
z_2 &= f_2 - L_{21}z_1 \\
z_3 &= f_3 - L_{31}z_1 - L_{32}z_2
\end{aligned}
\tag{6-67}
$$

From these and Eq. (6-66), it follows that

$$
\begin{aligned}
a_3 &= \frac{z_3}{U_{33}} \\[2mm]
a_2 &= \frac{z_2 - U_{23}a_3}{U_{22}} \\[2mm]
a_1 &= \frac{z_1 - U_{13}a_3 - U_{12}a_2}{U_{11}}
\end{aligned}
\tag{6-68}
$$

These steps may be generalized to the case of an $n \times n$ **K** matrix in the following manner:

$$z_1 = f_1 \tag{6-69a}$$

$$z_i = f_i - \sum_{j=1}^{i-1} L_{ij}z_j \qquad i = 2, 3, \ldots, n \tag{6-69b}$$

and

$$a_n = \frac{z_n}{U_{nn}} \tag{6-70a}$$

$$a_i = \frac{\left(z_i - \displaystyle\sum_{j=i+1}^{n} U_{ij} a_j \right)}{U_{ii}}, \qquad i = n-1, n-2, \ldots, 1 \tag{6-70b}$$

For rather obvious reasons, the steps in Eqs. (6-69) and (6-70) are referred to as the forward elimination and backward substitution steps, respectively. A numerical example should help to clarify the procedure.

Example 6-12

Solve the following system of algebraic equations by using forward elimination and backward substitution:

$$8a_1 + 4a_2 + 2a_3 = 34$$
$$4a_1 + 8a_2 + 4a_3 = 56$$
$$2a_1 + 4a_2 + 8a_3 = 46$$

Solution

Since the matrix $\mathbf{K}$ is the same as that in Example 6-11, the results of the triangular decomposition from that example may be used. The forward elimination step gives

$$z_1 = 34$$
$$z_2 = 56 - (0.5)(34) = 39$$
$$z_3 = 46 - (0.25)(34) - (0.5)(39) = 18$$

and the backward substitution step gives

$$a_3 = {}^{18}\!/_6 = 3$$
$$a_2 = \frac{39 - (3)(3)}{6} = 5$$
$$a_1 = \frac{34 - (2)(3) - (4)(5)}{8} = 1$$

Therefore, the solution to the original system of equations is given by

$$\mathbf{a} = \begin{bmatrix} 1 \\ 5 \\ 3 \end{bmatrix}$$

A quick check by direct substitution reveals that this is indeed the solution. ■

When the **K** matrix is very large, the triangular decomposition requires much more execution time than the forward elimination and backward substitution steps. Therefore, once the triangular decomposition step has been performed, the system **Ka** = **f** may be solved for several different **f**'s rather economically. This capability is very important in all practical finite element programs and is referred to as *resolution capability*.

Next we turn to the question of when the active zone solution method is guaranteed to give valid results. In order to answer such a question, we must first define what is meant by a symmetric, positive (or negative) definite matrix.

Classification of Symmetric Real Matrices [8]

First, the *principal minors* of a matrix **K** must be defined. If **K** is an $n \times n$ matrix, then the n principal minors of **K** are denoted as $\Delta_1, \Delta_2, \ldots, \Delta_n$ and are given by

$$\Delta_1 = K_{11} \qquad \Delta_2 = \det \begin{bmatrix} K_{11} & K_{12} \\ K_{21} & K_{22} \end{bmatrix} \qquad \Delta_3 = \det \begin{bmatrix} K_{11} & K_{12} & K_{13} \\ K_{21} & K_{22} & K_{23} \\ K_{31} & K_{32} & K_{33} \end{bmatrix} \quad \textbf{(6-71)}$$

and so forth, up to $\Delta_n = \det \mathbf{K}$.

With the help of the principal minors, symmetric real matrices can then be classified as (1) positive definite, (2) negative definite, (3) positive semidefinite, (4) negative semidefinite, and (5) indefinite, as summarized in Table 6-2. Note that a positive definite, symmetric real matrix is one whose principal minors are all strictly positive. On the other hand, note how the principal minors alternate between strictly negative and strictly positive for negative definite matrices.

Fortunately, the finite element formulations in this text result in positive (or negative) definite **K** matrices, except in a few cases as noted later. This is fortunate because the active zone equation solver presents no unusual difficulties in these cases and may be used as presented. When the **K** matrix is unsymmetric (see Chapter 8) or indefinite, some additional checks may be necessary to ensure that the equations can be solved. In these cases, certain rows and columns of **K** may

Table 6-2   Classification of Symmetric Real Matrices*

Classification of symmetric real matrix	Conditions on the principal minors
Positive definite	$\Delta_1 > 0, \Delta_2 > 0, \ldots, \Delta_n > 0$
Negative definite	$\Delta_1 < 0, \Delta_2 > 0, \Delta_3 < 0$, etc.
Positive semidefinite	$\Delta_1 \geq 0, \Delta_2 \geq 0, \ldots, \Delta_n \geq 0$
Negative semidefinite	$\Delta_1 \leq 0, \Delta_2 \geq 0, \Delta_3 \leq 0$, etc.
Indefinite	None of the above patterns

*From Brogan, W. L., *Modern Control Theory*, Quantum Publishers, New York, 1974.

have to be interchanged to effect a solution. Treatment of such cases is beyond the scope of this text; the interested reader may wish to consult references [9–11]. In what follows it is tacitly assumed that interchanges of rows and columns are not necessary.

Storage Considerations

It should be recalled that the assemblage stiffness matrix $\mathbf{K}$ is generally banded and symmetric. This is quite significant because it becomes possible to reduce significantly the storage requirements of such a matrix. If it is symmetric, then the entries below the main diagonal need not be stored. This alone reduces the storage requirement by a factor of nearly one-half. If the matrix is banded or if there are many leading zeros in the upper triangular portion of the matrix, then the storage requirements are reduced further.

Consider, for example, the 7×7 matrix $\mathbf{K}$ that is given by

$$\mathbf{K} = \begin{bmatrix} K_{11} & K_{12} & K_{13} & 0 & 0 & 0 & 0 \\ & K_{22} & K_{23} & 0 & 0 & K_{26} & 0 \\ & & K_{33} & K_{34} & 0 & K_{36} & 0 \\ & & & K_{44} & K_{45} & 0 & 0 \\ & & & & K_{55} & K_{56} & K_{57} \\ & \text{symmetric} & & & & K_{66} & K_{67} \\ & & & & & & K_{77} \end{bmatrix} \quad \textbf{(6-72)}$$

The half-bandwidth b_w in this case is readily seen to be five, as dictated by the second row. One way to store the coefficients in this matrix is by the so-called banded storage method.

Banded Storage Method

If an $n \times n$ matrix $\mathbf{K}$ has a half-bandwidth b_w, then the nonzero coefficients (and the imbedded zeros by row) are stored in a compacted $n \times b_w$ matrix. The matrix in Eq. (6-72), for example, would be stored in a 7×5 matrix as follows:

$$\begin{bmatrix} K_{11} & K_{12} & K_{13} & & \\ K_{22} & K_{23} & 0 & 0 & K_{26} \\ K_{33} & K_{34} & 0 & K_{36} & \\ K_{44} & K_{45} & & & \\ K_{55} & K_{56} & K_{57} & & \\ K_{66} & K_{67} & & & \\ K_{77} & & & & \end{bmatrix}$$

Note how each row from the diagonal entry is *slid* to the left, with imbedded zeros in the row also necessarily being stored. If the original matrix $\mathbf{K}$ is not banded, no reduction in storage requirements occurs. Nevertheless, in this example, the original matrix would require 49 (or 7×7) storage locations, whereas the banded storage would require 35 (or 7×5) locations. A more efficient storage method is presented next.

Skyline Storage Method

It is far more efficient to store the **K** matrix in a column vector form as illustrated below. The method is most easily understood by way of an example. Reconsider the 7×7 matrix given by Eq. (6-72). Let us store the nonzero coefficients (and imbedded zeros by column) in a column vector $A(i)$ as shown in Fig. 6-19. Note how each column in the **K** matrix is *stacked* in $A(i)$. Note further that leading zeros in the column need not be stored. Zeros that are imbedded in a column, however, must be stored.

i	$A(i)$
1	K_{11}
2	K_{12}
3	K_{22}
4	K_{13}
5	K_{23}
6	K_{33}
7	K_{34}
8	K_{44}
9	K_{45}
10	K_{55}
11	K_{26}
12	K_{36}
13	0
14	K_{56}
15	K_{66}
16	K_{57}
17	K_{67}
18	K_{77}

j	JDIAG(j)
1	1
2	3
3	6
4	8
5	10
6	15
7	18

Figure 6-19 Skyline storage method for **K** matrix in Eq. (6-72).

The JDIAG(j) array in Fig. 6-19 is used to find the location of each nonzero (or imbedded zero) entry in the column vector. Physically, for a given value of j $(1 \leq j \leq n)$, JDIAG(j) represents the number of entries stored in $A(i)$ up to and including the diagonal entry (i.e., K_{jj}). For example, K_{55} is the tenth entry in $A(i)$ and, hence, JDIAG(5) is 10. The location of K_{mn} in $A(i)$ is most easily found by using the following simple formula:

$$i = JDIAG(n) - |(n - m)| \tag{6-73}$$

For example, let us locate the position in the $A(i)$ array that is reserved for K_{36}. Here n is 6 and m is 3. Therefore, $i = $ JDIAG(6) $- |(6 - 3)| = 15 - 3 = 12$. The reader should note in Fig. 6-19 that K_{36} is indeed the twelfth entry in the $A(i)$ array. *Note*: n should always be the larger subscript (not necessarily the second subscript).

This same method can be used to store unsymmetric matrices with a symmetric profile. In this case, the lower coefficients are stored by row, with unity assigned to each diagonal entry in another column vector, e.g., the $C(i)$ array. Because the *profile* is assumed to be symmetric, the same JDIAG(j) array may be used. For obvious reasons, the JDIAG(j) array is frequently referred to as the *pointer array of the diagonal pivots*, or simply the *pointer array*.

Computer Implementation

Subroutines are readily available that incorporate the active zone solution and skyline storage methods. Two such FORTRAN subroutines are introduced in this section— subroutine ACTCOL for symmetric matrices and subroutine UACTCL for unsymmetric matrices [12]. These subroutines are given in Appendix C. In Chapter 8, it will be seen how an unsymmetric **K** matrix may arise. In such cases, subroutine UACTCL must be used. The subprogram header lines are given by

```
SUBROUTINE ACTCOL (A, B, JDIAG, NEQ, AFAC, BACK)
```

and

```
SUBROUTINE UACTCL (A, C, B, JDIAG, NEQ, AFAC, BACK)
```

where the parameters A, B, C, JDIAG, NEQ, AFAC, and BACK are defined in Table 6-3. In subroutine UACTCL, the diagonal entries in the C array (see Table 6-3) are set to unity (the actual diagonal entries are stored in the A array). The reader may wish to consult the original reference [12] for further details on the use of these subroutines. It should also be remarked that these subroutines require function DOT (also given in Appendix C), which performs a vector dot product.

Table 6-3   Variables Used in Equation Solution Subprograms ACTCOL and UACTCL*

A(NAD)	Upper triangular coefficients when called, replaced by **U** on return
B(NEQ)	Right-hand side vector at call (i.e., **f**), solution vector (i.e., **a**) upon return to calling program
C(NAD)	Lower triangular coefficients at call, **L** at return (used in UACTCL only)
JDIAG(NEQ)	Pointer array to determine location in A or C of diagonal pivots
NEQ	Number of equations to be solved
NAD	Length of A or C arrays: equal to JDIAG(NEQ)
AFAC	Logical variable: if true, triangular decomposition performed
BACK	Logical variable: if true, forward elimination and backward substitution performed

*Copyright © 1977 McGraw-Hill (UK), London. From Zienkiewicz, O. C., *The Finite Element Method*, 3d ed. Reproduced by permission of the publisher.

6-9   REMARKS

In this chapter, the compatibility and completeness requirements have been introduced. If the assumed form of the parameter function satisfies these requirements, convergence is guaranteed as the elements are reduced in size in some regular fashion.

The one-dimensional lineal element was reintroduced along with two new one-dimensional normalized coordinates: the serendipity coordinate and the length coordinates. The shape functions for this element were derived in terms of all three coordinate systems.

Among the two-dimensional elements considered in this chapter were the three-node triangular element and the four-node rectangular element. The shape functions were derived for the triangular element in terms of the global coordinates. Then the so-called area coordinates were introduced, and each shape function was seen to be equal to the corresponding area coordinate. This fact has important implications in the evaluation of the resulting integrals because special integration formulas may be used. The shape functions for the rectangular element were given in terms of the serendipity coordinates.

In a similar fashion, the shape functions for two three-dimensional elements were presented. For the four-node tetrahedral element the shape functions were derived in terms of the global coordinates using the standard procedure. Volume coordinates, which facilitate the evaluation of the volume integrals that will naturally arise, were introduced. The eight-cornered brick element was introduced along with its associated shape functions in the so-called serendipity coordinates.

A special type of problem was discussed that is not strictly one, two, or three dimensional. These problems are referrred to as axisymmetric problems and must meet two basic conditions: both the geometry and the external effects must be axisymmetric. It was shown how the two-dimensional elements may be used in such situations. This point will be demonstrated in more detail in Chapters 7 and 8.

As mentioned above, special integration formulas may be used when line, area, and volume integrals need to be evaluated. Such integrals may be readily evaluated if the lineal, triangular, and tetrahedral elements are used with the shape functions given in terms of the length, area, and volume coordinates. Examples were given, but the reader may expect to see many more applications of these formulas in the next several chapters.

The chapter was concluded with a more efficient method for the solution of the system of algebraic equations—the active zone equation solver. This method may be divided into three steps: (1) triangular decomposition, (2) forward elimination, and (3) backward substitution. Up to now, we have been using the relatively inefficient matrix inversion method. The active zone method takes advantage of the banded (and symmetric) nature of the assemblage system equations. Two special FORTRAN subroutines were introduced that very effectively perform the necessary tasks, thereby paving the way to our obtaining solutions to larger, more practical problems.

In Chapters 7 and 8, specific application areas are considered: stress analysis in Chapter 7 and thermal and fluid flow analysis in Chapter 8. These two chapters will pull together everything that has been covered up to this point. The reader may wish to review Chapter 5 before tackling the next chapter.

REFERENCES

1. Zienkiewicz, O. C., *The Finite Element Method*, McGraw-Hill (UK), London, 1977, pp. 287–288.
2. Huebner, K. H., *The Finite Element Method for Engineers*, Wiley, New York, 1975, pp. 79–81.
3. Desai, C. S., *Elementary Finite Element Method*, Prentice-Hall, Englewood Cliffs, N.J., 1979, pp. 43–45.
4. Martin, H. C., and G. F. Carey, *Introduction to Finite Element Analysis: Theory and Application*, McGraw-Hill, New York, pp. 52–53.
5. Micropaedia, "Serendib," *Encyclopaedia Britannica*, vol. IX, p. 67, 1981.
6. Zienkiewicz, O. C., *The Finite Element Method*, McGraw-Hill (UK), London, 1977, pp. 141–144.
7. Zienkiewicz, O. C., *The Finite Element Method*, McGraw-Hill (UK), London, 1977, p. 719.
8. Brogan, W. L., *Modern Control Theory*, Quantum Publishers, New York, 1974, p. 134.

  9. Wilkinson, J. H., and C. Reinsch, *Linear Algebra. Handbook for Automatic Compu-tation II*, Springer-Verlag, New York, 1971.
 10. Ralston, A., *A First Course in Numerical Analyses*, McGraw-Hill, New York, 1965.
 11. Fox, L., *An Introduction to Numerical Linear Algebra*, Oxford University Press, New York, 1965.
 12. Zienkiewicz, O. C., *The Finite Element Method*, McGraw-Hill (UK), London, 1977, pp. 740–746.

PROBLEMS

6-1   In a mass transfer model, we typically try to find the concentration distribution of some chemical species in a fluid or a solid. What specific name could be suggested for the parameter funciton in this case?

6-2   In electromagnetics, what could be suggested for specific name(s) of the parameter function(s)?

6-3   Show that the three-node triangular element satisfies the compatibility requirement if a parameter function of the form $\phi = c_1 + c_2x + c_3y$ is assumed for C^0-continuous problems.

6-4   Show that the four-node tetrahedral element satisfies the compatibility requirement if a parameter function of the form $\phi = c_1 + c_2x + c_3y + c_4z$ is assumed for C^0-continuous problems.

6-5   Show that the three-node triangular element satisfies the completeness requirement if a parameter function of the form $\phi = c_1 + c_2x + c_3y$ is assumed for C^0-continuous problems.

6-6   Show that the four-node tetrahedral element satisfies the completeness requirement if a parameter function of the form $\phi = c_1 + c_2x + c_3y + c_4z$ is assumed for C^0-continuous problems.

6-7   In Problem 6-5, which term (or terms) represents the rigid-body mode and which term (or terms) represents the constant strain?

6-8   In Problem 6-6, which term (or terms) represents the rigid-body mode and which term (or terms) represents the constant strain?

6-9   Show that the shape function for node j for the two-node lineal element (with C^0-continuity) is given by Eq. (6-11b) in terms of the serendipity coordinate r.

6-10   With the help of Eqs. (6-19) and (6-20), show that the shape functions for the three-node triangular element are given by Eqs. (6-21) for C^0-continuous problems. *Hint:* See Sec. 2-7 for the method to be used in obtaining the matrix inverse. Denote the entries in the inverted matrix as m_{11}, m_{12}, etc.

6-11   Show that the shape functions given by Eqs. (6-21) for the triangular element satisfy the three properties delineated in Sec. 5-6.

6-12   Show the lines of constant L_j and L_k on the three-node triangular element for $L_j = 0, \frac{1}{4}, \frac{1}{2}, \frac{3}{4},$ and 1 and for $L_k = 0, \frac{1}{4},$ etc.

6-13 From the definition of L_j given by Eq. (6-22b), show that L_j is the same as the shape function N_j given in Eqs. (6-21), and hence $N_j = L_j$.

6-14 From the definition of L_k given by Eq. (6-22c), show that L_k is the same as the shape function N_k given in Eqs. (6-21), and hence $N_k = L_k$.

6-15 Determine the explicit form of the shape function for node i for the rectangular element in terms of the global coordinates (x,y) and the nodal coordinates. *Hint:* See Eqs. (6-29) and (6-30).

6-16 Repeat Problem 6-15 for node j.

6-17 Repeat Problem 6-15 for node k.

6-18 Repeat Problem 6-15 for node m.

6-19 For the four-node rectangular C^0-continuous element,

    a. Plot N_i on face mi.
    b. Plot N_i on face ij.
    c. Plot N_i along the straight line connecting nodes i and k.

6-20 For the four-node rectangular C^0-continuous element,

    a. Plot N_j on face ij.
    b. Plot N_j on face jk.
    c. Plot N_j along the straight line connecting nodes j and m.

6-21 For the four-node rectangular C^0-continuous element,

    a. Plot N_k on face jk.
    b. Plot N_k on face km.
    c. Plot N_k along the straight line connecting nodes k and i.

6-22 For the four-node rectangular C^0-continuous element,

    a. Plot N_m on face mi.
    b. Plot N_m on face km.
    c. Plot N_m along the straight line connecting nodes m and j.

6-23 Evaluate the following integral for the four-node rectangular element:

$$\int_A N_i^2 h \, dx \, dy$$

where A denotes the area of the rectangle and h is a constant.

6-24 Evaluate the following integral for the four-node rectangular element:

$$\int_A N_j N_k h \, dx \, dy$$

where A denotes the area of the rectangle and h is a constant.

6-25 Evaluate the following integral for the four-node rectangular element:

$$\int_A N_k N_m h \, dx \, dy$$

where A is the area of the rectangle and h is a constant.

6-26   For the four-node tetrahedral element shown in Fig. 6-12, give all twelve possible element definitions in terms of the four global node numbers.

6-27   Determine expressions similar to those in Eq. (6-40e) for m_{12}, m_{22}, m_{32}, and m_{42} for the four-node tetrahedral element.

6-28   Determine expressions similar to those in Eq. (6-40e) for m_{13}, m_{23}, m_{33}, and m_{43} for the four-node tetrahedral element.

6-29   Determine expressions similar to those in Eq. (6-40e) for m_{14}, m_{24}, m_{34}, and m_{44} for the four-node tetrahedral element.

6-30   Show that N_i evaluates to unity at node i for the four-node tetrahedral element and that N_i is zero if evaluated at the coordinates of node j, node k, and node m.

6-31   From Problem 6-30 and Eqs. (6-40), give a plausible argument for why N_i must decrease linearly from unity at node i to zero at each of the remaining nodes.

6-32   Show that $\Sigma N_\beta = 1$ for the four shape functions given by Eqs. (6-40) for the four-node tetrahedral element.

6-33   With the help of the definition of L_j given by Eq. (6-41), show that L_j is the same as the shape function N_j given in Eqs. (6-40) for the four-node tetrahedral element.

6-34   With the help of the definition of L_k given by Eq. (6-41), show that L_k is the same as the shape function N_k given in Eqs. (6-40) for the four-node tetrahedral element.

6-35   With the help of the definition of L_m given by Eq. (6-41), show that L_m is the same as the shape function N_m given in Eqs. (6-40) for the four-node tetrahedral element.

6-36   What two conditions must a problem satisfy in order to be considered axisymmetric?

6-37   For the axisymmetric body shown in Fig. 6-17(a) and discretized in Fig. 6-17(b), what are the shape functions in terms of the radial and axial coordinates r and z.

6-38   Evaluate the following line integrals for the two-node lineal element:

a. $\int_l N_i^2 \, dl$ **b.** $\int_l N_i N_j \, dl$

c. $\int_l N_i^2 N_j^2 \, dl$ **d.** $\int_l N_i^2 N_j^4 \, dl$

6-39   Evaluate the following line integrals for the two-node lineal element:

a. $\int_l N_i x \, dl$ **b.** $\int_l N_i N_j x^2 \, dl$

6-40   Evaluate the following area integrals for the three-node triangular element:

a. $\int_A N_i N_k \, dA$ **b.** $\int_A N_i^3 \, dA$

c. $\int_A N_j N_k \, dA$ **d.** $\int_A N_i^3 N_j^2 N_k \, dA$

6-41   Evaluate the following area integrals for the three-node triangular element:

a. $\int_A N_i x \, dA$ **b.** $\int_A N_j y \, dA$

6-42   Evaluate the following volume integrals for the four-node tetrahedral element:

a. $\int_V N_i \, dV$ **b.** $\int_V N_j N_k \, dV$

c. $\int_V N_i N_j N_k N_m \, dV$ **d.** $\int_V N_i^2 N_m^2 \, dV$

6-43 Evaluate the following volume integrals for the four-node tetrahedral element:

a. $\int_V N_k^2 N_m^2 \, dV$ **b.** $\int_V N_k x \, dV$

6-44 For symmetric $n \times n$ **K** matrices, show that the entries in the lower triangular matrix **L** can be obtained directly from the upper triangular matrix **U**. In other words, show that Eq. (6-61) is valid for $n \times n$ symmetric **K** matrices.

6-45 Determine **L** and **U** for the following matrix:

$$\mathbf{K} = \begin{bmatrix} 10 & 5 & 2 \\ 5 & 20 & 5 \\ 2 & 5 & 10 \end{bmatrix}$$

Is **K** positive definite? If not, classify it.

6-46 Determine **L** and **U** for the following matrix:

$$\mathbf{K} = \begin{bmatrix} 20 & 10 & 5 \\ 10 & 30 & 10 \\ 5 & 10 & 20 \end{bmatrix}$$

Is **K** positive definite? If not, classify it.

6-47 Determine **L** and **U** for the following matrix:

$$\mathbf{K} = \begin{bmatrix} 20 & 15 & 10 \\ 10 & 30 & 15 \\ 15 & 10 & 20 \end{bmatrix}$$

Can Table 6-2 be used to classify this matrix? Why or why not?

6-48 With the results from Problem 6-45, solve the following system of equations:

$$10a_1 + 5a_2 + 2a_3 = 26$$
$$5a_1 + 20a_2 + 5a_3 = 60$$
$$2a_1 + 5a_2 + 10a_3 = 42$$

6-49 With the help of the results from Problem 6-46, solve the following system of equations:

$$20a_1 + 10a_2 + 5a_3 = 90$$
$$10a_1 + 30a_2 + 10a_3 = 200$$
$$5a_1 + 10a_2 + 20a_3 = 135$$

6-50 Solve the following set of equations by using the results from Problem 6-47:

$$20a_1 + 15a_2 + 10a_3 = 155$$
$$10a_1 + 30a_2 + 15a_3 = 210$$
$$15a_1 + 10a_2 + 20a_3 = 135$$

6-51 Consider the following symmetric matrix:

$$K = \begin{bmatrix} 20 & 10 & 5 & 0 & 0 & 0 \\ & 40 & 20 & 10 & 0 & 0 \\ & & 40 & 20 & 10 & 0 \\ & & & 40 & 20 & 10 \\ & \text{symmetric} & & & 40 & 20 \\ & & & & & 20 \end{bmatrix}$$

a. What is the half-bandwidth?
b. How many storage locations are required if this matrix is stored by the banded storage method? What is the resulting banded storage matrix?
c. How many storage locations are required if this matrix is stored by the skyline storage method (i.e., in column-vector form)? What is the column vector [i.e., the $A(i)$ array in ACTCOL]?
d. What is the JDIAG(j) array for part (c)?

6-52 Consider the following symmetric matrix:

$$K = \begin{bmatrix} 50 & 20 & 10 & 0 & 0 & 0 & 0 & 0 \\ & 75 & 50 & 0 & 25 & 0 & 0 & 0 \\ & & 80 & 20 & 0 & 50 & 0 & 0 \\ & & & 30 & 0 & 0 & 0 & 0 \\ & & & & 40 & 20 & 0 & 0 \\ & \text{symmetric} & & & & 50 & 5 & 10 \\ & & & & & & 2 & 5 \\ & & & & & & & 1 \end{bmatrix}$$

a. What is the half-bandwidth?
b. How many storage locations are required if this matrix is stored by the banded storage method? What is the resulting banded storage matrix?
c. How many storage locations are required if this matrix is stored by the skyline storage method (i.e., in column-vector form)? What is the column vector [i.e., the $A(i)$ array in ACTCOL]?
d. What is the JDIAG(j) array for part (c)?

6-53 Consider the unsymmetric matrix:

$$K = \begin{bmatrix} 50 & 20 & 0 & 4 & 0 & 0 & 0 & 0 & 0 \\ 30 & 100 & 80 & 20 & 5 & 0 & 0 & 0 & 0 \\ 5 & 20 & 130 & 40 & 20 & 10 & 0 & 0 & 0 \\ 0 & 90 & 100 & 150 & 100 & 80 & 40 & 0 & 0 \\ 0 & 50 & 75 & 80 & 180 & 100 & 60 & 40 & 0 \\ 0 & 0 & 0 & 90 & 150 & 200 & 90 & 85 & 40 \\ 0 & 0 & 0 & 10 & 75 & 80 & 210 & 215 & 35 \\ 0 & 0 & 0 & 0 & 5 & 50 & 100 & 220 & 190 \\ 0 & 0 & 0 & 0 & 0 & 6 & 10 & 55 & 160 \end{bmatrix}$$

a. What is the half-bandwidth?
b. What are the $A(i)$ and $C(i)$ arrays for use in subroutine UACTCL?
c. What is the JDIAG(i) array for part (b)?

7

Stress Analysis

7-1 INTRODUCTION

In this chapter, the finite element method is applied to several different classes of problems in stress analysis. The chapter begins with two-dimensional problems, since one-dimensional stress analysis was introduced in Chapter 5. This is followed by the finite element formulations to axisymmetric and three-dimensional problems. In each case, expressions for the element stiffness matrix and element nodal force vectors are derived from Eqs. (5-87) to (5-92). However, an alternate approach is taken in the study of the beam model. More specifically, the finite element characteristics for the beam are derived with the help of the Galerkin method (one of the weighted-residual methods from Chapter 4). Appropriate working expressions are then derived for the element stiffness matrix and element nodal force vectors for the beam element. The concept of substructuring is described, which makes the development of extremely large finite element models feasible. The chapter is concluded with a brief description of a two-dimensional stress analysis program.

7-2 TWO-DIMENSIONAL STRESS ANALYSIS

In this section, the plane stress and plane strain models are defined. The finite element characteristics are derived from Eqs. (5-87) to (5-92) for the case of a specific two-dimensional element. Each of the basic steps in the FEM solution to such problems is discussed.

Definitions of Plane Stress and Plane Strain

Plane stress and plane strain models are both two-dimensional idealizations of three-dimensional problems. The situations in which each is appropriate are discussed below.

Plane Stress

The plane stress model is appropriate when a thin plate is loaded uniformly across its thickness t and in a direction parallel to the lateral surfaces of the plate as shown in Fig. 7-1. Note that the thickness t of the plate need not be constant. The loading around the plate periphery may be a result of tensile, compressive, or shear forces. The definition of plane stress implies

$$\sigma_{zz} = 0 \qquad \sigma_{xz} = 0 \qquad \text{and} \qquad \sigma_{yz} = 0 \qquad \textbf{(7-1)}$$

both within and on the faces of the plate. Also, the body force b_z must be zero. The body forces b_x and b_y may each be functions of x and y. Therefore, there are three nonzero stress components σ_{xx}, σ_{yy}, and σ_{xy}, which are functions of x and y and which remain constant through the thickness of the plate (in the z direction) at any point. It should be pointed out that the strain in the z direction, ϵ_{zz}, is not necessarily zero for the state of plane stress [see Eq. (7-16)]. In addition, the body may have self-strains ϵ_{xx0}, ϵ_{yy0}, ϵ_{xy0}, and ϵ_{zz0} possibly because of a temperature change ΔT. Also, initial residual stresses σ_{xx0}, σ_{yy0}, and σ_{xy0} may exist.

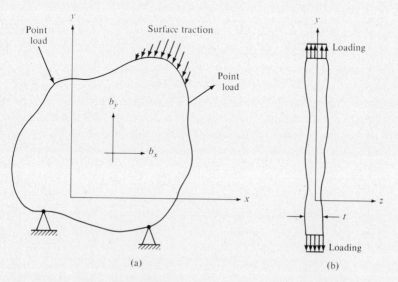

Figure 7-1   Plane stress (all loads are in the x-y plane only).

In summary, a problem may be classified as plane stress when the body to be analyzed is a relatively thin plate loaded only in the plane of the plate.

Plane Strain

The plain strain model is appropriate when the normal strain in the z direction, ϵ_{zz}, and the shear strains, ϵ_{xz} and ϵ_{yz}, may be assumed to be zero, or

$$\epsilon_{zz} = 0 \qquad \epsilon_{xz} = 0 \qquad \text{and} \qquad \epsilon_{yz} = 0 \qquad \textbf{(7-2)}$$

This situation is most likely to arise when a long prismatic member of constant cross section is held between two smooth fixed rigid planes as shown in Fig. 7-2. Although this is not a very practical situation, St. Venant's principle [1] allows us to use the plane strain model in the regions far from the ends that may not meet the above conditions.

Note the coordinate system shown in Fig. 7-2. External forces may have only x and y components and may be a function of x and y only. All cross sections are expected to have the same deformations. Body forces, b_x and b_y, per unit volume may each be functions of x and y only, and in the z direction we must have $b_z = 0$. As in plane stress, the body may have self-strains ϵ_{xx0}, ϵ_{yy0}, and ϵ_{xy0} possibly because of a temperature change ΔT and initial stresses σ_{xx0}, σ_{yy0}, σ_{xy0}, and σ_{zz0}. Note that in general σ_{zz} is not necessarily zero in plane strain and, therefore, we must allow for the possibility of nonzero σ_{zz0} [see Eq. (7-21)].

The Shape Function Matrix N

Before the shape function matrix can be derived, the two-dimensional region to be analyzed must be discretized into a suitable number of elements. From Chapter 6, two elements in particular can be used in this situation: the three-node triangular element and the four-node rectangular element. Because the triangular element can accommodate practically all two-dimensional regions, let us restrict the finite element formulation to only these elements, as shown in Fig. 7-3. In both plane stress and plane strain, each node has two degrees of freedom: the x and y components

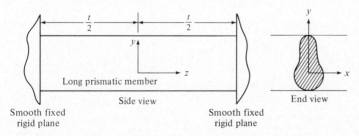

Figure 7-2 Plane strain (all loads are in the x-y plane only).

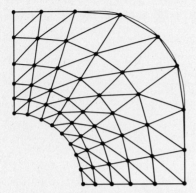

Figure 7-3 Typical two-dimensional region discretized into triangular elements.

of the displacements. Since each element has three nodes and each node has two degrees of freedom, the shape function matrix **N** is given by Eq. (5-61), which is repeated here for easy reference:

$$\mathbf{N} = \begin{bmatrix} N_i & 0 & N_j & 0 & N_k & 0 \\ 0 & N_i & 0 & N_j & 0 & N_k \end{bmatrix} \tag{5-61}$$

The three shape functions are given by Eqs. (6-21). The shape function matrix **N** is, therefore, the same for plane stress and plane strain and is given by Eqs. (5-61) and (6-21).

The Strain-Nodal Displacement Matrix B

Recall that the strain-nodal displacement matrix **B** is given by Eq. (5-76) or $\mathbf{B} = \mathbf{LN}$, where **L** is the linear operator matrix. It should further be recalled from Eq. (5-75) that $\boldsymbol{\epsilon} = \mathbf{Lu}$, where

$$\boldsymbol{\epsilon} = [\epsilon_{xx} \quad \epsilon_{yy} \quad \epsilon_{xy}]^T \tag{7-3}$$

and

$$\mathbf{u} = [u \quad v]^T \tag{7-4}$$

where u and v are the x and y components of the displacement field. The strain-displacement (or kinematic) relationships for both plane stress and plane strain are given by Eq. (5-14) for small deformations:

$$\epsilon_{xx} = \frac{\partial u}{\partial x} \qquad \epsilon_{yy} = \frac{\partial v}{\partial y} \qquad \epsilon_{xy} = \frac{\partial u}{\partial y} + \frac{\partial v}{\partial x} \tag{5-14}$$

It follows that the linear operator matrix **L** is given by

$$\mathbf{L} = \begin{bmatrix} \dfrac{\partial}{\partial x} & 0 \\[2ex] 0 & \dfrac{\partial}{\partial y} \\[2ex] \dfrac{\partial}{\partial y} & \dfrac{\partial}{\partial x} \end{bmatrix} \tag{7-5}$$

The matrix **B** is then easily determined to be

$$\mathbf{B} = \mathbf{LN} = \begin{bmatrix} \dfrac{\partial N_i}{\partial x} & 0 & \vline & \dfrac{\partial N_j}{\partial x} & 0 & \vline & \dfrac{\partial N_k}{\partial x} & 0 \\[2ex] 0 & \dfrac{\partial N_i}{\partial y} & \vline & 0 & \dfrac{\partial N_j}{\partial y} & \vline & 0 & \dfrac{\partial N_k}{\partial y} \\[2ex] \dfrac{\partial N_i}{\partial y} & \dfrac{\partial N_i}{\partial x} & \vline & \dfrac{\partial N_j}{\partial y} & \dfrac{\partial N_j}{\partial x} & \vline & \dfrac{\partial N_k}{\partial y} & \dfrac{\partial N_k}{\partial x} \end{bmatrix} \tag{7-6}$$

But the shape functions are given by Eqs. (6-21), so the matrix **B** is seen to be composed of the m_{ij}'s and is given by

$$\mathbf{B} = \begin{bmatrix} m_{21} & 0 & \vline & m_{22} & 0 & \vline & m_{23} & 0 \\ 0 & m_{31} & \vline & 0 & m_{32} & \vline & 0 & m_{33} \\ m_{31} & m_{21} & \vline & m_{32} & m_{22} & \vline & m_{33} & m_{23} \end{bmatrix} \tag{7-7}$$

Since the m_{ij}'s are known functions of the nodal coordinates [see Eq. (6-21d)], the **B** matrix is determined. The fact that this matrix is composed of only constant entries (for the three-node triangular element) has an important implication when the integrals defining the finite element characteristics are to be evaluated, as will be seen shortly.

The Constitutive Relationship

Recall that the constitutive relationship for a linear, elastic material is given by Eq. (5-24), which is repeated here for convenience:

$$\boldsymbol{\sigma} = \mathbf{D}(\boldsymbol{\epsilon} - \boldsymbol{\epsilon}_0) + \boldsymbol{\sigma}_0 \tag{5-24}$$

The material property matrix **D** depends on whether the material is isotropic or anisotropic. Only the isotropic case is considered here. The interested reader should consult Zienkiewicz's book [2] for the anisotropic form of the material property matrix. The matrix **D** and the vector $\boldsymbol{\epsilon}_0$ are different for plane stress and plane strain as indicated below.

Plane Stress

For an isotropic material in plane stress, we have by the generalized Hooke's law:

$$\epsilon_{xx} = \frac{\sigma_{xx}}{E} - \frac{\mu\sigma_{yy}}{E} + \epsilon_{xx0} \tag{7-8a}$$

$$\epsilon_{yy} = -\frac{\mu\sigma_{xx}}{E} + \frac{\sigma_{yy}}{E} + \epsilon_{yy0} \tag{7-8b}$$

$$\epsilon_{xy} = \frac{1}{G}\sigma_{xy} + \epsilon_{xy0} \tag{7-8c}$$

where μ is Poisson's ratio, E is the modulus of elasticity, and G is the shear modulus (one of the Lamé constants) related to μ and E by

$$G = \frac{E}{2(1 + \mu)} \tag{7-9}$$

If we solve Eqs. (7-8) for the three stresses and recognize that the initial stresses not shown in Eqs. (7-8) are simply additive to σ_{xx}, etc., we get

$$\sigma_{xx} = \frac{E}{1 - \mu^2}[(\epsilon_{xx} - \epsilon_{xx0}) + \mu(\epsilon_{yy} - \epsilon_{yy0})] + \sigma_{xx0} \tag{7-10a}$$

$$\sigma_{yy} = \frac{E}{1 - \mu^2}[\mu(\epsilon_{xx} - \epsilon_{xx0}) + (\epsilon_{yy} - \epsilon_{yy0})] + \sigma_{yy0} \tag{7-10b}$$

$$\sigma_{xy} = \frac{E}{1 - \mu^2}\left[\frac{1 - \mu}{2}(\epsilon_{xy} - \epsilon_{xy0})\right] + \sigma_{xy0} \tag{7-10c}$$

Equations (7-10) may now be cast into the form of Eq. (5-24) by defining the following [in addition to ϵ from Eq. (7-3)]:

$$\boldsymbol{\sigma} = [\sigma_{xx} \quad \sigma_{yy} \quad \sigma_{xy}]^T \tag{7-11}$$

$$\boldsymbol{\sigma}_0 = [\sigma_{xx0} \quad \sigma_{yy0} \quad \sigma_{xy0}]^T \tag{7-12}$$

$$\boldsymbol{\epsilon}_0 = [\epsilon_{xx0} \quad \epsilon_{yy0} \quad \epsilon_{xy0}]^T \tag{7-13}$$

and

$$\mathbf{D} = \frac{E}{1 - \mu^2}\begin{bmatrix} 1 & \mu & 0 \\ \mu & 1 & 0 \\ 0 & 0 & \dfrac{1 - \mu}{2} \end{bmatrix} \tag{7-14}$$

If the self-strain vector $\boldsymbol{\epsilon}_0$ is a result of a temperature change ΔT, then $\boldsymbol{\epsilon}_0$ is given by

$$\boldsymbol{\epsilon}_0 = [\alpha_t\,\Delta T \quad \alpha_t\,\Delta T \quad 0]^T \tag{7-15}$$

In Eq. (7-15), α_t is the coefficient of thermal expansion. Note that no shear strain is induced by a thermal excursion in an isotropic material. This is not the case in anisotropic materials.

It is noted that ϵ_{zz} is not necessarily zero and is given by

$$\epsilon_{zz} = -\frac{\mu}{E}(\sigma_{xx} + \sigma_{yy}) + \epsilon_{zz0} \qquad \textbf{(7-16)}$$

where ϵ_{zz0} is taken to be $\alpha_t \Delta T$ in the case of a thermal strain. However, this strain component need not be included in Eq. (7-3); ϵ_{zz} is simply determined from Eq. (7-16) after σ_{xx} and σ_{yy} have been found.

Plane Strain

The material property matrix **D** for the case of plane strain proves to be different from that for plane stress. For a linear, elastic isotropic material in plane strain, the appropriate form of Hooke's law is

$$\sigma_{xx} = (2G + \lambda)(\epsilon_{xx} - \epsilon_{xx0}) + \lambda(\epsilon_{yy} - \epsilon_{yy0}) + \sigma_{xx0} \qquad \textbf{(7-17a)}$$

$$\sigma_{yy} = \lambda(\epsilon_{xx} - \epsilon_{xx0}) + (2G + \lambda)(\epsilon_{yy} - \epsilon_{yy0}) + \sigma_{yy0} \qquad \textbf{(7-17b)}$$

$$\sigma_{xy} = G(\epsilon_{xy} - \epsilon_{xy0}) + \sigma_{xy0} \qquad \textbf{(7-17c)}$$

where the shear modulus G is given by Eq. (7-9) and λ is another one of the Lamé constants defined by

$$\lambda = \frac{\mu E}{(1 + \mu)(1 - 2\mu)} \qquad \textbf{(7-18)}$$

If $\boldsymbol{\sigma}$, $\boldsymbol{\epsilon}$, $\boldsymbol{\sigma}_0$, and $\boldsymbol{\epsilon}_0$ are defined as above, it follows from Eqs. (7-17) that the material property matrix **D** is given by

$$\mathbf{D} = \frac{E}{(1 + \mu)(1 - 2\mu)} \begin{bmatrix} 1 - \mu & \mu & 0 \\ \mu & 1 - \mu & 0 \\ 0 & 0 & \dfrac{1 - 2\mu}{2} \end{bmatrix} \qquad \textbf{(7-19)}$$

Therefore, the material property matrix for an isotropic material in plane strain is known in terms of the material constants E and μ.

If the initial or self-strains are a result of a temperature change ΔT, we have

$$\boldsymbol{\epsilon}_0 = [(1 + \mu)\alpha_t \Delta T \qquad (1 + \mu)\alpha_t \Delta T \qquad 0]^T \qquad \textbf{(7-20)}$$

Note the additional factor $1 + \mu$, which was not present for the case of plane stress [see Eq. (7-15)]. In Eq. (7-20), α_t is the coefficient of thermal expansion. Again it is pointed out that the shear strains as a result of a temperature change cannot develop in an isotropic material.

It should be noted that σ_{zz} is not necessarily zero for the case of plane strain and is given by

$$\sigma_{zz} = \mu(\sigma_{xx} + \sigma_{yy}) + \sigma_{zz_0} - E\epsilon_{zz_0} \qquad \text{(7-21)}$$

where ϵ_{zz_0} is taken to be $\alpha_t \, \Delta T$ in the case of a thermal strain.

This completes the determination of the constitutive relationship for the cases of plane stress and plane strain. It is emphasized that only linear, elastic, isotropic materials are considered here. The reader is referred to Zienkiewicz's book [2] for the case of anisotropic materials.

The Element Stiffness Matrix

The element stiffness matrix is given by Eq. (5-87) for all problems in stress analysis and is repeated here for convenience:

$$\mathbf{K}^e = \int_{V^e} \mathbf{B}^T \mathbf{D} \mathbf{B} \, dV \qquad \text{(5-87)}$$

Both **B** and **D** are composed only of constants. Moreover, the elemental volume dV is given by

$$dV = t \, dx \, dy \qquad \text{(7-22)}$$

where t is the thickness of the plate for plane stress or the length of the long prismatic member in plane strain. The thickness t is frequently taken to be unity for the plane strain case. Therefore, Eq. (5-87) becomes

$$\mathbf{K}^e = \mathbf{B}^T \mathbf{D} \mathbf{B} \int_{A^e} t \, dx \, dy \qquad \text{(7-23)}$$

For reasonably small elements, the thickness t may be taken as a constant average value (perhaps the value at the centroid or the average of the three values at the nodes). It then follows that t may be taken through the integral. We are left with $\int dx \, dy$, which is really the area A of the triangle and is given by Eq. (6-21e). Therefore, the element stiffness matrix is given by

$$\mathbf{K}^e = \mathbf{B}^T \mathbf{D} \mathbf{B} t A \qquad \text{(7-24)}$$

It should be noted that $\mathbf{K}^e$ is a 6×6 matrix because $\mathbf{B}^T$ is 6×3, **D** is 3×3, and **B** is 3×6. This is consistent with the fact that there are six nodal displacements (two at each of the three nodes of the triangle).

Example 7-1

Consider the element shown in Fig. 7-4. Determine the element stiffness matrix if nodes i, j, k are 2, 1, 3, respectively. The material is steel, which has an elastic modulus of 30×10^6 psi and a Poisson's ratio of 0.3. The thickness may be assumed to be 0.25 in. (constant). The nodal coordinates (in inches) are shown on the figure. All loads (not shown) may be assumed to be in the plane of the element; hence the plane stress case applies.

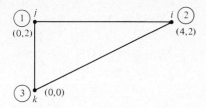

Figure 7-4   Element used in Example 7-1.

Solution

From Eq. (7-7), we see that we need six of the nine m_{ij}'s, which are easily calculated as shown below. The area A of the element is needed in the calculation of the m_{ij}'s and is determined from Eq. (6-21e) as follows:

$$A = \tfrac{1}{2} \det \begin{bmatrix} 1 & 4 & 2 \\ 1 & 0 & 2 \\ 1 & 0 & 0 \end{bmatrix} = 4 \text{ in.}^2$$

This result checks with the value of the area from the simple formula that is given by the product of one-half the base and height. The m_{ij}'s are calculated as follows:

$$m_{21} = \frac{y_j - y_k}{2A} = \frac{y_1 - y_3}{2A} = \frac{2 - 0}{2(4)} = 0.25 \text{ in.}^{-1}$$

$$m_{31} = \frac{x_k - x_j}{2A} = \frac{x_3 - x_1}{2A} = \frac{0 - 0}{2(4)} = 0$$

$$m_{22} = \frac{y_k - y_i}{2A} = \frac{y_3 - y_2}{2A} = \frac{0 - 2}{2(4)} = -0.25 \text{ in.}^{-1}$$

$$m_{32} = \frac{x_i - x_k}{2A} = \frac{x_2 - x_3}{2A} = \frac{4 - 0}{2(4)} = 0.50 \text{ in.}^{-1}$$

$$m_{23} = \frac{y_i - y_j}{2A} = \frac{y_2 - y_1}{2A} = \frac{2 - 2}{2(4)} = 0$$

$$m_{33} = \frac{x_j - x_i}{2A} = \frac{x_1 - x_2}{2A} = \frac{0 - 4}{2(4)} = -0.50 \text{ in.}^{-1}$$

It then follows that the **B** matrix is given by

$$\mathbf{B} = \begin{bmatrix} 0.25 & 0 & -0.25 & 0 & 0 & 0 \\ 0 & 0 & 0 & 0.50 & 0 & -0.50 \\ 0 & 0.25 & 0.50 & -0.25 & -0.50 & 0 \end{bmatrix}$$

The material property matrix $\mathbf{D}$ (for plane stress) is given by

$$\mathbf{D} = \frac{E}{1 - \mu^2}\begin{bmatrix} 1 & \mu & 0 \\ \mu & 1 & 0 \\ 0 & 0 & \dfrac{1 - \mu}{2} \end{bmatrix} = \frac{30 \times 10^6}{1 - 0.3^2}\begin{bmatrix} 1 & 0.3 & 0 \\ 0.3 & 1 & 0 \\ 0 & 0 & \dfrac{1 - 0.3}{2} \end{bmatrix}$$

or

$$\mathbf{D} = \begin{bmatrix} 33.0 & 9.89 & 0 \\ 9.89 & 33.0 & 0 \\ 0 & 0 & 11.5 \end{bmatrix} \times 10^6 \text{ lbf/in.}^2$$

The reader may now verify that product $\mathbf{B}^T\mathbf{D}\mathbf{B}tA$ gives the following stiffness matrix for this element:

$$\mathbf{K}^e = \begin{bmatrix} 2.06 & 0 & -2.06 & 1.24 & 0 & -1.24 \\ 0 & 0.72 & 1.44 & -0.72 & -1.44 & 0 \\ -2.06 & 1.44 & 4.94 & -2.68 & -2.88 & 1.24 \\ 1.24 & -0.72 & -2.68 & 8.97 & 1.44 & -8.25 \\ 0 & -1.44 & -2.88 & 1.44 & 2.88 & 0 \\ -1.24 & 0 & 1.24 & -8.25 & 0 & 8.25 \end{bmatrix} \times 10^6 \text{ lbf/in.}$$

The symmetry of $\mathbf{K}^e$ should be noted. The 2×2 submatrices, however, are not necessarily symmetric. Partitioning the element stiffness matrix in this way greatly facilitates the assemblage step.   ∎

The Element Nodal Force Vectors

Each of the five nodal force vectors given by Eqs. (5-88) to (5-92) may now be evaluated for plane stress and plane strain problems.

Self-Strain

The element nodal force vector as a result of the self-strains is given by Eq. (5-88), or

$$\mathbf{f}^e_{\epsilon_0} = \int_{V^e} \mathbf{B}^T\mathbf{D}\boldsymbol{\epsilon}_0 \, dV = \int_{A^e} \mathbf{B}^T\mathbf{D}\boldsymbol{\epsilon}_0 t \, dx \, dy \tag{7-25}$$

where, for thermal strains, $\boldsymbol{\epsilon}_0$ is given by Eq. (7-15) for plane stress and by Eq. (7-20) for plane strain. If we limit the present discussion to thermal strains and take the coefficient of thermal expansion α_t and the temperature change as constants within an element (and hence constant $\boldsymbol{\epsilon}_0$), then Eq. (7-25) becomes

$$\mathbf{f}^e_{\epsilon_0} = \mathbf{B}^T\mathbf{D}\boldsymbol{\epsilon}_0 tA \tag{7-26}$$

Note that the thickness t was also assumed to be constant. Whenever a property varies over the element, it may be evaluated at the centroid and subsequently treated as a constant. More sophisticated integration schemes are discussed in Chapter 9. Note that $\mathbf{f}_{\epsilon_0}^e$ is 6×1 because $\mathbf{B}^T$ is 6×3, $\mathbf{D}$ is 3×3, and $\boldsymbol{\epsilon}_0$ is 3×1. It is emphasized that $\mathbf{B}$ and $\mathbf{D}$ are known and are given by Eqs. (7-7) and (7-14) [or Eq. (7-19)], respectively.

Example 7-2

For the element in Example 7-1, determine the vector $\mathbf{f}_{\epsilon_0}^e$ if the coefficient of thermal expansion is 6.0×10^{-6} in./in.-°F and the temperature increases by 150°F.

Solution

First we calculate $\boldsymbol{\epsilon}_0$ from Eq. (7-15) since the plane stress case is applicable:

$$\boldsymbol{\epsilon}_0 = \begin{bmatrix} \alpha_t \, \Delta T \\ \alpha_t \, \Delta T \\ 0 \end{bmatrix} = \begin{bmatrix} (6.0 \times 10^{-6})(150) \\ (6.0 \times 10^{-6})(150) \\ 0 \end{bmatrix} = \begin{bmatrix} 900 \\ 900 \\ 0 \end{bmatrix} \times 10^{-6} \text{ in./in.}$$

Then, using the results for the $\mathbf{B}$ and $\mathbf{D}$ matrices from Example 7-1, we have

$$\mathbf{f}_{\epsilon_0}^e = \begin{bmatrix} 0.25 & 0 & 0 \\ 0 & 0 & 0.25 \\ -0.25 & 0 & 0.50 \\ 0 & 0.50 & -0.25 \\ 0 & 0 & -0.50 \\ 0 & -0.50 & 0 \end{bmatrix} \begin{bmatrix} 33.0 & 9.89 & 0 \\ 9.89 & 33.0 & 0 \\ 0 & 0 & 11.5 \end{bmatrix} \begin{bmatrix} 900 \\ 900 \\ 0 \end{bmatrix} (0.25)(4.0)$$

or

$$\mathbf{f}_{\epsilon_0}^e = \begin{bmatrix} 9{,}650 \\ 0 \\ -9{,}650 \\ 19{,}300 \\ 0 \\ -19{,}300 \end{bmatrix} \text{ lbf} \qquad \blacksquare$$

Prestresses

The element nodal force vector as a result of the prestresses σ_{xx0}, σ_{yy0}, and σ_{xy0} is easily determined from Eq. (5-89) as follows:

$$\mathbf{f}_{\sigma_0}^e = \int_{V^e} \mathbf{B}^T \boldsymbol{\sigma}_0 \, dV = \int_{A^e} \mathbf{B}^T \boldsymbol{\sigma}_0 t \, dx \, dy = \mathbf{B}^T \boldsymbol{\sigma}_0 t A \qquad (7\text{-}27)$$

Note that the prestresses are assumed to be constant within the element (or the values at the centroid are used). Once again it is noted that $\mathbf{f}_{\sigma_0}^e$ is of size 6×1.

Body Forces

It follows from Eq. (5-90) that the nodal force vector as a result of the body forces b_x and b_y is given by

$$\mathbf{f}_b^e = \int_{V^e} \mathbf{N}^T \mathbf{b} \; dV = \int_{A^e} \begin{bmatrix} N_i & 0 \\ 0 & N_i \\ \hline N_j & 0 \\ 0 & N_j \\ \hline N_k & 0 \\ 0 & N_k \end{bmatrix} \begin{bmatrix} b_x \\ b_y \end{bmatrix} t \; dx \; dy \qquad \textbf{(7-28)}$$

Noting that each shape function is given by its respective area coordinate [see Eqs. (6-25)], it follows that

$$\mathbf{f}_b^e = \begin{bmatrix} \int L_i b_x t \; dx \; dy \\ \int L_i b_y t \; dx \; dy \\ \hline \int L_j b_x t \; dx \; dy \\ \int L_j b_y t \; dx \; dy \\ \hline \int L_k b_x t \; dx \; dy \\ \int L_k b_y t \; dx \; dy \end{bmatrix} \qquad \textbf{(7-29)}$$

Integrating each term with the help of the integration formula given by Eq. (6-49) gives the very simple (and expected) result below. A typical term is evaluated in the following manner:

$$\int_A L_i b_x t \; dx \; dy = \frac{1!0!0!}{(1 + 0 + 0 + 2)!} b_x t 2A = \tfrac{1}{3} b_x t A$$

The complete result is given by

$$\mathbf{f}_b^e = \frac{tA}{3} \begin{bmatrix} b_x \\ b_y \\ \hline b_x \\ b_y \\ \hline b_x \\ b_y \end{bmatrix} \qquad \textbf{(7-30)}$$

which says, in effect, that the body forces in the x and y directions are allocated equally to each node. This would not necessarily be the case if b_x, b_y, or t were not constant within the element.

Surface Tractions

The element nodal force vector as a result of the surface tractions s_x and s_y as given by Eq. (5-91) may be evaluated by first noting that

$$\mathbf{f}_s^e = \int_{S^e} \mathbf{N}^T \mathbf{s} \, dS = \int_{S^e}
\begin{bmatrix}
N_i & 0 \\
0 & N_i \\
\hline
N_j & 0 \\
0 & N_j \\
\hline
N_k & 0 \\
0 & N_k
\end{bmatrix}
\begin{bmatrix}
s_x \\
s_y
\end{bmatrix} dS \tag{7-31}$$

Note that the integrals are to be evaluated around the element boundaries (in a counterclockwise direction) as denoted by the integration limit S^e. As we have seen in the examples presented thus far, the evaluation of such integrals for elements totally within the body never contributes to the assemblage nodal force vector. In this case, each internal force is equal and opposite on adjacent sides of the element, therefore, the net contribution of these internal forces to the assemblage nodal force vector is zero.

Figure 7-5(a) shows an element e with nodes i and j (but not k) on the global boundary. A surface traction is assumed to act on leg ij. Let us denote the length of this leg as l_{ij}. The integral may be evaluated rather easily if the shape functions are written in terms of the area coordinates [see Eq. (6-25)] and if dS is replaced by $t \, dl$. In addition, on leg ij, we have $N_k = L_k = 0$, and hence Eq. (7-31) becomes

$$\mathbf{f}_s^e =
\begin{bmatrix}
\int_{l_{ij}} L_i s_x t \, dl \\
\int_{l_{ij}} L_i s_y t \, dl \\
\hline
\int_{l_{ij}} L_j s_x t \, dl \\
\int_{l_{ij}} L_j s_y t \, dl \\
\hline
0 \\
0
\end{bmatrix} \tag{7-32}$$

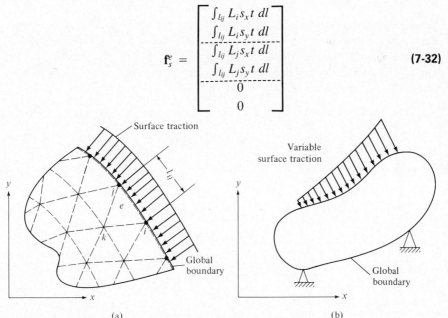

Figure 7-5 (a) Portion of the global boundary showing a surface traction on leg ij of element e and (b) a variable surface traction.

With the help of the integral formula for the length coordinates [i.e., Eq. (6-48)], and assuming s_x, s_y, and t to be constant over leg ij, the nodal force vector as a result of the surface tractions becomes

$$\mathbf{f}_s^e = \frac{t l_{ij}}{2} \begin{bmatrix} s_x \\ s_y \\ \hline s_x \\ s_y \\ \hline 0 \\ 0 \end{bmatrix} \qquad \text{for leg } ij \text{ on the global boundary} \qquad \textbf{(7-33)}$$

Expressions may be derived in a similar fashion, or set up by inspection if leg jk or ki happen to be on the portion of the global boundary over which a surface traction acts. If the surface traction is distributed as shown in Fig. 7-5(b), the integrals in Eq. (7-32) could be evaluated as illustrated in the problems or a numerical integration method from Chapter 9 may be used.

Example 7-3

Reconsider the element in Example 7-1. A distributed surface traction acts on the leg connecting nodes 1 and 3 (local nodes j and k), as shown in Fig. 7-6. Determine the corresponding element nodal force vector.

Solution

First, we need to determine s_x and s_y. Since the traction is directed only in the x direction, we have $s_y = 0$. The surface traction s_x is not constant over leg jk and, therefore, the effective s_x is taken to be the average of 1600 and 2000 lbf/in.^{2}, or $s_x = 1800$ lbf/in.^{2}. Noting that leg jk (and not leg ij) in this case is on the global boundary, we write the following using Eq. (7-33) as a general guide:

$$\mathbf{f}_s^e = \frac{t l_{jk}}{2} \begin{bmatrix} 0 \\ 0 \\ \hline s_x \\ s_y \\ \hline s_x \\ s_y \end{bmatrix} = \frac{(0.25)(2)}{2} \begin{bmatrix} 0 \\ 0 \\ \hline 1800 \\ 0 \\ \hline 1800 \\ 0 \end{bmatrix} = \begin{bmatrix} 0 \\ 0 \\ \hline 450 \\ 0 \\ \hline 450 \\ 0 \end{bmatrix} \text{ lbf}$$

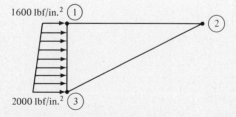

Figure 7-6 Element with imposed surface traction for Example 7-3.

Because an average value for the surface traction is used, the allocation to nodes 1 and 3 is equal. ∎

Point Loads

The nodal force vector as a result of a point load (PL) at (x_0, y_0) is given by Eq. (5-92), or

$$
\mathbf{f}^e_{PL} = \Sigma \mathbf{N}^T \mathbf{f}_p = \Sigma
\begin{bmatrix}
N_i & 0 \\
0 & N_i \\
\hline
N_j & 0 \\
0 & N_j \\
\hline
N_k & 0 \\
0 & N_k
\end{bmatrix}
\begin{bmatrix}
f_{px} \\
f_{py}
\end{bmatrix}
=
\begin{bmatrix}
N_i(x_0, y_0)f_{px} \\
N_i(x_0, y_0)f_{py} \\
\hline
N_j(x_0, y_0)f_{px} \\
N_j(x_0, y_0)f_{py} \\
\hline
N_k(x_0, y_0)f_{px} \\
N_k(x_0, y_0)f_{py}
\end{bmatrix}
\tag{7-34}
$$

where f_{px} and f_{py} are the magnitudes of the point load in the x and y directions, respectively. Note that if the point load is applied at, e.g., node i, then the corresponding shape function (N_i) is unity and the others are zero, thus giving the obvious intuitive result. Although the summation sign has been dropped above, it is implied. In other words, each point load acting within an element has a corresponding nodal force vector as given above. Given the coordinates of the point of application of the point load [i.e., (x_0, y_0)], we can determine the element in which the load occurs by evaluating N_i, N_j, and N_k at (x_0, y_0). When all three shape functions yield values between zero and unity (i.e., nonnegative and less than one), the element containing the point load is found.

Example 7-4

A point load acts at $(0.6, 1.6)$ for the element in Example 7-1. The x and y components of the point load are $+1500$ and -2300 lbf, respectively. Determine the corresponding nodal force vector.

Solution

The shape functions are given by Eqs. (6-21). Since they make use of all nine of the m_{ij}'s, we must compute m_{11}, m_{12}, and m_{13} (which were not needed in Example 7-1):

$$
m_{11} = \frac{x_j y_k - x_k y_j}{2A} = \frac{x_1 y_3 - x_3 y_1}{2A} = \frac{(0)(0) - (0)(2)}{2(4)} = 0
$$

$$
m_{12} = \frac{x_k y_i - x_i y_k}{2A} = \frac{x_3 y_2 - x_2 y_3}{2A} = \frac{(0)(4) - (4)(0)}{2(4)} = 0
$$

$$
m_{13} = \frac{x_i y_j - x_j y_i}{2A} = \frac{x_2 y_1 - x_1 y_2}{2A} = \frac{(4)(2) - (0)(2)}{2(4)} = 1
$$

Therefore, the shape functions may be evaluated at $x = 0.6$ and $y = 1.6$ as follows:

$$N_i = 0 + (0.25)(0.6) + (0)(1.6) = 0.15$$

$$N_j = 0 + (-0.25)(0.6) + (0.50)(1.6) = 0.65$$

$$N_k = 1 + (0)(0.6) + (-0.50)(1.6) = 0.20$$

Note that the sum of the three shape functions is unity as expected. It follows from Eq. (7-34) that $\mathbf{f}^e_{PL}$ is given by

$$
\mathbf{f}^e_{PL} =
\begin{bmatrix}
(0.15)(1500) \\
(0.15)(-2300) \\
\hline
(0.65)(1500) \\
(0.65)(-2300) \\
\hline
(0.20)(1500) \\
(0.20)(-2300)
\end{bmatrix}
=
\begin{bmatrix}
225 \\
-345 \\
\hline
975 \\
-1495 \\
\hline
300 \\
-460
\end{bmatrix}
$$

Note that most of the point load is allocated to the node j (global node number 1) because this node is the closest to the point load. In a similar fashion, node i (global node number 2) receives the lowest allocation since it is farthest from the point load. ∎

Composite Nodal Force Vector

The reader is reminded that the contributions from each of the nodal force vectors for a given element must be combined according to Eq. (5-86), which is repeated here for convenience:

$$\mathbf{f}^e = \mathbf{f}^e_{\epsilon_0} - \mathbf{f}^e_{\sigma_0} + \mathbf{f}^e_b + \mathbf{f}^e_s + \mathbf{f}^e_{PL} \tag{5-86}$$

The vector $\mathbf{f}^e$ (without subscripts) may be referred to as the *composite nodal force vector*. We are now in a position to discuss the assemblage step.

Assemblage Step

The assemblage of the element stiffness matrices $\mathbf{K}^e$ and the composite nodal force vector $\mathbf{f}^e$ follows the standard procedure illustrated in Chapters 3 and 4. Example 7-5 should help to clarify further the basic idea.

Example 7-5

Consider the mesh shown in Fig. 7-7, which contains only two elements. The elements are defined in terms of the global node numbers, as indicated below the

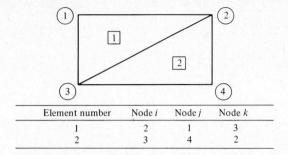

Element number	Node i	Node j	Node k
1	2	1	3
2	3	4	2

Figure 7-7 Mesh used in Example 7-5.

mesh. The 6×6 element stiffness matrices for the two elements may be written symbolically in terms of nine 2×2 submatrices as follows:

$$\mathbf{K}^{(1)} = \begin{bmatrix} \mathbf{K}^{(1)}_{22} & \mathbf{K}^{(1)}_{21} & \mathbf{K}^{(1)}_{23} \\ \mathbf{K}^{(1)}_{12} & \mathbf{K}^{(1)}_{11} & \mathbf{K}^{(1)}_{13} \\ \mathbf{K}^{(1)}_{32} & \mathbf{K}^{(1)}_{31} & \mathbf{K}^{(1)}_{33} \end{bmatrix}$$

and

$$\mathbf{K}^{(2)} = \begin{bmatrix} \mathbf{K}^{(2)}_{33} & \mathbf{K}^{(2)}_{34} & \mathbf{K}^{(2)}_{32} \\ \mathbf{K}^{(2)}_{43} & \mathbf{K}^{(2)}_{44} & \mathbf{K}^{(2)}_{42} \\ \mathbf{K}^{(2)}_{23} & \mathbf{K}^{(2)}_{24} & \mathbf{K}^{(2)}_{22} \end{bmatrix}$$

Show the assemblage stiffness matrix in terms of the 2×2 submatrices.

Solution

The assemblage stiffness matrix $\mathbf{K}^a$ is of size 8×8 and is given by

$$\mathbf{K}^a = \begin{bmatrix} \mathbf{K}^{(1)}_{11} & \mathbf{K}^{(1)}_{12} & \mathbf{K}^{(1)}_{13} & \mathbf{0} \\ \mathbf{K}^{(1)}_{21} & \mathbf{K}^{(1)}_{22} + \mathbf{K}^{(2)}_{22} & \mathbf{K}^{(1)}_{23} + \mathbf{K}^{(2)}_{23} & \mathbf{K}^{(2)}_{24} \\ \mathbf{K}^{(1)}_{31} & \mathbf{K}^{(1)}_{32} + \mathbf{K}^{(2)}_{32} & \mathbf{K}^{(1)}_{33} + \mathbf{K}^{(2)}_{33} & \mathbf{K}^{(2)}_{34} \\ \mathbf{0} & \mathbf{K}^{(2)}_{42} & \mathbf{K}^{(2)}_{43} & \mathbf{K}^{(2}_{44} \end{bmatrix}$$

Although it is not obvious from this result, the assemblage stiffness matrix is symmetric. Moreover, it is banded, although the bandedness is not too obvious because only two elements were considered here. It should be recalled that the assemblage stiffness matrix $\mathbf{K}^a$ is zeroed-out before the element stiffness matrix for the first element is added into it. ∎

Example 7-6

Reconsider the mesh given in Example 7-5. The 6×1 composite nodal force vectors for both elements may be written symbolically in terms of three 2×1 subvectors as follows:

$$
\mathbf{f}^{(1)} = \begin{bmatrix} \mathbf{f}_2^{(1)} \\ \hline \mathbf{f}_1^{(1)} \\ \hline \mathbf{f}_3^{(1)} \end{bmatrix}
\quad \text{and} \quad
\mathbf{f}^{(2)} = \begin{bmatrix} \mathbf{f}_3^{(2)} \\ \hline \mathbf{f}_4^{(2)} \\ \hline \mathbf{f}_2^{(2)} \end{bmatrix}
$$

Show the assemblage nodal force vector in terms of the 2×1 subvectors.

Solution

The assemblage nodal force vector is of size 8×1 and is given by

$$
\mathbf{f}^a = \begin{bmatrix} \mathbf{f}_1^{(1)} \\ \hline \mathbf{f}_2^{(1)} + \mathbf{f}_2^{(2)} \\ \hline \mathbf{f}_3^{(1)} + \mathbf{f}_3^{(2)} \\ \hline \mathbf{f}_4^{(2)} \end{bmatrix}
$$

In the two previous examples, the mesh was comprised of only two elements. In general, the assemblage step is performed for every element. This results in an assemblage stiffness matrix and an assemblage nodal force vector of sizes $2N \times 2N$ and $2N \times 1$, respectively, where N is the number of nodes used in the entire mesh. The reader is reminded that the half-bandwidth of $\mathbf{K}^a$ is given by Eq. (3-33). Like the assemblage stiffness matrix, the assemblage nodal force vector is zeroed-out before the nodal force vector for the first element is added into it. ∎

Prescribed Displacements

The prescribed displacements are imposed according to either of the two methods illustrated in Sec. 3-2. Method 1 is preferred because it results in a better conditioned stiffness matrix in large problems. After application of the prescribed displacements, the system of equations to be solved is given by

$$
\mathbf{Ka} = \mathbf{f} \tag{7-35}
$$

It should be recalled that the vector $\mathbf{a}$ is defined by

$$
\mathbf{a} = [u_1 \quad v_1 \;\vdots\; u_2 \quad v_2 \;\vdots\; \cdots \;\vdots\; u_N \quad v_N]^T \tag{7-36}
$$

where u_i and v_i are the displacements in the x and y directions at node i, and where N is the maximum number of nodes used.

Solution for the Nodal Displacements

Before application of the prescribed displacements, $\mathbf{K}^a$ is singular and does not possess an inverse. After the proper application of the prescribed displacements, $\mathbf{K}$ is nonsingular for well-posed problems. Thus, a unique solution for the nodal displacements is ensured. Subroutine ACTCOL from Sec. 6-8 may be used quite effectively to obtain the solution. It should be recalled that the subroutine utilizes a three-step procedure: triangular decomposition, forward elimination, and backward substitution.

The Element Resultants

Let us now indicate how we could obtain the strains and stresses within a typical element once the nodal displacements in the vector **a** have been found. Recall that the strains within the region are related to the nodal displacements on an element basis by

$$\boldsymbol{\epsilon} = \mathbf{B}\mathbf{a}^e \qquad (5\text{-}77)$$

Since both the matrix **B** for any element e and $\mathbf{a}^e$ are known (because **a** is known), it is clear that Eq. (5-77) can be used to determine the strains (ϵ_{xx}, ϵ_{yy}, and ϵ_{xy}) within element e. It should be noted that the resulting strains are really averages for the element and are generally associated with the centroid of the element. In terms of the m_{ij}'s [defined by Eqs. (6-21)] and the now known nodal displacements, the average strains (denoted by $\bar{\epsilon}_{xx}$, $\bar{\epsilon}_{yy}$, and $\bar{\epsilon}_{xy}$) within element e are given explicitly by

$$\bar{\epsilon}_{xx} = m_{21}u_i + m_{22}u_j + m_{23}u_k \qquad (7\text{-}37a)$$

$$\bar{\epsilon}_{yy} = m_{31}v_i + m_{32}v_j + m_{33}v_k \qquad (7\text{-}37b)$$

$$\bar{\epsilon}_{xy} = m_{31}u_i + m_{21}v_i + m_{32}u_j + m_{22}v_j + m_{33}u_k + m_{23}v_k \qquad (7\text{-}37c)$$

The average stresses ($\bar{\sigma}_{xx}$, $\bar{\sigma}_{yy}$, and $\bar{\sigma}_{xy}$) within element e may be evaluated from

$$\bar{\boldsymbol{\sigma}} = \mathbf{D}[\mathbf{B}\mathbf{a}^e - \boldsymbol{\epsilon}_0] + \boldsymbol{\sigma}_0 \qquad (7\text{-}38)$$

where each matrix or vector on the right-hand side is known. These average stresses are also generally associated with the centroid of the element. It should be recalled that in the case of plane stress, the strains in the z direction are not necessarily zero and can be computed from Eq. (7-16). Similarly, in the case of plane strain, the stresses in the z direction are not necessarily zero and may be determined from Eq. (7-21). Like the other stresses and strains, these too are average values and are usually associated with the centroid.

Remarks

The displacement function for the three-node triangle was taken to be linear [see Eq. (6-16a)]. Hence, the displacements are assumed to vary linearly within each element. The strains and stresses, on the other hand, being related to the derivatives of the displacements, must then be constant. Hence, this element is often referred to as the *constant strain triangle*.

Although the expressions for the finite element characteristics have been evaluated for the triangular element, the reader should have little difficulty in extending this development to the rectangular element. In Chapter 9, other, more powerful, two-dimensional elements are introduced.

7-3   AXISYMMETRIC STRESS ANALYSIS

In this section, the axisymmetric stress model is defined. The finite element characteristics are derived from Eqs. (5-87) to (5-92) for the case of the axisymmetric triangular element. Each of the basic steps in the FEM formulation and solution is discussed.

Definitions

An axisymmetric problem in stress analysis represents another example in which a three-dimensional problem may be idealized as a two-dimensional problem. Such problems must meet the two basic conditions delineated in Sec. 6-6. It should further be recalled that these problems lend themselves to a cylindrical coordinate system with coordinates (r,θ,z) as opposed to a cartesian (x,y,z) system. However, because of the axisymmetry, the θ coordinate need not be included. In other words, the geometry, material properties, loadings, etc., may be a function of r and z only (and not a function of θ). For example, in Fig. 7-8, note how the line load acts in an axisymmetric manner about the z axis. The displacements in the r and z directions are denoted by u and v, respectively. In this section, the triangular element discussed in Sec. 6-6 is used for illustrative purposes. The element is really a triangular donut or toroid with the nodes being represented as circles formed about the z axis, as shown in Fig. 7-9.

The Strain-Displacement Relationship

A fundamental difference exists between axisymmetric stress analysis and that of plane stress and plane strain; namely, a fourth component of the strain $\epsilon_{\theta\theta}$ (and hence stress $\sigma_{\theta\theta}$) must be explicitly considered in addition to the other three strains, ϵ_{rr}, ϵ_{zz}, and ϵ_{rz} (and stresses, σ_{rr}, σ_{zz}, and σ_{rz}). The circumferential strain at a

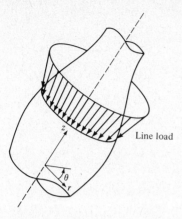

Figure 7-8   Axisymmetric body with an axisymmetric line load.

point within the axisymmetric body is caused by the radial displacement u at this same point. Therefore, the strain vector $\boldsymbol{\epsilon}$ has four components and is defined by

$$\boldsymbol{\epsilon} = [\epsilon_{rr} \quad \epsilon_{\theta\theta} \quad \epsilon_{zz} \quad \epsilon_{rz}]^{T} \tag{7-39}$$

These strains are related to the radial displacement u and the axial displacement v by the following strain-displacement relationships [3]:

$$\epsilon_{rr} = \frac{\partial u}{\partial r} \tag{7-40a}$$

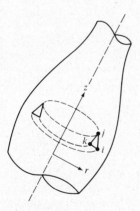

Figure 7-9   Axisymmetric body showing the axisymmetric triangular element (i.e., a triangular toroid).

$$\epsilon_{\theta\theta} = \frac{u}{r} \tag{7-40b}$$

$$\epsilon_{zz} = \frac{\partial v}{\partial z} \tag{7-40c}$$

$$\epsilon_{rz} = \frac{\partial u}{\partial z} + \frac{\partial v}{\partial r} \tag{7-40d}$$

Defining the displacement field vector **u** by $\mathbf{u} = [u \quad v]^T$, we may write Eqs. (7-40) in a very compact form as

$$\boldsymbol{\epsilon} = \mathbf{L}\mathbf{u} \tag{7-41}$$

where the linear operator matrix **L** is defined by

$$\mathbf{L} = \begin{bmatrix} \dfrac{\partial}{\partial r} & 0 \\[2ex] \dfrac{1}{r} & 0 \\[2ex] 0 & \dfrac{\partial}{\partial z} \\[2ex] \dfrac{\partial}{\partial z} & \dfrac{\partial}{\partial r} \end{bmatrix} \tag{7-42}$$

The Shape Function and Strain-Nodal Displacement Matrices

Let us define the nodal displacement vector $\mathbf{a}^e$ by

$$\mathbf{a}^e = [u_i \quad v_i \; \vdots \; u_j \quad v_j \; \vdots \; u_k \quad v_k]^T \tag{5-59}$$

since each of the three nodes i, j, k has two components of displacement. It follows from Sec. 5-6 that the shape function matrix is given by

$$\mathbf{N} = \begin{bmatrix} N_i & 0 & \vdots & N_j & 0 & \vdots & N_k & 0 \\ 0 & N_i & \vdots & 0 & N_j & \vdots & 0 & N_k \end{bmatrix} \tag{5-61}$$

where the shape functions themselves are given by Eqs. (6-21) with x and y replaced by r and z, respectively. Hence, the displacement field vector **u** may be related to the nodal displacement vector $\mathbf{a}^e$ in the usual manner:

$$\mathbf{u} = \mathbf{N}\mathbf{a}^e \tag{7-43}$$

It then follows from $\boldsymbol{\epsilon} = \mathbf{Lu} = \mathbf{LNa}^e = \mathbf{Ba}^e$ that

$$
\mathbf{B} = \mathbf{LN} =
\begin{bmatrix}
\dfrac{\partial N_i}{\partial r} & 0 & \dfrac{\partial N_j}{\partial r} & 0 & \dfrac{\partial N_k}{\partial r} & 0 \\[2mm]
\dfrac{N_i}{r} & 0 & \dfrac{N_j}{r} & 0 & \dfrac{N_k}{r} & 0 \\[2mm]
0 & \dfrac{\partial N_i}{\partial z} & 0 & \dfrac{\partial N_j}{\partial z} & 0 & \dfrac{\partial N_k}{\partial z} \\[2mm]
\dfrac{\partial N_i}{\partial z} & \dfrac{\partial N_i}{\partial r} & \dfrac{\partial N_j}{\partial z} & \dfrac{\partial N_j}{\partial r} & \dfrac{\partial N_k}{\partial z} & \dfrac{\partial N_k}{\partial r}
\end{bmatrix}
\tag{7-44}
$$

whereupon computing the derivatives we get

$$
\mathbf{B} =
\begin{bmatrix}
m_{21} & 0 & m_{22} & 0 & m_{23} & 0 \\[2mm]
\dfrac{N_i}{r} & 0 & \dfrac{N_j}{r} & 0 & \dfrac{N_k}{r} & 0 \\[2mm]
0 & m_{31} & 0 & m_{32} & 0 & m_{33} \\[2mm]
m_{31} & m_{21} & m_{32} & m_{22} & m_{33} & m_{23}
\end{bmatrix}
\tag{7-45}
$$

Note that N_i/r, N_j/r, and N_k/r are not constant but rather are functions of r and z. Nevertheless, the shape function matrix $\mathbf{N}$ and the strain-nodal displacement matrix $\mathbf{B}$ are both determined.

The Constitutive Relationship

As mentioned earlier, four stress components need to be considered. Hence, let us define

$$
\boldsymbol{\sigma} = [\sigma_{rr} \quad \sigma_{\theta\theta} \quad \sigma_{zz} \quad \sigma_{rz}]^T
\tag{7-46}
$$

Note the addition of the stress $\sigma_{\theta\theta}$, which acts in the θ direction (i.e., circumferentially). This definition is consistent with that given for the strain vector $\boldsymbol{\epsilon}$ in Eq. (7-39). Restricting the present discussion to isotropic materials only and assuming the self-strain is a result of a temperature change ΔT, we define

$$
\boldsymbol{\epsilon}_0 =
\begin{bmatrix}
\epsilon_{rr0} \\
\epsilon_{\theta\theta0} \\
\epsilon_{zz0} \\
\epsilon_{rz0}
\end{bmatrix}
=
\begin{bmatrix}
\alpha_t \, \Delta T \\
\alpha_t \, \Delta T \\
\alpha_t \, \Delta T \\
0
\end{bmatrix}
\tag{7-47}
$$

where α_t is the coefficient of thermal expansion. Note that a thermal shear strain cannot exist in an isotropic material. Let us also define the residual stress vector as

$$
\boldsymbol{\sigma}_0 = [\sigma_{rr0} \quad \sigma_{\theta\theta0} \quad \sigma_{zz0} \quad \sigma_{rz0}]^T
\tag{7-48}
$$

For a linear, elastic material, the constitutive equation is given by Eq. (5-24), or

$$\boldsymbol{\sigma} = \mathbf{D}(\boldsymbol{\epsilon} - \boldsymbol{\epsilon}_0) + \boldsymbol{\sigma}_0 \qquad (5\text{-}24)$$

Zienkiewicz [4] gives the corresponding material property matrix (for an isotropic material) as

$$\mathbf{D} = \frac{E}{(1 + \mu)(1 - 2\mu)} \begin{bmatrix} 1 - \mu & \mu & \mu & 0 \\ \mu & 1 - \mu & \mu & 0 \\ \mu & \mu & 1 - \mu & 0 \\ 0 & 0 & 0 & \frac{1}{2}(1 - 2\mu) \end{bmatrix}$$

$$(7\text{-}49)$$

where E and μ are the modulus of elasticity and Poisson's ratio, respectively. The material property matrix is seen once again to be symmetric.

The Element Stiffness Matrix

The general expression for the stiffness matrix is given by Eq. (5-87). If we substitute $2\pi r\, dr\, dz$ for the elemental volume dV, we get

$$\mathbf{K}^e = \int_{V^e} \mathbf{B}^T \mathbf{D} \mathbf{B}\, dV = 2\pi \int_{A^e} \mathbf{B}^T \mathbf{D} \mathbf{B} r\, dr\, dz \qquad (7\text{-}50)$$

Note that the volume integration reduces to an integration over the cross-sectional area of the triangular donut. In other words, we must evaluate the latter integral as an area integral. In effect, the half-plane of the body has been discretized with a number of triangular elements as shown in Fig. 7-10.

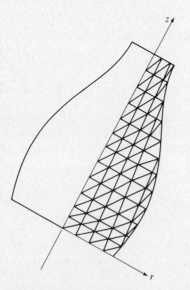

Figure 7-10 Axisymmetric body discretized into a number of three-node triangular elements.

The area integral to be evaluated in Eq. (7-50) is not as simple as the plane stress or plane strain counterpart because the integrand is now a function of r and z. The integration in this case is no longer trivial. At this point we could revert to quadrature or numerical integration to evaluate the area integral in Eq. (7-50), as explained in Chapter 9. However, let us evaluate the integral approximately by evaluating the integrand at the centroid of the triangle and then treating B and r as though they were constants. The coordinates of the area centroid are denoted by $\bar{r}$ and $\bar{z}$ and are given in terms of the nodal coordinates by

$$\bar{r} = \frac{r_i + r_j + r_k}{3} \tag{7-51}$$

and

$$\bar{z} = \frac{z_i + z_j + z_k}{3} \tag{7-52}$$

The resulting $\mathbf{B}$ matrix is denoted as $\overline{\mathbf{B}}$, where $\overline{\mathbf{B}} = \mathbf{B}(\bar{r}, \bar{z})$, denotes that $\mathbf{B}$ is evaluated at the coordinates of the centroid $\bar{r}$ and $\bar{z}$. Therefore, Eq. (7-50) may be written as

$$\mathbf{K}^e = 2\pi \, \overline{\mathbf{B}}^T \mathbf{D} \overline{\mathbf{B}} \bar{r} \int_{A^e} dr \, dz$$

Clearly, the remaining integral is simply the area A of the triangle given by Eq. (6-21e) with r and z replacing x and y, respectively. The resulting element stiffness matrix is given by

$$\mathbf{K}^e = 2\pi \, \overline{\mathbf{B}}^T \mathbf{D} \overline{\mathbf{B}} \bar{r} A \tag{7-53}$$

Note that $\mathbf{K}^e$ is of size 6×6 and is symmetric. The matrix and scalar multiplications in Eq. (7-53) would actually be performed in a computer program by calls to appropriate subroutines.

This simple approach is known to give convergent and valid results if it gives the exact value for the volume. For example, the volume of a triangular toroid is given exactly by $2\pi\bar{r}A$, which is what this approximate integration method yields. Therefore, this approximate integration method is guaranteed to give good results. The accuracy deteriorates as the elements increase in size.

The Nodal Force Vectors

Useful expressions for each of the five nodal force vectors are now derived for axisymmetric stress analysis performed with the three-node triangular element.

Self-Strain

The nodal force vector as a result of the self-strain is given by Eq. (5-88) and may be evaluated here by first noting that

$$\mathbf{f}_{\epsilon_0}^e = \int_{V^e} \mathbf{B}^T \mathbf{D} \boldsymbol{\epsilon}_0 \, dV = 2\pi \int_{A^e} \mathbf{B}^T \mathbf{D} \boldsymbol{\epsilon}_0 r \, dr \, dz \tag{7-54}$$

where the integrand is seen once again to be a function of r and z (because $\mathbf{B}$ is actually a function of r and z). Let us, therefore, take the approximate approach first by evaluating the integrand at the coordinates of the centroid $(\bar{r}, \bar{z})$ and then by treating the integrand as a constant to give

$$\mathbf{f}_{\epsilon_0}^e = 2\pi \bar{\mathbf{B}}^T \mathbf{D} \boldsymbol{\epsilon}_0 \bar{r} A \qquad (7\text{-}55)$$

where $\bar{\mathbf{B}} = \mathbf{B}(\bar{r}, \bar{z})$, as in the evaluation of the stiffness matrix. This nodal force vector is of size 6×1 as expected. If the self-strain is due to a temperature change, then we simply use Eq. (7-47), which explicitly includes the coefficient of thermal expansion α_t and the temperature change ΔT. If ΔT varies over the element in question, then the value at the centroid may be used and thus treated as a constant. This simple approach is consistent with the approximation made above.

Residual Stresses

The element nodal force vector as a result of the residual stresses is determined in the following manner beginning with Eq. (5-89):

$$\mathbf{f}_{\sigma_0}^e = \int_{V^e} \mathbf{B}^T \boldsymbol{\sigma}_0 \, dV = 2\pi \int_{A^e} \mathbf{B}^T \boldsymbol{\sigma}_0 r \, dr \, dz = 2\pi \bar{\mathbf{B}}^T \boldsymbol{\sigma}_0 \bar{r} A \qquad (7\text{-}56)$$

where $\boldsymbol{\sigma}_0$ is given by Eq. (7-48). If the element in question is not under a prestress, then $\boldsymbol{\sigma}_0$ is simply taken to be zero.

Body Forces

The body forces per unit volume in the radial and axial directions are denoted as b_r and b_z, respectively. The corresponding nodal force vector is evaluated by starting with Eq. (5-90) and noting that

$$\mathbf{f}_b^e = \int_{V^e} \mathbf{N}^T \mathbf{b} \, dV = 2\pi \int_{A^e} \mathbf{N}^T \mathbf{b} \, r \, dr \, dz \qquad (7\text{-}57)$$

where the body force vector $\mathbf{b}$ is defined by

$$\mathbf{b} = [b_r \quad b_z]^T \qquad (7\text{-}58)$$

for this axisymmetric analysis. Recall that $\mathbf{N}$ is a function of r and z. If the body forces are assumed to be constant over the element in question and the integrand is evaluated at $\bar{r}$ and $\bar{z}$, then Eq. (7-57) may be evaluated to give

$$\mathbf{f}_b^e = \frac{2\pi \bar{r} A}{3} \begin{bmatrix} b_r \\ b_z \\ \hline b_r \\ b_z \\ \hline b_r \\ b_z \end{bmatrix} \qquad (7\text{-}59)$$

Thus it is seen that one-third of the body force in each direction is allocated to each node. This result should be intuitively obvious since b_r and b_z were assumed to be constant.

If the axisymmetric body is oriented such that the z axis is collinear with direction of gravity, then $b_z = \rho g$ may be used, where ρ is the mass density and g is the acceleration due to gravity. If the body is rotating at an angular speed ω about the z axis, then we may take $b_r = \rho r \omega^2$. In this case, b_r is seen to be a function of r and we may, in effect, use $b_r = \rho \bar{r} \omega^2$.

Surface Tractions

The integral representing the nodal force vector as a result of the surface tractions is not a volume integral, but rather a surface integral. Equation (5-91) provides the starting point, which may be rewritten in terms of the elemental surface area on the boundary of the element as

$$\mathbf{f}_s^e = \int_{S^e} \mathbf{N}^T \mathbf{s} \, dS = 2\pi \int_l \mathbf{N}^T \mathbf{s} r \, dl \qquad \textbf{(7-60)}$$

where $\mathbf{s}$ is defined in terms of the surface tractions s_r and s_z by

$$\mathbf{s} = [s_r \quad s_z]^T \qquad \textbf{(7-61)}$$

and dl is the elemental length around the boundary of the element. If the leg of the triangular element in question is internal and not on the global boundary, then the corresponding nodal force vector as a result of the internal loads do not contribute to the assemblage nodal force vector. The reader will recall that these internal surface tractions are always equal and opposite and, hence, contribute nothing to the assemblage nodal force vector. In other words, only those legs on the global boundary over which surface tractions are imposed need to be considered.

For example, consider leg ij for the element shown in Fig. 7-11. One way to evaluate the integral in Eq. (7-60) is to evaluate the integrand at the midpoint of

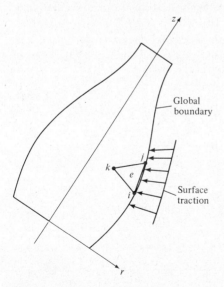

Figure 7-11   Typical triangular element with leg ij on the global boundary on which a surface traction is imposed. *Note:* The surface traction must be axisymmetric.

leg ij and then treat the integrand as a constant. Let us define $\bar{r}_{ij}$ as the radial coordinate at the point halfway between nodes i and j. Let us also note that N_k is zero on leg ij. With these assumptions, it can be shown that Eq. (7-60) simplifies to

$$\mathbf{f}_s^e = 2\pi\bar{r}_{ij}l_{ij} \begin{bmatrix} s_r/2 \\ s_z/2 \\ \hline s_r/2 \\ s_z/2 \\ \hline 0 \\ 0 \end{bmatrix} \qquad \text{for leg } ij \text{ on the global boundary} \qquad \textbf{(7-62)}$$

where l_{ij} is the length of leg ij. A more sophisticated way of evaluating this integral is illustrated in the problems where the special integration formula [see Eq. (6-48)] involving length coordinates is used to simplify the integration. The reader should be able to write, by inspection, the expression for $\mathbf{f}_s^e$ when leg jk or ki is on the global boundary.

It is interesting to note that if a body of revolution is loaded in a nonaxisymmetric manner, it may still be possible to perform an axisymmetric analysis by using a one-term Fourier series of sine and cosine functions to represent the loads and the displacements. The loads, however, must still be symmetric about a plane through the z axis. An example of such a case would be the wind load on a circular smoke stack. Huebner [5] and Wilson [6] give further details. Treatment of this situation is beyond the scope of this text.

Line Loads

For axisymmetric bodies the so-called point loads are actually line loads around the circumference of the body as shown in Fig. 7-8. It is customary to define the line load in the radial and axial directions on a per unit circumference basis. Let us define the *line load* per unit circumference in the radial and axial directions as F_r and F_z, respectively. The equivalent point loads are then given by

$$f_{p_r} = 2\pi r F_r \qquad \textbf{(7-63a)}$$

and

$$f_{p_z} = 2\pi r F_z \qquad \textbf{(7-63b)}$$

Let us also assume the load acts at coordinates (r_0, z_0). Beginning with Eq. (5-92), we can write

$$\mathbf{f}_{PL}^e = \Sigma \mathbf{N}^T \mathbf{f}_p = \Sigma \begin{bmatrix} N_i(r_0,z_0) & 0 \\ 0 & N_i(r_0,z_0) \\ \hline N_j(r_0,z_0) & 0 \\ 0 & N_j(r_0,z_0) \\ \hline N_k(r_0,z_0) & 0 \\ 0 & N_k(r_0,z_0) \end{bmatrix} \begin{bmatrix} 2\pi r_0 F_r \\ 2\pi r_0 F_z \end{bmatrix} \qquad \textbf{(7-64)}$$

If leg ij is the leg on which the load acts as shown in Fig. 7-12, then $N_k(r_0, z_0)$ is identically zero, and the nodal force vector as a result of a line load is given by

$$
\mathbf{f}_{PL}^e = 2\pi r_0
\begin{bmatrix}
N_i(r_0, z_0)F_r \\
N_i(r_0, z_0)F_z \\
\hdashline
N_j(r_0, z_0)F_r \\
N_j(r_0, z_0)F_z \\
\hdashline
0 \\
0
\end{bmatrix}
\tag{7-65}
$$

In Eq. (7-65) the summation sign has been dropped because it is understood that all line load contributions would be added in the usual manner. Similar expressions can be set up by inspection if leg jk or ki is on the global boundary. It should be obvious that the shape functions in Eq. (7-65) allocate the intuitively expected amounts of $2\pi r_0 F_r$ and $2\pi r_0 F_z$ to each of the nodes. For example, if leg ij is on the global boundary and if the line load acts one-third of the distance from node i to node j, then Eq. (7-65) would give

$$
\mathbf{f}_{PL}^e = 2\pi r_0 [\tfrac{2}{3}F_r \quad \tfrac{2}{3}F_z \mid \tfrac{1}{3}F_r \quad \tfrac{1}{3}F_z \mid 0 \quad 0]^T
$$

as expected. In these cases, however, the stress distribution in the immediate vicinity of the load would be very poorly represented unless hundreds of tiny elements were used in this region. It is generally recommended that nodes be placed at the locations of these line loads for improved accuracy. In this case, Eq. (7-65) still gives the correct result.

Remarks

The assemblage of the element stiffness matrices and nodal force vectors is performed in the usual manner. The discussion on the assemblage step in Sec. 7-2 is

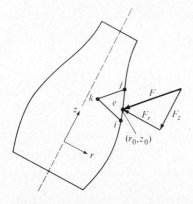

Figure 7-12 Typical triangular element with leg ij on the global boundary on which an axisymmetric line load is applied.

applicable here as well. Again Method 1 is recommended for the application of the prescribed displacements. Finally, the resulting system of equations embodied in $\mathbf{Ka} = \mathbf{f}$ may be solved with the help of subroutine ACTCOL, as explained in Sec. 6-8. Thus, the nodal displacements for every node are known and may be used to obtain the element strains and stresses as follows. The average strains within an element may be computed with the help of $\boldsymbol{\epsilon} = \mathbf{Ba}^e$ or

$$\bar{\boldsymbol{\epsilon}} = \bar{\mathbf{B}}\mathbf{a}^e \tag{7-66}$$

and the average stresses from

$$\bar{\boldsymbol{\sigma}} = \mathbf{D}(\bar{\mathbf{B}}\mathbf{a}^e - \boldsymbol{\epsilon}_0) + \boldsymbol{\sigma}_0 \tag{7-67}$$

These average stresses and strains are generally associated with the centroid of the element.

7-4   THREE-DIMENSIONAL STRESS ANALYSIS

Three-dimensional stress analysis encompasses all practical structural engineering problems since no approximations or simplifying assumptions are inherently made. The price paid for this generality is that of much larger computer storage requirements and obviously longer computational times, as shown in Table 7-1 where two problems are compared. Note that the computational time is roughly proportional to the number of unknowns and to the square of the half-bandwidth. Table 7-1 illustrates the enormity of the three-dimensional problem and has motivated the development of higher-order (distorted) elements, some of which are presented in Chapter 9.

The simplest three-dimensional element is the four-node tetrahedral element with nodes i, j, k, m as shown in Fig. 7-13. The finite element formulation of

Table 7-1   Comparison of Computational Times for Typical Two- and
Three-dimensional Problems*

	Two-dimensional problem	Three-dimensional problem
Number of nodes	$30 \times 30 = 900$	$30 \times 30 \times 30 = 27,000$
Number of equations (and unknowns)	$2 \times 900 = 1800$	$3 \times 27,000 = 81,000$
Variables in bandwidth	$2 \times 30 = 60$	$3 \times 30 \times 30 = 2700$
Relative computational time	1 unit	90,000 units

*The computational time is roughly proportional to the number of unknowns and to the square of the half-bandwidth.

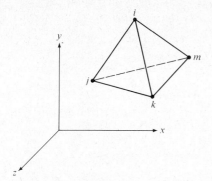

Figure 7-13   The four-node tetrahedral element.

problems in three-dimensional stress analysis is illustrated quite well with this element. Recall from Sec. 6-5 that two rules must be followed when specifying which nodes are associated with a particular element. The reader may want to review the pertinent material in Sec. 6-5 in general and Example 6-5 in particular.

If the reader understands the basic approach taken here, little difficulty will be encountered if an alternate formulation in terms of the eight-node brick is desired. The integrations in this case would be best performed numerically as illustrated in Chapter 9, where other, more practical, elements are introduced.

The Strain-Displacement Relationship

The strain-displacement or kinematic relationship for small deformations in three dimensions was given by Eq. (5-15) and is repeated here for easy reference:

$$\epsilon_{xx} = \frac{\partial u}{\partial x} \qquad \epsilon_{yy} = \frac{\partial v}{\partial y} \qquad \epsilon_{zz} = \frac{\partial w}{\partial z}$$

$$\epsilon_{xy} = \frac{\partial u}{\partial y} + \frac{\partial v}{\partial x} \qquad \epsilon_{yz} = \frac{\partial v}{\partial z} + \frac{\partial w}{\partial y} \qquad \epsilon_{zx} = \frac{\partial w}{\partial x} + \frac{\partial u}{\partial z} \qquad \textbf{(5-15)}$$

where u, v, and w are the displacements in the x, y, and z directions, respectively. Let us define the displacement field vector **u** as

$$\mathbf{u} = [u \quad v \quad w]^T \qquad \textbf{(5-69)}$$

and the strain vector as

$$\boldsymbol{\epsilon} = [\epsilon_{xx} \quad \epsilon_{yy} \quad \epsilon_{zz} \quad \epsilon_{xy} \quad \epsilon_{yz} \quad \epsilon_{zx}]^T \qquad \textbf{(5-26)}$$

Equation (5-15) may then be written in the usual compact form $\boldsymbol{\epsilon} = \mathbf{Lu}$, where

$$
\mathbf{L} = \begin{bmatrix} \dfrac{\partial}{\partial x} & 0 & 0 \\[2mm] 0 & \dfrac{\partial}{\partial y} & 0 \\[2mm] 0 & 0 & \dfrac{\partial}{\partial z} \\[2mm] \dfrac{\partial}{\partial y} & \dfrac{\partial}{\partial x} & 0 \\[2mm] 0 & \dfrac{\partial}{\partial z} & \dfrac{\partial}{\partial y} \\[2mm] \dfrac{\partial}{\partial z} & 0 & \dfrac{\partial}{\partial x} \end{bmatrix}
$$

(7-68)

The Shape Function and Strain-Nodal Displacement Matrices

Let us define the nodal displacement vector $\mathbf{a}^e$ as follows:

$$
\mathbf{a}^e = [u_i \quad v_i \quad w_i \mid u_j \quad v_j \quad w_j \mid u_k \quad v_k \quad w_k \mid u_m \quad v_m \quad w_m]^T \quad \textbf{(5-70)}
$$

since each of the four nodes i, j, k, m has three components of displacement. It follows from Sec. 5-6 that the shape function matrix $\mathbf{N}$ is given by

$$
\mathbf{N} = \begin{bmatrix} N_i & 0 & 0 & N_j & 0 & 0 & N_k & 0 & 0 & N_m & 0 & 0 \\ 0 & N_i & 0 & 0 & N_j & 0 & 0 & N_k & 0 & 0 & N_m & 0 \\ 0 & 0 & N_i & 0 & 0 & N_j & 0 & 0 & N_k & 0 & 0 & N_m \end{bmatrix} \quad \textbf{(5-72)}
$$

where the shape functions themselves are given by Eqs. (6-40). It then follows that the displacement field vector $\mathbf{u}$ is related to the nodal displacememnt vector $\mathbf{a}^e$ in the usual manner by

$$
\mathbf{u} = \mathbf{N}\mathbf{a}^e \quad \textbf{(7-69)}
$$

and from $\mathbf{B} = \mathbf{L}\mathbf{N}$ that

$$
\mathbf{B} = \begin{bmatrix} m_{21} & 0 & 0 & m_{22} & 0 & 0 & m_{23} & 0 & 0 & m_{24} & 0 & 0 \\ 0 & m_{31} & 0 & 0 & m_{32} & 0 & 0 & m_{33} & 0 & 0 & m_{34} & 0 \\ 0 & 0 & m_{41} & 0 & 0 & m_{42} & 0 & 0 & m_{43} & 0 & 0 & m_{44} \\ m_{31} & m_{21} & 0 & m_{32} & m_{22} & 0 & m_{33} & m_{23} & 0 & m_{34} & m_{24} & 0 \\ 0 & m_{41} & m_{31} & 0 & m_{42} & m_{32} & 0 & m_{43} & m_{33} & 0 & m_{44} & m_{34} \\ m_{41} & 0 & m_{21} & m_{42} & 0 & m_{22} & m_{43} & 0 & m_{23} & m_{44} & 0 & m_{24} \end{bmatrix} \quad \textbf{(7-70)}
$$

where the m_{ij}'s are known for any given element and may be computed from Eq. (6-40e). Clearly, the strain-nodal displacement matrix $\mathbf{B}$ is known for any given element.

The Constitutive Relationship

Assuming a linear, elastic material, we can write the stress-strain relationship in the usual form:

$$\sigma = D(\epsilon - \epsilon_0) + \sigma_0 \tag{5-24}$$

where in this case, in addition to ϵ from Eq. (5-26), we define

$$\sigma = [\sigma_{xx} \quad \sigma_{yy} \quad \sigma_{zz} \quad \sigma_{xy} \quad \sigma_{yz} \quad \sigma_{zx}]^T \tag{5-25}$$

$$\epsilon_0 = [\epsilon_{xx0} \quad \epsilon_{yy0} \quad \epsilon_{zz0} \quad \epsilon_{xy0} \quad \epsilon_{yz0} \quad \epsilon_{zx0}]^T \tag{5-27}$$

$$\sigma_0 = [\sigma_{xx0} \quad \sigma_{yy0} \quad \sigma_{zz0} \quad \sigma_{xy0} \quad \sigma_{yz0} \quad \sigma_{zx0}]^T \tag{5-28}$$

The corresponding material property matrix for linear, elastic isotropic materials is given by Zienkiewicz [7] as

$$D = \frac{E}{(1+\mu)(1-2\mu)} \begin{bmatrix} 1-\mu & \mu & \mu & 0 & 0 & 0 \\ \mu & 1-\mu & \mu & 0 & 0 & 0 \\ \mu & \mu & 1-\mu & 0 & 0 & 0 \\ 0 & 0 & 0 & \dfrac{1-2\mu}{2} & 0 & 0 \\ 0 & 0 & 0 & 0 & \dfrac{1-2\mu}{2} & 0 \\ 0 & 0 & 0 & 0 & 0 & \dfrac{1-2\mu}{2} \end{bmatrix} \tag{7-71}$$

If the self-strains ϵ_0 stem from a temperature change ΔT, then we have

$$\epsilon_0 = [\alpha_t \Delta T \quad \alpha_t \Delta T \quad \alpha_t \Delta T \quad 0 \quad 0 \quad 0]^T \tag{7-72}$$

where α_t is the coefficient of thermal expansion for the material comprising the element. Once again it is noted that only normal strains are induced by a temperature change in an isotropic material.

The Element Stiffness Matrix

Starting with Eq. (5-87) and recognizing that B and D are constant, we get

$$K^e = \int_{V^e} B^T D B \, dV = B^T D B V \tag{7-73}$$

where V, the volume of the tetrahedron, is given by Eq. (6-40f) and is a function of the nodal coordinates only. For the four-node tetrahedral element, the element stiffness matrix is of size 12×12 and is symmetric.

The Element Nodal Force Vectors

The element nodal force vector as a result of the self-strains is easily determined from Eq. (5-88):

$$\mathbf{f}^e_{\epsilon_0} = \int_{V^e} \mathbf{B}^T \mathbf{D} \boldsymbol{\epsilon}_0 \, dV = \mathbf{B}^T \mathbf{D} \boldsymbol{\epsilon}_0 V \tag{7-74}$$

If the self-strain is due to a temperature change ΔT in a material with a coefficient of thermal expansion α_t, then Eq. (7-72) should be used for $\boldsymbol{\epsilon}_0$. This element nodal force vector is of size 12×1 since the element has four nodes and three degrees of freedom per node.

The element nodal force vector as a result of the prestresses is computed from Eq. (5-89) to give

$$\mathbf{f}^e_{\sigma_0} = \int_{V^e} \mathbf{B}^T \boldsymbol{\sigma}_0 \, dV = \mathbf{B}^T \boldsymbol{\sigma}_0 V \tag{7-75}$$

which also is of size 12×1.

The element nodal force vector as a result of the body forces is computed by starting with Eq. (5-90). Writing the shape functions in terms of the volume coordinates and integrating the result with the help of Eq. (6-50), we get

$$\mathbf{f}^e_b = \int_{V^e} \mathbf{N}^T \mathbf{b} \, dV = \int_{V^e} \begin{bmatrix} L_i & 0 & 0 \\ 0 & L_i & 0 \\ 0 & 0 & L_i \\ L_j & 0 & 0 \\ 0 & L_j & 0 \\ 0 & 0 & L_j \\ L_j & 0 & 0 \\ 0 & L_k & 0 \\ 0 & 0 & L_k \\ L_m & 0 & 0 \\ 0 & L_m & 0 \\ 0 & 0 & L_m \end{bmatrix} \begin{bmatrix} b_x \\ b_y \\ b_z \end{bmatrix} dV = \frac{V}{4} \begin{bmatrix} b_x \\ b_y \\ b_z \\ b_x \\ b_y \\ b_z \\ b_x \\ b_y \\ b_z \\ b_x \\ b_y \\ b_z \end{bmatrix} \tag{7-76}$$

Note that the body forces per unit volume (b_x, b_y, and b_z) were assumed to be constant. As expected, one-fourth of b_x is allocated to the x degrees of freedom of each of the four nodes, etc.

The nodal force vector as a result of the surface tractions needs to be evaluated for only those elements that have one or more faces on the global boundary. Moreover, only those exposed faces with imposed surface tractions acting on them need to be considered. For example, consider the tetrahedron shown in Fig. 7-14, which is assumed to have face ijk on the global boundary and an imposed surface

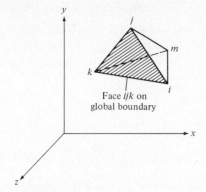

y

j

m

k

i

Face *ijk* on
global boundary

x

z

Figure 7-14 Tetrahedral element with face *ijk* on the global boundary with an imposed surface traction.

traction with components s_x, s_y, and s_z. On this face N_m is identically zero, and Eq. (5-91) simplifies to

$$\mathbf{f}_s^e = \int_{S^e} \mathbf{N}^T \mathbf{s}\, dS = \int_{A_{ijk}} \begin{bmatrix} N_i & 0 & 0 \\ 0 & N_i & 0 \\ 0 & 0 & N_i \\ \hline N_j & 0 & 0 \\ 0 & N_j & 0 \\ 0 & 0 & N_j \\ \hline N_k & 0 & 0 \\ 0 & N_k & 0 \\ 0 & 0 & N_k \\ \hline 0 & 0 & 0 \\ 0 & 0 & 0 \\ 0 & 0 & 0 \end{bmatrix} \begin{bmatrix} s_x \\ s_y \\ s_z \end{bmatrix} dA = \int_{A_{ijk}} \begin{bmatrix} N_i s_x \\ N_i s_y \\ N_i s_z \\ \hline N_j s_x \\ N_i s_y \\ N_j s_z \\ \hline N_k s_x \\ N_k s_y \\ N_k s_z \\ \hline 0 \\ 0 \\ 0 \end{bmatrix} dA \qquad \textbf{(7-77)}$$

where A_{ijk} is the area of the triangle *ijk* that comprises face *ijk* of the tetrahedron. Equation (7-77) may be integrated with the help of the integration formula for area coordinates [Eq. (6-49)] providing the shape functions are written in terms of the area coordinates [see Eq. (6-25)] to give the intuitively obvious result

$$\mathbf{f}_s^e = \frac{A_{ijk}}{3} \begin{bmatrix} s_x & s_y & s_z & \vdots & s_x & s_y & s_z & \vdots & s_x & s_y & s_z & \vdots & 0 & 0 & 0 \end{bmatrix}^T \qquad \textbf{(7-78)}$$

If any of the remaining three faces were on the global boundary, the corresponding nodal force vector could easily be set up using Eq. (7-78) as a guide.

The element nodal force vector as a result of a point load may be computed from Eq. (5-92) as follows. First, note that the external load must necessarily act on one of the four faces of the tetrahedron (or at a node), but only if the face (or node) is on the global boundary. For example, consider a point load acting somewhere on face ijk of the tetrahedral element in Fig. 7-14. Let the location of the point load be represented by the coordinates (x_0, y_0, z_0). Obviously, $N_m(x_0, y_0, z_0)$ is identically zero on this face. Therefore, in this case Eq. (5-92) reduces to

$$
\mathbf{f}_{\mathrm{PL}}^e = \Sigma \mathbf{N}^T \mathbf{f}_p =
\begin{bmatrix}
N_i & 0 & 0 \\
0 & N_i & 0 \\
0 & 0 & N_i \\
\hdashline
N_j & 0 & 0 \\
0 & N_j & 0 \\
0 & 0 & N_j \\
\hdashline
N_k & 0 & 0 \\
0 & N_k & 0 \\
0 & 0 & N_k \\
\hdashline
0 & 0 & 0 \\
0 & 0 & 0 \\
0 & 0 & 0
\end{bmatrix}
\begin{bmatrix}
f_{p_x} \\
f_{p_y} \\
f_{p_z}
\end{bmatrix}_{\substack{x=x_0 \\ y=y_0 \\ z=z_0}}
=
\begin{bmatrix}
N_i(x_0, y_0, z_0) f_{p_x} \\
N_i(x_0, y_0, z_0) f_{p_y} \\
N_i(x_0, y_0, z_0) f_{p_z} \\
\hdashline
N_j(x_0, y_0, z_0) f_{p_x} \\
N_j(x_0, y_0, z_0) f_{p_y} \\
N_j(x_0, y_0, z_0) f_{p_z} \\
\hdashline
N_k(x_0, y_0, z_0) f_{p_x} \\
N_k(x_0, y_0, z_0) f_{p_y} \\
N_k(x_0, y_0, z_0) f_{p_z} \\
\hdashline
0 \\
0 \\
0
\end{bmatrix}
\tag{7-79}
$$

Similar expressions for $\mathbf{f}_{\mathrm{PL}}^e$ can be set up by inspection if the point load acts on any of the other three faces.

Remarks

The assemblage, application of prescribed displacements, and solution of $\mathbf{Ka} = \mathbf{f}$ yield the three nodal displacements for every node. These displacements may be used to obtain strains and stresses for any element e from

$$
\bar{\boldsymbol{\epsilon}} = \mathbf{B}\mathbf{a}^e \tag{7-80}
$$

and

$$
\bar{\boldsymbol{\sigma}} = \mathbf{D}(\mathbf{B}\mathbf{a}^e - \boldsymbol{\epsilon}_0) + \boldsymbol{\sigma}_0 \tag{7-81}
$$

These strains and stresses are average values for the element and are usually associated with the volume centroid of the tetrahedron. Principal stresses can be computed with the help of Eqs. (5-12).

7-5 BEAMS

The finite element characteristics for the beam element based on the elementary theory of beams are derived in this section. This particular model proves to be very useful when statically indeterminate beams are encountered, although the model is equally applicable to the statically determinate case. It should be pointed out that in some cases the plane stress finite element characteristics may be used to model certain types of beams with greater accuracy and detail than the elementary theory can provide. For example, if a thin flat plate is cantilevered and loaded as shown in Fig. 7-15, a rather detailed two-dimensional stress distribution may be found within the beam using the plane stress formulation described in Sec. 7-2. In Fig. 7-15, the point loads P_1, P_2, and P_3 and the distributed load $p(x)$ must act in the plane of the plate for the plane stress model to be valid.

Review of the Elementary Theory

The following assumptions are consistent with the elementary theory of beams. Hooke's law applies in that the stresses are related to the strains in the usual linear fashion. The deflections are assumed to be small in comparison to the beam dimensions. Loadings (forces and moments) must be transverse to the beam, and longitudinal strains are assumed to be negligible.

A typical beam is shown in Fig. 7-16, where the point loads P_1 and P_2 act on the beam in the plane of the paper. In addition, a distributed load $p(x)$ and the externally applied moments M_1 and M_2 are shown. The distributed load $p(x)$ may be a function of x. The left end of the beam is said to be cantilevered while the right end is simply supported. The deflection and slope at the cantilevered end are zero, whereas the deflection and moment at the simply supported end are zero.

With the help of Fig. 7-17, the following sign conventions are adopted: A positive moment M results in compressive stresses for positive values of y; otherwise

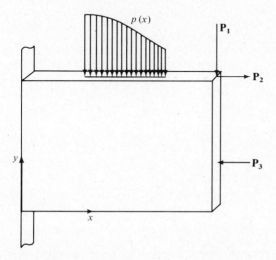

Figure 7-15   Beam for which the plane stress model is valid.

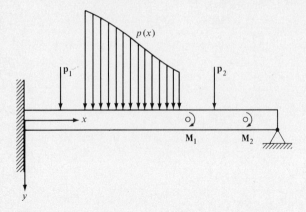

Figure 7-16 Typical beam with a distributed load per unit length $p(x)$, point loads $\mathbf{P}_1$ and $\mathbf{P}_2$, and applied moments $\mathbf{M}_1$ and $\mathbf{M}_2$.

the moment is negative. A shear force V is positive if directed in the positive direction on a positive face (a positive face is one in which the outward normal to the face is in the positive coordinate direction) or if directed in the negative direction on a negative face; otherwise the shear force is negative. This sign convention for shear force is consistent with that defined in Sec. 5-2. A distributed lateral force p is positive if directed in the positive y direction. Note that $p(x)$ is shown to be constant in Fig. 7-17 because as dx approaches zero, $p(x)$ approaches a constant from x to $x + dx$.

With these sign conventions, the following equations [8] relate the shear force V, the moment M, and the lateral load p:

$$\frac{dV}{dx} = -p \tag{7-82}$$

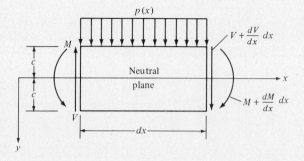

Figure 7-17 Infinitesimal beam element showing positive shear forces, moments, and distributed loading.

$$\frac{dM}{dx} = -V \tag{7-83}$$

$$M = EI \frac{d^2w}{dx^2} \tag{7-84}$$

where w represents the deflection in the y direction and EI is the product of the modulus of elasticity E and the moment of inertia I of the cross section about the axis going through the neutral plane of the beam. The stresses as a result of bending are zero in the neutral plane. The product EI is often referred to as the *flexural rigidity*. Recall that the moment of inertia is defined by

$$I = \int_A y^2 \, dA \tag{7-85}$$

where A is the cross-sectional area of the beam. Furthermore, the slope θ of the beam is related to the lateral deflection w by

$$\theta = \frac{dw}{dx} \tag{7-86}$$

If Eqs. (7-82) and (7-83) are combined, we obtain

$$\frac{d^2M}{dx^2} = p \tag{7-87}$$

If the last result is combined with Eq. (7-84), we get

$$\frac{d^2}{dx^2} \left(EI \frac{d^2w}{dx^2} \right) = p(x) \tag{7-88}$$

which is the governing equation for the deflection of the beam. Equation (7-88) allows for the fact that the flexural rigidity EI may vary along the length of the beam.

C^1-Continuous Shape Functions

The finite element characteristics for the beam model can be derived from Eq. (5-87), etc., provided that the generalized displacements are taken to be the deflections and slopes (w and dw/dx); the strains to be the curvatures of the neutral plane (d^2w/dx^2); and the stresses as the bending moments (M). However, let us take an alternate but completely equivalent approach by deriving the finite element characteristics by using the Galerkin weighted-residual method presented in Chapter 4.

Recall that the Galerkin method (in the finite element context) requires that we choose a suitable trial or basis function that is applied locally over only a portion of the complete x domain; i.e., over a typical finite element. Let us denote this trial displacement function or *parameter function* by w. Figure 7-18 shows a typical finite element e with nodes i and j over which w is to apply. However, we must ensure that we have not only interelement deflection continuity but slope continuity

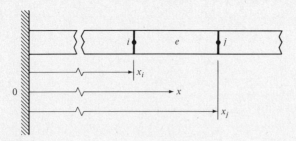

Figure 7-18   Typical beam element e with nodes i and j.

as well. In other words, at each element boundary (or node in this case) we insist that both the deflection and slope be continuous, or that the parameter function be C^1-continuous.

Recall further that problems were defined to be C^1-continuous when the weak formulation contains at most only second-order derivatives. The weak formulation corresponding to Eq. (7-88) may be obtained by either the variational approach or by the Galerkin weighted-residual method. The latter is illustrated in this section and indeed yields an integral equation that contains no derivatives of higher order than two. Therefore, the beam model requires C^1-continuous parameter functions, which in turn requires C^1-continuous shape functions as seen below.

Each element has two nodes, but four requirements must be met: the deflections and slopes at the two nodes must be continuous. Stated differently, at node i, we must have a deflection w_i and slope θ_i, while at node j we must have deflection w_j and slope θ_j. Therefore, four constants must be introduced in the assumed form of the parameter function. Not surprisingly, the following function meets the compatibility and completeness requirements delineated in Sec. 6-2:

$$w = c_1 + c_2 x + c_3 x^2 + c_4 x^3 \tag{7-89}$$

From Eq. (7-86) we see that the slope within an element must be taken to be

$$\theta = \frac{dw}{dx} = c_2 + 2c_3 x + 3c_4 x^2 \tag{7-90}$$

These last two equations provide the starting point for the derivation of the shape functions.

For a typical element e, the vector $\mathbf{a}^e$ is taken as

$$\mathbf{a}^e = [\; w_i \quad \theta_i \;\vdots\; w_j \quad \theta_j \;]^T \tag{7-91}$$

From Eqs. (7-89) and (7-90) it follows that

$$w_i = c_1 + c_2 x_i + c_3 x_i^2 + c_4 x_i^3$$

$$\theta_i = c_2 + 2c_3 x_i + 3c_4 x_i^2$$

$$w_j = c_1 + c_2 x_j + c_3 x_j^2 + c_4 x_j^3$$

$$\theta_j = c_2 + 2c_3 x_j + 3c_4 x_j^2$$

Let us rewrite these last four equations in matrix form as

$$
\mathbf{a}^e = \begin{bmatrix} w_i \\ \theta_i \\ w_j \\ \theta_j \end{bmatrix} = \begin{bmatrix} 1 & x_i & x_i^2 & x_i^3 \\ 0 & 1 & 2x_i & 3x_i^2 \\ 1 & x_j & x_j^2 & x_j^3 \\ 0 & 1 & 2x_j & 3x_j^2 \end{bmatrix} \begin{bmatrix} c_1 \\ c_2 \\ c_3 \\ c_4 \end{bmatrix}
\tag{7-92}
$$

If we solve for the vector of coefficients we get

$$
\begin{bmatrix} c_1 \\ c_2 \\ c_3 \\ c_4 \end{bmatrix} = \begin{bmatrix} 1 & x_i & x_i^2 & x_i^3 \\ 0 & 1 & 2x_i & 3x_i^2 \\ 1 & x_j & x_j^2 & x_j^3 \\ 0 & 1 & 2x_j & 3x_j^2 \end{bmatrix}^{-1} \begin{bmatrix} w_i \\ \theta_i \\ w_j \\ \theta_j \end{bmatrix}
\tag{7-93}
$$

Note that the inverse of the 4 × 4 matrix of nodal coordinates is indicated, but let us postpone getting this inverse for now. Instead let us rewrite Eq. (7-89) in matrix form as

$$
w = \begin{bmatrix} 1 & x & x^2 & x^3 \end{bmatrix} \begin{bmatrix} c_1 \\ c_2 \\ c_3 \\ c_4 \end{bmatrix}
\tag{7-94}
$$

If Eqs. (7-93) and (7-94) are combined by eliminating the vector of coefficients in the usual manner, we get

$$
w = \begin{bmatrix} 1 & x & x^2 & x^3 \end{bmatrix} \begin{bmatrix} 1 & x_i & x_i^2 & x_i^3 \\ 0 & 1 & 2x_i & 3x_i^2 \\ 1 & x_j & x_j^2 & x_j^3 \\ 0 & 1 & 2x_j & 3x_j^2 \end{bmatrix}^{-1} \begin{bmatrix} w_i \\ \theta_i \\ w_j \\ \theta_j \end{bmatrix}
\tag{7-95}
$$

It is very convenient from a mathematical point of view to work in terms of a local coordinate ξ defined by

$$
\xi = \frac{x - x_i}{x_j - x_i} = \frac{x - x_i}{L}
\tag{7-96}
$$

where L (not to be confused with the linear operator matrix $\mathbf{L}$, which is not used in this section) is defined by

$$
L = x_j - x_i
\tag{7-97}
$$

Clearly, L represents the length of the element.

Furthermore, let us write the parameter function w in terms of the shape functions and nodal deflections and slopes as follows:

$$
w = \mathbf{N}\mathbf{a}^e = N_{wi}w_i + N_{\theta i}\theta_i + N_{wj}w_j + N_{\theta j}\theta_j
\tag{7-98}
$$

Note that each node has two shape functions associated with it: one for the deflection and one for the slope. If the right-hand side of Eq. (7-95) is multiplied out (after performing the indicated inversion) and the variable x is eliminated in favor of the local coordinate ξ by using Eq. (7-96), we get

$$w = [\, 1 - 3\xi^2 + 2\xi^3 \,]w_i + [\, L(\xi - 2\xi^2 + \xi^3) \,]\theta_i$$
$$+ [\, 3\xi^2 - 2\xi^3 \,]w_j + [\, L(-\xi^2 + \xi^3) \,]\theta_j \quad \textbf{(7-99)}$$

Comparing Eqs. (7-99) and (7-98) reveals that the shape function matrix $\mathbf{N}$ in terms of the local coordinate ξ is given by

$$\mathbf{N} = \begin{bmatrix} N_{wi} \\ N_{\theta i} \\ N_{wj} \\ N_{\theta j} \end{bmatrix}^T_{\neq} \begin{bmatrix} 1 - 3\xi^2 + 2\xi^3 \\ L(\xi - 2\xi^2 + \xi^3) \\ 3\xi^2 - 2\xi^3 \\ L(-\xi^2 + \xi^3) \end{bmatrix}^T \quad \textbf{(7-100)}$$

The reader should verify that these shape functions satisfy the following conditions: at $x = x_i$ (where $\xi = 0$), we have $N_{wi} = 1$ and $N_{\theta i} = N_{wj} = N_{\theta j} = 0$; and at $x = x_j$ (where $\xi = 1$), we have $N_{wj} = 1$ and $N_{wi} = N_{\theta i} = N_{\theta j} = 0$. In addition, by Eq. (7-86) we must have

$$\theta = \frac{dw}{dx} = \frac{dw}{d\xi}\frac{d\xi}{dx} = \frac{1}{L}\frac{dw}{d\xi} = \frac{1}{L}\frac{d\mathbf{N}}{d\xi}\mathbf{a}^e \quad \textbf{(7-101)}$$

Therefore, the slope θ within an element is given by

$$\theta = \left(\frac{1}{L}\frac{dN_{wi}}{d\xi}\right)w_i + \left(\frac{1}{L}\frac{dN_{\theta i}}{d\xi}\right)\theta_i + \left(\frac{1}{L}\frac{dN_{wj}}{d\xi}\right)w_j + \left(\frac{1}{L}\frac{dN_{\theta j}}{d\xi}\right)\theta_j \quad \textbf{(7-102)}$$

Again the reader should verify that the derivatives of the shape functions satisfy the following conditions: at $x = x_i$ (where $\xi = 0$), we have

$$\left. \begin{aligned} \frac{1}{L}\frac{dN_{\theta i}}{d\xi} &= 1 \\ \frac{1}{L}\frac{dN_{wi}}{d\xi} = \frac{1}{L}\frac{dN_{wj}}{d\xi} &= \frac{1}{L}\frac{dN_{\theta j}}{d\xi} = 0 \end{aligned} \right\} \quad \text{at } \xi = 0 \quad \textbf{(7-103)}$$

and at $x = x_j$ (where $\xi = 1$), we have

$$\left. \begin{aligned} \frac{1}{L}\frac{dN_{\theta j}}{d\xi} &= 1 \\ \frac{1}{L}\frac{dN_{wi}}{d\xi} = \frac{1}{L}\frac{dN_{\theta i}}{d\xi} &= \frac{1}{L}\frac{dN_{wj}}{d\xi} = 0 \end{aligned} \right\} \quad \text{at } \xi = 1 \quad \textbf{(7-104)}$$

With these shape functions in hand, we are now in a position to determine the finite element characteristics for the beam element.

The Finite Element Characteristics

As stated earlier, the Galerkin weighted-residual method will be used to obtain the finite element characteristics for the beam element. Recall from Chapter 4 that the Galerkin method requires that

$$\sum_{e=1}^{n} \int_{x_i}^{x_j} \mathbf{N}^T \left\{ \frac{d^2}{dx^2} \left(EI \frac{d^2w}{dx^2} \right) - p \right\} dx = 0 \qquad \textbf{(7-105)}$$

Note that the term in the braces is really the residual that results when the assumed parameter function for the element is substituted into the governing equation, Eq. (7-88). The so-called weighting functions are precisely the shape functions as embodied in the shape function matrix $\mathbf{N}$.

Again it is more convenient to work in terms of the local coordinate ξ defined by Eq. (7-96). Therefore, $dx = L \, d\xi$ and Eq. (7-105) may be written in terms of the local coordinate ξ as follows:

$$\int_0^1 \mathbf{N}^T \left\{ \frac{1}{L} \frac{d^2}{d\xi^2} \left(\frac{EI}{L^2} \frac{d^2w}{d\xi^2} \right) - pL \right\} d\xi = 0 \qquad \textbf{(7-106)}$$

where the summation over the n elements has been intentionally dropped because we are seeking the element stiffness matrix and the element nodal force vectors. It should be a well-known fact by now that the summation represents the assemblage step, which is considered later.

Equation (7-106) may be written as follows:

$$\frac{1}{L} \int_0^1 \mathbf{N}^T \frac{d^2}{d\xi^2} \left(\frac{EI}{L^2} \frac{d^2w}{d\xi^2} \right) d\xi = \int_0^1 \mathbf{N}^T pL \, d\xi$$

If we integrate the term on the left-hand side of this last equation by parts, we get

$$\frac{1}{L} \mathbf{N}^T \frac{d}{d\xi} \left(\frac{EI}{L^2} \frac{d^2w}{d\xi^2} \right) \bigg|_0^1 - \frac{1}{L} \int_0^1 \frac{d\mathbf{N}^T}{d\xi} \frac{d}{d\xi} \left(\frac{EI}{L^2} \frac{d^2w}{d\xi^2} \right) d\xi = \int_0^1 \mathbf{N}^T pL \, d\xi \qquad \textbf{(7-107)}$$

Writing Eqs. (7-83) and (7-84) in terms of the local coordinate ξ, we get

$$\frac{1}{L} \frac{dM}{d\xi} = -V \qquad \textbf{(7-108)}$$

and

$$M = \frac{EI}{L^2} \frac{d^2w}{d\xi^2} \qquad \textbf{(7-109)}$$

Combining these last two results gives

$$\frac{1}{L} \frac{d}{d\xi} \left(\frac{EI}{L^2} \frac{d^2w}{d\xi^2} \right) = -V \qquad \textbf{(7-110)}$$

Therefore, with the help of Eq. (7-110), the *integrated term* in Eq. (7-107) becomes

$$
- V\mathbf{N}^T \Big|_{\xi=0}^{\xi=1} = - V(x_j) \begin{bmatrix} 0 \\ 0 \\ 1 \\ 0 \end{bmatrix} + V(x_i) \begin{bmatrix} 1 \\ 0 \\ 0 \\ 0 \end{bmatrix} = \begin{bmatrix} V(x_i) \\ 0 \\ - V(x_j) \\ 0 \end{bmatrix}
$$

Therefore, Eq. (7-107) becomes

$$
\begin{bmatrix} V(x_i) \\ 0 \\ - V(x_j) \\ 0 \end{bmatrix} - \frac{1}{L} \int_0^1 \frac{d\mathbf{N}^T}{d\xi} \frac{d}{d\xi} \left(\frac{EI}{L^2} \frac{d^2 w}{d\xi^2} \right) d\xi = \int_0^1 \mathbf{N}^T p L \, d\xi
$$

Integrating the integral on the left-hand side of this last equation by parts again gives

$$
\begin{bmatrix} V(x_i) \\ 0 \\ - V(x_j) \\ 0 \end{bmatrix} - \frac{1}{L} \frac{d\mathbf{N}^T}{d\xi} \frac{EI}{L^2} \frac{d^2 w}{d\xi^2} \Big|_{\xi=0}^{\xi=1} + \frac{1}{L} \int_0^1 \frac{d^2\mathbf{N}^T}{d\xi^2} \frac{EI}{L^2} \frac{d^2 w}{d\xi^2} \, d\xi = \int_0^1 \mathbf{N}^T p L \, d\xi
$$

$$(7\text{-}111)$$

Evaluating the integrated term in this last equation with the help of Eqs. (7-109) and (7-100), we get

$$
- \frac{1}{L} \frac{d\mathbf{N}^T}{d\xi} \frac{EI}{L^2} \frac{d^2 w}{d\xi^2} \Big|_{\xi=0}^{\xi=1} = - M \frac{1}{L} \frac{d\mathbf{N}^T}{d\xi} \Big|_{\xi=0}^{\xi=1}
$$

$$
= (-M) \begin{bmatrix} -6\xi + 6\xi^2/L \\ 1 - 4\xi + 3\xi^2 \\ 6\xi - 6\xi^2/L \\ -2\xi + 3\xi^2 \end{bmatrix} \Big|_{\xi=0}^{\xi=1}
$$

$$
= - M(x_j) \begin{bmatrix} 0 \\ 0 \\ 0 \\ 1 \end{bmatrix} + M(x_i) \begin{bmatrix} 0 \\ 1 \\ 0 \\ 0 \end{bmatrix} = \begin{bmatrix} 0 \\ M(x_i) \\ 0 \\ - M(x_j) \end{bmatrix}
$$

Therefore, Eq. (7-111) becomes

$$
\begin{bmatrix} V(x_i) \\ M(x_i) \\ - V(x_j) \\ - M(x_j) \end{bmatrix} + \frac{1}{L^3} \int_0^1 \frac{d^2\mathbf{N}^T}{d\xi^2} EI \frac{d^2 w}{d\xi^2} \, d\xi = \int_0^1 \mathbf{N}^T p L \, d\xi \qquad (7\text{-}112)
$$

With the help of Eq. (7-98) and noting $\mathbf{a}^e$ is not a function of ξ, we may write Eq. (7-112) as follows:

$$\left\{ \frac{1}{L^3} \int_0^1 \frac{d^2\mathbf{N}^T}{d\xi^2} EI \frac{d^2\mathbf{N}}{d\xi^2} \, d\xi \right\} \mathbf{a}^e = \begin{bmatrix} -V(x_i) \\ -M(x_i) \\ +V(x_j) \\ +M(x_j) \end{bmatrix} + \int_0^1 \mathbf{N}^T pL \, d\xi \qquad \textbf{(7-113)}$$

This last equation may be written concisely as

$$\mathbf{K}^e \mathbf{a}^e = \mathbf{f}^e \qquad \textbf{(7-114)}$$

where $\mathbf{f}^e$ is defined by

$$\mathbf{f}^e = \mathbf{f}^e_{VM} + \mathbf{f}^e_p \qquad \textbf{(7-115)}$$

and $\mathbf{K}^e$ is the element stiffness matrix given by

$$\mathbf{K}^e = \frac{1}{L^3} \int_0^1 \frac{d^2\mathbf{N}^T}{d\xi^2} EI \frac{d^2\mathbf{N}}{d\xi^2} \, d\xi \qquad \textbf{(7-116)}$$

In Eq. (7-115) $\mathbf{f}^e_{VM}$ is the element nodal force vector as a result of shear forces and moments and is defined by

$$\mathbf{f}^e_{VM} = \begin{bmatrix} -V(x_i) \\ -M(x_i) \\ \hdashline V(x_j) \\ M(x_j) \end{bmatrix} \qquad \textbf{(7-117)}$$

and $\mathbf{f}^e_p$ is the element nodal force vector as a result of the distributed load p and is defined by

$$\mathbf{f}^e_p = \int_0^1 \mathbf{N}^T pL \, d\xi \qquad \textbf{(7-118)}$$

Let us now evaluate $\mathbf{K}^e$ and $\mathbf{f}^e_p$. Working on $\mathbf{K}^e$ first, we note that

$$\frac{d^2\mathbf{N}^T}{d\xi^2} = \begin{bmatrix} -6 + 12\xi \\ L(-4 + 6\xi) \\ 6 - 12\xi \\ L(-2 + 6\xi) \end{bmatrix} \qquad \textbf{(7-119)}$$

If the right-hand side of Eq. (7-119) is substituted into Eq. (7-116), the integrand multiplied out, and the integral integrated and evaluated, we get

$$\mathbf{K}^e = \frac{EI}{L^3} \begin{bmatrix} 12 & 6L & \vdots & -12 & 6L \\ 6L & 4L^2 & \vdots & -6L & 2L^2 \\ \hdashline -12 & -6L & \vdots & 12 & -6L \\ 6L & 2L^2 & \vdots & -6L & 4L^2 \end{bmatrix} \qquad \textbf{(7-120)}$$

Note that EI was assumed to be constant for the element. If EI is variable, a suitable average value could be used.

Next the element nodal force vector $\mathbf{f}_p^e$ as a result of the distributed lateral loading p is evaluated. Let us take an average value of the distributed load (written $\bar{p}$) for the element, which is defined as

$$\bar{p} = \frac{p(x_i) + p(x_j)}{2} \qquad (7\text{-}121)$$

Then Eq. (7-118) becomes

$$\mathbf{f}_p^e = \bar{p}L \int_0^1 \mathbf{N}^T \, d\xi = \bar{p}L \begin{bmatrix} \int_0^1 (1 - 3\xi^2 + 2\xi^3) \, d\xi \\ \int_1^0 L(\xi - 2\xi^2 + \xi^3) \, d\xi \\ \hline \int_1^0 (3\xi^2 - 2\xi^3) \, d\xi \\ \int_1^0 L(-\xi^2 + \xi^3) \, d\xi \end{bmatrix} = \begin{bmatrix} \bar{p}L/2 \\ \bar{p}L^2/12 \\ \hline \bar{p}L/2 \\ -\bar{p}L^2/12 \end{bmatrix} \qquad (7\text{-}122)$$

Remarks

The assemblage step is routine and is performed in the manner illustrated in Chapter 3 for the two-dimensional truss model. Imposed external moments and point loads may be readily applied via $\mathbf{f}_{VM}^e$ providing that a node is placed at such points. The point loads are applied via the shear force $V(x_i)$ or $V(x_j)$. Application of the prescribed deflections and slopes is also routine, and Method 1 is again recommended. The solution of $\mathbf{Ka} = \mathbf{f}$ for the nodal deflections and slopes may be obtained either by the matrix inversion method or with the help of subroutine ACTCOL [9] given in Appendix C. These nodal deflections and slopes may be used to obtain the element resultants, as explained below.

In this case, the element resultants include the internal moments M, shear forces V, maximum longitudinal stresses σ_{max}, and maximum shear stresses τ_{max} at each point along the beam. The internal moments M may be evaluated from

$$M = \frac{EI}{L^2} \frac{d^2\mathbf{N}}{d\xi^2} \mathbf{a}^e \qquad (7\text{-}123)$$

and the shear forces V from

$$V = -\frac{1}{L} \frac{d}{d\xi} \left(\frac{EI}{L^2} \frac{d^2\mathbf{N}}{d\xi^2} \right) \mathbf{a}^e \qquad (7\text{-}124)$$

Note that these last two equations imply that M varies linearly over the element e, whereas V is constant. The maximum longitudinal stress (or bending stress) occurs in the outermost fibers of the beam (i.e., the upper or lower surface) and is given by

$$\sigma_{max} = \frac{Mc}{I} \tag{7-125}$$

where c is the distance from the neutral plane to the outermost fibers of the beam as shown in Fig. 7-17. The maximum shear stress τ_{max} occurs in the neutral plane and is given by

$$\tau_{max} = \frac{V \int_0^c y \, dA}{Ib} \tag{7-126}$$

where b is the width of the beam in the neutral plane. Note that in these equations, we would use the results from Eqs. (7-123) and (7-124) to provide M and V at any value of x.

Example 7-7 below illustrates the use of the material in this section and introduces a new simple element—the linear spring element. In this example, one point on the beam is supported by a spring with a known spring constant k_s. Up to now, only one type of element has been used in any one problem or model. Example 7-7 shows how we can easily model systems with a variety of elements. Large structural analysis programs such as NASTRAN [10] and STARDYNE [11] have hundreds of different types of built-in elements.

Example 7-7

Solve for the deflections and slopes at the ends and midspan of the beam shown in Fig. 7-19(a). The distributed load p is 4800 N/m and the point load P is 3000 N. The left end is cantilevered and the right end is attached to a linear, elastic spring with a spring constant k_s of 200 kN/m. The beam has a rectangular cross section with a width b of 3 cm and a height h of 4.31 cm and is made of steel with an elastic modulus E of 2.0×10^{11} N/m². Also determine the moment, shear force, and maximum bending and shear stresses at the middle of the distributed load. Assume $L_1 = L_2 = 1$ m.

Solution

First, we must determine the moment of inertia I for the beam:

$$I = \tfrac{1}{12}bh^3 = \tfrac{1}{12}(0.03)(0.0431)^3$$

or

$$I = 2.00 \times 10^{-7} \text{ m}^4$$

Therefore, the flexural rigidity is given by

$$EI = (2 \times 10^{11})(2 \times 10^{-7}) = 40.0 \times 10^3 \text{ Nm}^2$$

The element computations are summarized below. Note that only two equal-length beam elements are used (here $L = 1$ m) in addition to the spring element as shown in Fig. 7-19(b).

Element 1

From Eq. (7-120), we may compute

$$
\mathbf{K}^{(1)} = \frac{40 \times 10^3}{(1)^3}
\begin{bmatrix}
12 & 6(1) & -12 & 6(1) \\
6(1) & 4(1)^2 & -6(1) & 2(1)^2 \\
-12 & -6(1) & 12 & -6(1) \\
6(1) & 2(1)^2 & -6(1) & 4(1)^2
\end{bmatrix}
$$

$$
=
\begin{bmatrix}
480 & 240 & -480 & 240 \\
240 & 160 & -240 & 80 \\
-480 & -240 & 480 & -240 \\
240 & 80 & -240 & 160
\end{bmatrix} \times 10^3
$$

and from Eq. (7-122),

$$
\mathbf{f}_p^{(1)} =
\begin{bmatrix}
(4800)(1)/2 \\
(4800)(1)^2/12 \\
(4800)(1)/2 \\
-(4800)(1)^2/12
\end{bmatrix}
=
\begin{bmatrix}
2.4 \\
0.4 \\
2.4 \\
-0.4
\end{bmatrix} \times 10^3
$$

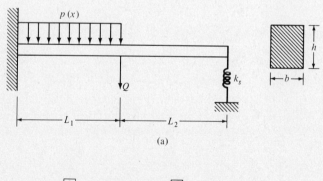

(a)

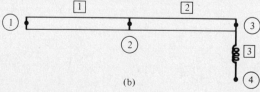

(b)

Figure 7-19 Beam used in Example 7-7. (a) Loading and restraints and (b) discretization for the FEM solution.

If the point load P is considered in the computation of $\mathbf{f}_{VM}^e$ for element 1, we get

$$
\mathbf{f}_{VM}^{(1)} = \begin{bmatrix} 0 \\ 0 \\ \hline 3.0 \\ 0 \end{bmatrix} \times 10^3
$$

Therefore, the composite nodal force vector for element 1 is given by

$$
\mathbf{f}^{(1)} = \begin{bmatrix} 2.4 \\ 0.4 \\ \hline 5.4 \\ -0.4 \end{bmatrix} \times 10^3
$$

Element 2

The stiffness computation for element 2 is identical to that of element 1. There is no distributed load acting on element 2, so $\mathbf{f}_p^{(2)}$ is zero. Moreover, the point load was considered in element 1, so $\mathbf{f}_{VM}^{(2)}$ is zero as well. (How would these computations be modified if the point load is considered in the computations for element 2? Is the same result obtained for the assemblage nodal force vector?)

Element 3

The stiffness relationship for the spring element is given by

$$
\mathbf{K}^{(3)} = k_s \begin{bmatrix} 1 & -1 \\ -1 & 1 \end{bmatrix} = \begin{bmatrix} 200 & -200 \\ -200 & 200 \end{bmatrix} \times 10^3
$$

This follows immediately from the direct approach illustrated in Chapter 3 for the truss element (AE/L for the truss element is analogous to the spring constant k_s).

The assemblage of these element stiffness matrices may be more readily understood if it is given symbolically as follows if we let $k = EI/L^3$:

$$
\begin{bmatrix}
12k & 6kL & -12k & 6kL & 0 & 0 & 0 \\
6kL & 4kL^2 & -6kL & 2kL^2 & 0 & 0 & 0 \\
-12k & -6kL & 24k & 0 & -12kL & 6kL & 0 \\
6kL & 2kL^2 & 0 & 8kL^2 & -6kL & 2kL^2 & 0 \\
0 & 0 & -12k & -6kL & 12k+k_s & -6kL & -k_s \\
0 & 0 & 6kL & 2kL^2 & -6kL & 4kL^2 & 0 \\
0 & 0 & 0 & 0 & -k_s & 0 & k_s
\end{bmatrix}
$$

where the vector of nodal unknowns $\mathbf{a}$ is implied to be

$$
\mathbf{a} = [\, w_1 \quad \theta_1 \mid w_2 \quad \theta_2 \mid w_3 \quad \theta_3 \mid w_4 \,]^T
$$

Note how the element stiffness matrix for the spring element is incorporated into the assemblage stiffness matrix. Numerically, we have

$$
\begin{bmatrix}
480 & 240 & -480 & 240 & 0 & 0 & 0 \\
240 & 160 & -240 & 80 & 0 & 0 & 0 \\
-480 & -240 & 960 & 0 & -480 & 240 & 0 \\
240 & 80 & 0 & 320 & -240 & 80 & 0 \\
0 & 0 & -480 & -240 & 680 & -240 & -200 \\
0 & 0 & 240 & 80 & -240 & 160 & 0 \\
0 & 0 & 0 & 0 & -200 & 0 & 200
\end{bmatrix}
\begin{bmatrix}
w_1 \\ \theta_1 \\ w_2 \\ \theta_2 \\ w_3 \\ \theta_3 \\ w_4
\end{bmatrix}
=
\begin{bmatrix}
2.4 \\ 0.4 \\ 5.4 \\ -0.4 \\ 0 \\ 0 \\ 0
\end{bmatrix}
$$

But $w_1 = 0$, $\theta_1 = 0$, and $w_4 = 0$ must be imposed, and if Method 1 is used from Chapter 3, the result is

$$
\begin{bmatrix}
1 & 0 & 0 & 0 & 0 & 0 & 0 \\
0 & 1 & 0 & 0 & 0 & 0 & 0 \\
0 & 0 & 960 & 0 & -480 & 240 & 0 \\
0 & 0 & 0 & 320 & -240 & 80 & 0 \\
0 & 0 & -480 & -240 & 680 & -240 & 0 \\
0 & 0 & 240 & 80 & -240 & 160 & 0 \\
0 & 0 & 0 & 0 & 0 & 0 & 1
\end{bmatrix}
\begin{bmatrix}
w_1 \\ \theta_1 \\ w_2 \\ \theta_2 \\ w_3 \\ \theta_3 \\ w_4
\end{bmatrix}
=
\begin{bmatrix}
0 \\ 0 \\ 5.4 \\ -0.4 \\ 0 \\ 0 \\ 0
\end{bmatrix}
$$

Solving for the nodal unknowns yields

$$w_1 = 0 \qquad w_2 = 0.01166 \text{ m} \qquad w_3 = 0.00680 \text{ m} \qquad w_4 = 0$$

$$\theta_1 = 0 \qquad \theta_2 = 0.00648 \text{ rad} \qquad \theta_3 = -0.01052 \text{ rad}$$

The moment in the beam at the middle of the distributed load (where $\xi = \frac{1}{2}$ for element 1) is determined from Eqs. (7-123) and (7-119) as follows:

$$
M = \left. \frac{EI}{L^2} \frac{d^2 \mathbf{N}^T}{d\xi^2} \right|_{\xi = 1/2} \mathbf{a}^{(1)} = \frac{EI}{L^2}(0w_1 - L\theta_1 + 0w_2 + L\theta_2)
$$

$$
= \frac{EI(\theta_2 - \theta_1)}{L} = \frac{(40 \times 10^3)(0.00648 - 0.)}{(1)}
$$

or

$$M = 259 \text{ N·m}$$

Therefore, the maximum bending stress at this location is

$$
\sigma_{max} = \frac{Mc}{I} = \frac{(259)(0.0431/2)}{2 \times 10^{-7}} = 27.9 \times 10^6 \text{ N/m}^2
$$

The shear force V at this point is computed from Eq. (7-124) as follows:

$$
V = \left. -\frac{1}{L} \frac{d}{d\xi} \frac{EI}{L^2} \frac{d^2 \mathbf{N}}{d\xi^2} \right|_{\xi = 1/2} \mathbf{a}^e
$$

$$V = \frac{-EI}{L^3}(12w_1 + 6L\theta_1 - 12w_2 + 6L\theta_2)$$

$$= -40 \times 10^3[12(0) + 6(1)(0) - 12(0.01166) + (6)(0.00648)]$$

or

$$V = 4040 \text{ N}$$

Therefore, the maximum shear stress at this location is

$$\tau_{max} = \frac{V\int_0^{h/2} y\, dA}{Ib} = \frac{Vb\,(y^2/2)}{Ib}\Bigg|_0^{h/2} = \frac{Vh^2}{8I}$$

or

$$\tau_{max} = \frac{(4040)(0.0431)^2}{8(2 \times 10^{-7})} = 4.7 \times 10^6 \text{ N/m}^2 \qquad \blacksquare$$

7-6   SUBSTRUCTURING

Occasionally, the structure to be analyzed results in a model that is too large to be solved even on the largest computers. In this section we will see how it is possible to analyze such a structure by breaking it into several substructures. For example, the Boeing 747 airliner was analyzed by breaking the airplane into four parts, or substructures [12]. This approach is referred to as *substructuring* and is explained below. Substructuring can also make it feasible to solve practical structural analysis problems on microcomputers with limited memory.

In an effort to make this discussion a little less abstract, consider the two-dimensional region shown in Fig. 7-20(a). The ultimate goal is to find the nodal displacements and the element resultants. The concept of substructuring requires that we divide the original region into two or more parts as shown in Fig. 7-20(b). Let us concentrate on the first substructure and temporarily disregard the second. The element characteristics for this substructure are determined and assembled in the usual manner to form $\mathbf{K}^a\mathbf{a}^a = \mathbf{f}^a$ for this substructure only. Actually, the assemblage step is not performed in exactly the usual manner; instead, we assemble $\mathbf{K}^a$ and $\mathbf{f}^a$ such that in partitioned form we have

$$\left[\begin{array}{c:c} \mathbf{K}_{11}^a & \mathbf{K}_{12}^a \\ \hdashline \mathbf{K}_{21}^a & \mathbf{K}_{22}^a \end{array}\right] \left[\begin{array}{c} \mathbf{a}_1^a \\ \hdashline \mathbf{a}_2^a \end{array}\right] = \left[\begin{array}{c} \mathbf{f}_1^a \\ \hdashline \mathbf{f}_2^a \end{array}\right] \qquad \textbf{(7-127)}$$

where $\mathbf{a}_1^a$ contains the displacements of the nodes that are on the boundary of the substructure, and $\mathbf{a}_2^a$ contains the displacements of the internal nodes. The displacements of the internal nodes may be eliminated as described below.

If the second matrix equation in Eq. (7-127) is written out and solved for $\mathbf{a}_2^a$, we get

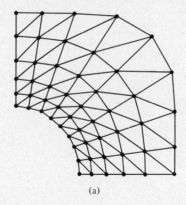

(a)

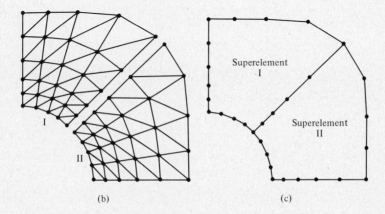

(b)   (c)

Figure 7-20 Two-dimensional region illustrating the concept of substructuring. (a) Original mesh, (b) divided into two substructures, and (c) creating two superelements.

$$\mathbf{a}_2^a = (\mathbf{K}_{22}^a)^{-1}[\mathbf{f}_2^a - \mathbf{K}_{21}^a\mathbf{a}_1^a] \tag{7-128}$$

But the first matrix equation in Eq. (7-127) implies

$$\mathbf{K}_{11}^a\mathbf{a}_1^a + \mathbf{K}_{12}^a\mathbf{a}_2^a = \mathbf{f}_1^a \tag{7-129}$$

Eliminating $\mathbf{a}_2^a$ with the help of Eq. (7-128) and rearranging the result gives

$$[\mathbf{K}_{11}^a - \mathbf{K}_{12}^a(\mathbf{K}_{22}^a)^{-1}\mathbf{K}_{21}^a]\mathbf{a}_1^a = \mathbf{f}_1^a - \mathbf{K}_{12}^a(\mathbf{K}_{22}^a)^{-1}\mathbf{f}_2^a \tag{7-130}$$

which is of the form

$$\mathbf{K}^E\mathbf{a}^E = \mathbf{f}^E \tag{7-131}$$

where the superscript E is used to denote the superelement created by eliminating the internal degrees of freedom. It can be shown that $\mathbf{K}^E$ is still symmetric. The process summarized by Eqs. (7-127) to (7-131) is referred to as *condensation* and may be performed on each substructure. The resulting superelements shown in Fig. 7-20(c) have nodes only on the boundaries. The stiffness matrix $\mathbf{K}^E$ and nodal force vector $\mathbf{f}^E$ for each superelement E are assembled in the usual manner to form the assemblage stiffness matrix $\mathbf{K}^A$ and assemblage nodal force vector $\mathbf{f}^A$. The prescribed displacements are also imposed in the usual manner (Method 1 from Chapter 3 is preferred). The resulting system of equations in $\mathbf{Ka} = \mathbf{f}$ is solved for the nodal displacements. However, the vector $\mathbf{a}$ contains only the displacements for nodes on the boundaries of the superelements. In order to obtain the nodal displacements of the internal nodes, we may use Eq. (7-128) for each substructure, where $\mathbf{a}_1^q$ is now known from the solution of $\mathbf{Ka} = \mathbf{f}$. With the nodal displacements now known for each of the original nodes, the element resultants may be obtained in the usual manner.

Note that Eq. (7-130) requires the inverse of $\mathbf{K}_{22}^q$, which can be a rather large matrix in practical problems. Subroutine ACTCOL [9] may be used to obtain this inverse by successively solving $\mathbf{K}_{22}^q \mathbf{x} = \boldsymbol{\delta}$, where $\boldsymbol{\delta}$ is taken to be $[1 \quad 0 \quad 0 \quad \cdots]^T$ to get the first column in $(\mathbf{K}_{22}^q)^{-1}$, then $[0 \quad 1 \quad 0 \quad 0 \quad \cdots]^T$ to get the second column in $(\mathbf{K}_{22}^q)^{-1}$, and so forth. The triangular decomposition of $\mathbf{K}_{22}^q$ needs to be performed only once (which is accomplished by setting AFAC to .FALSE. after the first decomposition). Since a large portion of the solution time is spent on the triangular decomposition phase, obtaining the inverse in this way is quite practical.

In addition to the obvious advantages of substructuring just mentioned, there are others. For example, the analysis of large structures can be performed by several different groups of engineers, with each group responsible for developing the finite element model for one of the substructures. Obviously, each group must use the same number of nodes at the same locations along the substructure interfaces. Moreover, those portions of the design, and hence the model, that are finalized need not be regenerated during each computer run. Only those portions of the design that are being changed need to be modeled during each run. This can result in an obvious economic advantage when large and complex models are used.

Substructuring may also be performed quite readily in all nonstructural applications, such as in thermal and fluid flow analyses. In addition, substructuring is not limited to static or steady-state analyses. After mastering Chapter 10, the reader should have little difficulty in applying substructuring to dynamic structural and transient thermal problems.

7-7 DEVELOPMENT OF A TWO-DIMENSIONAL STRESS ANALYSIS PROGRAM: PROGRAM STRESS

Some helpful hints and comments are given in this section so that the reader can develop a two-dimensional stress analysis program with further instructions from the instructor.[1] Specific comments are made with respect to the main program, mesh generation, and data storage.

---

[1] A two-dimensional stress analysis program and a user's manual are included in the Instructor/Solution's manual for this text.

The Main Program

The main program should be kept as brief as possible. An example is the following seven lines of code:

```
PROGRAM STRESS
COMMON XL(2500)
MAX = 2500
LCONSL = 3
CALL DRIVER (MAX,LCONSL)
CALL EXIT
END
```

The variable MAX represents the length of the XL array, the contents of which are described below. This variable should be assigned the value of the dimension of XL in the unlabeled COMMON. Much larger problems may be solved with the program by increasing this parameter to the memory limit of the computer being used. The variable LCONSL (as in the TRUSS program) represents the logical unit number for the console. Control is then transferred to subroutine DRIVER.

Mesh Generation

The same type of mesh generator discussed in Sec. 3-6 may be used in this program except that the nodes no longer need to be equally spaced. This is accomplished with the help of two spacing factors f_x and f_y (FX and FY in the program). These factors are used as described below.

Consider the starting node NI and the final node NF shown in Fig. 7-21(a). Additional nodes can be generated between these two end nodes if a nonzero value of NG is used. The nodes need not be equally spaced, however. Without loss of generality, let us number the nodes 1, 2, . . . , n, as shown in Fig. 7-21(b) and refer to the x and y coordinates of node 1 by x_1, y_1, etc. Then the spacing factors f_x and f_y may be defined by

$$f_x = \frac{x_3 - x_2}{x_2 - x_1} = \frac{x_4 - x_3}{x_3 - x_2} = \cdots = \frac{x_n - x_{n-1}}{x_{n-1} - x_{n-2}} \qquad (7\text{-}132)$$

and

$$f_y = \frac{y_3 - y_2}{y_2 - y_1} = \frac{y_4 - y_3}{y_3 - y_2} = \cdots = \frac{y_n - y_{n-1}}{y_{n-1} - y_{n-2}} \qquad (7\text{-}133)$$

Concentrating on the x direction for now, we may write

$$x_2 - x_1 = x_2 - x_1 \qquad (7\text{-}134a)$$

$$x_3 - x_2 = f_x(x_2 - x_1) \qquad (7\text{-}134b)$$

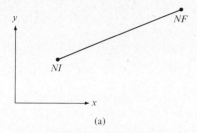
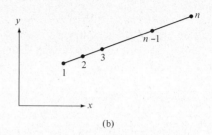

Figure 7-21 (a) Starting node *NI* and ending node *NF* are used to generate additional nodes. (b) Line along which *n* nodes are generated (not necessarily equally spaced).

$$x_4 - x_3 = f_x(x_3 - x_2) = f_x^2(x_2 - x_1) \tag{7-134c}$$

. . . .

$$x_n - x_{n-1} = f_x(x_{n-1} - x_{n-2}) = f_x^{n-2}(x_2 - x_1) \tag{7-134d}$$

Adding these results gives

$$x_n - x_1 = [1 + f_x + f_x^2 + \cdots + f_x^{n-2}](x_2 - x_1) \tag{7-135}$$

from which it follows that

$$x_2 = x_1 + \frac{x_n - x_1}{1 + \sum\limits_{i=1}^{n-2} f_x^i} \tag{7-136}$$

Since x_1 and x_n may be input (as XI and XF, respectively) and since f_x may be input (as FX), then x_2 can be found from Eq. (7-136). Note that *n* is the total number of nodes in the generating sequence (including NI and NF). Obviously x_3 can be computed from Eq. (7-134b), or

$$x_3 = x_2 + f_x(x_2 - x_1) \tag{7-137}$$

and so forth. It should also be obvious that in the *y* direction the same results hold except that each x_i is replaced by y_i and f_x is replaced by f_y. Note that if f_x (or f_y) is greater than unity, the nodes are spaced farther apart in moving from NI to NF; if f_x (or f_y) is less than unity (but greater than zero), the spacing between two consecutive nodes decreases in moving from NI to NF.

Data Storage

The data for the nodal coordinates, elements, boundary conditions, material properties, surface tractions, point loads, and so forth, should be stored in the XL array. In addition, the assemblage stiffness matrix (in column vector form), the assemblage nodal force vector, and the diagonal pointer array (JDIAG from Sec. 6-8) should also be stored in the XL array. The partitions between each of these sections should float, and if fewer nodes are used, for example, more materials may be present. The number of storage locations required must not exceed MAX. If it does, an error message should be displayed on the console and execution should be terminated.

The numbering of the nodes is critical if the program does not renumber the nodes to minimize the bandwidth of the assemblage stiffness matrix. However, the program should store only the banded portion of these matrices, which reduces the storage requirements drastically over storing the full matrices. Nodes on the object being analyzed are always numbered consecutively, from 1 to the maximum number of nodes. The bandwidth is minimized when the maximum difference between any two nodes on each element is minimized. Figure 7-22 shows that this is accomplished quite simply by numbering the nodes in the direction of fewer nodes.

Note that in Fig. 7-22(a) the nodes are numbered in the direction of the smaller number of nodes. In Fig. 7-22(b) the nodes are numbered to the right and in the direction of the greater number of nodes. The storage requirements for the mesh in the latter are higher than for the mesh in the former. In Fig. 7-22(c) the nodes are numbered in an alternating sweeping fashion, which more than doubles the storage requirements over that required in Fig. 7-22(a). Finally, in Fig. 7-22(d) the element on the lower left has nodes 1, 2, and 21; the implication is that the stiffness matrix is no longer banded, which in turn means higher storage requirements and increased execution times. Although the program need not store the zero terms in

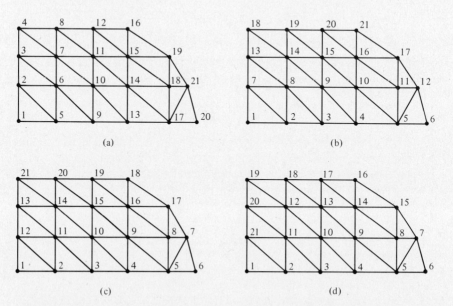

Figure 7-22 (a) Proper node numbering scheme. (b) Less-desirable node numbering scheme. (c) Nodes should never be numbered in an alternating sweeping fashion. (d) Worst node numbering scheme—results in an unbanded stiffness matrix.

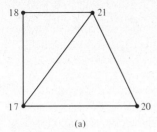
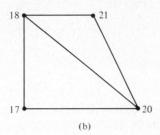

Figure 7-23   (a) Preferred way of dividing a quadrilateral into two triangles and (b) less-desirable way of forming two triangles.

the stiffness matrix outside the bandwidth, it should store everything within the bandwidth (including leading zeros).

Numbering the elements is *not* critical, but some sort of regular numbering scheme usually allows the use of an automatic element generation feature. Note that in forming the elements shown in the lower right-hand corner of Fig. 7-22(a), the quadrilateral formed by nodes 17, 20, 21, and 18 is divided into two triangles by using the shorter diagonal (nodes 17 and 21) as opposed to the longer diagonal (nodes 18 and 20). This is summarized by Fig. 7-23. Regular triangles generally give better results than obtuse or needle-shaped triangles.

7-8   REMARKS

This chapter illustrates how the finite element method is used in static, linear stress analysis. The formulations for plate bending and nonlinear problems are conspicuously absent. For these applications, the reader is referred to more advanced books on this subject by Zienkiewicz [13], Bathe [14], and Ugural [15].

REFERENCES

1. Ugural, A. C., and S. K. Fenster, *Advanced Strength and Applied Elasticity*, American Elsevier, New York, 1975, pp. 57–58.
2. Zienkiewicz, O. C., *The Finite Element Method*, McGraw-Hill (UK), London, 1977, pp. 99–101.
3. Ugural, A. C., and S. K. Fenster, *Advanced Strength and Applied Elasticity*, American Elsevier, New York, 1975, pp. 75–77.
4. Zienkiewicz, O. C., *The Finite Element Method*, McGraw-Hill (UK), London, 1977, p. 125.
5. Huebner, K. H., *The Finite Element Method for Engineers*, Wiley, New York, 1975, pp. 218–219.
6. Wilson, E. L., "Structural Analysis of Axisymmetric Solids," *AIAA Journal*, vol. 3, no. 12, pp. 2267–2274, December 1965.

7. Zienkiewicz, O. C., *The Finite Element Method*, McGraw-Hill (UK), London, 1977, p. 140.

8. Popov, E. P., *Introduction to Mechanics of Solids*, Prentice-Hall, Englewood Cliffs, N.J., 1968, pp. 32–36.

9. Zienkiewicz, O. C., *The Finite Element Method*, McGraw-Hill (UK), London, 1977, pp. 740–741.

10. MSC/NASTRAN, User's Manual, MacNeal-Schwendler Corporation, Los Angeles, 1979.

11. STARDYNE for Scope 3.4 Operating System, User Information Manual, Cybernet Services, Control Data Corporation, Minneapolis, 1978.

12. Hansen, S. D., et al., "Analysis of the 747 Aircraft Wing-Body Intersection," *Proceedings of the 2nd Conference on Matrix Methods in Structural Mechanics*, Wright-Patterson Air Force Base, Dayton, Ohio, October 15–17, 1968.

13. Zienkiewicz, O. C., *The Finite Element Method*, McGraw-Hill (UK), London, 1977, pp. 226–267, 450–526.

14. Bathe, K., *Finite Element Procedures in Engineering Analysis*, Prentice-Hall, Englewood Cliffs, N.J., 1982, pp. 251–260.

15. Ugural, A. C., *Stresses in Plates and Shells*, McGraw-Hill, New York, 1981, pp. 126–136.

PROBLEMS

Note: The properties in Appendix A should be used unless stated otherwise.

7-1 Show that the B matrix for the three-node triangular element in plane stress or plane strain is given by Eq. (7-7).

7-2 Derive the B matrix for the cases of plane stress and plane strain for the four-node rectangular element. How does this matrix differ from that for the three-node triangular element?

7-3 For an isotropic material, how is the shear modulus related to the modulus of elasticity and Poisson's ratio?

7-4 Why is the normal strain ε_{zz} not necessarily zero for a thin plate in plane stress? Give a specific example.

7-5 Why is the normal stress σ_{zz} not necessarily zero in a long bar in plane strain? Give a specific example.

7-6 Repeat Example 7-1 for the case of plane strain. Since the member is now assumed to be long in the longitudinal direction, take t to be unity. Use as much of Example 7-1 as possible.

7-7 Consider the triangular element shown in Fig. P7-7. The plate from which the element is extracted is made of cast iron and has a thickness of 0.5 in. The nodal coordinates are $x_i = 2.0$, $y_i = 1.5$, $x_j = 1.7$, $y_j = 3.0$, $x_k = 0.6$, and $y_k = 1.8$ in. Determine the element stiffness matrix if all external forces act in the plane of the plate (and hence in the plane of the element).

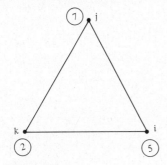

Figure P7-7

7-8 Repeat Problem 7-7 for the case of plane strain. Take t to be unity.

7-9 Consider the triangular element shown in Fig. P7-9. The plate from which the element is extracted is made of brass and has a thickness of 1 cm. The nodal coordinates are $x_i = 5$, $y_i = 6$, $x_j = 4$, $y_j = 4$, $x_k = 6$, and $y_k = 4$ cm. Determine the element stiffness matrix if all external forces act in the plane of the plate (and hence in the plane of the element).

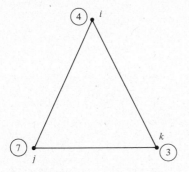

Figure P7-9

7-10 Repeat Problem 7-9 for the case of plane strain. Take t to be unity.

7-11 Consider the triangular element shown in Fig. P7-11. The element is extracted from a thin plate of thickness 0.5 cm. The material is hot rolled, low carbon steel. The nodal coordinates are $x_i = 0$, $y_i = 0$, $x_j = 0$, $y_j = -1$, $x_k = 2$, and $y_k = -1$ cm. Determine the element stiffness matrix if all external forces act in the plane of the plate (and hence in the plane of the element).

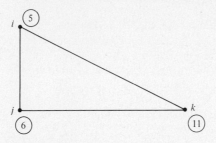

Figure P7-11

7-12 Repeat Problem 7-11 for the case of plane strain. Take t to be unity.

7-13 Consider the triangular element shown in Fig. P7-13. The element is extracted from a thin plate of thickness 0.75 in. The material is hard drawn copper. The nodal coordinates are $x_i = 0$, $y_i = 0$, $x_j = 1$, $y_j = 2$, $x_k = -1$, and $y_k = 2$ in. Determine the element stiffness matrix if all external forces act in the plane of the plate (and hence in the plane of the element).

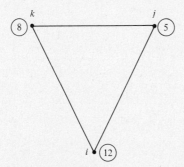

Figure P7-13

7-14 Repeat Problem 7-13 for the case of plane strain. Take t to be unity.

7-15 Show that the element stiffness matrix for plane stress and plane strain as given by Eq. (7-24) for the three-node triangular element is always symmetric.

7-16 Show that the element stiffness matrix for plane stress and plane strain for the four-node rectangular element is always symmetric.

7-17 What is the size of the element stiffness matrix for the case of plane stress (or plane strain) if the four-node rectangular element is used? Explain your answer.

7-18 Determine the element stiffness matrix for the element shown in Fig. P7-18. The element is extracted from an aluminum (6061 alloy) plate with a thickness of

1.25 cm. All external forces are in the plane of the plate. The coordinates of the nodes are $x_i = 4$, $y_i = 2$, $x_j = 4$, $y_j = 3$, $x_k = 2$, $y_k = 3$, $x_m = 2$, and $y_m = 2$ cm. Perform the integrations by evaluating the integrands at the element centroid and treating the integrand as a constant.

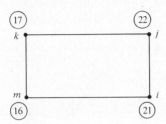

Figure P7-18

7-19 Repeat Problem 7-18 for the case of plane strain. Take t to be unity.

7-20 Determine the element stiffness matrix for the element shown in Fig. P7-20. The element is extracted from a brass plate with a thickness of 0.375 in. All external forces are in the plane of the plate. The coordinates of the nodes are $x_i = 5$, $y_i = 2$, $x_j = 5$, $y_j = 4$, $x_k = 2$, $y_k = 4$, $x_m = 2$, and $y_m = 2$ in. Perform the integrations by evaluating the integrands at the element centroid and treating the integrand as a constant.

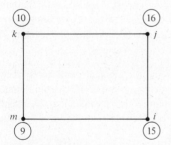

Figure P7-20

7-21 Repeat Problem 7-20 for the case of plane strain. Take t to be unity.

7-22 What size is the element nodal force vector from a self-strain if the four-node rectangular element is used in plane stress or plane strain formulations?

7-23 Determine a general expression for the element nodal force vector from a uniform self-strain for the rectangular element in plane stress. Perform the integrations by evaluating the integrand at the element centroid and treating the integrands as though they were constants.

7-24 Repeat Problem 7-23 for the case of plane strain.

7-25 For the element in Problem 7-7, determine the element nodal force vector as a result of a self-strain if the temperature increases by 50°F. Assume plane stress.

7-26 For the element in Problem 7-7, determine the element nodal force vector as a result of a self-strain if the temperature increases by 50°F. Assume plane strain and take t to be unity.

7-27 For the element in Problem 7-9, determine the element nodal force vector as a result of a self-strain if the temperature increases by 30°C. Assume plane stress.

7-28 For the element in Problem 7-9, determine the element nodal force vector as a result of a self-strain if the temperature increases by 30°C. Assume plane strain and take t to be unity.

7-29 For the element in Problem 7-11, determine the element nodal force vector as a result of a self-strain if the temperature decreases by 40°C. Assume plane stress.

7-30 For the element in Problem 7-11, determine the element nodal force vector as a result of a self-strain if the temperature decreases by 40°C. Assume plane strain and take t to be unity.

7-31 For the element in Problem 7-13, determine the element nodal force vector as a result of a self-strain if the temperature decreases by 75°F. Assume plane stress.

7-32 For the element in Problem 7-13, determine the element nodal force vector as a result of a self-strain if the temperature decreases by 75°F. Assume plane strain and take t to be unity.

7-33 For the element in Problem 7-18, determine the element nodal force vector as a result of a self-strain if the temperature decreases by 35°C. Assume plane stress and perform the integrations by evaluating the integrands at the element centroid and treating the integrands as though they were constants.

7-34 For the element in Problem 7-18, determine the element nodal force vector as a result of a self-strain if the temperature decreases by 35°C. Assume plane strain with t taken as unity and perform the integrations by evaluating the integrands at the element centroid and treating the integrands as though they were constants.

7-35 For the element in Problem 7-20, determine the element nodal force vector as a result of a self-strain if the temperature decreases by 80°F. Assume plane stress and perform the integrations by evaluating the integrands at the element centroid and treating the integrands as though they were constants.

7-36 For the element in Problem 7-20, determine the element nodal force vector as a result of a self-strain if the temperature decreases by 80°F. Assume plane strain with t taken as unity and perform the integrations by evaluating the integrands at the element centroid and treating the integrands as though they were constants.

7-37 Determine a general expression for the element nodal force vector from a uniform prestress for the rectangular element in plane stress or plane strain. Perform the integrations by evaluating the integrand at the element centroid and treating the integrands as though they were constants.

7-38 For the element in Problem 7-7, determine the element nodal force vector as a result of the following prestresses: $\sigma_{xx0} = 1000$, $\sigma_{yy0} = -750$, and $\sigma_{xy0} = 500$ psi.

7-39 For the element in Problem 7-9, determine the element nodal force vector as a result of the following prestresses: $\sigma_{xx0} = -2000$, $\sigma_{yy0} = 1200$, and $\sigma_{xy0} = -1500$ N/cm^2.

7-40 For the element in Problem 7-11, determine the element nodal force vector as a result of the following prestresses: $\sigma_{xx0} = 1300$, $\sigma_{yy0} = -800$, and $\sigma_{xy0} = 0$ N/cm^2.

7-41 For the element in Problem 7-13, determine the element nodal force vector as a result of the following prestresses: $\sigma_{xx0} = 4200$, $\sigma_{yy0} = -2800$, and $\sigma_{xy0} = -1400$ psi.

7-42 For the element in Problem 7-18, determine the element nodal force vector as a result of the following prestresses: $\sigma_{xx0} = -3400$, $\sigma_{yy0} = 1800$, and $\sigma_{xy0} = -2200$ N/cm^2. Perform the integrations by evaluating the integrands at the element centroid and treating the integrands as though they were constants.

7-43 For the element in Problem 7-20, determine the element nodal force vector as a result of the following prestresses: $\sigma_{xx0} = -400$, $\sigma_{yy0} = 800$, and $\sigma_{xy0} = -1200$ psi.

Perform the integrations by evaluating the integrands at the element centroid and treating the integrands as though they were constants.

7-44 Determine a general expression for the element nodal force vector from a uniform body force for the rectangular element in plane stress or plane strain. Perform the integrations by evaluating the integrand at the element centroid and treating the integrands as though they were constants.

7-45 For the element in Problem 7-7, determine the element nodal force vector as a result of a body force with the following components: $b_x = 100$ and $b_y = 0$ lbf/in^3.

7-46 For the element in Problem 7-9, determine the element nodal force vector as a result of a body force with the following components: $b_x = 500$ and $b_y = 700$ N/cm^3.

7-47 Assume that the element in Problem 7-11 is oriented such that the y axis is in the direction opposite that of gravity. If there are no other body forces present, determine the element nodal force vector. The acceleration due to gravity is 9.81 m/s^2.

7-48 Assume that the element in Problem 7-13 is oriented such that the x axis is in the same direction as that of gravity. If there are no other body forces present, determine the element nodal force vector. The acceleration due to gravity is 32.2 ft/s^2.

7-49 For the element in Problem 7-18, determine the element nodal force vector as a result of a body force with the following components: $b_x = 500$ and $b_y = 0$ N/cm^3. Perform the integrations by evaluating the integrands at the element centroid and treating the integrands as though they were constants.

7-50 Assume that the element in Problem 7-20 is oriented such that the y axis is in the direction opposite that of gravity. If there are no other body forces present, determine the element nodal force vector. The acceleration due to gravity is 32.2 ft/sec^2. Perform the integrations by evaluating the integrands at the element centroid and treating the integrands as though they were constants.

7-51 Determine a general expression for the element nodal force vector from a uniform surface traction on face ij for the rectangular element in plane stress or plane strain.

Perform the integrations by evaluating the integrand at the center of face ij and treating the integrands as though they were constants.

7-52   For the element in Problem 7-7, determine the element nodal force vector as a result of a surface traction on leg ij with the following components: $s_x = 1000$ and $s_y = 200$ psi.

7-53   For the element in Problem 7-9, determine the element nodal force vector as a result of a surface traction on leg jk with the following components: $s_x = 1500$ and $s_y = 800$ N/cm^2.

7-54   For the element in Problem 7-11, determine the element nodal force vector as a result of a surface traction on leg ki with the following components: $s_x = 1200$ and $s_y = 750$ N/cm^2.

7-55   For the element in Problem 7-13, determine the element nodal force vector as a result of a surface traction on leg jk with the following components: $s_x = 4200$ and $s_y = 1800$ psi.

7-56   For the element in Problem 7-18, determine the element nodal force vector as a result of a surface traction on face ij with the following components: $s_x = 1500$ and $s_y = 500$ N/cm^2. Perform the integrations by evaluating the integrands at the centroid of face ij and treating the integrands as though they were constants.

7-57   For the element in Problem 7-20, determine the element nodal force vector as a result of a surface traction on face mi with the following components: $s_x = 1100$ and $s_y = 1500$ psi. Perform the integrations by evaluating the integrands at the centroid of face mi and treating the integrands as though they were constants.

7-58   It has been assumed in the derivation leading to Eq. (7-33) in the text that the surface traction is uniform over the leg of the triangle in question. Derive an alternate expression for the nodal force vector for a linearly varying surface traction on leg ij of the three-node triangular element shown in Fig. P7-58. Proceed by assuming the following forms for s_x and s_y:

$$s_x = N_i s_{xi} + N_j s_{xj}$$

$$s_y = N_i s_{yi} + N_j s_{yj}$$

where s_{xi} is the x component of the surface traction at node i, etc. Write the shape functions in terms of the length coordinates (defined on leg ij) and use Eq. (6-48) to perform the integrations.

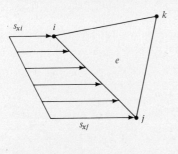
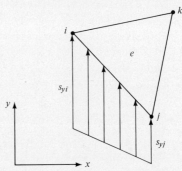

Figure P7-58

7-59 Repeat Problem 7-58 for a uniformly varying surface traction on leg jk. At nodes j and k, the x and y components of the surface traction are s_{xj} and s_{xk} and s_{yj} and s_{yk}, respectively.

7-60 Repeat Problem 7-59 for a uniformly varying surface traction on leg ki. At nodes k and i, the x and y components of the surface traction are s_{xk} and s_{xi}, and s_{yk} and s_{yi}, respectively.

7-61 For the element in Problem 7-7, determine the element nodal force vector as a result of a point load that has the following components: $f_{px} = 2000$ and $f_{py} = 1200$ lbf. The point load is located at $x_0 = 1.5$ and $y_0 = 2.0$ in.

7-62 For the element in Problem 7-9, determine the element nodal force vector as a result of a point load with the following components: $f_{px} = 1000$ and $f_{py} = 500$ N. The point load is located at $x_0 = 5$ and $y_0 = 5$ cm.

7-63 For the element in Problem 7-11, determine the element nodal force vector as a result of a point load with the following components: $f_{px} = 2400$ and $f_{py} = -1500$ N. The point load is located at $x_0 = 2$ and $y_0 = -1$ cm.

7-64 For the element in Problem 7-13, determine the element nodal force vector as a result of a point load with the following components: $f_{px} = -1250$ and $f_{py} = -1800$ lbf. The point load is located at $x_0 = 1$ and $y_0 = 2$ in.

7-65 For the element in Problem 7-18, determine the element nodal force vector as a result of a point load with the following components: $f_{px} = 2330$ and $f_{py} = -1500$ N. The point load is located at $x_0 = 3.0$ and $y_0 = 2.5$ cm.

7-66 For the element in Problem 7-18, determine the element nodal force vector as a result of a point load with the following components: $f_{px} = -2330$ and $f_{py} = 1500$ N. The point load is located at $x_0 = 4$ and $y_0 = 2$ cm.

7-67 For the element in Problem 7-20, determine the element nodal force vector as a result of a point load with the following components: $f_{px} = -2000$ and $f_{py} = 1000$ lbf. The point load is located at $x_0 = 4$ and $y_0 = 3$ in.

7-68 For the element in Problem 7-20, determine the element nodal force vector as a result of a point load with the following components: $f_{px} = 2500$ and $f_{py} = -3000$ lbf. The point load is located at $x_0 = 2$ and $y_0 = 4$ in.

7-69 Using the symbolic notation from Example 7-5, give the expression for the assemblage stiffness matrix for the discretized two-dimensional region in Fig. P7-69. What is the half-bandwidth? Can the half-bandwidth be reduced? If so, explain how.

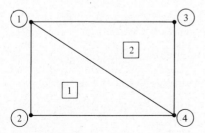

| Element | Nodes | | |
number	i	j	k
1	4	1	2
2	1	4	3

Figure P7-69

7-70 Using the symbolic notation from Example 7-6, give the expression for the assemblage nodal force vector for the discretized two-dimensional region in Fig. P7-69.

7-71 Using the symbolic notation from Example 7-5, give the expression for the assemblage stiffness matrix for the discretized two-dimensional region in Fig. P7-71. What is the half-bandwidth? Can the half-bandwidth be reduced? If so, explain how.

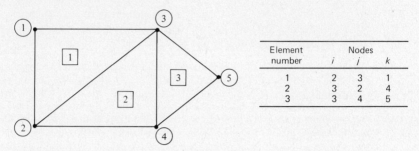

Element	Nodes		
number	*i*	*j*	*k*
1	2	3	1
2	3	2	4
3	3	4	5

Figure P7-71

7-72 Using the symbolic notation from Example 7-6, give the expression for the assemblage nodal force vector for the discretized two-dimensional region in Fig. P7-71.

7-73 Using the symbolic notation from Example 7-5, give the expression for the assemblage stiffness matrix for the discretized two-dimensional region in Fig. P7-73. What is the half-bandwidth? Can the half-bandwidth be reduced? If so, explain how.

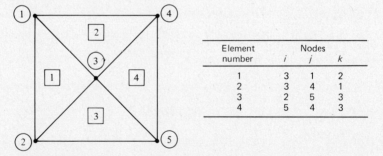

Element	Nodes		
number	*i*	*j*	*k*
1	3	1	2
2	3	4	1
3	2	5	3
4	5	4	3

Figure P7-73

7-74 Using the symbolic notation from Example 7-6, give the expression for the assemblage nodal force vector for the discretized two-dimensional region in Fig. P7-73.

7-75 By extending the symbolic notation from Example 7-5, give the expression for the assemblage stiffness matrix for the discretized two-dimensional region in Fig. P7-75. What is the half-bandwidth? Can the half-bandwidth be reduced? If so, explain how.

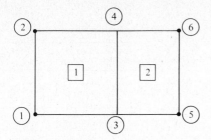

Element	Nodes			
number	*i*	*j*	*k*	*m*
1	3	4	2	1
2	5	6	4	3

Figure P7-75

7-76 Be extending the symbolic notation from Example 7-6, give the expression for the assemblage nodal force vector for the discretized two-dimensional region in Fig. P7-75.

7-77 By extending the symbolic notation from Example 7-5, give the expression for the assemblage stiffness matrix for the discretized two-dimensional region in Fig. P7-77. What is the half-bandwidth? Can the half-bandwidth be reduced? If so, explain how.

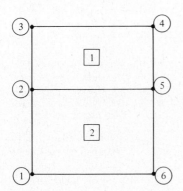

Element	Nodes			
number	*i*	*j*	*k*	*m*
1	5	4	3	2
2	6	5	2	1

Figure P7-77

7-78 By extending the symbolic notation from Example 7-6, give the expression for the assemblage nodal force vector for the discretized two-dimensional region in Fig. P7-77.

7-79 By extending the symbolic notation from Example 7-5, give the expression for the assemblage stiffness matrix for the discretized two-dimensional region in Fig. P7-79. Note the use of two different types of elements. What is the half-bandwidth? Can the half-bandwidth be reduced? If so, explain how.

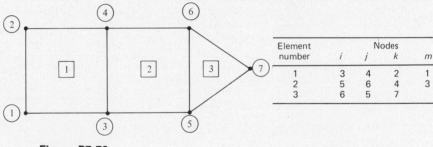

Figure P7-79

7-80 By extending the symbolic notation from Example 7-6, give the expression for the assemblage nodal force vector for the discretized two-dimensional region in Fig. P7-79. Note the use of two different types of elements.

7-81 For the element in Problem 7-7 for a particular loading condition, the following values of nodal displacements are obtained: $u_i = 0.00105$, $v_i = 0.0$, $u_j = -0.00055$, $v_j = 0.00200$, $u_k = 0.00150$, and $v_k = -0.00405$ in. Determine the element strains and stresses that correspond to these displacements if the element is in a state of plane stress. Assume both the self-strains and prestresses are zero.

7-82 Repeat Problem 7-81 for the case of plane strain. Take t to be unity.

7-83 For the element in Problem 7-9 for a particular loading condition, the following values of nodal displacements are obtained: $u_i = 0.00215$, $v_i = -0.00300$, $u_j = 0.0$, $v_j = 0.00560$, $u_k = -0.00520$, and $v_k = 0.00650$ cm. Determine the element strains and stresses that correspond to these displacements if the element is in a state of plane stress. Assume both the self-strains and prestresses are zero.

7-84 Repeat Problem 7-83 for the case of plane strain. Take t to be unity.

7-85 For the element in Problem 7-11 for a particular loading condition, the following values of nodal displacement are obtained: $u_i = -0.01250$, $v_i = 0.02300$, $u_j = 0.01500$, $v_j = -0.01550$, $u_k = 0.01750$, and $v_k = 0.0$ cm. Determine the element strains and stresses that correspond to these displacements if the element is in a state of plane stress. Assume the self-strains and prestresses from Problems 7-29 and 7-40.

7-86 Repeat Problem 7-85 for the case of plane strain. Take t to be unity. Assume the self-strains and prestresses from Problems 7-30 and 7-40.

7-87 For the element in Problem 7-13 for a particular loading condition, the following values of nodal displacements are obtained: $u_i = 0.02150$, $v_i = 0.02350$, $u_j = -0.01575$, $v_j = 0.02550$, $u_k = 0.02500$, and $v_k = 0.0$ in. Determine the element strains and stresses that correspond to these displacements if the element is in a state of plane stress. Assume the self-strains and prestresses from Problems 7-31 and 7-41.

7-88 Repeat Problem 7-87 for the case of plane strain. Take t to be unity. Assume the self-strains and prestresses from Problems 7-32 and 7-41.

7-89 For the element in Problem 7-18 for a particular loading condition, the following values of nodal displacements are obtained: $u_i = 0.02250$, $v_i = -0.01250$, $u_j = 0.02555$, $v_j = 0.02350$, $u_k = 0.02500$, $v_k = -0.02340$, $u_m = 0.01235$, and $v_m = -0.02500$ cm. Determine the element strains and stresses at the element centroid that correspond to these displacements if the element is in a state of plane stress. Assume both the self-strains and prestresses are zero.

7-90 Repeat Problem 7-89 for the case of plane strain. Take t to be unity.

7-91 For the element in Problem 7-20 for a particular loading condition, the following values of nodal displacements are obtained: $u_i = -0.01250$, $v_i = -0.03050$, $u_j = 0.02500$, $v_j = 0.03150$, $u_k = -0.02500$, $v_k = 0.03350$, $u_m = 0.0$, and $v_m = 0.03500$ in. Determine the element strains and stresses at the element centroid that correspond to these displacements if the element is in a state of plane stress. Assume both the self-strains and prestresses are zero.

7-92 Repeat Problem 7-91 for the case of plane strain. Take t to be unity.

7-93 Derive the **B** matrix for the case of axisymmetric stress analysis for the four-node rectangular element.

7-94 Consider the triangular element shown in Fig. P7-7. The body from which the element is extracted is made of hard drawn copper. The nodal coordinates are $r_i = 2.0$, $z_i = 1.5$, $r_j = 1.7$, $z_j = 3.0$, $r_k = 0.6$, and $z_k = 1.8$ in. Determine the element stiffness matrix assuming the body is a body of revolution and is loaded axisymmetrically.

7-95 Consider the triangular element shown in Fig. P7-9. The body from which the element is extracted is made of cast iron. The nodal coordinates are $r_i = 5$, $z_i = 6$, $r_j = 4$, $z_j = 4$, $r_k = 6$, and $z_k = 4$ cm. Assuming the body is a body of revolution loaded axisymmetrically, determine the element stiffness matrix.

7-96 Consider the triangular element shown in Fig. P7-11. The element is extracted from a body of revolution and is loaded axisymmetrically. The material is aluminum (6061 alloy). The nodal coordinates are $r_i = 10$, $z_i = 10$, $r_j = 10$, $z_j = 9$, $r_k = 12$, and $z_k = 9$ cm. Determine the element stiffness matrix.

7-97 Consider the triangular element shown in Fig. P7-13. The element is extracted from a body of revolution and is loaded axisymmetrically. The material is cast iron. The nodal coordinates are $r_i = 20$, $z_i = 20$, $r_j = 21$, $z_j = 22$, $r_k = 19$, and $z_k = 22$ in. Determine the element stiffness matrix.

7-98 Show that the element stiffness matrix for axisymmetric stress analysis as given by Eq. (7-53) for the three-node triangular element is always symmetric.

7-99 Show that the element stiffness matrix for axisymmetric stress analysis for the four-node rectangular element is always symmetric.

7-100 What is the size of the element stiffness matrix for the case of axisymmetric stress analysis if the four-node rectangular element is used? Justify your answer.

7-101 Determine the element stiffness matrix for the element shown in Fig. P7-18. The element is extracted from a cast iron body of revolution that is loaded axisymmetrically. The coordinates of the nodes are $r_i = 34$, $z_i = 32$, $r_j = 34$, $z_j = 33$,

$r_k = 32$, $z_k = 33$, $r_m = 32$, and $z_m = 32$ cm. Perform the integrations by evaluating the integrands at the element centroid and treating the integrands as constants.

7-102 Determine the element stiffness matrix for the element shown in Fig. P7-20. The element is extracted from a hot rolled, low carbon steel body of revolution that is loaded axisymmetrically. The coordinates of the nodes are $r_i = 55$, $z_i = 52$, $r_j = 55$, $z_j = 54$, $r_k = 52$, $z_k = 54$, $r_m = 52$, and $z_m = 52$ in. Perform the integrations by evaluating the integrands at the element centroid and treating the integrands as constants.

7-103 What size is the element nodal force vector from a self-strain if the four-node rectangular element is used in axisymmetric stress analysis?

7-104 Determine a general expression for the element nodal force vector from a uniform self-strain for the rectangular element in axisymmetric stress formulations. Perform the integrations by evaluating the integrand at the element centroid and treating the integrands as though they were constants.

7-105 For the element in Problem 7-94, determine the element nodal force vector as a result of a self-strain if the temperature increases by 70°F.

7-106 For the element in Problem 7-95, determine the element nodal force vector as a result of a self-strain if the temperature increases by 25°C.

7-107 For the element in Problem 7-96, determine the element nodal force vector as a result of a self-strain if the temperature decreases by 60°C.

7-108 For the element in Problem 7-97, determine the element nodal force vector as a result of a self-strain if the temperature increases by 55°F.

7-109 For the element in Problem 7-101, determine the element nodal force vector as a result of a self-strain if the temperature increases by 35°C. Perform the integrations by evaluating the integrands at the element centroid and treating the integrands as though they were constants.

7-110 For the element in Problem 7-102, determine the element nodal force vector as a result of a self-strain if the temperature decreases by 72°F. Perform the integrations by evaluating the integrands at the element centroid and treating the integrands as though they were constants.

7-111 Determine a general expression for the element nodal force vector from a uniform prestress for the triangular element in axisymmetric stress analysis by performing the exact integrations. *Hint*: Use area coordinates to represent r in the following manner:

$$r = L_i r_i + L_j r_j + L_k r_k$$

and then apply Eq. (6-49).

7-112 Determine a general expression for the element nodal force vector from a uniform prestress for the rectangular element in axisymmetric stress analysis. Perform the integrations by evaluating the integrand at the element centroid and treating the integrands as though they were constants.

7-113   For the element in Problem 7-94, determine the element nodal force vector as a result of the following prestresses: $\sigma_{rr_0} = -800$, $\sigma_{\theta\theta_0} = 1750$, $\sigma_{zz_0} = 1000$, and $\sigma_{rz_0} = -500$ psi.

7-114   For the element in Problem 7-95, determine the element nodal force vector as a result of the following prestresses: $\sigma_{rr_0} = 2000$, $\sigma_{\theta\theta_0} = -3200$, $\sigma_{zz_0} = 1500$, and $\sigma_{rz_0} = -1800$ N/cm^2.

7-115   For the element in Problem 7-96, determine the element nodal force vector as a result of the following prestresses: $\sigma_{rr_0} = 2300$, $\sigma_{\theta\theta_0} = 0$, $\sigma_{zz_0} = 1550$, and $\sigma_{rz_0} = -2100$ N/cm^2.

7-116   For the element in Problem 7-97, determine the element nodal force vector as a result of the following prestresses: $\sigma_{rr_0} = 1200$, $\sigma_{\theta\theta_0} = -800$, $\sigma_{zz_0} = 900$, $\sigma_{rz_0} = 400$ psi.

7-117   For the element in Problem 7-101, determine the element nodal force vector as a result of the following prestresses: $\sigma_{rr_0} = -2400$, $\sigma_{\theta\theta_0} = -1500$, $\sigma_{zz_0} = 2000$, and $\sigma_{rz_0} = 500$ N/cm^2. Perform the integrations by evaluating the integrands at the element centroid and treating the integrands as though they were constants.

7-118   For the element in Problem 7-102, determine the element nodal force vector from the following prestresses: $\sigma_{rr_0} = 1400$, $\sigma_{\theta\theta_0} = -2800$, $\sigma_{zz_0} = 900$ and $\sigma_{rz_0} = 1925$ psi. Perform the integrations by evaluating the integrands at the element centroid and treating the integrands as though they were constants.

7-119   Determine a general expression for the element nodal force vector from a uniform body force for the triangular element in axisymmetric stress problems by performing the exact integrations. *Hint*: See Problem 7-111.

7-120   Determine a general expression for the element nodal force vector from a uniform body force for the rectangular element in axisymmetric stress problems. Perform the integrations by evaluating the integrand at the element centroid and treating the integrands as though they were constants.

7-121   For the element in Problem 7-94, determine the element nodal force vector as a result of a body force with the following components: $b_r = 500$ and $b_z = 125$ lbf/in^3.

7-122   For the element in Problem 7-95, determine the element nodal force vector as a result of a body force with the following components: $b_r = 1500$ and $b_z = 500$ N/cm^3.

7-123   Assume that the element in Problem 7-96 is oriented such that the z axis is in the direction opposite that of gravity. If the body is rotated at 12 rad/sec about the z axis, determine the element nodal force vector. The acceleration due to gravity is 9.81 m/s^2. State all assumptions made.

7-124   Assume that the element in Problem 7-97 is oriented such that the z axis is in the same direction as that of gravity. If the body is rotated at 15 rad/sec about the z axis, determine the element nodal force vector. The acceleration due to gravity is 32.2 ft/s^2. State all assumptions made.

7-125   For the element in Problem 7-101, determine the element nodal force vector from a body force with the following components: $b_r = 1500$ and $b_z = 450$ N/cm^3. Perform the integrations by evaluating the integrands at the element centroid and treating the integrands as though they were constants.

7-126   Assume that the element in Problem 7-102 is oriented such that the z axis is in the direction opposite that of gravity. If the body is rotated at 20 rad/sec about the z axis, determine the element nodal force vector. The acceleration due to gravity is 32.2 ft/sec^2. Perform the integrations by evaluating the integrands at the element centroid and treating the integrands as though they were constants.

7-127   Determine a general expression for the element nodal force vector from a uniform surface traction on face ij for the rectangular element in a state of axisymmetric stress. Perform the integrations by evaluating the integrand at the center of face ij and treating the integrands as though they were constants.

7-128   Determine a general expression for the element nodal force vector from a uniform surface traction on face km for the rectangular element in a state of axisymmetric stress. Perform the integrations by evaluating the integrand at the center of face km and treating the integrands as though they were constants.

7-129   For the element in Problem 7-94, determine the element nodal force vector as a result of a surface traction on leg ij with the following components: $s_r = -1000$ and $s_z = 500$ psi.

7-130   For the element in Problem 7-95, determine the element nodal force vector as a result of a surface traction on leg ki with the following components: $s_r = -1500$ and $s_z = -800$ N/cm^2.

7-131   For the element in Problem 7-96, determine the element nodal force vector as a result of a surface traction on leg jk with the following components: $s_r = -1200$ and $s_z = 925$ N/cm^2.

7-132   For the element in Problem 7-97, determine the element nodal force vector as a result of a surface traction on leg ki with the following components: $s_r = -4200$ and $s_z = 0$ psi.

7-133   For the element in Problem 7-101, determine the element nodal force vector as a result of a surface traction on face ij with the following components: $s_r = -2350$ and $s_z = 500$ N/cm^2. Perform the integrations by evaluating the integrands at the center of face ij and treating the integrands as though they were constants.

7-134   For the element in Problem 7-102, determine the element nodal force vector as a result of a surface traction on face mi with the following components: $s_r = -1600$ and $s_z = 800$ psi. Perform the integrations by evaluating the integrands at the center of face mi and treating the integrands as though they were constants.

7-135   It has been assumed in the derivation leading to Eq. (7-62) in the text that the surface traction is uniform over the leg of the triangle in question. Derive an alternate expression for the nodal force vector for a *linearly varying* surface traction on leg ij of the three-node triangular element, shown in Fig. P7-135. Proceed by assuming the following forms for s_r and s_z:

$$s_r = N_i s_{ri} + N_j s_{rj}$$

$$s_z = N_i s_{zi} + N_j s_{zj}$$

where s_{ri} is the r component of the surface traction at node i, etc. In addition, represent the variable r in the integrands in terms of the nodal values of r (i.e., r_i and r_j) and the length coordinates L_i and L_j. Finally, write the shape functions in terms of the length coordinates (defined on leg ij) and use Eq. (6-48) to perform the integrations.

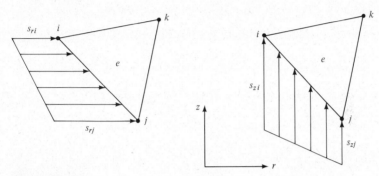

Figure P7-135

7-136 Repeat Problem 7-135 for a uniformly varying surface traction on leg jk. At nodes j and k, the r and z components of the surface traction are s_{rj} and s_{rk}, and s_{zj} and s_{zk}, respectively.

7-137 Repeat Problem 7-136 for a uniformly varying surface traction on leg ki. At nodes k and i, the r and z components of the surface traction are s_{rk} and s_{ri}, and s_{zk} and s_{zi}, respectively.

7-138 For the element in Problem 7-94, determine the element nodal force vector as a result of a point load that has the following components: $f_{pr} = 1200$ and $f_{pz} = -750$ lbf/in. The point load is located at $r_0 = 1.2$ and $z_0 = 2.2$ in.

7-139 For the element in Problem 7-95, determine the element nodal force vector as a result of a point load with the following components: $f_{pr} = -1000$ and $f_{pz} = 500$ N/cm. The point load is located at $r_0 = 5.5$ and $z_0 = 4.3$ cm.

7-140 For the element in Problem 7-96, determine the element nodal force vector as a result of a point load with the following components: $f_{pr} = 2400$ and $f_{pz} = 1500$ N/cm. The point load is located at $r_0 = 11$ and $z_0 = 9$ cm.

7-141 For the element in Problem 7-97, determine the element nodal force vector as a result of a point load with the following components: $f_{pr} = -2250$ and $f_{pz} = -2800$ lbf/in. The point load is located at $r_0 = 20$ and $z_0 = 21.5$ in.

7-142 For the element in Problem 7-101, determine the element nodal force vector as a result of a point load with the following components: $f_{pr} = -3150$ and $f_{pz} = 1500$ N/cm. The point load is located at $r_0 = 33.0$ and $z_0 = 32.5$ cm.

7-143 For the element in Problem 7-101, determine the element nodal force vector as a result of a point load with the following components: $f_{pr} = -3150$ and $f_{pz} = 1500$ N/cm. The point load is located at $r_0 = 34$ and $z_0 = 32$ cm.

7-144 For the element in Problem 7-102, determine the element nodal force vector as a result of a point load with the following components: $f_{pr} = -2575$ and $f_{pz} = 1260$ lbf/in. The point load is located at $r_0 = 54$ and $z_0 = 53$ in.

7-145 For the element in Problem 7-102, determine the element nodal force vector as a result of a point load with the following components: $f_{pr} = -3500$ and $f_{pz} = 800$ lbf/in. The point load is located at $r_0 = 52$ and $z_0 = 54$ in.

7-146 Using the symbolic notation from Example 7-5, give the expression for the assemblage stiffness matrix for the discretized axisymmetric region in Fig. P7-69.

7-147 Using the symbolic notation from Example 7-6, give the expression for the assemblage nodal force vector for the discretized axisymmetric region in Fig. P7-69.

7-148 Using the symbolic notation from Example 7-5, give the expression for the assemblage stiffness matrix for the discretized axisymmetric region in Fig. P7-71.

7-149 Using the symbolic notation from Example 7-6, give the expression for the assemblage nodal force vector for the discretized axisymmetric region in Fig. P7-71.

7-150 Using the symbolic notation from Example 7-5, give the expression for the assemblage stiffness matrix for the discretized axisymmetric region in Fig. P7-73.

7-151 Using the symbolic notation from Example 7-6, give the expression for the assemblage nodal force vector for the discretized axisymmetric region in Fig. P7-73.

7-152 By extending the symbolic notation from Example 7-5, give the expression for the assemblage stiffness matrix for the discretized axisymmetric region in Fig. P7-75.

7-153 By extending the symbolic notation from Example 7-6, give the expression for the assemblage nodal force vector for the discretized axisymmetric region in Fig. P7-75.

7-154 By extending the symbolic notation from Example 7-5, give the expression for the assemblage stiffness matrix for the discretized axisymmetric region in Fig. P7-77.

7-155 By extending the symbolic notation from Example 7-6, give the expression for the assemblage nodal force vector for the discretized axisymmetric region in Fig. P7-77.

7-156 By extending the symbolic notation from Example 7-5, give the expression for the assemblage stiffness matrix for the discretized axisymmetric region in Fig. P7-79. Note the use of two different types of elements.

7-157 By extending the symbolic notation from Example 7-6, give the expression for the assemblage nodal force vector for the discretized axisymmetric region in Fig. P7-79. Note the use of two different types of elements.

7-158 For the element in Problem 7-94 for a particular loading condition, the following values of nodal displacements are obtained: $u_i = 0.00235$, $v_i = 0.0$, $u_j = -0.00555$, $v_j = 0.00260$, $u_k = 0.00150$, and $v_k = -0.00645$ in. Determine the element strains and stresses that correspond to these displacements. Assume that both the element self-strains and prestresses are zero.

7-159   For the element in Problem 7-95 for a particular loading condition, the following values of nodal displacements are obtained: $u_i = 0.01215$, $v_i = -0.01300$, $u_j = 0.0$, $v_j = 0.00560$, $u_k = -0.00555$, and $v_k = -0.00750$ cm. Determine the element strains and stresses that correspond to these displacements. Assume that both the self-strains and prestresses are zero.

7-160   For the element in Problem 7-96 for a particular loading condition, the following values of nodal displacements are obtained: $u_i = -0.01250$, $v_i = -0.02350$, $u_j = 0.03500$, $v_j = 0.01550$, $u_k = -0.01775$, and $v_k = 0.01010$ cm. Determine the element strains and stresses that correspond to these displacements. Assume that both the self-strains and prestresses are zero.

7-161   For the element in Problem 7-97 for a particular loading condition, the following values of nodal displacements are obtained: $u_i = -0.02150$, $v_i = -0.02350$, $u_j = 0.01875$, $v_j = 0.02550$, $u_k = 0.02500$, and $v_k = 0.0$ in. Determine the element strains and stresses that correspond to these displacements. Assume that both the self-strains and prestresses are zero.

7-162   For the element in Problem 7-101 for a particular loading condition, the following values of nodal displacements are obtained: $u_i = -0.02750$, $v_i = 0.01250$, $u_j = 0.04555$, $v_j = 0.03350$, $u_k = 0.02555$, $v_k = -0.02350$, $u_m = -0.01235$, and $v_m = 0.02500$ cm. Determine the element strains and stresses at the element centroid that correspond to these displacements. Assume that both the self-strains and prestresses are zero.

7-163   For the element in Problem 7-102 for a particular loading condition, the following values of nodal displacements are obtained: $u_i = -0.02250$, $v_i = -0.02050$, $u_j = 0.01500$, $v_j = -0.03150$, $u_k = 0.02530$, $v_k = -0.03350$, $u_m = 0.0$, and $v_m = 0.03500$ in. Determine the element strains and stresses at the element centroid that correspond to these displacements. Assume that both the self-strains and prestresses are zero.

7-164   The body from which a tetrahedral element is extracted is made of bronze. The nodal coordinates of the element are $x_i = 2.0$, $y_i = 1.5$, $z_i = 0.0$, $x_j = 1.7$, $y_j = 3.0$, $z_j = -0.2$, $x_k = 1.5$, $y_k = 2.0$, $z_k = 1.7$, $x_m = 0.6$, $y_m = 1.8$, and $z_m = 0.1$ in. Determine the element stiffness matrix.

7-165   The body from which a tetrahedral element is extracted is made of hard drawn copper. The nodal coordinates of the element are $x_i = 5$, $y_i = 6$, $z_i = 0$, $x_j = 4$, $y_j = 4$, $z_j = 0$, $x_k = 5$, $y_k = 5$, $z_k = 4$, $x_m = 6$, $y_m = 4$, and $z_m = 0$ cm. Determine the element stiffness matrix.

7-166   Show that the element stiffness matrix for three-dimensional stress analysis as given by Eq. (7-73) for the four-node tetrahedral element is always symmetric.

7-167   Show that the element stiffness matrix for three-dimensional stress analysis for the eight-node brick element is always symmetric.

7-168   What is the size of the element stiffness matrix for the case of three-dimensional stress analysis if the eight-node brick element is used?

7-169   Determine the element stiffness matrix for the eight-node brick element defined below. The element is extracted from a hot rolled, high carbon steel structure. The coordinates of the nodes are $x_1 = 34$, $y_1 = 32$, $z_1 = 20$, $x_2 = 34$, $y_2 = 33$, $z_2 = 20$, $x_3 = 32$, $y_3 = 33$, $z_3 = 20$, $x_4 = 32$, $y_4 = 32$, $z_4 = 20$, $x_5 = 34$, $y_5 = 32$, $z_5 = 18$, $x_6 = 34$, $y_6 = 33$, $z_6 = 18$, $x_7 = 32$, $y_7 = 33$, $z_7 = 18$, $x_8 = 32$, $y_8 = 32$, and $z_8 = 18$ in. Perform the integrations by evaluating the integrands at the element centroid and treating the integrands as constants.

7-170   Determine the element stiffness matrix for the eight-node brick element defined below. The element is extracted from an aluminum (6061 alloy) structure. The coordinates of the nodes are $x_1 = 45$, $y_1 = 32$, $z_1 = 12$, $x_2 = 45$, $y_2 = 34$, $z_2 = 12$, $x_3 = 42$, $y_3 = 34$, $z_3 = 12$, $x_4 = 42$, $y_4 = 32$, $z_4 = 12$, $x_5 = 45$, $y_5 = 32$, $z_5 = 9$, $x_6 = 45$, $y_6 = 34$, $z_6 = 9$, $x_7 = 42$, $y_7 = 34$, $z_7 = 9$, $x_8 = 42$, $y_8 = 32$, and $z_8 = 9$ cm. Perform the integrations by evaluating the integrands at the element centroid and treating the integrands as constants.

7-171   What size is the element nodal force vector from a self-strain if the eight-node brick element is used in three-dimensional stress analysis?

7-172   Determine a general expression for the element nodal force vector from a uniform self-strain for the brick element in three-dimensional stress formulations. Perform the integrations by evaluating the integrands at the element centroid and treating the integrands as though they were constants.

7-173   For the element in Problem 7-164, determine the element nodal force vector as a result of a self-strain if the temperature increases by 50°F.

7-174   For the element in Problem 7-165, determine the element nodal force vector as a result of a self-strain if the temperature decreases by 72°C.

7-175   For the element in Problem 7-169, determine the element nodal force vector as a result of a self-strain if the temperature decreases by 48°F. Perform the integrations by evaluating the integrands at the element centroid and treating the integrands as though they were constants.

7-176   For the element in Problem 7-170, determine the element nodal force vector as a result of a self-strain if the temperature increases by 37°C. Perform the integrations by evaluating the integrands at the element centroid and treating the integrands as though they were constants.

7-177   Determine a general expression for the element nodal force vector from a uniform prestress for the brick element in three-dimensional stress analysis. Perform the integrations by evaluating the integrand at the element centroid and treating the integrands as though they were constants.

7-178   Determine a general expression for the element nodal force vector from a uniform body force for the brick element in three-dimensional stress problems. Perform the integrations by evaluating the integrand at the element centroid and treating the integrands as though they were constants.

7-179   For the element in Problem 7-164, determine the element nodal force vector as a result of a body force with the following components: $b_x = 100$, $b_y = 125$, and $b_z = 75$ lbf/in³.

7-180   For the element in Problem 7-165, determine the element nodal force vector as a result of a body force with the following components: $b_x = 70$, $b_y = 100$, and $b_z = 40$ N/cm^3.

7-181   For the element in Problem 7-169, determine the element nodal force vector from a body force with the following components: $b_x = 30$, $b_y = 50$, and $b_z = 20$ lbf/in.^{3}. Perform the integrations by evaluating the integrands at the element centroid and treating the integrands as though they were constants.

7-182   Assume that the element in Problem 7-170 is oriented such that the z axis is in the direction opposite that of gravity. If no other body forces exist, determine the corresponding element nodal force vector. The acceleration due to gravity is 9.81 m/s^2. Perform the integrations by evaluating the integrands at the element centroid and treating the integrands as though they were constants.

7-183   Derive a procedure that could be used to evaluate the area of a typical face of the tetrahedral element. This area is need for use in Eq. (7-78).

7-184   Determine a general expression for the element nodal force vector from a uniform surface tractions on face 1-2-3-4 for the brick element in a state of three-dimensional stress. Perform the integrations by evaluating the integrand at the center of the face and treating the integrands as though they were constants.

7-185   Determine a general expression for the element nodal force vector from a uniform surface traction on face 5-6-2-1 for the brick element in a state of three-dimensional stress. Perform the integrations by evaluating the integrand at the center of the face and treating the integrands as though they were constants.

7-186   For the element in Problem 7-164, determine the element nodal force vector as a result of a surface traction on face ijk with the following components: $s_x = -1000$, $s_y = 500$, and $s_z = 750$ psi.

7-187   For the element in Problem 7-165, determine the element nodal force vector as a result of a surface traction on face ikm with the following components: $s_x = -1500$, $s_y = -800$, and $s_z = 300$ N/cm^2.

7-188   For the element in Problem 7-169, determine the element nodal force vector as a result of a surface traction on face 5-6-7-8 with the following components: $s_x = 1600$, $s_y = -925$, and $s_z = -400$ psi. Perform the integrations by evaluating the integrands at the centroid of the face and treating the integrands as though they were constants.

7-189   For the element in Problem 7-170, determine the element nodal force vector as a result of a surface traction on face 8-7-3-4 with the following components: $s_x = -3200$, $s_y = 0$, and $s_z = 2500$ N/cm^2. Perform the integrations by evaluating the integrands at the centroid of the face and treating the integrands as though they were constants.

7-190   It has been assumed in the derivation leading to Eq. (7-78) in the text that the surface traction is uniform over the face of the tetradedral element in question. Derive an alternate expression for the nodal force vector for a *linearly varying* surface traction on face ijk of a four-node tetrahedral element. Proceed by assuming the following forms for s_x, s_y, and s_z:

$$s_x = N_i s_{xi} + N_j s_{xj} + N_k s_{xk}$$

$$s_y = N_i s_{yi} + N_j s_{yj} + N_k s_{yk}$$

$$s_z = N_i s_{zi} + N_j s_{zj} + N_k s_{zk}$$

where s_{xi} is the x component of the surface traction at node i, etc. Write the shape functions in terms of the area coordinates (defined on face ijk) and use Eq. (6-49) to perform the integrations.

7-191   Repeat Problem 7-190 for a linearly varying surface traction on leg jkm.

7-192   For the element in Problem 7-165, determine the element nodal force vector as a result of a point load that has the following components: $f_{px} = 600$, $f_{py} = -750$, and $f_{pz} = 500$ N. The point load is located at $x_0 = 5$, $y_0 = 5$, and $z_0 = 4$ cm.

7-193   For the element in Problem 7-169, determine the element nodal force vector as a result of a point load with the following components: $f_{px} = -1500$, $f_{py} = 500$, and $f_{pz} = 880$ lbf. The point load is located at $x_0 = 33.5$, $y_0 = 32.5$, and $z_0 = 20.0$ in.

7-194   For the element in Problem 7-164 for a particular loading condition, the following values of nodal displacements are obtained: $u_i = -0.01250$, $v_i = -0.02350$, $w_i = 0.02220$, $u_j = 0.03500$, $v_j = 0.01550$, $w_j = -0.01245$, $u_k = -0.01775$, $v_k = 0.01010$, $w_k = 0.0$, $u_m = -0.01135$, $v_m = 0.00990$, and $w_m = 0.0$ in. Determine the element strains and stresses that correspond to these displacements. Assume that both the self-strains and prestresses are zero.

7-195   For the element in Problem 7-165 for a particular loading condition, the following values of nodal displacements are obtained: $u_i = 0.02250$, $v_i = 0.02050$, $w_i = -0.05520$, $u_j = -0.03575$, $v_j = 0.01750$, $w_j = 0.03245$, $u_k = 0.01775$, $v_k = -0.01750$, $w_k = 0.07850$, $u_m = 0.02435$, $v_m = -0.01990$, and $w_m = 0.05450$ cm. Determine the element strains and stresses that correspond to these displacements. Assume that both the self-strains and prestresses are zero.

7-196   With the help of Fig. 7-17, verify Eq. (7-82) by doing a force balance on the infinitesimal beam element in the y direction.

7-197   With the help of Fig. 7-17, verify Eq. (7-83) by doing a moment balance about some convenient point on the infinitesimal beam element.

7-198   Derive the variational principle that corresponds to Eq. (7-88). What is the highest-order derivative present in the functional? Please explain. *Hint:* Refer to Chapter 4 (and use integration by parts twice).

7-199   Verify that the shape functions given by Eq. (7-100) for the beam element satisfy the following conditions:
a. At $x = x_i$ (where $\xi = 0$):

$$N_{wi} = 1$$

$$N_{\theta i} = N_{wj} = N_{\theta j} = 0$$

b. At $x = x_j$ (where $\xi = 1$):

$$N_{wj} = 1$$

$$N_{wi} = N_{\theta i} = N_{\theta j} = 0$$

7-200 Verify that the shape functions given by Eq. (7-100) for the beam element satisfy the conditions given by Eqs. (7-103) and (7-104).

7-201 Derive the finite element characteristics for the beam model [i.e., Eqs. (7-116) to (7-118)] from the variational principle from Problem 7-198. Are the resulting integral expressions the same as those derived with the Galerkin method? Please explain. It is not necessary to evaluate the integrals.

7-202 Consider the beam element shown in Fig. P7-202. The beam from which the element is extracted is made of hot rolled, low carbon steel. The beam is rectangular in cross-section with a width w of 4 cm and a height h of 8 cm. The coordinates of nodes i and j are $x_i = 5$ and $x_j = 6$ cm. Determine the element stiffness matrix.

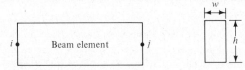

Figure P7-202

7-203 Consider the beam element shown in Fig. P7-202. The beam from which the element is extracted is made of aluminum (6061 alloy). The beam is rectangular in cross-section with a width w of 3 in. and a height h of 4 in. The coordinates of nodes i and j are $x_i = 32$ and $x_j = 36$ in. Determine the element stiffness matrix.

7-204 Consider the beam element shown in Fig. P7-204. The I-beam from which the element is extracted is made of hot rolled, high carbon steel. The cross-section of the beam is also shown in Fig. P7-204 with the dimensions $w = 4$ in., $h = 8$ in., and $t = 1$ in. The coordinates of nodes i and j are $x_i = 42$ and $x_j = 48$ in. Determine the element stiffness matrix.

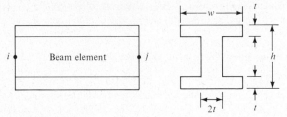

Figure P7-204

7-205 Consider the beam element shown in Fig. P7-204. The I-beam from which the element is extracted is made of cast iron. The cross-section of the beam is shown in Fig. P7-204 with the dimensions $w = 6$ cm, $h = 10$ cm, and $t = 2$ cm. The coordinates of nodes i and j are $x_i = 20$ and $x_j = 24$ cm. Determine the element stiffness matrix.

7-206 A solid circular bar made from brass is to be used as a beam in a certain application. The diameter of the bar is 3 cm. The coordinates of nodes i and j are $x_i = 32$ and $x_j = 36$ in. Determine the element stiffness matrix.

7-207 A solid circular bar made from hot rolled, low carbon steel is to be used as a beam in a certain application. The diameter of the bar is 1 in. The coordinates of nodes i and j are $x_i = 16$ and $x_j = 18$ in. Determine the element stiffness matrix.

7-208 Consider the beam element from Problem 7-202. Determine the nodal force vector for the element if a downward force of 1000 N and a counterclockwise moment of 800 N·cm act at node i.

7-209 Consider the beam element from Problem 7-203. Determine the nodal force vector for the element if a downward force of 1200 lbf and a counterclockwise moment of 800 lbf·in act at node i.

7-210 Consider the beam element from Problem 7-204. Determine the nodal force vector for the element if an upward force of 1500 lbf and a clockwise moment of 1800 lbf·in act at node i.

7-211 Consider the beam element from Problem 7-205. Determine the nodal force vector for the element if an upward force of 500 N and a clockwise moment of 750 N·cm act at node j.

7-212 Consider the beam element from Problem 7-206. If an upward force of 350 N and a counterclockwise moment of 400 N·cm act at node i, and if a downward force of 425 N acts at node j, determine the nodal force vector for the element.

7-213 Consider the beam element from Problem 7-207. If an upward force of 425 lbf and a clockwise moment of 675 lbf·in act at node i, and if a counterclockwise moment of 425 lbf·in acts at node j, determine the nodal force vector for the element.

7-214 For the beam element from Problem 7-202, determine the nodal force vector for the element if a uniformly distributed load of 1500 N/cm acts in the downward direction.

7-215 For the beam element from Problem 7-203, determine the nodal force vector for the element if a uniformly distributed load of 1200 lbf/in. acts in the downward direction.

7-216 For the beam element from Problem 7-204, determine the nodal force vector for the element if a uniformly distributed load of 1650 lbf/in. acts in the upward direction.

7-217 For the beam element from Problem 7-205, determine the nodal force vector for the element if a uniformly distributed load of 850 N/cm acts in the upward direction.

7-218 For the beam element from Problem 7-206, determine the nodal force vector for the element if a uniformly distributed load of 785 N/cm acts in the downward direction.

7-219 For the beam element from Problem 7-207, determine the nodal force vector for the element if a uniformly distributed load of 685 lbf/in. acts in the upward direction.

7-220 Using as much of Example 7-7 as is possible, resolve for the nodal deflections and slopes if the spring is removed. Assume only two elements and that the right end of the beam is simply supported.

7-221 Assuming only two elements and using as much of Example 7-7 as is possible, resolve for the nodal deflections and slopes if the left end of the beam is simply supported.

7-222 Using as much of Example 7-7 as is possible, resolve for the nodal deflections and slopes if the both ends of the beam are simply supported (the spring is removed). Assume only two elements.

7-223   Assuming only two elements and using as much of Example 7-7 as is possible, resolve for the nodal deflections and slopes if both ends of the beam are cantilevered (the spring is removed).

7-224   Reconsider the beam in Example 7-7. The distributed load p is now 3000 lbf/ft and the point load P is 1000 lbf. The spring is removed (so k_s should be taken as zero). The beam is rectangular in cross-section with a width b of 1.5 in. and a height h of 3 in. The beam is made of hot rolled, high carbon steel and is 3 ft long. Using only two equal-length elements, determine the deflections and slopes at the ends and midspan of the beam. In addition, determine the moment, shear force, maximum bending stress, and maximum shear stress at the middle of the distributed load.

7-225   Modify the TRUSS program in Appendix B so that it may be used to solve for the nodal deflections and slopes of a beam. Allow for up to 30 elements with up to 5 different materials. The input to the program should be modeled after the input to the TRUSS program. For each material, read in the following parameters: the elastic modulus E, the moment of inertia I, the distributed loading p, and the distance c_{max} from the neutral plane to the outermost fibers (needed in the stress calculations). Treat the prescribed deflections and slopes with positive boundary condition flags, and the imposed forces and moments with negative flags (see the TRUSS program). At the center of each element, the following results should be given in the output: the bending moment, the shear force, the maximum bending stress, and the maximum shear stress. These results should be calculated in a postprocessor subroutine.

7-226   Using the computer program from Problem 7-225 or one furnished by the instructor, solve for the nodal deflections and slopes for the beam from Problem 7-220. In addition, determine the element resultants at the midpoint of each element. Use 2, 4, and 8 elements.

7-227   Using the computer program from Problem 7-225 or one furnished by the instructor, solve for the nodal deflections and slopes for the beam from Problem 7-222. In addition, determine the element resultants at the midpoint of each element. Use 2, 4, and 8 elements.

7-228   Using the computer program from Problem 7-225 or one furnished by the instructor, solve for the nodal deflections and slopes for the beam from Problem 7-223. In addition, determine the element resultants at the midpoint of each element. Use 2, 4, and 8 elements.

7-229   Using the computer program from Problem 7-225 or one furnished by the instructor, solve for the nodal deflections and slopes for the beam in Problem 7-224. In addition, determine the element resultants at the midpoint of each element. Use 2, 4, and 8 elements.

7-230   Explain in your own words what is meant by substructuring and why is it so vital in large finite element models. Do not hesitate to use some equations in your explanation.

7-231   Using the stress analysis program furnished by the instructor, solve for the two-dimensional stress distribution in the thin plate shown in Fig. P7-231. Note that the plate has a circular hole in it, and therefore a stress concentration exists in the vicinity of the hole. Verify that the stress concentration factor is 3 for this condition. Assume the following: $D = 0.5$ in., $w = 6$ in., $h = 8$ in., and $s = 1000$ psi.

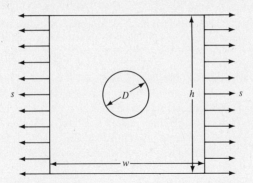

Figure P7-231

The plate is made of hot rolled, high carbon steel and has a thickness of 0.375 in. Note that only one-fourth of the plate needs to be. modeled because of the two-axis symmetry. Use at least 100 elements.

7-232 Using the stress analysis program furnished by the instructor, solve for the two-dimensional stress distribution in the thin plate shown in Fig. P7-231. Note that the plate has a circular hole in it, and therefore a stress concentration exists in the vicinity of the hole. Verify that the stress concentration factor is 3 for this condition. Assume the following: $D = 2$ cm, $w = 12$ cm, $h = 16$ cm, and $s = 1500$ N/cm^2. The plate is made of hot rolled, high carbon steel and has a thickness of 0.75 cm. Note that only one-fourth of the plate needs to be modeled because of the two-axis symmetry. Use at least 100 elements.

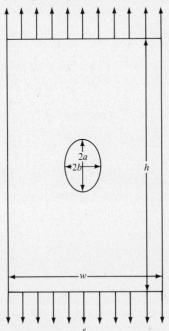

Figure P7-233

7-233   Using the stress analysis program furnished by the instructor, solve for the two-dimensional stress distribution in the thin plate shown in Fig. P7-233. Note that the plate has an elliptical hole in it, and therefore a stress concentration exists in the vicinity of the hole. Determine the value of the stress concentration factor for this condition. Assume the following: $a = 2$ cm, $b = 1$ cm, $w = 10$ cm, $h = 20$ cm, and $s = 1800$ N/cm^2. The plate is made of brass and has a thickness of 0.5 cm. Note that only one-fourth of the plate needs to be modeled because of the two-axis symmetry. Use at least 100 elements.

7-234   Using the stress analysis program furnished by the instructor, solve for the two-dimensional stress distribution in the thin plate shown in Fig. P7-233. Note that the plate has an elliptical hole in it, and therefore a stress concentration exists in the vicinity of the hole. Determine the value of the stress concentration factor for this condition. Assume the following: $a = 1$ in., $b = 0.5$ in., $w = 5$ in., $h = 10$ in., and $s = 1400$ psi. The plate is made of brass and has a thickness of 0.25 in. Note that only one-fourth of the plate needs to be modeled because of the two-axis symmetry. Use at least 100 elements.

7-235   Using the stress analysis program furnished by the instructor, solve for the two-dimensional stress distribution in the thin plate shown in Fig. P7-235. Note that the plate has a square hole in it, and therefore a stress concentration exists in the vicinity of the hole. Determine the value of the stress concentration factor for this condition. Assume the following: $d = 0.5$ in., $w = 10$ in., $h = 10$ in., and $s = 2000$ psi. The plate is made of cast iron and has a thickness of 0.5 in. Note that only one-fourth of the plate needs to be modeled because of the two-axis symmetry. Use at least 100 elements.

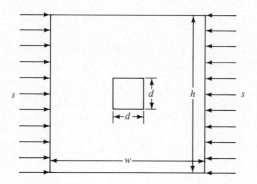

Figure P7-235

7-236   Using the stress analysis program furnished by the instructor, solve for the two-dimensional stress distribution in the thin plate shown in Fig. P7-235. Note that the plate has a square hole in it, and therefore a stress concentration exists in the vicinity of the hole. Determine the value of the stress concentration factor for this condition. Assume the following: $d = 1$ cm, $w = 20$ cm, $h = 20$ cm, and $s = 2500$ N/cm^2. The plate is made of cast iron and has a thickness of 0.5 cm. Note that only one-fourth of the plate needs to be modeled because of the two-axis symmetry. Use at least 100 elements.

8

Steady-State Thermal and Fluid Flow Analysis

8-1 INTRODUCTION

In this chapter several nonstructural applications are considered. Heat conduction is given fairly extensive treatment. One-, two-, and three-dimensional problems are formulated; these include convection, thermal radiation to a large enclosure, prescribed heat fluxes, insulation, and prescribed temperatures. Axisymmetric problems are also formulated. In all cases, the thermal conductivity may be dependent on temperature. Only isotropic materials, however, are considered in the formal development. Several problems involving fluid flow are also formulated. The first is a problem involving convective energy transport. The second is that of two-dimensional potential flow. This is followed by a general formulation to steady, two-dimensional, incompressible viscous fluid flow. The chapter is concluded with a description of a steady, two-dimensional heat transfer program.

Before these formulations are developed, however, it is necessary to introduce the reader to one of the simpler nonlinear solution methods. This is necessitated by the fact that if any of the thermal properties is temperature-dependent, the resulting system of equations for the nodal temperatures is nonlinear. It will also be seen that the system of equations in some fluid flow problems is nonlinear as well.

8-2 THE DIRECT ITERATION METHOD

From both implementation and mathematical points of view, the direct iteration method is the simplest of the nonlinear solution methods. However, this method

371

has two main drawbacks. The first is that it does not always converge, and the second is that the system matrix equation must be in the special form

$$[K(a)]a = f \tag{8-1}$$

as opposed to the more general form

$$g(a) = f \tag{8-2}$$

Nonetheless, this method is worth studying because of the simplicity it affords.

To solve Eq. (8-1) for the nodal unknowns a, we start the solution process by *guessing* appropriate values for a. These values are then used to compute K. The system $Ka = f$ is then solved for the vector a. If the *new* a, denoted by a_{i+1}, agrees with the *old* a, denoted by a_i, to within an acceptable tolerance, then the solution process is stopped; otherwise, a new K matrix is formed using the newly calculated values in the vector a and the process is repeated. This method is summarized in Fig. 8-1. Let us now illustrate this method with two examples.

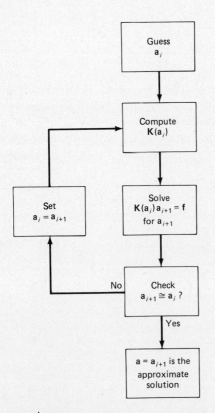

Figure 8-1 The direct iteration method.

Example 8-1

Use the direct iteration method to solve the following nonlinear equation:

$$x^2 + 6x - 5 = 0$$

Solution

The equation must be cast into the form given by Eq. (8-1), or

$$(x + 6)x = 5$$

In terms of the indices i and $i + 1$, we have

$$(x_i + 6)x_{i+1} = 5$$

For an initial guess of $x_0 = 10$, the iterations are as follows:

i	x_i	x_{i+1}
0	10.	0.3125
1	0.3125	0.7921
2	0.7921	0.7362
3	0.7362	0.7423
4	0.7423	0.7416
5	0.7416	0.7417
6	0.7417	0.7417

It is seen that convergence to four-place accuracy is attained in six iterations. The reader should take note that the second root is -6.7416 (obtained from the quadratic formula). In nonlinear problems, the root that is taken to be *the solution* is generally based on physical reasoning (e.g., perhaps x cannot be negative). ■

Example 8-2

Use the direct iteration method to obtain a set of roots to the following system of nonlinear equations:

$$x^2 + 6x + xy^2 = 23$$
$$6x + xy = 10$$

Solution

These equations must be cast into the form of Eq. (8-1), or

$$\begin{bmatrix} x + 6 & xy \\ 6 & x \end{bmatrix} \begin{bmatrix} x \\ y \end{bmatrix} = \begin{bmatrix} 23 \\ 10 \end{bmatrix}$$

Using the indices i and $i + 1$, we have

$$\begin{bmatrix} x_i + 6 & x_i y_i \\ 6 & x_i \end{bmatrix} \begin{bmatrix} x_{i+1} \\ y_{i+1} \end{bmatrix} = \begin{bmatrix} 23 \\ 10 \end{bmatrix}$$

which may be solved by the matrix-inversion method once an initial guess for x and y is made. Alternatively, such a system could be solved with the active zone equation solver from Sec. 6-8. For initial guesses of $x_0 = 2$ and $y_0 = 2$, the iterations are summarized as follows:

i	x_i	y_i	x_{i+1}	y_{i+1}
0	2.000	2.000	−.7500	7.250
1	−.7500	7.250	1.294	−2.980
2	1.294	−2.980	2.097	−1.997
3	2.097	−1.997	2.140	−1.354
4	2.140	−1.354	2.247	−1.626
5	2.247	−1.626	2.181	−1.373
6	2.181	−1.373	2.237	−1.569
7	2.237	−1.569	2.192	−1.409
8	2.192	−1.409	2.228	−1.537
9	2.228	−1.537	2.199	−1.433
10	2.199	−1.433	2.222	−1.517
11	2.222	−1.517	2.203	−1.449
12	2.203	−1.449	2.219	−1.503
13	2.219	−1.503	2.206	−1.459
14	2.206	−1.459	2.216	−1.495
15	2.216	−1.495	2.208	−1.466
16	2.208	−1.466	2.215	−1.489
17	2.215	−1.489	2.209	−1.470
18	2.209	−1.470	2.214	−1.486

The iterations were stopped when two successive iterations were within 1% of each other. ∎

It is emphasized that nonlinear equations in general have more than one solution. For example, try to verify that $x = 1$ and $y = 4$ is also a solution to the problem posed in Example 8-2. Caution must be exercised when attempting to obtain the solutions to nonlinear problems. For all problems formulated in this chapter, the direct iteration method tends to give only the physically realizable solution, providing the initial guesses are reasonable.

8-3 ONE-DIMENSIONAL HEAT CONDUCTION

Recall that a one-dimensional finite element formulation to a specific heat conduction problem was presented in Sec. 4-10. In the problem presented there, convection from the lateral surface of the body was included, but not thermal radiation. In this section, the one-dimensional heat conduction problem is extended to allow for radiation from the lateral surface of the body. Convection, radiation, and heat fluxes are also included on the ends of the body, i.e., on the boundaries. One application of such a problem is an extended surface (or fin) convecting to a fluid and/or radiating to some other large body or to space. In the interest of completeness, heat generation is also included. A schematic of the one-dimensional heat conduction model is shown in Fig. 8-2(a).

The Governing Equation

The governing equation for the temperature T is given by

$$\frac{d}{dx}\left(kA\frac{dT}{dx}\right) - hP(T - T_a) - \varepsilon\sigma P(T^4 - T_r^4) + QA = 0 \qquad \textbf{(8-3)}$$

where k is the thermal conductivity of the materials, h is the convective heat transfer coefficient, ε is the emissivity of the surface of the body, P is the perimeter, A is the cross-sectional area, T_a is the ambient fluid temperature far removed from the

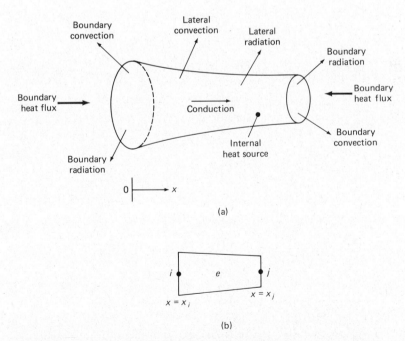

Figure 8-2   (a) Schematic of one-dimensional heat transfer problem and (b) typical element. Note that the boundary heat fluxes may act at a distance (such as the sun).

body, T_r is the receiver temperature for the radiation, Q is the heat generation rate per unit volume, and σ is the Stefan-Boltzmann constant, given by

$$\sigma = 5.670 \times 10^{-8} \text{ W/m}^2\text{-K}^4$$

or

$$\sigma = 0.1712 \times 10^{-8} \text{ Btu/hr-ft}^2\text{-}°\text{R}^4$$

Except for σ, each of these parameters may be a function of x. In addition, the thermal properties (k, h, and ε) may be temperature-dependent. Note that an absolute temperature scale must be used if radiation is present.

The FEM Formulation

Recall from Chapter 4 that two different approaches may be taken to obtain the finite element formulations once the governing differential equation is known: the variational and weighted-residual approaches. The former is not as convenient as the latter since it requires the intermediate step of determining the corresponding functional. The Galerkin weighted residual method is particularly well-suited to nonstructural problems and is used here.

Recall that if an approximation to the temperature T *on an element basis* is substituted into Eq. (8-3), the equation will not be satisfied exactly. In general, a residual R^e results. The weighted residual method is stated mathematically by

$$\sum_{e=1}^{M} \int_{V^e} \mathbf{W}^T R^e \, dV = 0 \tag{8-4}$$

where M elements are assumed and the matrix $\mathbf{W}^T$ is the transpose of the so-called weighting function matrix. For the specific case of the Galerkin method, Eq. (8-4) becomes

$$\sum_{e=1}^{M} \int_{V^e} \mathbf{N}^T R^e \, dV = 0 \tag{8-5}$$

where $\mathbf{N}$ represents the shape function matrix. Since the terms in Eq. (8-3) represent rate of energy transfer *per unit length*, the form of Eq. (8-5) to be applied to Eq. (8-3) is

$$\sum_{e=1}^{M} \int_{l^e} \mathbf{N}^T R^e \, dx = 0 \tag{8-6}$$

However, let us agree to drop the summation sign because we are really seeking the finite element characteristics for a typical element. The summation represents the assemblage process and is no longer delineated since this step is routine.

In what follows, the lineal element from Sec. 6-3 is used. As shown in Fig. 8-2(b), a typical element e connects nodes i and j with coordinates x_i and x_j and temperatures T_i and T_j, respectively. This implies a linear temperature distribution in each element. Therefore, Eq. (8-6) may be written as

$$\int_{le} \mathbf{N}^T \left[\frac{d}{dx} \left(kA \frac{dT}{dx} \right) - hP(T - T_a) - \varepsilon \sigma P(T^4 - T_r^4) + QA \right] dx = \mathbf{0} \quad \textbf{(8-7)}$$

If the first term is integrated by parts, we get

$$\mathbf{N}^T kA \frac{dT}{dx} \bigg|_{x_i}^{x_j} - \int_{le} \frac{d\mathbf{N}^T}{dx} kA \frac{dT}{dx} dx - \int_{le} \mathbf{N}^T hPT \, dx + \int_{le} \mathbf{N}^T hPT_a \, dx$$

$$- \int_{le} \mathbf{N}^T \varepsilon \sigma PT^4 \, dx + \int_{le} \mathbf{N}^T \varepsilon \sigma PT_r^4 \, dx + \int_{le} \mathbf{N}^T QA \, dx = \mathbf{0} \quad \textbf{(8-8)}$$

The first term in Eq. (8-8) is related to the heat flux q_x from conduction in the x direction. By Fourier's law of heat conduction, we have

$$q_x = -k \frac{dT}{dx} \quad \textbf{(8-9)}$$

This heat flux represents the heat transfer rate per unit cross-sectional area in the x direction. The minus sign is used to give a positive heat flux in the direction of decreasing temperature. The first term in Eq. (8-8) will contribute to the assemblage equations for only the two elements at each end of the body; all internal contributions to the assemblage equations cancel each other during the assemblage step. Therefore, this term needs to be evaluated for elements 1 and M only, assuming the elements are numbered consecutively from one end of the body to the other.

Let us assume that the heat transfer to or from the ends is given by any combination of the following three conditions: (1) convection with heat transfer coefficients h_i and h_j to a fluid at ambient temperature T_{ai} and T_{aj}, (2) radiation with surface emissivities ε_i and ε_j and receiver temperatures T_{ri} and T_{rj}, and (3) heat fluxes q_i and q_j (possibly acting at a distance such as sun). The subscripts i and j are used to denote the local node numbers i and j. Let us denote the heat flux from convection and radiation as q_{cv} and q_r, respectively. The imposed heat flux will be denoted as q_s. These boundary conditions are shown in Fig. 8-3 for one end of the body. An energy balance on this end of the body gives

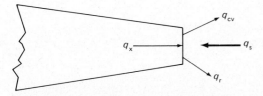

Figure 8-3 Typical boundary in the one-dimensional heat conduction problem showing the heat fluxes considered in the finite element formulation.

$$q_x + q_s = q_{cv} + q_r \tag{8-10a}$$

or

$$q_x = q_{cv} + q_r - q_s \tag{8-10b}$$

Note that q_s is defined to be *positive* if the heat flux is imposed *toward* the surface. From Newton's law of cooling at node j, we have

$$q_{cv} = h_j(T - T_{aj}) \tag{8-11}$$

and from the Stefan-Boltzmann law,

$$q_r = \varepsilon_j\sigma(T^4 - T_{rj}^4) \tag{8-12}$$

Since $q_s = q_j$ at node j, we have for the integrated term at $x = x_j$,

$$
\begin{aligned}
\mathbf{N}^T kA \frac{dT}{dx}\bigg|_{x=x_j} &= \mathbf{N}^T(-q_x A)\big|_{x=x_j} \\
&= \mathbf{N}^T A_j(-q_{cv} - q_r + q_s)\big|_{x=x_j} \\
&= (-\mathbf{N}^T h_j A_j T + \mathbf{N}^T h_j A_j T_{aj} - \mathbf{N}^T \varepsilon_j A_j \sigma T^4 \\
&\quad + \mathbf{N}^T \varepsilon_j A_j \sigma T_{rj}^4 + \mathbf{N}^T A_j q_j)\big|_{x=x_j}
\end{aligned}
\tag{8-13}
$$

A similar expression results for the integrated term at node i, namely,

$$
\begin{aligned}
-\mathbf{N}^T kA \frac{dT}{dx}\bigg|_{x=x_i} &= (-\mathbf{N}^T h_i A_i T + \mathbf{N}^T h_i A_i T_{ai} \\
&\quad - \mathbf{N}^T \varepsilon_i A_i \sigma T^4 + \mathbf{N}^T \varepsilon_i A_i \sigma T_{ri}^4 + \mathbf{N}^T A_i q_i)\big|_{x=x_i}
\end{aligned}
\tag{8-14}
$$

Note the sign of the term on the left; with the minus sign included here, the integrated term is now given by the *sum* of the right-hand sides of Eqs. (8-13) and (8-14).

At this point, the parameter function T must be related to the nodal temperatures in the usual manner by

$$T = \mathbf{N}\mathbf{a}^e \tag{8-15}$$

For convenience, let us write T^4 as follows:

$$T^4 = T^3 T = (\mathbf{N}\mathbf{a}^e)^3 \mathbf{N}\mathbf{a}^e \tag{8-16}$$

With the help of Eqs. (8-13) to (8-16), we may write Eq. (8-8) in the form

$$\mathbf{K}^e \mathbf{a}^e = \mathbf{f}^e \tag{8-17}$$

The composite element stiffness matrix $\mathbf{K}^e$ is comprised of five element stiffness or *conductance* matrices, or

$$\mathbf{K}^e = \mathbf{K}_x^e + \mathbf{K}_{cv}^e + \mathbf{K}_r^e + \mathbf{K}_{cvB}^e + \mathbf{K}_{rB}^e \tag{8-18}$$

and the composite element nodal force vector $\mathbf{f}^e$ is comprised of six other element nodal force vectors, or

$$\mathbf{f}^e = \mathbf{f}_{cv}^e + \mathbf{f}_r^e + \mathbf{f}_Q^e + \mathbf{f}_{cvB}^e + \mathbf{f}_{rB}^e + \mathbf{f}_{qB}^e \tag{8-19}$$

Note the use of the B in the subscripts on those terms that arise from the boundary conditions. The expressions for the *element characteristics* are given by

$$\mathbf{K}_x^e = \int_{l^e} \frac{d\mathbf{N}^T}{dx} kA \frac{d\mathbf{N}}{dx} dx \tag{8-20a}$$

$$\mathbf{K}_{cv}^e = \int_{l^e} \mathbf{N}^T hP\mathbf{N} \, dx \tag{8-20b}$$

$$\mathbf{K}_r^e = \int_{l^e} \mathbf{N}^T \varepsilon\sigma P(\mathbf{Na}^e)^3 \mathbf{N} \, dx \tag{8-20c}$$

$$\mathbf{K}_{cvB}^e = \mathbf{N}^T h_i A_i \mathbf{N}\big|_{x=x_i} + \mathbf{N}^T h_j A_j \mathbf{N}\big|_{x=x_j} \tag{8-20d}$$

$$\mathbf{K}_{rB}^e = \mathbf{N}^T \varepsilon_i A_i \sigma(\mathbf{Na}^e)^3 \mathbf{N}\big|_{x=x_i} + \mathbf{N}^T \varepsilon_j A_j \sigma(\mathbf{Na}^e)^3 \mathbf{N}\big|_{x=x_j} \tag{8-20e}$$

$$\mathbf{f}_{cv}^e = \int_{l^e} \mathbf{N}^T hPT_a \, dx \tag{8-21a}$$

$$\mathbf{f}_r^e = \int_{l^e} \mathbf{N}^T \varepsilon\sigma PT_r^4 \, dx \tag{8-21b}$$

$$\mathbf{f}_Q^e = \int_{l^e} \mathbf{N}^T QA \, dx \tag{8-21c}$$

$$\mathbf{f}_{cvB}^e = \mathbf{N}^T h_i A_i T_{ai}\big|_{x=x_i} + \mathbf{N}^T h_j A_j T_{aj}\big|_{x=x_j} \tag{8-21d}$$

$$\mathbf{f}_{rB}^e = \mathbf{N}^T \varepsilon_i A_i \sigma T_{ri}^4\big|_{x=x_i} + \mathbf{N}^T \varepsilon_j A_j \sigma T_{rj}^4\big|_{x=x_j} \tag{8-21e}$$

and

$$\mathbf{f}_{qB}^e = \mathbf{N}^T q_i A_i\big|_{x=x_i} + \mathbf{N}^T q_j A_j\big|_{x=x_j} \tag{8-21f}$$

Although it has been stated at the outset that the two-node lineal element from Sec. 6-3 is to be used here, the above expressions for the element characteristics are quite general. They can be applied readily to all other one-dimensional elements, some of which are presented in Chapter 9. Several of these expressions will now be evaluated by way of examples.

Example 8-3

Evaluate the element stiffness matrix $\mathbf{K}_x^e$ for the lineal element from Sec. 6-3.

Solution

Recognizing that we need the derivatives of the shape functions in the expression for $\mathbf{K}_x^e$ given by Eq. (8-20a), we first write

$$\frac{dN_i}{dx} = \frac{d}{dx}\left(\frac{x_j - x}{x_j - x_i}\right) = -\frac{1}{L}$$

and

$$\frac{dN_j}{dx} = \frac{d}{dx}\left(\frac{x - x_i}{x_j - x_i}\right) = +\frac{1}{L}$$

where L is the element length. Substituting these results into Eq. (8-20a) gives

$$\mathbf{K}_x^e = \int_{x_i}^{x_j} \begin{bmatrix} -\dfrac{1}{L} \\[2mm] \dfrac{1}{L} \end{bmatrix} kA \begin{bmatrix} -\dfrac{1}{L} & \dfrac{1}{L} \end{bmatrix} dx$$

or

$$\mathbf{K}_x^e = \frac{kA}{L} \begin{bmatrix} 1 & -1 \\ -1 & 1 \end{bmatrix} \tag{8-22}$$

where it has been assumed that k and A are constant. If this is not the case, then suitable average values may be used [e.g., values at $\bar{x} = (x_i + x_j)/2$]. ∎

Example 8-4

Try to evaluate the element stiffness matrix $\mathbf{K}_r^e$. Use the lineal element from Sec. 6-3.

Solution

For mathematical convenience, let us use the serendipity form of the shape functions given by Eqs. (6-11), or

$$N_i = \tfrac{1}{2}(1 - r) \tag{6-11a}$$

and

$$N_j = \tfrac{1}{2}(1 + r) \tag{6-11b}$$

where

$$r = \frac{2(x - \bar{x})}{x_j - x_i} = \frac{x - \bar{x}}{L/2} \tag{6-10}$$

and $\bar{x} = (x_i + x_j)/2$ defines the centroid of the element. With these definitions, we attempt to evaluate $\mathbf{K}_r^e$ as follows:

$$\mathbf{K}_r^e = \int_{-1}^{+1} \mathbf{N}^T \varepsilon \sigma P (\mathbf{N} \mathbf{a}^e)^3 \mathbf{N} \frac{L}{2}\, dr$$

$$= \int_{-1}^{+1} \frac{\varepsilon \sigma P L}{8} \begin{bmatrix} 1 - r \\ 1 + r \end{bmatrix} [1 - r \quad 1 + r] \tag{8-23}$$

$$\cdot\, [\tfrac{1}{2}(1 - r)T_i + \tfrac{1}{2}(1 + r)T_j]^3\, dr$$

This integral is most easily evaluated using Gauss-Legendre quadrature, discussed in Chapter 9. For now, we shall leave $\mathbf{K}_r^e$ in this form. ∎

From Example 8-4, it should be noted that the matrix $\mathbf{K}_f^e$ contains the (unknown) nodal temperatures (T_i and T_j) and, therefore, the resulting assemblage system equations will be nonlinear. The direct iteration method from Sec. 8-2 may be used to obtain the solution. However, if the major stiffness contribution is from the matrix $\mathbf{K}_f^e$, the direct iteration method is likely to be divergent. In this case, overrelaxation or underrelaxation may be needed to obtain convergence. A detailed discussion of these refinements to the direct iteration method is beyond the intended scope of this text.

Example 8-5

Evaluate the element stiffness matrix $\mathbf{K}_{cvB}^e$ if the lineal element from Sec. 6-3 is used.

Solution

Recall that one of the properties of the shape functions is given by

$$N_i(x_i) = 1 \qquad N_i(x_j) = 0$$
$$N_j(x_i) = 0 \qquad N_j(x_j) = 1 \tag{8-24}$$

Therefore, $\mathbf{K}_{cvB}^e$ may be evaluated as

$$\mathbf{K}_{cvB}^e = \begin{bmatrix} 1 \\ 0 \end{bmatrix} h_i A_i [1 \quad 0] + \begin{bmatrix} 0 \\ 1 \end{bmatrix} h_j A_j [0 \quad 1]$$

or

$$\mathbf{K}_{cvB}^e = \begin{bmatrix} h_i A_i & 0 \\ 0 & h_j A_j \end{bmatrix} \tag{8-25}$$

∎

It is emphasized that the stiffness matrix given by Eq. (8-25) will contribute to the assemblage stiffness matrix only if the element is on the boundary (that is, at one of the ends). Naturally, convection must be present at this end also. For example, if element 1 connects nodes 1 and 2, then

$$\mathbf{K}_{cvB}^{(1)} = \begin{bmatrix} h_1 A_1 & 0 \\ 0 & 0 \end{bmatrix}$$

Similarly for element M that connects nodes $n - 1$ and n we have

$$\mathbf{K}_{cvB}^{(M)} = \begin{bmatrix} 0 & 0 \\ 0 & h_n A_n \end{bmatrix}$$

All other $\mathbf{K}_{cvB}^e$'s may be taken as the 2×2 null matrix.

Example 8-6

Evaluate the element nodal force vector $\mathbf{f}_{cv}^e$. Use the lineal element from Sec. 6-3.

Solution

It is convenient to evaluate the integral in Eq. (8-21a) by using length coordinates and the integration formula given by Eq. (6-48):

$$\mathbf{f}_{cv}^e = \int_{l^e} \begin{bmatrix} L_i \\ L_j \end{bmatrix} hPT_a \, dx$$

$$= \begin{bmatrix} \int_{l^e} L_i \, hPT_a \, dx \\ \int_{l^e} L_j \, hPT_a \, dx \end{bmatrix}$$

$$= \begin{bmatrix} \dfrac{1!0!}{(1 + 0 + 1)!} LhPT_a \\ \dfrac{0!1!}{(0 + 1 + 1)!} LhPT_a \end{bmatrix} \tag{8-26}$$

or

$$\mathbf{f}_{cv}^e = \frac{hPLT_a}{2} \begin{bmatrix} 1 \\ 1 \end{bmatrix} \tag{8-27}$$

Not surprisingly, one-half of the total $hPLT_a$ is allocated to each node. Note that L (without subscripts) represents the element length and is not to be confused with the length coordinates. ∎

Example 8-7

Evaluate the element nodal force vector $\mathbf{f}_{cvB}^e$. Use the two-node lineal element.

Solution

Referring back to Example 8-5 and using Eq. (8-21d), it follows that

$$\mathbf{f}_{cvB}^e = \begin{bmatrix} 1 \\ 0 \end{bmatrix} h_i A_i T_{ai} + \begin{bmatrix} 0 \\ 1 \end{bmatrix} h_j A_j T_{aj}$$

or

$$\mathbf{f}_{cvB}^e = \begin{bmatrix} h_i A_i T_{ai} \\ h_j A_j T_{aj} \end{bmatrix} \tag{8-28}$$

∎

Again it is emphasized that this particular element nodal force vector will contribute to the assemblage stiffness matrix only if the element has a node on the global boundary (assuming convection there). For example, if element 1 connects nodes 1 and 2, then

$$\mathbf{f}_{cvB}^{(1)} = \begin{bmatrix} h_1 A_1 T_{a1} \\ 0 \end{bmatrix} \tag{8-29a}$$

Similarly for element M that connects nodes $n - 1$ and n we have

$$\mathbf{f}_{cvB}^{(M)} = \begin{bmatrix} 0 \\ h_n A_n T_{an} \end{bmatrix} \tag{8-29b}$$

In effect, all other $\mathbf{f}_{cvB}^e$'s may be taken to be the 2×1 null vector.

Example 8-8

Reevaluate the element stiffness matrix $\mathbf{K}_x^e$ by assuming the cross-sectional area A varies linearly from node i to node j over the lineal element.

Solution

It proves to be very convenient to write the linearly varying cross-sectional area A as

$$A(x) = N_i(x)A_i + N_j(x)A_j$$

where A_i and A_j are the cross-sectional areas at nodes i and j. Note that the shape functions from Sec. 6-3 are linear themselves and, in effect, provide a convenient *interpolation polynomial*. Therefore, we have

$$\mathbf{K}_x^e = \int_{le} \begin{bmatrix} -\dfrac{1}{L} \\ \dfrac{1}{L} \end{bmatrix} k(N_i A_i + N_j A_j) \begin{bmatrix} -\dfrac{1}{L} & \dfrac{1}{L} \end{bmatrix} dx$$

where L is the element length. In terms of the length coordinates L_i and L_j, this becomes

$$\mathbf{K}_x^e = \int_{le} \begin{bmatrix} 1 & -1 \\ -1 & 1 \end{bmatrix} \frac{k}{L^2} (L_i A_i + L_j A_j) \, dx$$

Using the integration formula given by Eq. (6-48), this reduces to the following simple result:

$$\mathbf{K}_x^e = \frac{k\overline{A}}{L} \begin{bmatrix} 1 & -1 \\ -1 & 1 \end{bmatrix} \tag{8-30a}$$

where $\overline{A}$ is defined by

$$\overline{A} = \frac{A_i + A_j}{2} \tag{8-30b}$$

∎

The reader is cautioned about generalizing this result. For example, if $\mathbf{K}_{cv}^e$ is evaluated in a similar fashion by assuming the perimeter varies linearly over the lineal element, the result is

$$\mathbf{K}_{cv}^e = \frac{hL}{12} \begin{bmatrix} 3P_i + P_j & P_i + P_j \\ P_i + P_j & P_i + 3P_j \end{bmatrix} \tag{8-31}$$

This is quite different from the result obtained when the perimeter is evaluated at the element centroid.

Example 8-9

Reconsider Example 4-11 from Sec. 4-10. Recall that the tip of the fin was assumed to be insulated. Resolve for the temperature distribution for the case of convection at the tip. Assume that the tip is in contact with a boiling fluid at 10°C with a heat transfer coefficient of 4000 W/m²-°C.

Solution

The element stiffness matrix

$$\mathbf{K}_{cvB}^{(2)} = \begin{bmatrix} 0 & 0 \\ 0 & h_j A_j \end{bmatrix} = \begin{bmatrix} 0 & 0 \\ 0 & (4000)(1.2566 \times 10^{-5}) \end{bmatrix}$$

needs to be added to $\mathbf{K}^{(2)}$ in Example 4-11, or

$$\mathbf{K}^{(2)} = \begin{bmatrix} 0.50893 & -0.49951 \\ -0.49951 & 0.50893 \end{bmatrix} + \begin{bmatrix} 0 & 0 \\ 0 & 0.05026 \end{bmatrix}$$

$$= \begin{bmatrix} 0.50893 & -0.49951 \\ -0.49951 & 0.55919 \end{bmatrix}$$

The assemblage stiffness matrix $\mathbf{K}^a$ becomes

$$\mathbf{K}^a = \begin{bmatrix} 0.50893 & -0.49951 & 0 \\ -0.49951 & 1.01786 & -0.49951 \\ 0 & -0.49951 & 0.55919 \end{bmatrix} \text{W/°C}$$

In addition, the element nodal force vector

$$\mathbf{f}_{cvB}^{(2)} = \begin{bmatrix} 0 \\ h_j A_j T_{aj} \end{bmatrix} = \begin{bmatrix} 0 \\ (4000)(1.2566 \times 10^{-5})(10) \end{bmatrix}$$

needs to be added to $\mathbf{f}^{(2)}$ in Example 4-11, or

$$\mathbf{f}^{(2)} = \begin{bmatrix} 0.23561 \\ 0.23561 \end{bmatrix} + \begin{bmatrix} 0 \\ 0.50264 \end{bmatrix} = \begin{bmatrix} 0.23561 \\ 0.73825 \end{bmatrix}$$

The assemblage nodal force vector $\mathbf{f}^a$ becomes

$$\mathbf{f}^a = \begin{bmatrix} 0.23561 \\ 0.47122 \\ 0.73825 \end{bmatrix}$$

After application of the prescribed temperature of 85°C at node 1 (using Method 1 from Sec. 3-2), we get

$$\begin{bmatrix} 1 & 0 & 0 \\ 0 & 1.01786 & -0.49951 \\ 0 & -0.49951 & 0.55919 \end{bmatrix} \begin{bmatrix} T_1 \\ T_2 \\ T_3 \end{bmatrix} = \begin{bmatrix} 85 \\ 42.930 \\ 0.73825 \end{bmatrix}$$

Solving for the nodal temperatures gives

$$T_1 = 85°C \qquad T_2 = 76.25°C \qquad T_3 = 69.43°C$$

As expected, the tip temperature is much lower than that obtained in Example 4-11, where the tip was assumed to be insulated. Section 4-10 should be consulted for a discussion of how the element resultants may be obtained. ∎

8-4 THE GREEN-GAUSS THEOREM

The Green-Gauss theorem is essentially a multidimensional version of integration by parts, the latter of which is given by Eq. (4-32). We found this equation to be useful when we developed the finite element characteristics in one-dimensional nonstructural problems. When dealing with two- and three-dimensional problems, we will again need to use integration by parts. However, the Green-Gauss theorem is far easier to apply and is mathematically more rigorous.

The Green-Gauss theorem may be derived from the divergence theorem. It may be recalled that the *divergence theorem* states that the integral of the divergence of a vector over a volume V is precisely equal to the integral of the flux of the same vector through the closed surface S that bounds the volume V. Stated mathematically, the three-dimensional form of the divergence theorem is given by

$$\int_V \nabla \cdot \mathbf{q} \, dV = \int_S \mathbf{q} \cdot \mathbf{n} \, dS \tag{8-32}$$

where $\mathbf{q}$ is any vector, dV is an infinitesimal volume element, dS is an infinitesimal surface element bounding the volume V, and $\mathbf{n}$ is a unit vector that is always normal to the closed surface S and always directed outward from the body, as shown in Fig. 8-4(a). The divergence of $\mathbf{q}$ is written $\nabla \cdot \mathbf{q}$ and in cartesian coordinates is given by

$$\nabla \cdot \mathbf{q} = \frac{\partial q_x}{\partial x} + \frac{\partial q_y}{\partial y} + \frac{\partial q_z}{\partial z} \tag{8-33}$$

where q_x, q_y, and q_z are the x, y, and z components of $\mathbf{q}$. As Eq. (8-33) implies, the divergence of a vector is a scalar.

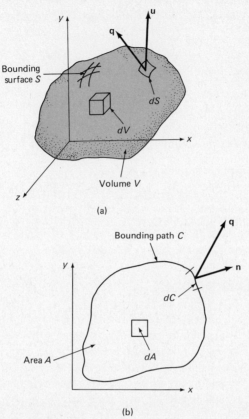

Figure 8-4 Definition of terms used in the divergence theorem in (a) three dimensions and (b) two dimensions.

Equation (8-32) may also be used to represent the two-dimensional form of the divergence theorem if V is interpreted to be an area A, S to be the path C enclosing the area, and $\mathbf{n}$ to be the normal unit vector pointing outward from the bounding path in the plane of the area A. In this case Eq. (8-32) is written

$$\int_A \nabla \cdot \mathbf{q} \, dA = \int_C \mathbf{q} \cdot \mathbf{n} \, dC \tag{8-34}$$

The integration around the path C must be performed in the counterclockwise direction. Figure 8-4(b) should help to clarify this notation.

We are now in a position to derive the *Green-Gauss theorem*. Let us write the vector $\mathbf{q}$ as the product of a scalar β and another vector $\mathbf{p}$, or

$$\mathbf{q} = \beta \mathbf{p} \tag{8-35}$$

Substituting this expression for $\mathbf{q}$ in Eq. (8-32) yields

$$\int_V \nabla \cdot (\beta \mathbf{p}) \, dV = \int_S \beta \mathbf{p} \cdot \mathbf{n} \, dS \tag{8-36}$$

The following identity [1] proves to be useful:

$$\nabla \cdot (\beta \mathbf{p}) = \beta \nabla \cdot \mathbf{p} + \nabla \beta \cdot \mathbf{p} \tag{8-37}$$

where $\nabla \beta$ denotes the gradient of the scalar β. In cartesian coordinates, $\nabla \beta$ is given in three dimensions by

$$\nabla \beta = \frac{\partial \beta}{\partial x}\mathbf{i} + \frac{\partial \beta}{\partial y}\mathbf{j} + \frac{\partial \beta}{\partial z}\mathbf{k} \tag{8-38a}$$

and in two dimensions by

$$\nabla \beta = \frac{\partial \beta}{\partial x}\mathbf{i} + \frac{\partial \beta}{\partial y}\mathbf{j} \tag{8-38b}$$

Note that the gradient of a scalar is a vector. Returning to Eq. (8-36) and using Eq. (8-37), we may write the following:

$$\int_V \beta \nabla \cdot \mathbf{p} \, dV = \int_S \beta \mathbf{p} \cdot \mathbf{n} \, dS - \int_V \nabla \beta \cdot \mathbf{p} \, dV \tag{8-39}$$

for the three-dimensional case and

$$\int_A \beta \nabla \cdot \mathbf{p} \, dA = \int_C \beta \mathbf{p} \cdot \mathbf{n} \, dC - \int_A \nabla \beta \cdot \mathbf{p} \, dA \tag{8-40}$$

for the two-dimensional case. Since p_x is independent of p_y and p_z, and vice versa, it follows that

$$\int_V \beta \frac{\partial p_x}{\partial x} \, dV = \int_S \beta p_x n_x \, dS - \int_V \frac{\partial \beta}{\partial x} p_x \, dV \tag{8-41a}$$

$$\int_V \beta \frac{\partial p_y}{\partial y} \, dV = \int_S \beta p_y n_y \, dS - \int_V \frac{\partial \beta}{\partial y} p_y \, dV \tag{8-41b}$$

and

$$\int_V \beta \frac{\partial p_z}{\partial z} \, dV = \int_S \beta p_z n_z \, dS - \int_V \frac{\partial \beta}{\partial z} p_z \, dV \tag{8-41c}$$

In the two-dimensional case, p_x is independent of p_y, and vice versa; so Eq. (8-40) implies

$$\int_A \beta \frac{\partial p_x}{\partial x} \, dA = \int_C \beta p_x n_x \, dC - \int_A \frac{\partial \beta}{\partial x} p_x \, dA \tag{8-42a}$$

and

$$\int_A \beta \frac{\partial p_y}{\partial y} \, dA = \int_C \beta p_y n_y \, dC - \int_A \frac{\partial \beta}{\partial y} p_y \, dA \tag{8-42b}$$

In the above, dV is generally taken to be $dx \, dy \, dz$ and dA to be $dx \, dy$; p_x, p_y, and p_z are the x, y, and z components of $\mathbf{p}$; and n_x, n_y, and n_z are the *direction cosines* of the outward normal unit vector with respect to the coordinate axes. Figure

8-5 shows that n_x is the cosine of the angle α_1 between the vector **n** (the outward normal unit vector) and the x axis. Similarly, n_y is the cosine of the angle α_2. Hence, n_x and n_y are the respective direction cosines.

Equations (8-42) and (8-41) are the two- and three-dimensional forms of the Green-Gauss theorem. Specific examples that illustrate the use of Eqs. (8-41) and (8-42) are not given here because several applications are forthcoming in the remainder of this chapter.

8-5   SIMPLE TWO-DIMENSIONAL HEAT CONDUCTION

Consider a long bar of uniform cross section as shown in Fig. 8-6(a). It is assumed that each cross section through the bar is no different from any other both geometrically and thermally. Note that only imposed heat fluxes and prescribed temperatures are allowed. Perfect insulation is a special case of zero heat flux. Let us allow for a heat source or internal energy generation Q per unit volume and unit time that is at most a function of x and y only. This three-dimensional body may therefore be analyzed as though it were two-dimensional. The governing equation for the steady-state temperature T for a typical cross section far from the ends is given by

$$\frac{\partial}{\partial x}\left(k\frac{\partial T}{\partial x}\right) + \frac{\partial}{\partial y}\left(k\frac{\partial T}{\partial y}\right) + Q = 0 \tag{8-43}$$

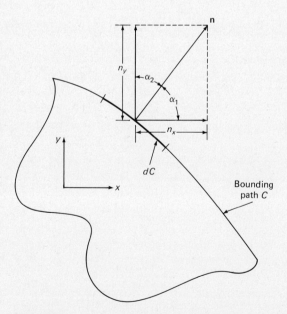

Figure 8-5   Two-dimensional region illustrating the unit normal **n** (perpendicular to the boundary C) and the angles α_1 and α_2. Note that $\cos \alpha_1 = n_x$ and $\cos \alpha_2 = n_y$.

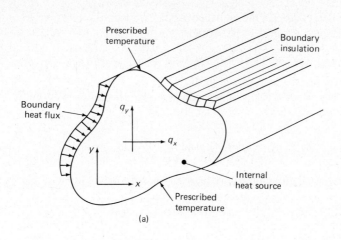

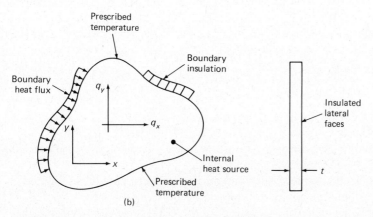

Figure 8-6   Schematic of simple two-dimensional heat conduction problem: (a) infinitely long body and (b) thin plate with insulated lateral faces.

Note that this same equation describes the plate of constant thickness shown in Fig. 8-6(b), providing the lateral faces of the plate are well-insulated.

In the next several sections, the heat flux in a direction normal to some surface, usually the global boundary, will be needed. A heat flux is generally a vector and represents the heat transfer rate per unit area. In isotropic materials, the heat flux is always normal to isothermal surfaces. A surface is isothermal if the temperature is constant on that surface. In any event, the net heat flux through a surface from conduction must be taken in the direction normal to the surface. Clearly, no contribution can be made by the tangential heat flux (if any exists). Let us denote the net (normal) heat flux vector as $\mathbf{q}_n$, the magnitude of which is simply q_n. In terms of the outward normal unit vector $\mathbf{n}$, $\mathbf{q}_n$ may be written

$$\mathbf{q}_n = q_n \mathbf{n} \tag{8-44}$$

In terms of the heat fluxes in the global coordinate directions, this same net (normal) heat flux from conduction may be written

$$\mathbf{q}_n = q_x \mathbf{i} + q_y \mathbf{j} \tag{8-45}$$

where q_x and q_y are the heat fluxes in the x and y directions. This is illustrated in Fig. 8-7. Invoking Fourier's law of heat conduction for heterogeneous (but isotropic) materials, we have

$$q_x = -k\frac{\partial T}{\partial x} \quad \text{and} \quad q_y = -k\frac{\partial T}{\partial y} \tag{8-46}$$

It should be recalled from elementary heat transfer that the minus signs are included so that a positive heat flux vector always points in the direction of decreasing temperature. From Eqs. (8-44) to (8-46), we conclude that

$$\mathbf{q}_n = -k\frac{\partial T}{\partial x}\mathbf{i} - k\frac{\partial T}{\partial y}\mathbf{j} \tag{8-47}$$

If the dot product of both sides of this last result is taken with the vector $\mathbf{n}$, we get

$$q_n = -k\frac{\partial T}{\partial x}n_x - k\frac{\partial T}{\partial y}n_y \tag{8-48}$$

where $n_x = \mathbf{i} \cdot \mathbf{n}$ and $n_y = \mathbf{j} \cdot \mathbf{n}$ are the cosines of the angles formed by the x and y axes with the vector $\mathbf{n}$; that is, n_x and n_y are the direction cosines. Equation (8-48) is very important. It will be used frequently in what follows and may be extended by inspection to the three-dimensional case. It is emphasized that q_n represents the net (normal) heat flux from conduction leaving a surface.

In the next section some of the concepts of variational calculus introduced in Chapter 4 are extended to two dimensions. In Sec. 8-6, these concepts are applied to Eq. (8-43) in order to derive the expressions for the finite element characteristics.

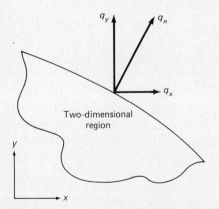

Figure 8-7 Boundary of two-dimensional region showing net (normal) heat flux $\mathbf{q}_n$ and the components in the x and y directions.

This is followed by the derivation of these expressions with the Galerkin method in Sec. 8-7. The latter will be seen to be much simpler and applicable to a wider variety of problems. Therefore, after Sec. 8-6, the Galerkin method will be used throughout the remainder of the book.

8-6   VARIATIONAL FORMULATIONS IN TWO DIMENSIONS

Recall from Chapter 4 that if the governing differential equation contains second-order derivatives, the functional contains only first-order derivatives. For Eq. (8-43) it seems reasonable to assume a functional F of the form

$$F = F(x,y,T,T_x,T_y) \tag{8-49}$$

where the subscripts on T indicate derivatives, i.e.,

$$T_x = \frac{\partial T}{\partial x} \qquad \text{and} \qquad T_y = \frac{\partial T}{\partial y} \tag{8-50}$$

The integral to be extremized is then given by

$$I = \int_A F(x,y,T,T_x,T_y) \, dx \, dy \tag{8-51}$$

Using a procedure that is completely analogous to that used in Sec. 4-5, let us derive the corresponding Euler-Lagrange equation. It should further be recalled that the Euler-Lagrange equation is really the governing differential equation for the problem.

The Euler-Lagrange Equation: Geometric and Natural Boundary Conditions

We proceed by taking the first variation of Eq. (8-51) and setting it to zero, or

$$\delta I = \delta \int_A F(x,y,T,T_xT_y) \, dx \, dy = 0 \tag{8-52}$$

From one of the commutative properties, we have

$$\delta \int_A F \, dx \, dy = \int_A \delta F \, dx \, dy \tag{8-53}$$

and by an extension of Eq. (4-37), we write

$$\delta F = \frac{\partial F}{\partial T} \delta T + \frac{\partial F}{\partial T_x} \delta T_x + \frac{\partial F}{\partial T_y} \delta T_y \tag{8-54}$$

Another of the cummutative properties allows us to write

$$\frac{\partial F}{\partial T_x} \delta T_x = \frac{\partial F}{\partial T_x} \delta \left(\frac{\partial T}{\partial x} \right) = \frac{\partial F}{\partial T_x} \frac{\partial}{\partial x} (\delta T) \tag{8-55a}$$

and, similarly

$$\frac{\partial F}{\partial T_y} \delta T_y = \frac{\partial F}{\partial T_y} \frac{\partial}{\partial y} (\delta T) \tag{8-55b}$$

Therefore, Eq. (8-52) becomes

$$\delta I = \int_A \left[\frac{\partial F}{\partial T} \delta T + \frac{\partial F}{\partial T_x} \frac{\partial}{\partial x} (\delta T) + \frac{\partial F}{\partial T_y} \frac{\partial}{\partial y} (\delta T) \right] dx \, dy = 0 \tag{8-56}$$

Normally, we would use integration by parts at this point. However, since the problem is two-dimensional, we use the Green-Gauss theorem instead. For example, let us examine the second term which is in effect an integral of the form

$$\int p_x \frac{\partial \beta}{\partial x} \, dA$$

where p_x is analogous to $\partial F/\partial T_x$ and β to δT. From Eq. (8-42a), it follows that

$$\int_A \frac{\partial F}{\partial T_x} \frac{\partial}{\partial x} (\delta T) \, dx \, dy = \int_C \frac{\partial F}{\partial T_x} n_x \, \delta T \, dC - \int_A \frac{\partial}{\partial x} \left(\frac{\partial F}{\partial T_x} \right) \delta T \, dx \, dy \tag{8-57}$$

In a similar fashion, the third term in Eq. (8-56) may be written as

$$\int_A \frac{\partial F}{\partial T_y} \frac{\partial}{\partial y} (\delta T) \, dx \, dy = \int_C \frac{\partial F}{\partial T_y} n_y \, \delta T \, dC - \int_A \frac{\partial}{\partial y} \left(\frac{\partial F}{\partial T_y} \right) \delta T \, dx \, dy \tag{8-58}$$

Therefore, Eq. (8-56) becomes

$$\delta I = \int_A \left[\frac{\partial F}{\partial T} - \frac{\partial}{\partial x} \left(\frac{\partial F}{\partial T_x} \right) - \frac{\partial}{\partial y} \left(\frac{\partial F}{\partial T_y} \right) \right] \delta T \, dx \, dy$$

$$+ \int_C \left[\frac{\partial F}{\partial T_x} n_x + \frac{\partial F}{\partial T_y} n_y \right] \delta T \, dC = 0 \tag{8-59}$$

Continuing further with the analogies from Chapter 4, we conclude that the *Euler-Lagrange equation* is given by

$$\frac{\partial F}{\partial T} - \frac{\partial}{\partial x} \left(\frac{\partial F}{\partial T_x} \right) - \frac{\partial}{\partial y} \left(\frac{\partial F}{\partial T_y} \right) = 0 \tag{8-60}$$

and the *natural boundary condition* by

$$\frac{\partial F}{\partial T_x} n_x + \frac{\partial F}{\partial T_y} n_y = 0 \tag{8-61}$$

This last result will be shown to be related to the condition of a zero temperature gradient; hence no conduction on that part of the boundary can occur (i.e., it is perfectly insulated). Also, the second integral in Eq. (8-59) is zero on those portions of the global boundary where the temperature is prescribed because if T is prescribed, we have $\delta T = 0$. This last condition is the so-called *geometric boundary condition*.

Example 8-10

For the problem described by the governing equation given by Eq. (8-43):
(a) Determine the functional F if the method illustrated in Example 4-3 is used.
(b) Determine the variational principle if the method illustrated in Example 4-4 is used, and if an imposed heat flux q_{sb} is assumed to act on the boundary.

Solution

(a) We first write Eq. (8-43) in a form where a one-to-one correspondence of terms can be made more readily with the Euler-Lagrange equation, or

$$Q - \frac{\partial}{\partial x}\left(-k\frac{\partial T}{\partial x}\right) - \frac{\partial}{\partial y}\left(-k\frac{\partial T}{\partial y}\right) = 0 \qquad \text{(8-62)}$$

from which it is concluded that

$$\frac{\partial F}{\partial T} = Q \qquad \text{(8-63a)}$$

$$\frac{\partial F}{\partial T_x} = -k\frac{\partial T}{\partial x} = -kT_x \qquad \text{(8-63b)}$$

$$\frac{\partial F}{\partial T_y} = -k\frac{\partial T}{\partial y} = -kT_y \qquad \text{(8-63c)}$$

If Eqs. (8-63) are integrated, we get

$$F = QT + f(T_x, T_y) \qquad \text{(8-64a)}$$

$$F = -\tfrac{1}{2}kT_x^2 + g(T, T_y) \qquad \text{(8-64b)}$$

and

$$F = -\tfrac{1}{2}kT_y^2 + h(T, T_x) \qquad \text{(8-64c)}$$

where f, g, and h must be such that F itself is given by

$$F = QT - \tfrac{1}{2}kT_x^2 - \tfrac{1}{2}kT_y^2 \qquad \text{(8-65a)}$$

or

$$F = QT - \tfrac{1}{2}k\left(\frac{\partial T}{\partial x}\right)^2 - \tfrac{1}{2}k\left(\frac{\partial T}{\partial y}\right)^2 \qquad \text{(8-65b)}$$

Actually a constant may be added to the expression for the functional F. However, since it would have no effect whatsoever on the extremization process to be illustrated below, it is not included in Eqs. (8-65).

(b) From Example 4-4, we begin by writing

$$\delta I = \int_A \left[Q + \frac{\partial}{\partial x}\left(k\frac{\partial T}{\partial x}\right) + \frac{\partial}{\partial y}\left(k\frac{\partial T}{\partial y}\right) \right] \delta T \, dx \, dy = 0 \qquad \text{(8-66)}$$

The two terms containing second derivatives may be rewritten with the help of the Green-Gauss theorem [see Eqs. (8-42)] if we note that

$$\beta = \delta T \qquad p_x = k\frac{\partial T}{\partial x} \qquad p_y = k\frac{\partial T}{\partial y} \qquad \text{(8-67)}$$

For example,

$$\int_A \frac{\partial}{\partial x}\left(k\frac{\partial T}{\partial x}\right)\delta T \, dx \, dy = \int_C k\frac{\partial T}{\partial x}n_x \, \delta T \, dC - \int_A \frac{\partial}{\partial x}(\delta T)k\frac{\partial T}{\partial x} \, dx \, dy \qquad \text{(8-68)}$$

and

$$\frac{\partial}{\partial x}(\delta T) = \delta\left(\frac{\partial T}{\partial x}\right) \qquad \text{(8-69)}$$

and

$$k\frac{\partial T}{\partial x}\frac{\partial}{\partial x}(\delta T) = k\frac{\partial T}{\partial x}\delta\left(\frac{\partial T}{\partial x}\right) = \tfrac{1}{2}k\delta\left(\frac{\partial T}{\partial x}\right)^2 \qquad \text{(8-70)}$$

A similar result is obtained for the other term. The final result is given by

$$\delta I = \int_A\left[Q\,\delta T - \tfrac{1}{2}k\delta\left(\frac{\partial T}{\partial x}\right)^2 - \tfrac{1}{2}k\delta\left(\frac{\partial T}{\partial y}\right)^2\right]dx\,dy$$

$$+ \int_C\left(k\frac{\partial T}{\partial x}n_x + k\frac{\partial T}{\partial y}n_y\right)\delta T \, dC = 0 \qquad \text{(8-71)}$$

This implies that I itself is given by

$$I = \int_A\left[QT - \tfrac{1}{2}k\left(\frac{\partial T}{\partial x}\right)^2 - \tfrac{1}{2}k\left(\frac{\partial T}{\partial y}\right)^2\right]dx\,dy - \int_C q_n T \, dC \qquad \text{(8-72)}$$

Integrations around the boundary C are always performed in a counterclockwise direction as shown in Figure 8-8. If we assume that an imposed heat flux q_{sB} acts on the boundary (transferring energy toward the two-dimensional region) as shown in Fig. 8-9, then an energy balance at this point on the boundary gives

$$I = \int_A\left[QT - \tfrac{1}{2}k\left(\frac{\partial T}{\partial x}\right)^2 - \tfrac{1}{2}k\left(\frac{\partial T}{\partial y}\right)^2\right]dx\,dy + \int_C q_{sB} T \, dC \qquad \text{(8-73)}$$

This result is used as a starting point in the next section in the finite element formulation to the problem. ∎

Example 8-11

Using the functional from Example 8-10, determine the so-called natural boundary condition for the problem described by the governing equation given by Eq. (8-43).

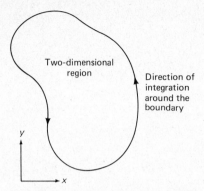

Figure 8-8 The boundary integrals for a two-dimensional region are evaluated by integrating in a counterclockwise direction as shown.

Solution

With the help of Eq. (8-65a), the natural boundary condition expressed in Eq. (8-61) becomes

$$-k\frac{\partial T}{\partial x}n_x - k\frac{\partial T}{\partial y}n_y = 0 \qquad \textbf{(8-74)}$$

From Eq. (8-48) it is concluded that the part of the boundary C over which Eq. (8-74) holds is insulated. The net heat flux q_n from conduction is zero in this case. ■

From Examples 8-10 and 8-11 and Eqs. (8-73) and (8-74) it is concluded that the variational principle corresponding to Eq. (8-43) is given by

$$I = \int_A \left[QT - \tfrac{1}{2}k\left(\frac{\partial T}{\partial x}\right)^2 - \tfrac{1}{2}k\left(\frac{\partial T}{\partial y}\right)^2 \right] dx\, dy \qquad \textbf{(8-75)}$$

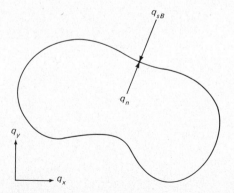

Figure 8-9 Two-dimensional region showing the heat fluxes from conduction and that which is imposed on a typical part of the global boundary.

providing the boundary conditions are either natural or geometric on various parts
of the global boundary.

The Finite Element Formulation

The functional I given by Eq. (8-73) holds over the entire area A. It is convenient
to perform the integrations on an element basis by noting that

$$I = \sum_{e=1}^{M} I^e \tag{8-76}$$

where M elements are assumed. Instead of extremizing the original integral I, we
take the derivative of I^e with respect to the nodal unknowns in the vector $\mathbf{a}^e$. Recall
that the parameter function for the temperature T may be written in terms of the
nodal temperatures by writing

$$T = \mathbf{N}\mathbf{a}^e \tag{8-77}$$

where $\mathbf{N}$ is the shape function matrix. For the three-node triangular element, for
example, $\mathbf{N}$ is given by

$$\mathbf{N} = [N_i \quad N_j \quad N_k] \tag{8-78}$$

where the shape functions themselves are given by Eqs. (6-21). In this case, $\mathbf{a}^e$
contains the three nodal temperatures or

$$\mathbf{a}^e = \begin{bmatrix} T_i \\ T_j \\ T_k \end{bmatrix} \tag{8-79}$$

Taking the derivative of I^e with respect to $\mathbf{a}^e$ and remembering that Eq. (4-119)
must be used, we have

$$\frac{dI^e}{d\mathbf{a}^e} = \int_{A^e} \left[Q\mathbf{N}^T - k\frac{\partial \mathbf{N}^T}{\partial x}\frac{\partial \mathbf{N}}{\partial x}\mathbf{a}^e - k\frac{\partial \mathbf{N}^T}{\partial y}\frac{\partial \mathbf{N}}{\partial y}\mathbf{a}^e \right] dx \, dy$$

$$+ \int_{C^e} q_{sB}\mathbf{N}^T \, dC = 0 \tag{8-80}$$

This last result may be written in the standard form

$$\mathbf{K}^e\mathbf{a}^e = \mathbf{f}^e \tag{8-81}$$

by defining

$$\mathbf{K}^e = \mathbf{K}^e_{xx} + \mathbf{K}^e_{yy} \tag{8-82}$$

and

$$\mathbf{f}^e = \mathbf{f}^e_Q + \mathbf{f}^e_{qB} \tag{8-83}$$

where in turn

$$\mathbf{K}_{xx}^e = \int_{A^e} \frac{\partial \mathbf{N}^T}{\partial x} k \frac{\partial \mathbf{N}}{\partial x} \, dx \, dy \qquad \text{(8-84a)}$$

$$\mathbf{K}_{yy}^e = \int_{A^e} \frac{\partial \mathbf{N}^T}{\partial y} k \frac{\partial \mathbf{N}}{\partial y} \, dx \, dy \qquad \text{(8-84b)}$$

$$\mathbf{f}_Q^e = \int_{A^e} \mathbf{N}^T Q \, dx \, dy \qquad \text{(8-85a)}$$

and

$$\mathbf{f}_{qB}^e = \int_{C^e} \mathbf{N}^T q_{sB} \, dC \qquad \text{(8-85b)}$$

In Sec. 8-8, several of these integrals are evaluated for the triangular element.

8-7 THE GALERKIN METHOD IN TWO DIMENSIONS

In direct contrast to the previous section, the Galerkin weighted-residual method is applied to Eq. (8-43) in a very straightforward manner. We begin the formulation on an element basis from the outset by forming the integral of the weighted residual and setting that result to zero where the weighting functions are the shape functions, or

$$\int_{A^e} \mathbf{N}^T \left[\frac{\partial}{\partial x} \left(k \frac{\partial T}{\partial x} \right) + \frac{\partial}{\partial y} \left(k \frac{\partial T}{\partial y} \right) + Q \right] dx \, dy = \mathbf{0} \qquad \text{(8-86)}$$

Applying the Green-Gauss theorem to the first two terms by noting

$$\beta = \mathbf{N}^T \qquad p_x = k \frac{\partial T}{\partial x} \qquad p_y = k \frac{\partial T}{\partial y}$$

we get

$$\int_{C^e} \mathbf{N}^T k \frac{\partial T}{\partial x} n_x \, dC - \int_{A^e} \frac{\partial \mathbf{N}^T}{\partial x} k \frac{\partial T}{\partial x} \, dx \, dy + \int_{C^e} \mathbf{N}^T k \frac{\partial T}{\partial y} n_y \, dC$$

$$- \int_{A^e} \frac{\partial \mathbf{N}^T}{\partial y} k \frac{\partial T}{\partial y} \, dx \, dy + \int_{A^e} \mathbf{N}^T Q \, dx \, dy = \mathbf{0}$$

Combining the two boundary integrals into one integral and using Eq. (8-48) gives

$$\int_{A^e} \frac{\partial \mathbf{N}^T}{\partial x} k \frac{\partial T}{\partial x} \, dx \, dy + \int_{A^e} \frac{\partial \mathbf{N}^T}{\partial y} k \frac{\partial T}{\partial y} \, dx \, dy = \int_{A^e} \mathbf{N}^T Q \, dx \, dy - \int_{C^e} \mathbf{N}^T q_n \, dC$$

If a heat flux q_{sB} is assumed to be imposed as shown in Fig. 8-9, we have

$$q_n = -q_{sB} \qquad \text{(8-87)}$$

from an energy balance at the point of application. Furthermore, using $T = \mathbf{N}\mathbf{a}^e$ (and noting that $\mathbf{a}^e$ is independent of x and y) gives

$$\mathbf{K}^e \mathbf{a}^e = \mathbf{f}^e \tag{8-88}$$

where

$$\mathbf{K}^e = \mathbf{K}_{xx}^e + \mathbf{K}_{yy}^e \tag{8-89}$$

and

$$\mathbf{f}^e = \mathbf{f}_Q^e + \mathbf{f}_{qB}^e \tag{8-90}$$

where in turn

$$\mathbf{K}_{xx}^e = \int_{A^e} \frac{\partial \mathbf{N}^T}{\partial x} k \frac{\partial \mathbf{N}}{\partial x} \, dx \, dy \tag{8-84a}$$

$$\mathbf{K}_{yy}^e = \int_{A^e} \frac{\partial \mathbf{N}^T}{\partial y} k \frac{\partial \mathbf{N}}{\partial y} \, dx \, dy \tag{8-84b}$$

$$\mathbf{f}_Q^e = \int_{A^e} \mathbf{N}^T Q \, dx \, dy \tag{8-85a}$$

$$\mathbf{f}_{qB}^e = \int_{C^e} \mathbf{N}^T q_{sB} \, dC \tag{8-85b}$$

These expressions are identical to those derived via the variational approach. Obviously, the Galerkin method requires significantly fewer steps. Several of these integrals are evaluated in the next section for the triangular element for which the shape function matrix $\mathbf{N}$ is given by

$$\mathbf{N} = [N_i \quad N_j \quad N_k]$$

where the shape functions themselves are given by Eqs. (6-21). The vector of nodal unknowns $\mathbf{a}^e$ is comprised of the three nodal temperatures in this case.

Since the Galerkin method is simple, the variational approach is abandoned in the remaining part of the chapter (and the book). The Galerkin method will be used exclusively from now on.

8-8   GENERAL TWO-DIMENSIONAL HEAT CONDUCTION

In this section the simple heat conduction problem defined in Sec. 8-5 is generalized as shown in Fig. 8-10. Note that the two-dimensional region is shown as a relatively thin plate of variable thickness. As in the simple model, a volumetric heat source is also to be included. Both lateral and boundary heat transfer effects are to be modeled including convection, simple radiation, and imposed heat fluxes. Simple radiation is defined to be radiation to or from a large enclosure. The imposed heat fluxes may act at a distance such as the sun and, therefore, may be present in addition to convection and radiation. Prescribed temperatures on the global boundary are also allowed.

The intent in this section is to derive the expressions for the element characteristics in general, i.e., without regard to the type of element. As mentioned in

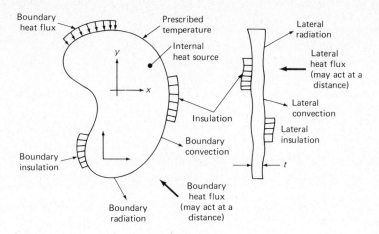

Figure 8-10 Schematic of general two-dimensional heat conduction problem.

Sec. 8-7, the Galerkin method will be used. Then several of these expressions will be evaluated by way of examples for the three-node triangular element. However, before these expressions can be derived, the governing equation that describes this general problem is needed.

The Governing Equation

The governing differential equation for the temperature T is derived quite easily if an energy balance is done on an elemental volume $t\, dx\, dy$, where t is the thickness of the plate. It can be shown that the governing equation is given by

$$\frac{\partial}{\partial x}\left(kt\frac{\partial T}{\partial x}\right) + \frac{\partial}{\partial y}\left(kt\frac{\partial T}{\partial y}\right) - h(T - T_a) - \varepsilon\sigma(T^4 - T_r^4) + q_s + Qt = 0 \quad \textbf{(8-91)}$$

where k is the thermal conductivity, h is the sum of the convective heat transfer coefficients on the two lateral faces, ε is the sum of the two surface emissivities on the two lateral faces, q_s is the sum of the heat fluxes imposed on the two lateral faces, Q is the heat generation per unit time and volume, T_a is the ambient temperature of the fluid in contact with the lateral faces, T_r is the temperature of the enclosure receiving the radiation from the lateral surfaces, and σ is the Stefan-Boltzmann constant (given in Sec. 8-3). Note that q_s is positive if it is directed as shown in Fig. 8-10.

As mentioned earlier, boundary convection, radiation, and heat fluxes are to be included also. These boundary conditions are considered in the next section where we derive the element characteristics. The reason for this is that it is far easier to apply these conditions on an element basis. Prescribed temperatures will also be handled quite routinely after the assemblage step.

It should be noted that if a particular heat transfer effect is negligible or not present, the corresponding term in Eq. (8-91) is simply omitted. For example, let us say that in a given problem we have negligible lateral radiation and no imposed lateral heat flux. In this case, we may taken ε and q_s both to be zero in a general computer program. It will be seen in the next section that this will eliminate the corresponding contributions to the element stiffness matrices and/or nodal force vectors.

The FEM Formulation

The expressions for the finite element characteristics may be derived without actually specifying the type of element at this point. However, later in this section the emphasis will be on the three-node triangular element from Sec. 6-4. Recall that in this case the shape function matrix $\mathbf{N}$ is given by

$$\mathbf{N} = [N_i \quad N_j \quad N_k] \tag{8-92}$$

where the shape functions themselves are given by Eqs. (6-21). A typical element e has nodes i, j, and k, specified in a counterclockwise order with temperatures T_i, T_j, and T_k. On an element basis, the Galerkin method requires

$$\int_{A^e} \mathbf{N}^T \left[\frac{\partial}{\partial x} \left(kt \frac{\partial T}{\partial x} \right) + \frac{\partial}{\partial y} \left(kt \frac{\partial T}{\partial y} \right) - h(T - T_a) \right.$$

$$\left. - \varepsilon\sigma(T^4 - T_r^4) + q_s + Qt \right] dx\, dy = 0 \tag{8-93}$$

It is emphasized that this integral applies to a typical element e and the integrations are to be performed over the area A^e of the element. Note that each term in Eq. (8-93) has units of energy per unit time. If the Green-Gauss theorem is applied to the two terms containing second-order derivatives, we get

$$\int_{C^e} \mathbf{N}^T kt \frac{\partial T}{\partial x} n_x\, dC - \int_{A^e} \frac{\partial \mathbf{N}^T}{\partial x} kt \frac{\partial T}{\partial x}\, dx\, dy$$

$$+ \int_{C^e} \mathbf{N}^T kt \frac{\partial T}{\partial y} n_y\, dC - \int_{A^e} \frac{\partial \mathbf{N}^T}{\partial y} kt \frac{\partial T}{\partial y}\, dx\, dy$$

$$- \int_{A^e} \mathbf{N}^T hT\, dx\, dy + \int_{A^e} \mathbf{N}^T hT_a\, dx\, dy$$

$$- \int_{A^e} \mathbf{N}^T \varepsilon\sigma T^4\, dx\, dy + \int_{A^e} \mathbf{N}^T \varepsilon\sigma T_r^4\, dx\, dy$$

$$+ \int_{A^e} \mathbf{N}^T q_s\, dx\, dy + \int_{A^e} \mathbf{N}^T Qt\, dx\, dy = 0 \tag{8-94}$$

However, the two integrals around the element boundary may be combined with the help of Eq. (8-48) to give

$$\int_{C^e} \mathbf{N}^T \left(k \frac{\partial T}{\partial x} n_x + k \frac{\partial T}{\partial y} n_y \right) t\, dC = \int_{C^e} \mathbf{N}^T (-q_n) t\, dC \tag{8-95}$$

where q_n represents the heat flux from conduction in the direction of the outward normal to the boundary as shown in Fig. 8-7.

At this point it is convenient to consider the nongeometric boundary conditions (i.e., all except the prescribed temperatures). As mentioned earlier, convection, radiation, and imposed heat fluxes are to be included. Perfect insulation is a special case of zero heat flux. Let us perform an energy balance (on an area basis) on the global boundary shown in Fig. 8-11 such that we may write

$$q_n + q_{sB} = q_{cvB} + q_{rB} \qquad \textbf{(8-96)}$$

where q_{cvB}, q_{rB}, and q_{sB} represent the convective, radiation, and imposed heat fluxes. Using this result to eliminate q_n in Eq. (8-95) gives

$$\int_{C_e} \mathbf{N}^T(-q_n)t\, dC = \int_{C_e} \mathbf{N}^T(q_{sB} - q_{cvB} - q_{rB})t\, dC \qquad \textbf{(8-97)}$$

But from Newton's law of cooling, we have

$$q_{cvB} = h_B(T - T_{aB}) \qquad \textbf{(8-98)}$$

and by application of the Stefan-Boltzmann law, we have

$$q_{rB} = \varepsilon_B \sigma(T^4 - T_{rB}^4) \qquad \textbf{(8-99)}$$

where h_B is the convective heat transfer coefficient on the element boundary, ε_B is the surface emissivity of the boundary, T_{aB} is the ambient fluid temperature, and T_{rB} is the temperature of the enclosure receiving the radiation from the boundary. Strictly speaking, the T in Eqs. (8-98) and (8-99) should be the temperature at the element boundary, but the shape function matrix in Eq. (8-95) will automatically take care of this as shown later.

With the help of Eqs. (8-98) and (8-99), we may write Eq. (8-97) as

$$\int_{C_e} \mathbf{N}^T(-q_n)t\, dC = \int_{C_e} \mathbf{N}^T q_{sB}t\, dC - \int_{C_e} \mathbf{N}^T h_B t T\, dC$$

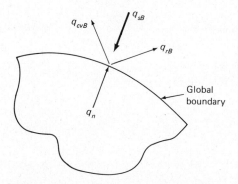

Figure 8-11 Typical part of the global boundary in the two-dimensional heat conduction problem showing the heat fluxes considered in the finite element formulation.

$$+ \int_{C^e} \mathbf{N}^T h_B t T_{aB} \ dC - \int_{C^e} \mathbf{N}^T \varepsilon_B \sigma t T^4 \ dC$$

$$+ \int_{C^e} \mathbf{N}^T \varepsilon_B \sigma t T_{rB}^4 \ dC \qquad (8\text{-}100)$$

Recalling that

$$T = \mathbf{N} \mathbf{a}^e \qquad (8\text{-}101)$$

where $\mathbf{N}$ is the shape function matrix and $\mathbf{a}^e$ is the vector of the nodal temperatures for element e. As in Sec. 8-3, it is convenient to write T^4 as follows:

$$T^4 = T^3 T = (\mathbf{N} \mathbf{a}^e)^3 \mathbf{N} \mathbf{a}^e \qquad (8\text{-}102)$$

It follows that Eq. (8-94) may be written as

$$\mathbf{K}^e \mathbf{a}^e = \mathbf{f}^e \qquad (8\text{-}103)$$

where

$$\mathbf{K}^e = \mathbf{K}_{xx}^e + \mathbf{K}_{yy}^e + \mathbf{K}_{cv}^e + \mathbf{K}_r^e + \mathbf{K}_{cvB}^e + \mathbf{K}_{rB}^e \qquad (8\text{-}104)$$

and

$$\mathbf{f}^e = \mathbf{f}_{cv}^e + \mathbf{f}_r^e + \mathbf{f}_q^e + \mathbf{f}_Q^e + \mathbf{f}_{cvB}^e + \mathbf{f}_{rB}^e + \mathbf{f}_{qB}^e \qquad (8\text{-}105)$$

The element stiffness matrices are in turn given by

$$\mathbf{K}_{xx}^e = \int_{A^e} \frac{\partial \mathbf{N}^T}{\partial x} kt \frac{\partial \mathbf{N}}{\partial x} \ dx \ dy \qquad (8\text{-}106a)$$

$$\mathbf{K}_{yy}^e = \int_{A^e} \frac{\partial \mathbf{N}^T}{\partial y} kt \frac{\partial \mathbf{N}}{\partial y} \ dx \ dy \qquad (8\text{-}106b)$$

$$\mathbf{K}_{cv}^e = \int_{A^e} \mathbf{N}^T h \mathbf{N} \ dx \ dy \qquad (8\text{-}106c)$$

$$\mathbf{K}_r^e = \int_{A^e} \mathbf{N}^T \varepsilon \sigma (\mathbf{N} \mathbf{a}^e)^3 \mathbf{N} \ dx \ dy \qquad (8\text{-}106d)$$

$$\mathbf{K}_{cvB}^e = \int_{C^e} \mathbf{N}^T h_B t \mathbf{N} \ dC \qquad (8\text{-}106e)$$

$$\mathbf{K}_{rB}^e = \int_{C^e} \mathbf{N}^T \varepsilon_B \sigma t (\mathbf{N} \mathbf{a}^e)^3 \mathbf{N} \ dC \qquad (8\text{-}106f)$$

and the element nodal force vectors by

$$\mathbf{f}_{cv}^e = \int_{A^e} \mathbf{N}^T h T_a \ dx \ dy \qquad (8\text{-}107a)$$

$$\mathbf{f}_r^e = \int_{A^e} \mathbf{N}^T \varepsilon \sigma T_r^4 \ dx \ dy \qquad (8\text{-}107b)$$

$$\mathbf{f}_q^e = \int_{A^e} \mathbf{N}^T q_s \ dx \ dy \qquad (8\text{-}107c)$$

$$\mathbf{f}_Q^e = \int_{A^e} \mathbf{N}^T Q t \ dx \ dy \tag{8-107d}$$

$$\mathbf{f}_{cvB}^e = \int_{C^e} \mathbf{N}^T h_B t T_{aB} \ dC \tag{8-107e}$$

$$\mathbf{f}_{rB}^e = \int_{C^e} \mathbf{N}^T \varepsilon_B \sigma t T_{rB}^4 \ dC \tag{8-107f}$$

$$\mathbf{f}_{qB}^e = \int_{C^e} \mathbf{N}^T q_{sB} t \ dC \tag{8-107g}$$

For the three-node triangular element, the stiffness matrices and nodal force vectors are each of sizes 3×3 and 3×1, respectively. It can be seen almost by inspection that each of these stiffness matrices is symmetric. Note the use of the subscript B on those terms that arise from the boundary conditions. In these cases, the integrations are to be performed around the boundaries of each element. However, if all legs of the element are internal (i.e., within the body) and not on the global boundary, the corresponding stiffness matrices (i.e., $\mathbf{K}_{cvB}^e$ and $\mathbf{K}_{rB}^e$) and the nodal force vectors (i.e., $\mathbf{f}_{cvB}^e$, $\mathbf{f}_{rB}^e$, and $\mathbf{f}_{qB}^e$) are simply taken to be null matrices and vectors, respectively. This is illustrated in Example 8-17.

Example 8-12

Evaluate the stiffness matrix from conduction in the x direction by using the three-node triangular element from Sec. 6-4.

Solution

Recall that the shape function matrix is given by Eq. (8-92) for the triangular element. From the definition of $\mathbf{K}_{xx}^e$ given by Eq. (8-106a) we see that the first derivatives of the shape functions are needed with respect to x. From Eqs. (6-21), we note that

$$\frac{\partial N_i}{\partial x} = m_{21} \qquad \frac{\partial N_j}{\partial x} = m_{22} \qquad \frac{\partial N_k}{\partial x} = m_{23}$$

Therefore, we may evaluate $\mathbf{K}_{xx}^e$ as follows:

$$\mathbf{K}_{xx}^e = \int_{A^e} \begin{bmatrix} m_{21} \\ m_{22} \\ m_{23} \end{bmatrix} kt[m_{21} \quad m_{22} \quad m_{23}] \ dx \ dy$$

$$= ktA \begin{bmatrix} m_{21}^2 & m_{21}m_{22} & m_{21}m_{23} \\ m_{22}m_{21} & m_{22}^2 & m_{22}m_{23} \\ m_{23}m_{21} & m_{23}m_{22} & m_{23}^2 \end{bmatrix} \tag{8-108}$$

The m_{ij}'s and area A are a function only of the nodal coordinates and may be computed with the help of Eqs. (6-21). In Eq. (8-108), the integrations were performed by treating k and t as constants. If these parameters are not constant, the above expression still holds, but k and t then represent the values at the element centroid. The accuracy of this approximation improves as the number of elements used in the discretization is increased. ∎

The reader may show in a completely analogous manner that the stiffness matrix from conduction in the y direction is given by

$$\mathbf{K}_{yy}^e = ktA \begin{bmatrix} m_{31}^2 & m_{31}m_{32} & m_{31}m_{33} \\ m_{32}m_{31} & m_{32}^2 & m_{32}m_{33} \\ m_{33}m_{31} & m_{33}m_{32} & m_{33}^2 \end{bmatrix} \tag{8-109}$$

for the triangular element. Note that both $\mathbf{K}_{xx}^e$ and $\mathbf{K}_{yy}^e$ are symmetric as expected.

Example 8-13

Evaluate the element stiffness matrix from lateral convection for the three-node triangular element.

Solution

From the definition of $\mathbf{K}_{cv}^e$ given by Eq. (8-106c), we see that the integrand contains the shape functions directly (not the derivatives). Therefore, it is more convenient to replace each shape function N_β with its corresponding *area coordinate* L_β from Eq. (6-25). The integrals may then be evaluated with the help of the special integration formula given by Eq. (6-49). From Eqs. (8-106c) and (6-25), we have

$$\mathbf{K}_{cv}^e = \int_{A^e} \begin{bmatrix} L_i \\ L_j \\ L_k \end{bmatrix} [L_i \quad L_j \quad L_k] h \, dx \, dy = \int_{A^e} \begin{bmatrix} L_i^2 & L_iL_j & L_iL_k \\ L_jL_i & L_j^2 & L_jL_k \\ L_kL_i & L_kL_j & L_k^2 \end{bmatrix} h \, dx \, dy$$

Using Eq. (8-49), we may evaluate a typical integral as follows:

$$\int_{A^e} hL_i^2 \, dx \, dy = h \int_{A^e} L_i^2 \, dx \, dy$$

$$= h \frac{2!0!0!}{(2+0+0+2)!} 2A = \tfrac{1}{6}hA$$

Another integral is evaluated as

$$\int_{A^e} hL_iL_j \, dx \, dy = h \int_{A^e} L_iL_j \, dx \, dy$$

$$= h \frac{1!1!0!}{(1+1+0+2)!} 2A = \tfrac{1}{12}hA$$

and so forth. Clearly $\mathbf{K}_{cv}^e$ must be given by

$$\mathbf{K}_{cv}^e = \frac{hA}{12} \begin{bmatrix} 2 & 1 & 1 \\ 1 & 2 & 1 \\ 1 & 1 & 2 \end{bmatrix} \tag{8-110}$$

Recall that h is the sum of the two convective heat transfer coefficients on the two lateral faces of the plate. Moreover, if h varies with x and y, the value of h at the centroid of element should be used in Eq. (8-110). ∎

Example 8-14

Try to evaluate the element stiffness matrix as a result of radiation from the lateral faces. Use the three-node triangular element.

Solution

Inspection of the expression for $\mathbf{K}_r^e$ given in Eq. (8-106d) reveals that $(\mathbf{N}\mathbf{a}^e)$ is needed. Using area coordinates for the shape functions and noting that

$$\mathbf{a}^e = [T_i \quad T_j \quad T_k]^T$$

we have

$$(\mathbf{N}\mathbf{a}^e)^3 = (L_i T_i + L_j T_j + L_k T_k)^3 \tag{8-111}$$

Since this is a scalar, we may write

$$\mathbf{K}_r^e = \int_{A^e} \varepsilon\sigma \begin{bmatrix} L_i \\ L_j \\ L_k \end{bmatrix} [L_i \quad L_j \quad L_k](L_i T_i + L_j T_j + L_k T_k)^3 \, dx \, dy \tag{8-112}$$

The indicated multiplications could be carried out and Eq. (6-49) used to integrate the various terms. However, this approach is not very practical, and a numerical integration should be performed. Such integration methods are covered in detail in Chapter 9. The unknown temperatures appear in this matrix and the resulting problem is nonlinear. The direct iteration method from Sec. 8-2 may be used in this case. ∎

Examples 8-12 to 8-14 have illustrated how the integrals are evaluated over the area of the element (i.e., over A^e). In the expressions for $\mathbf{K}_{cvB}^e$ and $\mathbf{K}_{rB}^e$, we note that the integrations are to be performed around the element boundaries (in a counterclockwise direction). However, there is no convection or radiation (assuming an opaque body) between two internal and adjacent elements. Therefore, these two matrices need to be evaluated only for those elements with at least one leg on the global boundary B as shown in Fig. 8-12 for the triangular element. Note that leg ij happens to be on the global boundary. The next example illustrates how the corresponding stiffness matrix from convection is evaluated in this case.

Example 8-15

Evaluate $\mathbf{K}_{cvB}^e$ for the element shown in Fig. 8-12 with leg ij on the global boundary.

Solution

From Eq. (8-106e), we note that the shape functions themselves are needed in the integrand. This generally means that area coordinates should be used to facilitate the evaluation of the integral in this two-dimensional application. Therefore, Eq. (8-106e) becomes

$$\mathbf{K}_{cvB}^e = \int_{C^e} \begin{bmatrix} L_i \\ L_j \\ L_k \end{bmatrix} h_B t [L_i \quad L_j \quad L_k] \, dC$$

But on leg ij, we have $L_k = 0$ and $dC = dl$, or

$$\mathbf{K}_{cvB}^e = \int_{l_{ij}} h_B t \begin{bmatrix} L_i^2 & L_i L_j & 0 \\ L_j L_i & L_j^2 & 0 \\ 0 & 0 & 0 \end{bmatrix} dl$$

Note that the integration is to be performed only on leg ij, the length of which is denoted as l_{ij}. The area coordinates have in effect degenerated to length coordinates. Equation (6-48) may now be used to perform the integrations. For example,

$$\int_{l_{ij}} h_B t L_i^2 \, dl = h_B t \frac{2!0!}{(2 + 0 + 1)!} l_{ij} = \frac{2 h_B t l_{ij}}{6}$$

and

$$\int_{l_{ij}} h_B t L_i L_j \, dl = h_B t \frac{1!1!}{(1 + 1 + 1)!} l_{ij} = \frac{h_B t l_{ij}}{6}$$

The complete result is given by

$$\mathbf{K}_{cvB}^e = \frac{h_B t l_{ij}}{6} \begin{bmatrix} 2 & 1 & 0 \\ 1 & 2 & 0 \\ 0 & 0 & 0 \end{bmatrix} \quad \begin{array}{l} \text{for leg } ij \text{ on the} \\ \text{global boundary} \end{array} \qquad \textbf{(8-113a)}$$

■

The results from Example 8-15 can be generalized. If nodes j and k are on the global boundary, we have

$$\mathbf{K}_{cvB}^e = \frac{h_B t l_{jk}}{6} \begin{bmatrix} 0 & 0 & 0 \\ 0 & 2 & 1 \\ 0 & 1 & 2 \end{bmatrix} \quad \begin{array}{l} \text{for leg } jk \text{ on the} \\ \text{global boundary} \end{array} \qquad \textbf{(8-113b)}$$

whereas if nodes k and i are on the global boundary, we have

$$\mathbf{K}_{cvB}^e = \frac{h_B t l_{ki}}{6} \begin{bmatrix} 2 & 0 & 1 \\ 0 & 0 & 0 \\ 1 & 0 & 2 \end{bmatrix} \quad \begin{array}{l} \text{for leg } ki \text{ on the} \\ \text{global boundary} \end{array} \qquad \textbf{(8-113c)}$$

where l_{jk} and l_{ki} denote the lengths of legs jk and ki, respectively. Strictly speaking, the thickness t in Eqs. (8-113) represents the average thickness of the element along the leg in question.

The reader should try to show that the stiffness matrix given by Eq. (8-106f) may be written as

$$\mathbf{K}_{rB}^e = \int_{l_{ij}} \varepsilon_B \sigma t \begin{bmatrix} L_i \\ L_j \\ 0 \end{bmatrix} [L_i \quad L_j \quad 0](L_i T_i + L_j T_j)^3 \, dl \tag{8-114}$$

for the element shown in Fig. 8-12. This integral may be evaluated with the help of Eq. (6-48). It may also be evaluated numerically using one of the techniques presented in Chapter 9.

Let us now turn to the nodal force vectors given by Eqs. (8-107). Note that $\mathbf{f}_{cv}^e$, $\mathbf{f}_r^e$, $\mathbf{f}_q^e$, and $\mathbf{f}_Q^e$ are to be evaluated by integrating over the element area A^e, whereas $\mathbf{f}_{cvB}^e$, $\mathbf{f}_{rB}^e$, and $\mathbf{f}_{qB}^e$ are to be evaluated on the element boundaries. The former are readily evaluated if area coordinates are used with the following results:

$$\mathbf{f}_{cv}^e = \frac{hAT_a}{3} \begin{bmatrix} 1 \\ 1 \\ 1 \end{bmatrix} \tag{8-115}$$

$$\mathbf{f}_r^e = \frac{\varepsilon \sigma A T_r^4}{3} \begin{bmatrix} 1 \\ 1 \\ 1 \end{bmatrix} \tag{8-116}$$

$$\mathbf{f}_q^e = \frac{q_s A}{3} \begin{bmatrix} 1 \\ 1 \\ 1 \end{bmatrix} \tag{8-117}$$

$$\mathbf{f}_Q^e = \frac{QAt}{3} \begin{bmatrix} 1 \\ 1 \\ 1 \end{bmatrix} \tag{8-118}$$

The result for $\mathbf{f}_Q^e$ in Eq. (8-118) deserves special attention. This nodal force vector results from a volumetric heat source, generally considered to be distributed

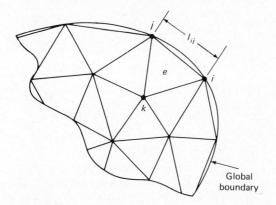

Figure 8-12 Typical element with at least one leg of the triangular element on the global boundary.

within the body. If Q is not uniform, then the value at the element centroid is to be used in Eq. (8-118). Let us derive an expression for the nodal force vector if Q happens to represent a point source at (x_0, y_0). With the help of the δ-function defined by Eq. (4-85), we may represent the nodal force vector for the point source as

$$\mathbf{f}_Q^e = \int_{A^e} \begin{bmatrix} N_i \\ N_j \\ N_k \end{bmatrix} Q't \, \delta(x - x_0) \, \delta(y - y_0) \, dx \, dy \qquad \textbf{(8-119)}$$

where Q' is the energy generation per unit time and *per unit thickness*. From the definition of the δ function, the above integral evaluates to zero everywhere within the element except at (x_0, y_0). Therefore, it follows that $\mathbf{f}_Q^e$ is given by

$$\mathbf{f}_Q^e = Q't \begin{bmatrix} N_i(x_0, y_0) \\ N_j(x_0, y_0) \\ N_k(x_0, y_0) \end{bmatrix} \qquad \textbf{(8-120)}$$

If more than one point source is present in the element, additional nodal force vectors similar to Eq. (8-120) would simply be included in the analysis.

From a practical point of view, it is best to situate a node at each point source. Not only does this give more accurate results, but it also allows us to consider the point sources after the assemblage is complete. For example, if a point source of Q' is located at *global node I*, then $Q't$ is simply *added* to the Ith position in the assemblage nodal force vector. If the point source is within an element (not at a node), we must use Eq. (8-120) at the element level. Section 8-14 discusses one method of implementing Eq. (8-120) in a computer program.

Let us now turn to the nodal force vectors that are to be evaluated on the element boundary. The results are

$$\mathbf{f}_{cvB}^e = \frac{h_B t l_{ij} T_{aB}}{2} \begin{bmatrix} 1 \\ 1 \\ 0 \end{bmatrix} \qquad \begin{array}{l} \text{for leg } ij \text{ on the} \\ \text{global boundary} \end{array} \qquad \textbf{(8-121)}$$

$$\mathbf{f}_{rB}^e = \frac{\varepsilon_B \sigma t l_{ij} T_{rB}^4}{2} \begin{bmatrix} 1 \\ 1 \\ 0 \end{bmatrix} \qquad \begin{array}{l} \text{for leg } ij \text{ on the} \\ \text{global boundary} \end{array} \qquad \textbf{(8-122)}$$

$$\mathbf{f}_{qB}^e = \frac{q_{sB} t l_{ij}}{2} \begin{bmatrix} 1 \\ 1 \\ 0 \end{bmatrix} \qquad \begin{array}{l} \text{for leg } ij \text{ on the} \\ \text{global boundary} \end{array} \qquad \textbf{(8-123)}$$

Similar expressions may be set up by inspection if leg jk or ki is on the global boundary.

The assemblage of the element stiffness matrices and nodal force vectors is routine (see Example 8-17). The prescribed temperature boundary conditions, if any exist, are applied with either of the two methods illustrated in Sec. 3-2. It has been noted that if radiation heat transfer is included, the problem becomes nonlinear. The problem is nonlinear also if temperature-dependent properties are to be used.

In these cases, the direct iteration method from Sec. 8-2 may be used to solve the system of equations for the nodal temperatures.

The Element Resultants

The heat fluxes q_x and q_y in the x and y directions, respectively, may be computed for every element by applying Fourier's law of heat conduction and writing

$$q_x = -k\frac{\partial T}{\partial x} = -k\frac{\partial \mathbf{N}}{\partial x}\mathbf{a}^e \qquad \text{(8-124a)}$$

and

$$q_y = -k\frac{\partial T}{\partial y} = -k\frac{\partial \mathbf{N}}{\partial y}\mathbf{a}^e \qquad \text{(8-124b)}$$

If the triangular element is used, we get

$$\bar{q}_x = -k(m_{21}T_i + m_{22}T_j + m_{23}T_k) \qquad \text{(8-125a)}$$

and

$$\bar{q}_y = -k(m_{31}T_i + m_{32}T_j + m_{33}T_k) \qquad \text{(8-125b)}$$

where the m_{ij}'s are calculated by Eqs. (6-21) and T_i, T_j, and T_k represent the nodal temperatures. Recall that these temperatures are known from the solution of $\mathbf{Ka} = \mathbf{f}$. Note that a bar ($^-$) is used on q_x and q_y to indicate that these are actually average heat fluxes for the element. This is a consequence of the fact that the shape functions are linear (and the derivatives are therefore constants). These average heat fluxes are usually associated with the centroid of the element.

Example 8-16

Consider the element shown in Fig. 8-13. Note that a point heat source with a strength of 500 Btu/hr per inch of thickness is located at $x_0 = 1$ in. and $y_0 = 0.2$ in. The nodal coordinates are shown in Fig. 8-13. The plate thickness is 0.5 in. The element is defined as having nodes 1, 3, and 4, in this order.

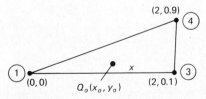

Figure 8-13   Element in Example 8-16 with the nodal coordinates shown in inches and a point heat source (denoted as x) at (1, 0.2).

Solution

From Eq. (8-120) we see that the shape functions need to be evaluated at (x_0, y_0). The shape functions for the triangular element are given by Eqs. (6-21). Each of the m_{ij}'s needs to be computed with the help of Eq. (6-21) with the results summarized as follows:

$$m_{11} = 1.0 \qquad m_{12} = 0. \qquad m_{13} = 0.$$

$$m_{21} = -0.50 \qquad m_{22} = 0.5625 \qquad m_{23} = -0.0625$$

$$m_{31} = 0. \qquad m_{32} = -1.25 \qquad m_{33} = 1.25$$

In these computations the area A of the element was needed and was computed to be 0.8 in.^{2} from Eq. (6-21e). The shape functions may now be evaluated at the location of the heat source as follows:

$$N_i = m_{11} + m_{21}x_0 + m_{31}y_0$$
$$= 1.0 + (-0.50)(1.0) + (0.00)(0.2) = 0.5000$$
$$N_j = m_{12} + m_{22}x_0 + m_{32}y_0$$
$$= 0.0 + (0.5625)(1.0) + (-1.25)(0.2) = 0.3125$$
$$N_k = m_{13} + m_{23}x_0 + m_{33}y_0$$
$$= 0.0 + (-0.0625)(1.0) + (1.25)(0.2) = 0.1875$$

Note that the sum of the shape functions evaluated at (x_0, y_0) is unity. From Eq. (8-120) we compute $\mathbf{f}_Q^e$ as follows:

$$\mathbf{f}_Q^e = (500)(0.5) \begin{bmatrix} 0.5000 \\ 0.3125 \\ 0.1875 \end{bmatrix} = \begin{bmatrix} 125 \\ 78 \\ 47 \end{bmatrix} \text{ Btu/hr}$$

Note that nodes 1 and 3 are closer to the source than node 4 and, therefore, most of the source is automatically allocated to nodes 1 and 3. It should also be noted that the sum of the values allocated to each node is 250 Btu/hr, the total amount available. ∎

Example 8-17

The temperature distribution is needed in the device shown in Fig. 8-14(a). This device is very long in the direction normal to the plane of the paper. The upper and lower surfaces convect to fluids at 36 and 24°C, through convective heat transfer coefficients of 500 and 1000 W/m²-°C, respectively. The outside boundary is held at a fixed temperature of 35°C. The device is comprised of two materials that are fused together at $x = \pm 3$ cm. The materials are bronze and aluminum with thermal conductivities of 52 and 186 W/m-°C, respectively. A volumetric heat source with a strength of 5 W/cm³ is present in the bronze only. Determine the temperature distribution within and the heat fluxes through the device. Neglect the effects of thermal radiation.

Solution

Because of the symmetry about the centerline, only one-half of the region needs to be modeled as shown in Fig. 8-14(b). The boundary condition on this line of symmetry is equivalent to that of perfect insulation (i.e., the natural boundary condition). A problem such as this one is usually analyzed with hundreds of elements. However, in order to show each step in the finite element solution process,

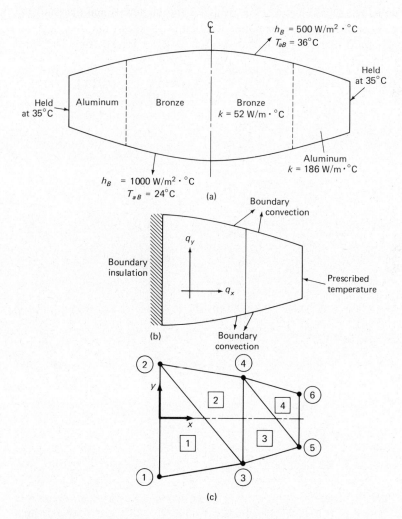

Figure 8-14 (a) Schematic of device in Example 8-17. (b) Region to be analyzed with heat transfer mechanisms shown. (c) Discretization of the two-dimensional region.

let us use only four triangular elements as shown in Fig. 8-14(c). The nodal co-ordinates are obtained from a scaled drawing of the cross section and are summarized in Table 8-1 along with the element definitions. The bandwidth of the assemblage stiffness matrix is minimized by numbering the nodes as shown. Note that two material set flags are used, one for each material. The element stiffness matrices and nodal force vectors are calculated as shown below. The assemblage stiffness matrix and assemblage nodal force vector are also determined after each element is processed.

Element 1

Element 1 is defined as having nodes 1, 3, and 2 (in this order) and material set 1. The nodal coordinates are summarized in Table 8-1. Recall that the element stiffness matrix for conduction in the x direction is given by Eq. (8-108). The area of the element is computed from Eq. (6-21e) as follows:

$$A = \tfrac{1}{2} \det \begin{bmatrix} 1 & x_1 & y_1 \\ 1 & x_3 & y_3 \\ 1 & x_2 & y_2 \end{bmatrix} = \tfrac{1}{2} \det \begin{bmatrix} 1 & 0.000 & -0.020 \\ 1 & 0.030 & -0.017 \\ 1 & 0.000 & 0.020 \end{bmatrix}$$

or

$$A = 0.00060 \text{ m}^2$$

The three m_{ij}'s that are needed are computed from Eq. (6-21d) as follows:

$$m_{21} = \frac{y_j - y_k}{2A} = \frac{y_3 - y_2}{2A}$$

Table 8-1 Node and Element Data for Example 8-17

Node number	x, cm	y, cm
1	0.0	−2.0
2	0.0	2.0
3	3.0	−1.7
4	3.0	1.7
5	5.0	−1.0
6	5.0	1.0

Element number	Nodes			Material set number
	i	j	k	
1	1	3	2	1
2	2	3	4	1
3	3	5	4	2
4	5	6	4	2

$$= \frac{-0.017 - 0.020}{2(0.00060)} = -30.833 \text{ m}^{-1}$$

$$m_{22} = \frac{y_k - y_i}{2A} = \frac{y_2 - y_1}{2A}$$

$$= \frac{0.020 - (-0.020)}{2(0.00060)} = 33.333 \text{ m}^{-1}$$

$$m_{23} = \frac{y_i - y_j}{2A} = \frac{y_1 - y_3}{2A}$$

$$= \frac{-0.020 - (-.017)}{2(0.00060)} = -2.500 \text{ m}^{-1}$$

The thermal conductivity k for the material comprising this element is 52 W/m-°C. Therefore, for $\mathbf{K}_{xx}^e$ we get

$$\mathbf{K}_{xx}^{(1)} = (52.)(1.0)(0.0006) \begin{bmatrix} (-30.833)^2 & (-30.833)(33.333) & (-30.833)(-2.500) \\ (33.333)(-30.833) & (33.333)^2 & (33.333)(-2.500) \\ (-2.500)(-30.833) & (-2.500)(33.333) & (-2.500)^2 \end{bmatrix}$$

or

$$\mathbf{K}_{xx}^{(1)} = \begin{bmatrix} 29.662 & -32.067 & 2.405 \\ -32.067 & 34.667 & -2.600 \\ 2.405 & -2.600 & 0.195 \end{bmatrix} \text{ W/°C}$$

Note that the thickness is taken to be 1 m since the body is very long in the direction perpendicular to the two-dimensional region being analyzed. Any value for the thickness t may be used as long as the same value is used in each element. Similarly, in the y direction we need

$$m_{31} = \frac{x_k - x_j}{2A} = \frac{x_2 - x_3}{2A} = \frac{0.0 - 0.030}{2(0.00060)}$$

$$= -25.000 \text{ m}^{-1}$$

$$m_{32} = \frac{x_i - x_k}{2A} = \frac{x_1 - x_2}{2A} = \frac{0.0 - 0.0}{2(0.00060)}$$

$$= 0.$$

$$m_{33} = \frac{x_j - x_i}{2A} = \frac{x_3 - x_1}{2A} = \frac{0.030 - 0.0}{2(0.00060)}$$

$$= 25.000 \text{ m}^{-1}$$

The element stiffness matrix for conduction in the y direction is calculated from Eq. (8-109) as follows:

$$\mathbf{K}_{yy}^{(1)} = (52.)(1.0)(0.0006) \begin{bmatrix} (-25.0)^2 & (-25.0)(0.0) & (-25.0)(25.0) \\ (0.0)(-25.0) & (0.0)^2 & (0.0)(25.0) \\ (25.0)(-25.0) & (25.0)(0.0) & (25.0)^2 \end{bmatrix}$$

or

$$\mathbf{K}_{yy}^{(1)} = \begin{bmatrix} 19.500 & 0.000 & -19.500 \\ 0.000 & 0.000 & 0.000 \\ -19.500 & 0.000 & 19.500 \end{bmatrix} \text{W/°C}$$

Leg ij or 1–3 is on the global boundary and undergoes convection to a fluid at a temperature T_{aB} of 24°C. The convective heat transfer coefficient h_B is 1000 W/m²-°C. From Eq. (8-113a), we see that we also need the length of leg 1–3. This length is easily computed from

$$l_{ij} = \sqrt{(x_j - x_i)^2 + (y_j - y_i)^2} = \sqrt{(x_3 - x_1)^2 + (y_3 - y_1)^2}$$
$$= \sqrt{(0.030 - 0.0)^2 + [-0.017 - (-0.020)]^2}$$

or

$$l_{ij} = 0.03015 \text{ m}$$

The element stiffness matrix from convection from leg ij is calculated as follows:

$$\mathbf{K}_{cvB}^{(1)} = \frac{(1000.)(1.0)(0.03015)}{6} \begin{bmatrix} 2 & 1 & 0 \\ 1 & 2 & 0 \\ 0 & 0 & 0 \end{bmatrix}$$

or

$$\mathbf{K}_{cvB}^{(1)} = \begin{bmatrix} 10.050 & 5.025 & 0.000 \\ 5.025 & 10.050 & 0.000 \\ 0.000 & 0.000 & 0.000 \end{bmatrix} \text{W/°C}$$

There are no other contributions to the element stiffness matrix because the boundary radiation is neglected and the lateral effects are zero (because a two-dimensional slice out of a long body is being analyzed). Therefore, the composite stiffness matrix for element 1 is given by Eq. (8-104) or, in this case, as

$$\mathbf{K}^{(1)} = \mathbf{K}_{xx}^{(1)} + \mathbf{K}_{yy}^{(1)} + \mathbf{K}_{cvB}^{(1)}$$

or

$$\mathbf{K}^{(1)} = \begin{bmatrix} 59.212 & -27.042 & -17.095 \\ -27.042 & 44.717 & -2.600 \\ -17.095 & -2.600 & 19.695 \end{bmatrix}$$

The assemblage stiffness matrix after processing one element is given by

$$\mathbf{K}^a = \begin{bmatrix} 59.212 & -17.095 & -27.042 & 0.000 & 0.000 & 0.000 \\ -17.095 & 19.695 & -2.600 & 0.000 & 0.000 & 0.000 \\ -27.042 & -2.600 & 44.717 & 0.000 & 0.000 & 0.000 \\ 0.000 & 0.000 & 0.000 & 0.000 & 0.000 & 0.000 \\ 0.000 & 0.000 & 0.000 & 0.000 & 0.000 & 0.000 \\ 0.000 & 0.000 & 0.000 & 0.000 & 0.000 & 0.000 \end{bmatrix}$$

Recall that $\mathbf{K}^a$ is zeroed out before adding the results from the first element.

The element nodal force vectors are now calculated as follows. The volumetric heat generation Q of 5 W/cm^3 or 5×10^6 W/m^3 is present in element 1. From Eq. (8-118) we get

$$\mathbf{f}_Q^{(1)} = \frac{(5 \times 10^6)(1.0)(0.0006)}{3}\begin{bmatrix} 1 \\ 1 \\ 1 \end{bmatrix} = \begin{bmatrix} 1000. \\ 1000. \\ 1000. \end{bmatrix} W$$

Leg ij or 1–3 undergoes convection and the corresponding nodal force vector is given by Eq. (8-121) or

$$\mathbf{f}_{cvB}^{(1)} = \frac{(1000)(1.0)(0.03015)(24)}{2}\begin{bmatrix} 1 \\ 1 \\ 0 \end{bmatrix} = \begin{bmatrix} 361.8 \\ 361.8 \\ 0.0 \end{bmatrix} W$$

All other contributions to the composite nodal force vector are zero, and from Eq. (8-105) we have

$$\mathbf{f}^{(1)} = \mathbf{f}_Q^{(1)} + \mathbf{f}_{cvB}^{(1)}$$

or

$$\mathbf{f}^{(1)} = \begin{bmatrix} 1361.8 \\ 1361.8 \\ 1000.0 \end{bmatrix} W$$

The assemblage nodal force vector after processing one element is given by

$$\mathbf{f}^a = \begin{bmatrix} 1361.8 \\ 1000.0 \\ 1361.8 \\ 0.0 \\ 0.0 \\ 0.0 \end{bmatrix} W$$

Element 2

Element 2 is defined as having nodes 2, 3, and 4 (in this order) and material set 1 with $k = 52$ W/m-°C and $Q = 5 \times 10^6$ W/m^3. Moreover, leg ki or 4–2 is on the global boundary and undergoes convection with $h_B = 500$ W/m^2-°C and $T_{aB} = 36$°C. The results are summarized below.

$$A = \frac{1}{2}\det\begin{bmatrix} 1 & x_2 & y_2 \\ 1 & x_3 & y_3 \\ 1 & x_4 & y_4 \end{bmatrix}$$

$$= \frac{1}{2}\det\begin{bmatrix} 1 & 0.000 & 0.020 \\ 1 & 0.030 & -0.017 \\ 1 & 0.030 & 0.017 \end{bmatrix} = 0.00051 \text{ m}^2$$

or

$$m_{21} = -33.333 \qquad m_{22} = -2.941 \qquad m_{23} = 36.274$$

$$m_{31} = 0.000 \qquad m_{32} = -29.412 \qquad m_{33} = 29.412$$

$$\mathbf{K}_{xx}^{(2)} = \begin{bmatrix} 29.467 & 2.600 & -32.067 \\ 2.600 & 0.229 & -2.829 \\ -32.067 & -2.829 & 34.896 \end{bmatrix}$$

$$\mathbf{K}_{yy}^{(2)} = \begin{bmatrix} 0.000 & 0.000 & 0.000 \\ 0.000 & 22.941 & -22.941 \\ 0.000 & -22.941 & 22.941 \end{bmatrix}$$

$$l_{ki} = 0.0315 \text{ m}$$

$$\mathbf{K}_{cvB}^{(2)} = \begin{bmatrix} 5.025 & 0.000 & 2.512 \\ 0.000 & 0.000 & 0.000 \\ 2.512 & 0.000 & 5.025 \end{bmatrix}$$

$$\mathbf{K}^{(2)} = \mathbf{K}_{xx}^{(2)} + \mathbf{K}_{yy}^{(2)} + \mathbf{K}_{cvB}^{(2)}$$

$$= \begin{bmatrix} 34.492 & 2.600 & -29.554 \\ 2.600 & 23.171 & -25.771 \\ -29.554 & -25.771 & 62.862 \end{bmatrix}$$

$$\mathbf{K}^a = \begin{bmatrix} 59.212 & -17.095 & -27.042 & 0.000 & 0.000 & 0.000 \\ -17.095 & 54.187 & 0.000 & -29.554 & 0.000 & 0.000 \\ -27.042 & 0.000 & 67.887 & -25.771 & 0.000 & 0.000 \\ 0.000 & -29.554 & -25.771 & 62.862 & 0.000 & 0.000 \\ 0.000 & 0.000 & 0.000 & 0.000 & 0.000 & 0.000 \\ 0.000 & 0.000 & 0.000 & 0.000 & 0.000 & 0.000 \end{bmatrix}$$

$$\mathbf{f}_Q^{(2)} = \begin{bmatrix} 850.0 \\ 850.0 \\ 850.0 \end{bmatrix}$$

$$\mathbf{f}_{cvB}^{(2)} = \begin{bmatrix} 271.3 \\ 0.0 \\ 271.3 \end{bmatrix}$$

$$\mathbf{f}^{(2)} = \mathbf{f}_Q^{(2)} + \mathbf{f}_{cvB}^{(2)} = \begin{bmatrix} 1121.3 \\ 850.0 \\ 1121.3 \end{bmatrix}$$

$$\mathbf{f}^a = \begin{bmatrix} 1361.8 \\ 2121.3 \\ 2211.8 \\ 1121.3 \\ 0.0 \\ 0.0 \end{bmatrix}$$

Element 3

Element 3 is defined as having nodes 3, 5, and 4 (in this order) and material set 2 with $k = 186$ W/m-°C and $Q = 0$. Leg ij or 3–5 is on the global boundary and undergoes convection with $h_B = 1000$ W/m²-°C and $T_{aB} = 24$°C. The results are summarized below.

$$A = 0.00034 \text{ m}^3$$

$$m_{21} = -39.706 \qquad m_{22} = 50.000 \qquad m_{23} = -10.294$$

$$m_{31} = -29.412 \qquad m_{32} = 0.000 \qquad m_{33} = 29.412$$

$$\mathbf{K}_{xx}^{(3)} = \begin{bmatrix} 99.701 & -125.550 & 25.849 \\ -125.550 & 158.100 & -32.550 \\ 25.849 & -32.550 & 6.701 \end{bmatrix}$$

$$\mathbf{K}_{yy}^{(3)} = \begin{bmatrix} 54.706 & 0.000 & -54.706 \\ 0.000 & 0.000 & 0.000 \\ -54.706 & 0.000 & 54.706 \end{bmatrix}$$

$$l_{ij} = 0.02119 \text{ m}$$

$$\mathbf{K}_{cvB}^{(3)} = \begin{bmatrix} 7.063 & 3.532 & 0.000 \\ 3.532 & 7.063 & 0.000 \\ 0.000 & 0.000 & 0.000 \end{bmatrix}$$

$$\mathbf{K}^{(3)} = \mathbf{K}_{xx}^{(3)} + \mathbf{K}_{yy}^{(3)} + \mathbf{K}_{cvB}^{(3)}$$

$$= \begin{bmatrix} 161.471 & -122.018 & -28.857 \\ -122.018 & 165.163 & -32.550 \\ -28.857 & -32.550 & 61.407 \end{bmatrix}$$

$$\mathbf{K}^a = \begin{bmatrix} 59.212 & -17.095 & -27.042 & 0.000 & 0.000 & 0.000 \\ -17.095 & 54.187 & 0.000 & -29.554 & 0.000 & 0.000 \\ -27.042 & 0.000 & 229.358 & -54.628 & -122.018 & 0.000 \\ 0.000 & -29.554 & -54.628 & 124.270 & -32.550 & 0.000 \\ 0.000 & 0.000 & -122.018 & -32.550 & 165.163 & 0.000 \\ 0.000 & 0.000 & 0.000 & 0.000 & 0.000 & 0.000 \end{bmatrix}$$

$$\mathbf{f}_Q^{(3)} = \begin{bmatrix} 0 \\ 0 \\ 0 \end{bmatrix}$$

$$\mathbf{f}_{cvB}^{(3)} = \begin{bmatrix} 254.3 \\ 254.3 \\ 0.0 \end{bmatrix}$$

$$\mathbf{f}^{(3)} = \mathbf{f}_Q^{(3)} + \mathbf{f}_{cvB}^{(3)} = \begin{bmatrix} 254.3 \\ 254.3 \\ 0.0 \end{bmatrix}$$

$$\mathbf{f}^a = \begin{bmatrix} 1361.8 \\ 2121.3 \\ 2466.1 \\ 1121.3 \\ 254.3 \\ 0.0 \end{bmatrix}$$

Element 4

Element 4 is defined as having nodes 5, 6, and 4 (in this order) and material set 2 with $k = 186$ W/m-°C and $Q = 0$. Leg jk or 6–4 is on the global boundary and undergoes convection with $h_B = 500$ W/m²-°C and $T_{aB} = 36$°C. The results are summarized below.

$$A = 0.00020 \text{ m}^2$$

$$m_{21} = -17.500 \qquad m_{22} = 67.500 \qquad m_{23} = -50.000$$

$$m_{31} = -50.000 \qquad m_{32} = 50.000 \qquad m_{33} = 0.000$$

$$\mathbf{K}_{xx}^{(4)} = \begin{bmatrix} 11.393 & -43.943 & 32.550 \\ -43.943 & 169.492 & -125.550 \\ 32.550 & -125.550 & 93.000 \end{bmatrix}$$

$$\mathbf{K}_{yy}^{(4)} = \begin{bmatrix} 93.000 & -93.000 & 0.000 \\ -93.000 & 93.000 & 0.000 \\ 0.000 & 0.000 & 0.000 \end{bmatrix}$$

$$l_{jk} = 0.02119 \text{ m}$$

$$\mathbf{K}_{cvB}^{(4)} = \begin{bmatrix} 0.000 & 0.000 & 0.000 \\ 0.000 & 3.532 & 1.766 \\ 0.000 & 1.766 & 3.532 \end{bmatrix}$$

$$\mathbf{K}^{(4)} = \mathbf{K}_{xx}^{(4)} + \mathbf{K}_{yy}^{(4)} + \mathbf{K}_{cvB}^{(4)}$$

$$= \begin{bmatrix} 104.392 & -136.942 & 32.550 \\ -136.942 & 266.024 & -123.784 \\ 32.550 & -123.784 & 96.532 \end{bmatrix}$$

$$\mathbf{K}^a = \begin{bmatrix} 59.212 & -17.095 & -27.042 & 0.000 & 0.000 & 0.000 \\ -17.095 & 54.187 & 0.000 & -29.554 & 0.000 & 0.000 \\ -27.042 & 0.000 & 229.358 & -54.628 & -122.018 & 0.000 \\ 0.000 & -29.554 & -54.628 & 220.801 & 0.000 & -123.784 \\ 0.000 & 0.000 & -122.018 & 0.000 & 269.556 & -136.942 \\ 0.000 & 0.000 & 0.000 & -123.784 & -136.942 & 266.024 \end{bmatrix}$$

$$\mathbf{f}_Q^{(4)} = \begin{bmatrix} 0 \\ 0 \\ 0 \end{bmatrix}$$

$$\mathbf{f}_{cvB}^{(4)} = \begin{bmatrix} 0.0 \\ 190.7 \\ 190.7 \end{bmatrix}$$

$$\mathbf{f}^{(4)} = \mathbf{f}_Q^{(4)} + \mathbf{f}_{cvB}^{(4)} = \begin{bmatrix} 0.0 \\ 190.7 \\ 190.7 \end{bmatrix}$$

$$\mathbf{f}^a = \begin{bmatrix} 1361.8 \\ 2121.3 \\ 2466.1 \\ 1312.0 \\ 254.3 \\ 190.7 \end{bmatrix}$$

Before applying the prescribed temperature boundary conditions on nodes 5 and 6, we have $\mathbf{K}^a\mathbf{a} = \mathbf{f}^a$, or

$$\begin{bmatrix} 59.212 & -17.095 & -27.042 & 0.000 & 0.000 & 0.000 \\ -17.095 & 54.187 & 0.000 & -29.554 & 0.000 & 0.000 \\ -27.042 & 0.000 & 229.358 & -54.628 & -122.018 & 0.000 \\ 0.000 & -29.554 & -54.628 & 220.801 & 0.000 & -123.784 \\ 0.000 & 0.000 & -122.018 & 0.000 & 269.556 & -136.942 \\ 0.000 & 0.000 & 0.000 & -123.784 & -136.942 & 266.024 \end{bmatrix} \begin{bmatrix} T_1 \\ T_2 \\ T_3 \\ T_4 \\ T_5 \\ T_6 \end{bmatrix}$$

$$= \begin{bmatrix} 1361.8 \\ 2121.3 \\ 2466.1 \\ 1312.0 \\ 254.3 \\ 190.7 \end{bmatrix}$$

After imposing the two prescribed temperatures with Method 1 from Sec. 3-2, we get

$$\begin{bmatrix} 59.212 & -17.095 & -27.042 & 0.000 & 0.000 & 0.000 \\ -17.095 & 54.187 & 0.000 & -29.554 & 0.000 & 0.000 \\ -27.042 & 0.000 & 229.358 & -54.628 & 0.000 & 0.000 \\ 0.000 & -29.554 & -54.628 & 220.801 & 0.000 & 0.000 \\ 0.000 & 0.000 & 0.000 & 0.000 & 1.000 & 0.000 \\ 0.000 & 0.000 & 0.000 & 0.000 & 0.000 & 1.000 \end{bmatrix} \begin{bmatrix} T_1 \\ T_2 \\ T_3 \\ T_4 \\ T_5 \\ T_6 \end{bmatrix} = \begin{bmatrix} 1361.8 \\ 2121.3 \\ 6736.7 \\ 5644.5 \\ 35.0 \\ 35.0 \end{bmatrix}$$

The solution is given by

$$T_1 = 71.3°C \qquad T_3 = 49.6°C \qquad T_5 = 35.0°C$$

$$T_2 = 88.8°C \qquad T_4 = 49.7°C \qquad T_6 = 35.0°C$$

The average heat fluxes through the elements may be computed from Eqs. (8-125). For example, for element 1 we have

$$\bar{q}_x^{(1)} = -52. \, [(-30.833)(71.3) + (33.333)(49.6) + (-2.500)(88.8)]$$

$$= 39,900 \text{ W/m}^2$$

and

$$\begin{aligned}\overline{q}_y^{(1)} &= -52. \, [(-25.000)(71.3) + (0.0)(49.6) + (25.000)(88.8)] \\ &= -22,800 \text{ W/m}^2\end{aligned}$$

or

$$\overline{q}_x^{(1)} = +3.99 \text{ W/cm}^2$$

and

$$\overline{q}_y^{(1)} = -2.28 \text{ W/cm}^2$$

The negative sign indicates that the heat is flowing in the negative y direction. Both of these signs seem to be intuitively correct (why?). The results for the other three elements are as follows:

$$\overline{q}_x^{(2)} = 6.76 \text{ W/cm}^2 \quad q_x^{(3)} = 13.6 \text{ W/cm}^2 \quad q_x^{(4)} = 13.7 \text{ W/cm}^2$$

$$\overline{q}_y^{(2)} = -0.01 \text{ W/cm}^2 \; q_y^{(3)} = -0.05 \text{ W/cm}^2 \; q_y^{(4)} = 0.0 \text{ W/cm}^2$$

Recall that these average heat fluxes are generally taken to be the local values at the respective element centroids. Because of the crude mesh used, these results are very approximate.   ■

One way to improve the results for the element resultants such as the heat fluxes is to average the results of two adjacent triangular elements as shown in Fig. 8-15. For example, if the results for elements 1 and 2 are averaged, we get

$$\overline{q}_x = \frac{\overline{q}_x^{(1)} + \overline{q}_x^{(2)}}{2} = \frac{3.99 + 6.76}{2} = 5.38 \text{ W/cm}^2$$

and

$$\overline{q}_y = \frac{\overline{q}_y^{(1)} + \overline{q}_y^{(2)}}{2} = \frac{(-2.28) + (-0.01)}{2} = -1.14 \text{ W/cm}^2$$

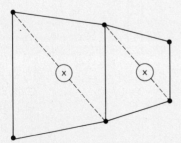

Figure 8-15   The method of quadrilateral averages. Note how each pair of triangles forms a quadrilateral. The element resultants for each pair of triangles are averaged and assigned to the centroid of the quadrilateral. Although this is illustrated here in a heat transfer application, the same technique is used in stress analysis.

whereas for elements 3 and 4, we get

$$\overline{q}_x = \frac{13.6 \, + \, 13.7}{2} \, = \, 13.65 \text{ W/cm}^2$$

and

$$\overline{q}_y = \frac{-0.05 \, + \, 0.0}{2} \, = \, -0.025 \text{ W/cm}^2$$

This method of combining the element resultants for two adjacent triangular elements is known as the method of *quadrilateral averages*. Note that a quadrilateral is formed by the two triangles. The combined element resultants from two triangles that form a quadrilateral are then associated with the centroid of the resulting quadrilateral as shown in Fig. 8-15.

8-9 AXISYMMETRIC HEAT CONDUCTION

Three-dimensional heat conduction may be modeled as a two-dimensional idealization if axisymmetry exists with respect to both the geometry and the thermal loads imposed on the body. A body of revolution is obviously geometrically axisymmetric as shown in Fig. 8-16. Each thermal load must also be axisymmetric, however, in order to take the temperature T as a function of only the radial coordinate r and the axial coordinate z. When conditions such as these exist, the problem is said to be one of axisymmetric heat conduction.

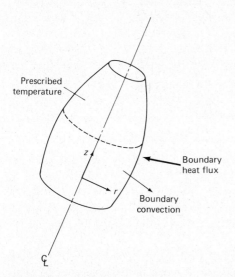

Figure 8-16 Schematic of axisymmetric heat conduction problem. Note that the body is a body of revolution about the z axis, and the thermal loading is axisymmetric as well.

Only the isotropic case is considered in the formal development; that is, the thermal conductivity k may be a function of r and z (for a heterogeneous body) and even the temperature T, but it may not be dependent on direction. Clearly, axisymmetric heat transfer cannot occur in a fully anisotropic body. Only if the body is stratified in such a way that the principal values of thermal conductivity occur in the r and z directions (with thermal conductivities k_r and k_z, respectively) can the problem be considered axisymmetric. The formulation to be presented could easily be extended to include this special anisotropic case.

Three types of boundary conditions are considered: convection, imposed heat fluxes, and prescribed temperatures. It should be apparent by now that convection and imposed heat fluxes are inherently included in the FEM formulation. Prescribed temperature boundary conditions, on the other hand, are handled in the usual manner after the assemblage step. Thermal radiation from the surface of the axisymmetric body is not included, but could easily be added by following the approach taken in Secs. 8-3 and 8-8.

The Governing Equation

The governing equation for axisymmetric steady-state heat conduction in an isotropic, heterogeneous body with a volumetric heat source is given by

$$\frac{1}{r}\frac{\partial}{\partial r}\left(rk\frac{\partial T}{\partial r}\right) + \frac{\partial}{\partial z}\left(k\frac{\partial T}{\partial z}\right) + Q = 0 \qquad \textbf{(8-126)}$$

where T is the temperature, k is the thermal conductivity, and Q is the heat generation per unit time and per unit volume. The global coordinates are denoted as r and z in the radial and axial directions, respectively. Note that Eq. (8-126) is actually the heat conduction equation in cylindrical coordinates with the term representing the circumferential conduction set to zero.

The mathematical statements of the boundary conditions for convection and imposed heat fluxes are more conveniently given later during the FEM formulation.

The Finite Element Formulation

It is not really necessary at this point to choose the particular type of axisymmetric element to be used in the discretization. However, after the general expressions for the element characteristics have been derived, the emphasis will be on the triangular donut-shaped element. This element and the rectangular donut-shaped element are shown in Fig. 8-17 for a typical axisymmetric body. Additional axisymmetric elements are presented in Chapter 9.

The Galerkin method will be used on an element basis to derive the expressions for the element characteristics. We begin by forming the volume integral of the product of the residual and transpose of the shape function matrix **N** and setting the result to zero, or

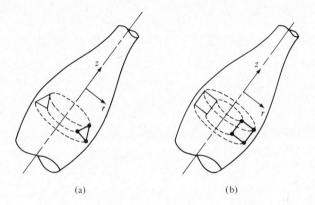

Figure 8-17   Typical axisymmetric elements: (a) the triangular donut-shaped element and (b) the rectangular donut-shaped element.

$$\int_{V^e} \mathbf{N}^T \left[\frac{1}{r} \frac{\partial}{\partial r} \left(kr \frac{\partial T}{\partial r} \right) + \frac{\partial}{\partial z} \left(k \frac{\partial T}{\partial z} \right) + Q \right] dV = \mathbf{0} \qquad \textbf{(8-127)}$$

where it is appropriate to take

$$dV = 2\pi r \, dr \, dz \qquad \textbf{(8-128)}$$

for the elemental volume dV. It is emphasized that the expression in the brackets in Eq. (8-127) is the residual that results when the approximation for the temperature T on an element basis (i.e., $T = \mathbf{N}\mathbf{a}^e$) is substituted into Eq. (8-126). Note that each term in Eq. (8-127) has dimensions of energy per unit time. If Eqs. (8-127) and (8-128) are combined and if the result is split into three separate integrals, we get

$$2\pi \int_{A^e} \mathbf{N}^T \frac{\partial}{\partial r} \left(kr \frac{\partial T}{\partial r} \right) dr \, dz + 2\pi \int_{A^e} \mathbf{N}^T \frac{\partial}{\partial z} \left(k \frac{\partial T}{\partial z} \right) r \, dr \, dz$$

$$+ 2\pi \int_{A^e} \mathbf{N}^T Qr \, dr \, dz = \mathbf{0} \qquad \textbf{(8-129)}$$

The 2π may be cancelled from each term, but we will continue to carry it. Note that the integration limit has been changed from V^e to A^e. Recall from Sec. 6-6 that we will, in effect, be analyzing the *half-plane* as shown in Fig. 6-17 for a typical axisymmetric body. For all practical purposes, the problem has been made two-dimensional with the thickness t (in the two-dimensional problem) being replaced by $2\pi r$! However, let us explicitly show how the expressions for the element characteristics can be derived when the integrands are functions of one or more of the global coordinates. Because r is independent of z, we should note that

$$\frac{\partial}{\partial z}\left(k\frac{\partial T}{\partial z}\right) r = \frac{\partial}{\partial z}\left(kr\frac{\partial T}{\partial z}\right) \tag{8-130}$$

Furthermore, if we apply the Green-Gauss theorem to the two terms in Eq. (8-129) that involve second-order derivatives, we get

$$2\pi\int_{C^e}\mathbf{N}^T kr\frac{\partial T}{\partial r}n_r\,dC - 2\pi\int_{A^e}\frac{\partial\mathbf{N}^T}{\partial r}kr\frac{\partial T}{\partial r}\,dr\,dz$$

$$+ 2\pi\int_{C^e}\mathbf{N}^T kr\frac{\partial T}{\partial z}n_z\,dC$$

$$- 2\pi\int_{A^e}\frac{\partial\mathbf{N}^T}{\partial z}kr\frac{\partial T}{\partial z}\,dr\,dz$$

$$+ 2\pi\int_{A^e}\mathbf{N}^T Qr\,dr\,dz = 0 \tag{8-131}$$

where C^e denotes that the integrations are to be performed around the boundary of the element (in a counterclockwise direction). The direction cosines n_r and n_z are shown in Fig. 8-18(a) for a typical triangular element. In a manner completely analogous to Eq. (8-48), we may write

$$q_n = -k\frac{\partial T}{\partial r}n_r - k\frac{\partial T}{\partial z}n_z \tag{8-132}$$

where q_n represents the net normal heat flux from conduction from within the element to the boundary of the element as shown in Fig. 8-18(b). As mentioned earlier both convection and imposed heat fluxes are to be considered. It is convenient at this point to perform an energy balance on the global boundary as shown in Fig. 8-18(b). If q_{sB} and q_{cvB} denote the imposed and convective heat fluxes, respectively, then we may write

$$q_n + q_{sB} = q_{cvB} \tag{8-133a}$$

or

$$q_n = q_{cvB} - q_{sB} \tag{8-133b}$$

We must now invoke Newton's law of cooling, or

$$q_{cvB} = h_B(T - T_{aB}) \tag{8-134}$$

where h_B is the convective heat transfer coefficient and T_{aB} is the temperature of the fluid far removed from the boundary. If Eqs. (8-131) to (8-134) are combined, we get

$$-2\pi\int_{C^e}\mathbf{N}^T h_B r(T - T_{aB})\,dC + 2\pi\int_{C^e}\mathbf{N}^T q_{sB} r\,dC$$

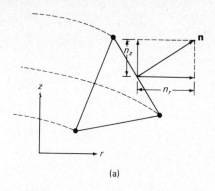

(a)

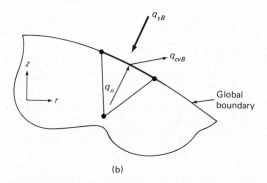

(b)

Figure 8-18   (a) Normal unit vector n shown for a typical triangular axisymmetric element. (b) Typical part of the global boundary in the axisymmetric heat conduction problem showing the heat fluxes considered in the finite element formulation.

$$- 2\pi \int_{A^e} \frac{\partial \mathbf{N}^T}{\partial r} kr \frac{\partial T}{\partial r} dr\, dz - 2\pi \int_{A^e} \frac{\partial \mathbf{N}^T}{\partial z} kr \frac{\partial T}{\partial z} dr\, dz \qquad \textbf{(8-135)}$$

$$+ 2\pi \int_{A^e} \mathbf{N}^T Qr\, dr\, dz = 0$$

But the temperature T within an element e is given by

$$T = \mathbf{N}\mathbf{a}^e \qquad \textbf{(8-136)}$$

where $\mathbf{N}$ is the shape function matrix and $\mathbf{a}^e$ is the vector of nodal temperatures for element e. For the three-node triangular element, $\mathbf{N}$ is given by

$$\mathbf{N} = [N_i \quad N_j \quad N_k] \qquad \textbf{(8-137)}$$

and

$$\mathbf{a}^e = [T_i \quad T_j \quad T_k]^T \qquad \textbf{(8-138)}$$

where the shape functions themselves are given by Eqs. (6-21) with x and y replaced by r and z. If the temperature T in Eq. (8-135) is eliminated by using Eq. (8-136), we get

$$\mathbf{K}^e \mathbf{a}^e = \mathbf{f}^e \tag{8-139}$$

where $\mathbf{K}^e$ may be referred to as the composite element stiffness matrix and $\mathbf{f}^e$ as the composite element nodal force vector, both defined as

$$\mathbf{K}^e = \mathbf{K}_{rr}^e + \mathbf{K}_{zz}^e + \mathbf{K}_{cvB}^e \tag{8-140}$$

and

$$\mathbf{f}^e = \mathbf{f}_Q^e + \mathbf{f}_{qB}^e + \mathbf{f}_{cvB}^e \tag{8-141}$$

The element stiffness matrices themselves are defined by

$$\mathbf{K}_{rr}^e = 2\pi \int_{A^e} \frac{\partial \mathbf{N}^T}{\partial r} \, kr \, \frac{\partial \mathbf{N}}{\partial r} \, dr \, dz \tag{8-142a}$$

$$\mathbf{K}_{zz}^e = 2\pi \int_{A^e} \frac{\partial \mathbf{N}^T}{\partial z} \, kr \, \frac{\partial \mathbf{N}}{\partial z} \, dr \, dz \tag{8-142b}$$

and

$$\mathbf{K}_{cvB}^e = 2\pi \int_{C^e} \mathbf{N}^T h_B r \, \mathbf{N} \, dC \tag{8-142c}$$

and the element nodal force vectors by

$$\mathbf{f}_Q^e = 2\pi \int_{A^e} \mathbf{N}^T Q r \, dr \, dz \tag{8-143a}$$

$$\mathbf{f}_{qB}^e = 2\pi \int_{C^e} \mathbf{N}^T q_{sB} r \, dC \tag{8-143b}$$

and

$$\mathbf{f}_{cvB}^e = 2\pi \int_{C^e} \mathbf{N}^T h_B r T_{aB} \, dC \tag{8-143c}$$

It is emphasized that the element characteristics given by Eqs. (8-142) and (8-143) are quite general and may be applied to any suitable axisymmetric element. Let us illustrate the use of the triangular axisymmetric element by way of several examples.

Example 8-18

Evaluate the element stiffness matrix from conduction in the radial direction for the triangular axisymmetric element.

Solution

From Eq. (8-142a), we see that the derivatives of each of the shape functions are needed with respect to r. Recall that the shape functions for this element are given by Eqs. (6-21) with r and z replacing x and y, respectively, or

$$N_i = m_{11} + m_{21}r + m_{31}z \tag{8-144a}$$

$$N_j = m_{12} + m_{22}r + m_{32}z \tag{8-144b}$$

$$N_k = m_{13} + m_{23}r + m_{33}z \tag{8-144c}$$

where the m_{ij}'s are given by Eq. (6-21d) with r_i and z_i replacing x_i and y_i, etc. Therefore, Eq. (8-142a) becomes

$$\mathbf{K}_{rr}^e = 2\pi \int_{A^e} \begin{bmatrix} m_{21} \\ m_{22} \\ m_{23} \end{bmatrix} [m_{21} \quad m_{22} \quad m_{23}] \, kr \, dr \, dz \tag{8-145}$$

Since the m_{ij}'s are constants and the thermal conductivity is assumed to be constant (in each element), Eq. (8-145) may be written

$$\mathbf{K}_{rr}^e = 2\pi \begin{bmatrix} m_{21} \\ m_{22} \\ m_{23} \end{bmatrix} [m_{21} \quad m_{22} \quad m_{23}] \int_{A^e} r \, dr \, dz \tag{8-146}$$

It proves to be very convenient at this point to represent the radial coordinate r over the element as follows:

$$r = N_i r_i + N_j r_j + N_k r_k \tag{8-147a}$$

where r_i, r_j, and r_k denote the radial coordinates of nodes i, j, and k. If the shape functions are written in terms of the area coordinates [see Eq. (6-25)], we get

$$r = L_i r_i + L_j r_j + L_k r_k \tag{8-147b}$$

Note that at node i, the radial coordinate r becomes r_i, etc. Moreover, the area coordinates (like the shape functions in this case) vary linearly with r over the element. It seems quite reasonable, therefore, to use Eqs. (8-147b) to represent r. The integral in Eq. (8-146) may be written as

$$\int_{A^e} r \, dr \, dz = r_i \int_{A^e} L_i \, dr \, dz + r_j \int_{A^e} L_j \, dr \, dz$$
$$+ r_k \int_{A^e} L_k \, dr \, dz \tag{8-148}$$

Using the integration formula given by Eq. (6-49), we may evaluate a typical integral as follows:

$$\int_{A^e} L_i \, dr \, dz = \frac{1!0!0!}{(1 + 0 + 0 + 2)!} \, 2A = \tfrac{1}{3}A$$

It should now be apparent that

$$\int_{A^e} r \, dr \, dz = \tfrac{1}{3}(r_i + r_j + r_k)A \tag{8-149}$$

where A, the area of the element, may be calculated with the help of Eq. (6-21e) if r_i and z_i are used in place of x_i and y_i, etc. Let us defined $\bar{r}$ as

$$\bar{r} = \tfrac{1}{3}(r_i + r_j + r_k) \tag{8-150}$$

so that the final result for $\mathbf{K}_{rr}^e$ may be given as

$$\mathbf{K}_{rr}^e = 2\pi k\bar{r}A \begin{bmatrix} m_{21}^2 & m_{21}m_{22} & m_{21}m_{23} \\ m_{22}m_{21} & m_{22}^2 & m_{22}m_{23} \\ m_{23}m_{21} & m_{23}m_{22} & m_{23}^2 \end{bmatrix} \qquad \textbf{(8-151)}$$

The reader should compare the approach taken here with that used in Sec. 7-3 for axisymmetric stress analysis. ■

It can be shown in a similar manner that the element stiffness matrix from conduction in the axial direction is given by

$$\mathbf{K}_{zz}^e = 2\pi k\bar{r}A \begin{bmatrix} m_{31}^2 & m_{31}m_{32} & m_{31}m_{33} \\ m_{32}m_{31} & m_{32}^2 & m_{32}m_{33} \\ m_{33}m_{31} & m_{33}m_{32} & m_{33}^2 \end{bmatrix} \qquad \textbf{(8-152)}$$

for the triangular element. In Eqs. (8-151) and (8-152) k is interpreted to be the average value of the thermal conductivity over the element if it varies from point to point. The element stiffness matrix from boundary convection is evaluated in the next example for the triangular element.

Example 8-19

Evaluate $\mathbf{K}_{cvB}^e$ given by Eq. (8-142c) for the element shown in Fig. 8-19.

Solution

From Fig. 8-19 we note that leg ij of element e is on the global boundary B. In order to facilitate the evaluation of Eq. (8-142c), we should represent the shape functions in terms of the area coordinates. In this case, the integral is to be evaluated on leg ij where $L_k = 0$ (and $N_k = 0$). Therefore, on leg ij the shape function matrix may be written

$$\mathbf{N} = [L_i \quad L_j \quad 0] \qquad \textbf{(8-153)}$$

and the matrix $\mathbf{K}_{cvB}^e$ may be written

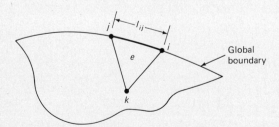

Figure 8-19   Typical element with at least one leg of the axisymmetric triangular element on the global boundary.

$$\mathbf{K}_{cvB}^e = 2\pi \int_{l_{ij}} \begin{bmatrix} L_i \\ L_j \\ 0 \end{bmatrix} [L_i \quad L_j \quad 0] \, h_B r \, dl \tag{8-154}$$

Let us represent r on leg ij as follows:

$$r = N_i r_i + N_j r_j = L_i r_i + L_j r_j \tag{8-155}$$

Thus, Eq. (8-154) becomes

$$\mathbf{K}_{cvB}^e = 2\pi h_B \int_{l_{ij}} \begin{bmatrix} L_i^2 & L_i L_j & 0 \\ L_j L_i & L_j^2 & 0 \\ 0 & 0 & 0 \end{bmatrix} (L_i r_i + L_j r_j) \, dl \tag{8-155}$$

where h_B is assumed to be constant over leg ij. It should be noted that there are two different types of terms that need to be evaluated. The first is evaluated as

$$\int_{l_{ij}} L_i^2 (L_i r_i) \, dl = r_i \int_{l_{ij}} L_i^3 \, dl$$

$$= r_i \frac{3!0!}{(3 + 0 + 1)!} l_{ij}$$

$$= \tfrac{1}{4} r_i l_{ij} \tag{8-156a}$$

and the second as

$$\int_{l_{ij}} L_i L_j (L_i r_i) \, dl = r_i \int_{l_{ij}} L_i^2 L_j \, dl$$

$$= r_i \frac{2!1!}{(2 + 1 + 1)!} l_{ij}$$

$$= \tfrac{1}{12} r_i l_{ij} \tag{8-156b}$$

where l_{ij} represents the length of leg ij. The final result for $\mathbf{K}_{cvB}^e$ is given by

$$\mathbf{K}_{cvB}^e = \frac{2\pi h_B l_{ij}}{12} \begin{bmatrix} 3r_i + r_j & r_i + r_j & 0 \\ r_i + r_j & r_i + 3r_j & 0 \\ 0 & 0 & 0 \end{bmatrix} \text{(for leg } ij \text{ on } B) \tag{8-157}$$

The reader should be able to derive the corresponding result if leg jk or ki happens to be on the global boundary. ∎

The three element nodal force vectors may be evaluated in a similar fashion with the results

$$\mathbf{f}_Q^e = \frac{2\pi Q A}{12} \begin{bmatrix} 2r_i + r_j + r_k \\ r_i + 2r_j + r_k \\ r_i + r_j + 2r_k \end{bmatrix} \tag{8-158}$$

$$\mathbf{f}_{cvB}^e = \frac{2\pi h_B T_{aB} l_{ij}}{6} \begin{bmatrix} 2r_i + r_j \\ r_i + 2r_j \\ 0 \end{bmatrix} \text{(for leg } ij \text{ on } B) \tag{8-159}$$

and

$$
\mathbf{f}_{qB}^e = \frac{2\pi q_{sB} l_{ij}}{6}
\begin{bmatrix}
2r_i + r_j \\
r_i + 2r_j \\
0
\end{bmatrix}
\qquad \text{(for leg } ij \text{ on } B) \qquad \textbf{(8-160)}
$$

Equations (8-159) and (8-160) are valid if leg ij is on the global boundary. Similar expressions could be written by inspection or easily derived for legs jk and ki. The reader should derive the expressions for $\mathbf{f}_Q^e$ in the case of a line source Q' at (r_0, z_0), where Q' is the heat generation rate per unit circumference (i.e., at $r = r_0$ and $z = z_0$).

This completes the evaluation of each of the element stiffness matrices and each of the nodal force vectors. The assemblage equations are formed in the usual manner. The prescribed temperature boundary conditions, if any are present, would be applied next. The solution for the temperatures would then follow and the heat flows through the body could be evaluated. If the thermal conductivity k or convective heat transfer coefficient h_B is dependent on the temperature, the direct iteration method from Sec. 8-2 may be used to solve for the nodal temperatures.

If a nonsymmetrical thermal load is imposed on the body, it may be possible to perform an axisymmetric analysis by using a one-term Fourier series of sine and cosine functions to represent the thermal loads and temperatures. The loads, however, must still be symmetric about a plane through the z axis. Huebner [2] and Wilson [3] give additional details. Treatment of this aspect of axisymmetric heat conduction is beyond the scope of this book.

8-10   THREE-DIMENSIONAL HEAT CONDUCTION

In this section the three-dimensional heat conduction problem is formulated. Figure 8-20 shows an arbitrary three-dimensional body in an xyz coordinate system. The boundary conditions that will be considered on the surface of the body are the following: convection, imposed heat fluxes (possibly acting at a distance such as the sun), and prescribed temperatures. Recall that perfect insulation is a special case of zero heat flux. A volumetric heat source will also be included. Thermal radiation from the surface is not considered but could easily be added.

In Fig. 8-20, the heat fluxes from conduction in the x, y, and z directions are denoted by q_x, q_y, and q_z. These heat fluxes may be related to the temperature T within the body by Fourier's law of heat conduction for an isotropic body in three dimensions, or

$$
q_x = -k\frac{\partial T}{\partial x} \qquad q_y = -k\frac{\partial T}{\partial y} \qquad q_z = -k\frac{\partial T}{\partial z} \qquad \textbf{(8-161)}
$$

where k is the thermal conductivity. The minus signs in Eq. (8-161) are necessary to ensure that each heat flux is positive if directed in the direction of decreasing temperature.

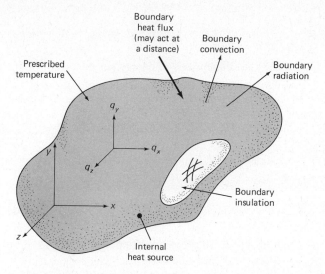

Figure 8-20   Schematic of three-dimensional heat conduction problem.

The Governing Equation

The governing equation for general three-dimensional heat conduction in a heterogeneous, isotropic body is given by

$$\frac{\partial}{\partial x}\left(k\,\frac{\partial T}{\partial x}\right) + \frac{\partial}{\partial y}\left(k\,\frac{\partial T}{\partial y}\right) + \frac{\partial}{\partial z}\left(k\,\frac{\partial T}{\partial z}\right) + Q = 0 \qquad \textbf{(8-162)}$$

where Q is the heat generation rate per unit volume. The thermal conductivity k may vary throughout the body and may even be a function of temperature. The mathematical form of the boundary conditions is given during the FEM formulation.

The Finite Element Formulation

As in the other heat conduction formulations, it is not necessary to choose the type of element to be used in the discretization in order to determine the expressions for the element characteristics. The reader may recall that in Sec. 6-5 two particular three-dimensional elements were presented, namely, the four-node tetrahedral and eight-node brick elements. After the general expressions for the element characteristics have been derived, the emphasis will be on the tetrahedral element. Additional three-dimensional elements are presented in Chapter 9.

The Galerkin method will be used on an element basis to derive the expressions for the element characteristics. We begin by writing

$$\int_{V^e} \mathbf{N}^T \left[\frac{\partial}{\partial x}\left(k\frac{\partial T}{\partial x} \right) + \frac{\partial}{\partial y}\left(k\frac{\partial T}{\partial y} \right) + \frac{\partial}{\partial z}\left(k\frac{\partial T}{\partial z} \right) + Q \right] dx\,dy\,dz = 0 \quad \textbf{(8-163)}$$

Note that each term in Eq. (8-163) has dimensions of energy per unit time. Using the Green-Gauss theorem on the three terms involving second-order derivatives gives

$$\int_{S^e} \mathbf{N}^T k \frac{\partial T}{\partial x} n_x\,dS - \int_{V^e} \frac{\partial \mathbf{N}^T}{\partial x} k \frac{\partial T}{\partial x} dx\,dy\,dz + \int_{S^e} \mathbf{N}^T k \frac{\partial T}{\partial y} n_y\,dS$$

$$- \int_{V^e} \frac{\partial \mathbf{N}^T}{\partial y} k \frac{\partial T}{\partial y} dx\,dy\,dz + \int_{S^e} \mathbf{N}^T k \frac{\partial T}{\partial z} n_z\,dS$$

$$- \int_{V^e} \frac{\partial \mathbf{N}^T}{\partial z} k \frac{\partial T}{\partial z} dx\,dy\,dz + \int_{V^e} \mathbf{N}^T Q\,dx\,dy\,dz = 0 \quad \textbf{(8-164)}$$

Note that three surface integrations arise. These integrals are identified by the S^e on the integrals and must be evaluated over the face (of each element) that is on the global boundary. However, at this point it is convenient to combine these three surface integrals into one integral by noting that

$$q_n = -k\frac{\partial T}{\partial x} n_x - k \frac{\partial T}{\partial y} n_y - k \frac{\partial T}{\partial z} n_z \quad \textbf{(8-165)}$$

where q_n is the net (normal) heat flux from conduction, i.e., in the direction of the unit normal $\mathbf{n}$ to the surface as shown in Fig. 8-21. Equation (8-165) is the three-dimensional form of Eq. (8-48).

Now it is convenient to consider the convective and imposed heat flux boundary conditions. An energy balance on the surface of the body in Fig. 8-21 gives

$$q_n + q_{sB} = q_{cvB} \quad \textbf{(8-166a)}$$

or

$$q_n = q_{cvB} - q_{sB} \quad \textbf{(8-166b)}$$

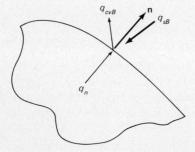

Figure 8-21 Typical part of the global boundary in the three-dimensional heat conduction problem showing the heat fluxes considered in the finite element formulation.

Recall that the heat flux q_{cvB} is proportional to the temperature difference between the body and the ambient or

$$q_{cvB} = h_B(T - T_{aB}) \tag{8-167}$$

where h_B is the convective heat transfer coefficient and T_{aB} is the temperature of fluid far removed from the body. If Eqs. (8-164) to (8-167) are combined, we get

$$- \int_{S^e} \mathbf{N}^T h_B T \, dS + \int_{S^e} \mathbf{N}^T h_B T_{aB} \, dS + \int_{S^e} \mathbf{N}^T q_{sB} \, dS$$

$$- \int_{V^e} \frac{\partial \mathbf{N}^T}{\partial x} k \frac{\partial T}{\partial x} \, dx \, dy \, dz - \int_{V^e} \frac{\partial \mathbf{N}^T}{\partial y} k \frac{\partial T}{\partial y} \, dx \, dy \, dz$$

$$- \int_{V^e} \frac{\partial \mathbf{N}^T}{\partial z} k \frac{\partial T}{\partial z} \, dx \, dy \, dz + \int_{V^e} \mathbf{N}^T Q \, dx \, dy \, dz = 0 \tag{8-168}$$

The temperature T within an element may be represented by

$$T = \mathbf{N}\mathbf{a}^e \tag{8-169}$$

where the shape function matrix $\mathbf{N}$ for the four-node tetrahedral element is given by

$$\mathbf{N} = [N_i \quad N_j \quad N_k \quad N_m] \tag{8-170}$$

and the vector of nodal temperatures $\mathbf{a}^e$ by

$$\mathbf{a}^e = [T_i \quad T_j \quad T_k \quad T_m]^T \tag{8-171}$$

The shape functions themselves are given by Eqs. (6-40) in this case. If the temperature T in Eq. (8-168) is eliminated by using Eq. (8-169), we get

$$\mathbf{K}^e \mathbf{a}^e = \mathbf{f}^e \tag{8-172}$$

where $\mathbf{K}^e$ is the composite element stiffness matrix and $\mathbf{f}^e$ is the composite element nodal force vector, or

$$\mathbf{K}^e = \mathbf{K}_{xx}^e + \mathbf{K}_{yy}^e + \mathbf{K}_{zz}^e + \mathbf{K}_{cvB}^e \tag{8-173}$$

and

$$\mathbf{f}^e = \mathbf{f}_Q^e + \mathbf{f}_{qB}^e + \mathbf{f}_{cvB}^e \tag{8-174}$$

The element stiffness matrices are defined as

$$\mathbf{K}_{xx}^e = \int_{V^e} \frac{\partial \mathbf{N}^T}{\partial x} k \frac{\partial \mathbf{N}}{\partial x} \, dx \, dy \, dz \tag{8-175a}$$

$$\mathbf{K}_{yy}^e = \int_{V^e} \frac{\partial \mathbf{N}^T}{\partial y} k \frac{\partial \mathbf{N}}{\partial y} \, dx \, dy \, dz \tag{8-175b}$$

$$\mathbf{K}_{zz}^e = \int_{V^e} \frac{\partial \mathbf{N}^T}{\partial z} k \frac{\partial \mathbf{N}}{\partial z} \, dx \, dy \, dz \tag{8-175c}$$

and

$$\mathbf{K}_{cvB}^e = \int_{S^e} \mathbf{N}^T h_B \mathbf{N} \, dS \tag{8-175d}$$

whereas the element nodal force vectors are defined as

$$\mathbf{f}_Q^e = \int_{V^e} \mathbf{N}^T Q \, dx \, dy \, dz \tag{8-176a}$$

$$\mathbf{f}_{qB}^e = \int_{S^e} \mathbf{N}^T q_{sB} \, dS \tag{8-176b}$$

and

$$\mathbf{f}_{cvB}^e = \int_{S^e} \mathbf{N}^T h_B T_{aB} \, dS \tag{8-176c}$$

The next step is to evaluate these integrals for the particular three-dimensional element used in the discretization. For example, if the four-node tetrahedral element is used, it can be shown that the results for the three element stiffness matrices from conduction are given by

$$\mathbf{K}_{xx}^e = kV \begin{bmatrix} m_{21}^2 & m_{21}m_{22} & m_{21}m_{23} & m_{21}m_{24} \\ m_{22}m_{21} & m_{22}^2 & m_{22}m_{23} & m_{22}m_{24} \\ m_{23}m_{21} & m_{23}m_{22} & m_{23}^2 & m_{23}m_{24} \\ m_{24}m_{21} & m_{24}m_{22} & m_{24}m_{23} & m_{24}^2 \end{bmatrix} \tag{8-177}$$

$$\mathbf{K}_{yy}^e = kV \begin{bmatrix} m_{31}^2 & m_{31}m_{32} & m_{31}m_{33} & m_{31}m_{34} \\ m_{32}m_{31} & m_{32}^2 & m_{32}m_{33} & m_{32}m_{34} \\ m_{33}m_{31} & m_{33}m_{32} & m_{33}^2 & m_{33}m_{34} \\ m_{34}m_{31} & m_{34}m_{32} & m_{34}m_{33} & m_{34}^2 \end{bmatrix} \tag{8-178}$$

and

$$\mathbf{K}_{zz}^e = kV \begin{bmatrix} m_{41}^2 & m_{41}m_{42} & m_{41}m_{43} & m_{41}m_{44} \\ m_{42}m_{41} & m_{42}^2 & m_{42}m_{43} & m_{42}m_{44} \\ m_{43}m_{41} & m_{43}m_{42} & m_{43}^2 & m_{43}m_{44} \\ m_{44}m_{41} & m_{44}m_{42} & m_{44}m_{43} & m_{44}^2 \end{bmatrix} \tag{8-179}$$

where the m_{ij}'s and volume V are given by Eqs. (6-40).

The element stiffness matrix from boundary convection deserves special attention. Note that the integral for $\mathbf{K}_{cvB}^e$ is a surface integral, not a volume integral. The surface referred to here is the surface of the element. The tetrahedral element has four surfaces and, strictly speaking, $\mathbf{K}_{cvB}^e$ should be evaluated over each of these faces. However, if all faces are internal and not on the global boundary, we must take $\mathbf{K}_{cvB}^e$ to be the 4×4 null matrix since there is no convection on the internal faces. Moreover, if a face is on the global boundary but no convection is present, then again $\mathbf{K}_{cvB}^e$ is taken to be the 4×4 null matrix. A nontrivial case is shown in Fig. 8-22, where face ijk of the element e is on the part of the global boundary that undergoes convection.

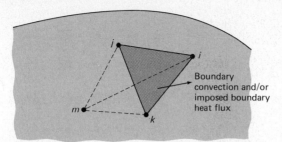

Figure 8-22   Typical element with at least one face of the tetrahedral element on the global boundary.

Example 8-20

Evaluate the element stiffness matrix $\mathbf{K}^e_{cvB}$ for the element shown in Fig. 8-22.

Solution

From Eq. (8-175d) we see that the shape functions themselves appear in the integrand. It is convenient to represent the shape functions in terms of the volume coordinates [see Eq. (6-43)]. Note that $L_m = 0$ on face ijk. Therefore, Eq. (8-175d) becomes

$$\mathbf{K}^e_{cvB} = \int_{A_{ijk}} \begin{bmatrix} L_i^2 & L_iL_j & L_iL_k & 0 \\ L_jL_i & L_j^2 & L_jL_k & 0 \\ L_kL_i & L_kL_j & L_k^2 & 0 \\ 0 & 0 & 0 & 0 \end{bmatrix} h_B \, dS \qquad \textbf{(8-180)}$$

where A_{ijk} denotes the area of face ijk. Note that this face is a triangle and the volume coordinates have degenerated to the three area coordinates. It seems appropriate, therefore, to use the special integration formula given by Eq. (6-49). For example, a typical term in the matrix may be evaluated as follows:

$$\int_{A_{ijk}} h_B L_i L_j \, dS = h_B \frac{1!1!0!}{(1 + 1 + 0 + 2)!} 2A_{ijk} = \frac{h_B A_{ijk}}{12}$$

The final result is given by

$$\mathbf{K}^e_{cvB} = \frac{h_B A_{ijk}}{12} \begin{bmatrix} 2 & 1 & 1 & 0 \\ 1 & 2 & 1 & 0 \\ 1 & 1 & 2 & 0 \\ 0 & 0 & 0 & 0 \end{bmatrix} \quad \begin{array}{l} \text{for face } ijk \text{ on the} \\ \text{global boundary } B \end{array} \qquad \textbf{(8-181)}$$

It is emphasized that this result is valid if face ijk of the tetrahedral element is on the global boundary B. The reader should be able to evaluate $\mathbf{K}^e_{cvB}$ if any of the other faces happen to be on the global boundary.   ■

The nodal force vector $\mathbf{f}^e_Q$ may be evaluated with the help of Eq. (6-50) if the shape functions are written in terms of the volume coordinates. The result is

$$\mathbf{f}_Q^e = \frac{QV}{4} \begin{bmatrix} 1 \\ 1 \\ 1 \\ 1 \end{bmatrix} \tag{8-182}$$

where the volume V is given by Eq. (6-40f). The heat source Q is assumed to be constant in arriving at Eq. (8-182); if Q is nonuniformly distributed in the body, then Q should be taken to represent a suitable average value over the element or the value of Q at the element centroid may be used. Note that one-fourth of the total heat source QV is allocated to each of the four nodes. Refer to Problem 8-176 for the case of a point heat source.

The same arguments used in the evaluation of $\mathbf{K}_{cvB}^e$ apply to the evaluation of $\mathbf{f}_{qB}^e$ and $\mathbf{f}_{cvB}^e$ with the results

$$\mathbf{f}_{qB}^e = \frac{q_{sB}A_{ijk}}{3} \begin{bmatrix} 1 \\ 1 \\ 1 \\ 0 \end{bmatrix} \qquad \text{for face } ijk \text{ on the global boundary } B \tag{8-183}$$

and

$$\mathbf{f}_{cvB}^e = \frac{h_B A_{ijk} T_{aB}}{3} \begin{bmatrix} 1 \\ 1 \\ 1 \\ 0 \end{bmatrix} \qquad \text{for face } ijk \text{ on the global boundary } B \tag{8-184}$$

Note that Eqs. (8-183) and (8-184) hold if face ijk is on the global boundary. These results can be extended by inspection if faces jkm, ijm, or ikm happen to be on the global boundary.

This completes the evaluation of the element stiffness matrices and the nodal force vectors. The assemblage step could now be done in the usual manner. The solution is also standard, providing the thermal properties are not a function of temperature; otherwise the direct iteration method presented in Sec. 8-2 could be used. The element heat fluxes $\bar{q}_x$, $\bar{q}_y$, and $\bar{q}_z$ may also be calculated using the general approach taken in Sec. 8-8. These heat fluxes would represent average values and are associated with the centroid of the tetrahedron.

8-11   TWO-DIMENSIONAL POTENTIAL FLOW

The flow of an incompressible and frictionless fluid (also known as an ideal fluid) is referred to as *potential flow*. No real fluid is truly frictionless. Outside the boundary layer, however, frictional effects may be neglected. In addition, if the fluid may be assumed to be incompressible, then the assumption of potential flow is valid. The flow is also then said to be *irrotational*. It may be recalled from gas dynamics that the flow of a gas may be considered to be incompressible for Mach numbers of 0.3 or less. A few of the many applications of potential flow are aerodynamics (including flow around airfoils), groundwater flow, flow through large enclosures (such as a reservoir), and flow around corners.

Only two-dimensional incompressible flow is modeled in the formal development. However, both the velocity potential and stream function formulations are developed. In both types of formulations, the *continuity equation* is given by

$$\frac{\partial u}{\partial x} + \frac{\partial v}{\partial y} = 0 \qquad\qquad \textbf{(8-185)}$$

and the *irrotational flow condition* by

$$\frac{\partial u}{\partial y} - \frac{\partial v}{\partial x} = 0 \qquad\qquad \textbf{(8-186)}$$

where u and v represent the x and y components of the fluid velocity. The velocity potential formulation will be seen to satisfy the irrotational flow condition exactly, whereas the stream function formulation will be seen to satisfy the continuity equation exactly.

The development in this book is limited further to external flows around symmetric bodies such as the one shown in Fig. 8-23. For unsymmetrical bodies such as airfoils, an additional condition called the Kutta or Kutta-Joukowski condition must be satisfied. This condition requires that the downstream stagnation point be located at the downstream edge. This can only happen if there is *circulation* around the body, thus causing *lift*. The reader is referred to several excellent books on fluid mechanics [4–6] for more information on this aspect of potential flow.

Velocity Potential Formulation

The irrotational flow condition given by Eq. (8-186) is satisfied exactly if the velocity potential ϕ is defined by

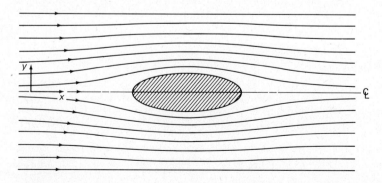

Figure 8-23 Two-dimensional potential flow around a symmetric body.

$$u = -\frac{\partial \phi}{\partial x} \qquad \text{(8-187a)}$$

and

$$v = -\frac{\partial \phi}{\partial y} \qquad \text{(8-187b)}$$

If these expressions for the velocity components are substituted into the continuity equation given by Eq. (8-185), the result is the two-dimensional form of *Laplace's equation*, or

$$\frac{\partial^2 \phi}{\partial x^2} + \frac{\partial^2 \phi}{\partial y^2} = 0 \qquad \text{(8-188)}$$

The Galerkin method on an element basis gives

$$\int_{A^e} \mathbf{N}^T \left[\frac{\partial^2 \phi}{\partial x^2} + \frac{\partial^2 \phi}{\partial y^2} \right] dx \, dy = \mathbf{0} \qquad \text{(8-189)}$$

where the shape function matrix $\mathbf{N}$ is dependent on the type of element used. We will derive the expressions for the element characteristics in general and then apply these results to the triangular element in particular.

If the Green-Gauss theorem is used, we get

$$\int_{C^e} \mathbf{N}^T \frac{\partial \phi}{\partial x} n_x \, dC - \int_{A^e} \frac{\partial \mathbf{N}^T}{\partial x} \frac{\partial \phi}{\partial x} dx \, dy$$

$$+ \int_{C^e} \mathbf{N}^T \frac{\partial \phi}{\partial y} n_y \, dC - \int_{A^e} \frac{\partial \mathbf{N}^T}{\partial y} \frac{\partial \phi}{\partial y} dx \, dy = \mathbf{0} \qquad \text{(8-190)}$$

Let us use Eqs. (8-187) to introduce the two velocity components into the formulation or

$$-\int_{C^e} \mathbf{N}^T u n_x \, dC - \int_{A^e} \frac{\partial \mathbf{N}^T}{\partial x} \frac{\partial \phi}{\partial x} dx \, dy$$

$$- \int_{C^e} \mathbf{N}^T v n_y \, dC - \int_{A^e} \frac{\partial \mathbf{N}^T}{\partial y} \frac{\partial \phi}{\partial y} dx \, dy = \mathbf{0} \qquad \text{(8-191)}$$

At this point it is convenient to represent the velocity potential ϕ on an element basis by writing

$$\phi = \mathbf{N} \mathbf{a}^e \qquad \text{(8-192)}$$

where the vector $\mathbf{a}^e$ contains the values of the potential at the nodes for element e. Since $\mathbf{a}^e$ is not a function of x or y, we can combine Eqs. (8-191) and (8-192) to get

$$\mathbf{K}^e \mathbf{a}^e = \mathbf{f}^e \qquad \text{(8-193)}$$

where the composite element stiffness matrix $\mathbf{K}^e$ is defined by

$$\mathbf{K}^e = \mathbf{K}^e_{xx} + \mathbf{K}^e_{yy} \qquad \text{(8-194)}$$

and the composite element nodal force vector $\mathbf{f}^e$ is defined by

$$\mathbf{f}^e = \mathbf{f}^e_{uB} + \mathbf{f}^e_{vB} \tag{8-195}$$

The element stiffness matrices $\mathbf{K}^e_{xx}$ and $\mathbf{K}^e_{yy}$ are defined by

$$\mathbf{K}^e_{xx} = \int_{A^e} \frac{\partial \mathbf{N}^T}{\partial x} \frac{\partial \mathbf{N}}{\partial x} \, dx \, dy \tag{8-196a}$$

and

$$\mathbf{K}^e_{yy} = \int_{A^e} \frac{\partial \mathbf{N}^T}{\partial y} \frac{\partial \mathbf{N}}{\partial y} \, dx \, dy \tag{8-196b}$$

whereas the element nodal force vectors $\mathbf{f}^e_{uB}$ and $\mathbf{f}^e_{vB}$ are defined by

$$\mathbf{f}^e_{uB} = - \int_{C^e} \mathbf{N}^T u n_x \, dC \tag{8-197a}$$

and

$$\mathbf{f}^e_{vB} = - \int_{C^e} \mathbf{N}^T v n_y \, dC \tag{8-197b}$$

Equations (8-196) and (8-197) are quite general in that they may be applied to any two-dimensional element. For example, if the three-node triangular element is used, $\mathbf{K}^e_{xx}$ and $\mathbf{K}^e_{yy}$ evaluate to

$$\mathbf{K}^e_{xx} = A \begin{bmatrix} m^2_{21} & m_{21}m_{22} & m_{21}m_{23} \\ m_{22}m_{21} & m^2_{22} & m_{22}m_{23} \\ m_{23}m_{21} & m_{23}m_{22} & m^2_{23} \end{bmatrix} \tag{8-198}$$

and

$$\mathbf{K}^e_{yy} = A \begin{bmatrix} m^2_{31} & m_{31}m_{32} & m_{31}m_{33} \\ m_{32}m_{31} & m^2_{32} & m_{32}m_{33} \\ m_{33}m_{31} & m_{33}m_{32} & m^2_{33} \end{bmatrix} \tag{8-199}$$

The integrals for $\mathbf{f}^e_{uB}$ and $\mathbf{f}^e_{vB}$ in Eqs. (8-197) need to be evaluated on the element boundaries only. However, each direction cosine takes on opposite signs on legs shared by two elements. In effect, the vectors $\mathbf{f}^e_{uB}$ and $\mathbf{f}^e_{vB}$ on an internal leg for one element cancel the corresponding vectors on the same leg for the adjacent element. The cancellation occurs at the assemblage step. Therefore, Eqs. (8-197) need to be evaluated only for those elements with at least one leg on the global boundary. For example, consider the triangular element shown in Fig. 8-24(a). Note that leg ij is on the global boundary. In Example 8-21, $\mathbf{f}^e_{uB}$ is evaluated for the case when u is assumed to vary linearly over the leg as shown in Fig. 8-24(b).

Example 8-21

Evaluate $\mathbf{f}^e_{uB}$ for the element shown in Fig. 8-24(b). Note that u is assumed to vary linearly over the leg such that at nodes i and j, the x components of velocity are u_i and u_j, respectively.

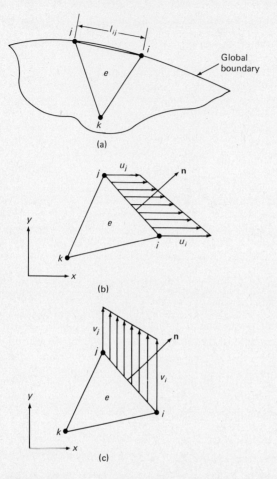

Figure 8-24 (a) Typical element with at least one leg of the triangular element on the global boundary. (b) Linearly varying x component of velocity on leg ij of triangle. (c) Linearly varying y component of velocity on leg ij of triangle.

Solution

On leg ij we have $N_k = L_k = 0$, and so we may write

$$u = L_i u_i + L_j u_j \qquad \text{(8-200)}$$

since this gives the desired linear variation of u over the leg. Equation (8-197a) becomes

$$\mathbf{f}^e_{uB} = -\int_{l_{ij}} \begin{bmatrix} L_i \\ L_j \\ 0 \end{bmatrix} (L_i u_i + L_j u_j) n_x \, dl \qquad \text{(8-201)}$$

where l_{ij} is the length of leg ij. If L_i and L_j are interpreted to be length coordinates, Eq. (6-48) may be used to integrate Eq. (8-201) to give

$$\mathbf{f}^e_{uB} = -\frac{l_{ij}n_x}{6} \begin{bmatrix} 2u_i + u_j \\ u_i + 2u_j \\ 0 \end{bmatrix} \qquad (8\text{-}202)$$

■

The reader should show that if v is assumed to vary linearly over leg ij as shown in Fig. 8-24(c), the result is

$$\mathbf{f}^e_{vB} = -\frac{l_{ij}n_y}{6} \begin{bmatrix} 2v_i + v_j \\ v_i + 2v_j \\ 0 \end{bmatrix} \qquad (8\text{-}203)$$

where v_i and v_j are the y components of velocity at nodes i and j, respectively. Similar expressions could be derived or written by inspection if leg jk or ki happens to be on the global boundary.

One method for determining the two direction cosines n_x and n_y for any leg of the triangular element is now presented. Let us arbitrarily concentrate on leg ij and define the vector $\mathbf{r}_{ij}$ as the vector running from node i to node j, as shown in Fig. 8-25. Therefore, we have

$$\mathbf{r}_{ij} = (x_j - x_i)\,\mathbf{i} + (y_j - y_i)\,\mathbf{j} \qquad (8\text{-}204)$$

where x_i and y_i are the coordinates of node i; x_j and y_j are the coordinates of node j; and $\mathbf{i}$ and $\mathbf{j}$ are the unit vectors along the x and y axes, respectively. The unit vector $\mathbf{n}$ is defined to be the outward normal to leg ij such that

$$\mathbf{n} = n_x\mathbf{i} + n_y\,\mathbf{j} \qquad (8\text{-}205)$$

where the direction cosines n_x and n_y are to be determined. The reader may recall that the cross product of two vectors results in another vector whose direction is determined by the right-hand screw rule moving from the first vector to the second

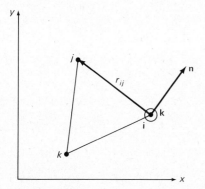

Figure 8-25 Vectors $\mathbf{r}_{ij}\mathbf{k}$ and $\mathbf{n}$ needed to determine n_x and n_y, the direction cosines.

through the smaller angle. If this is applied to the situation shown in Fig. 8-25, we may write

$$\mathbf{n} = \frac{\mathbf{r}_{ij} \times \mathbf{k}}{|\mathbf{r}_{ij} \times \mathbf{k}|} \tag{8-206}$$

where $\mathbf{k}$ is the unit vector in the z direction. The cross product of two vectors $\mathbf{a}$ and $\mathbf{b}$ is most easily evaluated from

$$\mathbf{a} \times \mathbf{b} = \det \begin{bmatrix} \mathbf{i} & \mathbf{j} & \mathbf{k} \\ a_x & a_y & a_z \\ b_x & b_y & b_z \end{bmatrix}$$

$$= (a_y b_z - a_z b_y)\,\mathbf{i} + (a_z b_x - a_x b_z)\,\mathbf{j} + (a_x b_y - a_y b_x)\,\mathbf{k} \tag{8-207}$$

where a_x, a_y, and a_z are the x, y, and z components of $\mathbf{a}$, etc. When Eqs. (8-205) and (8-206) are compared after the cross products are evaluated with the help of Eq. (8-207), the result is

$$n_x = \frac{y_j - y_i}{l_{ij}} \tag{8-208a}$$

and

$$n_y = -\frac{x_j - x_i}{l_{ij}} \tag{8-208b}$$

where l_{ij} is the length of leg ij and is given by

$$l_{ij} = \sqrt{(x_j - x_i)^2 + (y_j - y_i)^2} \tag{8-209}$$

Similar results may be obtained if leg jk or ki is on the global boundary.

Example 8-22

Determine the direction cosines n_x and n_y on leg ij for the element shown in Fig. 8-26.

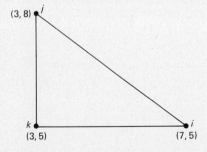

Figure 8-26   Triangular element in Example 8-22.

Solution

Using the nodal coordinates shown in Fig. 8-26, we get

$$n_x = \frac{8 - 5}{\sqrt{(3 - 7)^2 + (8 - 5)^2}} = \frac{3}{5} = 0.6$$

$$n_y = -\frac{3 - 7}{5} = 0.8$$

Note that the sum of the squares of the direction cosines is unity as it should be.

∎

On occasion it proves to be convenient to combine $\mathbf{f}_{uB}^e$ and $\mathbf{f}_{vB}^e$ into one term by noting that the normal velocity V_n may be written in terms of u and v as

$$V_n = un_x + vn_y \tag{8-210}$$

Equations (8-197) may be combined and $\mathbf{f}_{VB}^e$ may be defined as

$$\mathbf{f}_{VB}^e = \mathbf{f}_{uB}^e + \mathbf{f}_{vB}^e = -\int_{C^e} \mathbf{N}^T V_n \, dC \tag{8-211}$$

It also follows that

$$V_n = -\frac{\partial \phi}{\partial n} \tag{8-212}$$

where n represents the coordinate in the direction of the outward normal.

The Element Resultants

This completes the velocity potential formulation of two-dimensional, steady potential flow problems. The assemblage of each element stiffness matrix and nodal force vector is routine. Note that imposed velocities are not considered to be geometric boundary conditions; instead, nodal velocities are imposed via Eqs. (8-202) and (8-203) for the triangular element and by Eqs. (8-197) in general. The geometric boundary conditions are the prescribed velocity potentials. These may be applied with either Method 1 or 2 from Sec. 3-2. The resulting system of linear equations may be solved either by the matrix inversion method or by the active zone equation solver. The average velocities $\bar{u}$ and $\bar{v}$ for a typical triangular element e could then be determined from

$$\bar{u} = -\frac{\partial \phi}{\partial x} = -\frac{\partial \mathbf{N}}{\partial x} \mathbf{a}^e = -m_{21}\phi_i - m_{22}\phi_j - m_{23}\phi_k \tag{8-213a}$$

$$\bar{v} = -\frac{\partial \phi}{\partial y} = -\frac{\partial \mathbf{N}}{\partial y} \mathbf{a}^e = -m_{31}\phi_i - m_{32}\phi_j - m_{33}\phi_k \tag{8-213b}$$

where the m_{ij}'s are computed from Eqs. (6-21) and ϕ_i, ϕ_j, and ϕ_k are the now known nodal values of the potential. These average velocities are generally associated with the centroid of the triangular element. The method of quadrilateral averages could be used to improve these results.

Application: Flow around a Long Cylinder

Let us consider a long cylinder positioned transversely in a flow field between two flat plates as shown in Fig. 8-27(a). Because the cylinder is assumed to be long, end effects may be neglected; hence, the flow field is two-dimensional. Only the region outside of the boundary layer is considered so that viscous effects may be neglected. Let us also assume the fluid to be incompressible. Clearly, the potential flow formulation applies and the resulting streamlines and potential lines are shown in Fig. 8-27(b).

Because of the two-axis symmetry, only one-fourth of the region needs to be modeled as shown in Fig. 8-28. Note that the inlet velocity U enters the finite

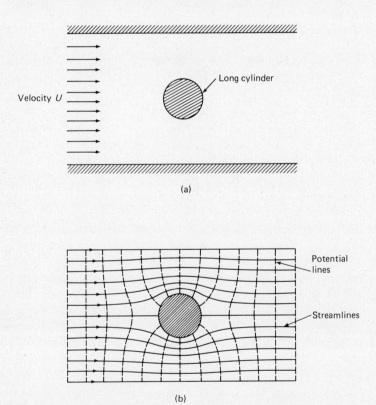

(a)

(b)

Figure 8-27 Application of two-dimensional, incompressible potential flow. (a) Long cylinder placed transversely in a uniform flow field between two flat plates and (b) the associated flow field around the cylinder. Lines of constant potential are always perpendicular to the streamlines.

element formulation via $\mathbf{f}_{uB}^e$ (or $\mathbf{f}_{vB}^e$) for those elements with a leg along side a-e. Also note that $\mathbf{f}_{uB}^e$ and $\mathbf{f}_{vB}^e$ (or $\mathbf{f}_{VB}^e$) are identically zero along sides a-b and d-e. Since the flow cannot penetrate the solid cylinder, V_n must be zero along the cylinder bc and hence $\mathbf{f}_{vB}^e$ is zero here. Finally, along side c-d, an arbitrary value of ϕ (a constant) must be chosen. This last boundary condition is not explicitly included in the finite element formulation and is imposed in the usual manner after the assemblage equations are formed.

Stream Function Formulation

The continuity equation given by Eq. (8-185) is satisfied exactly if the stream function ψ is defined by

$$u = \frac{\partial \psi}{\partial y}$$

(8-214a)

and

$$v = -\frac{\partial \psi}{\partial x}$$

(8-214b)

If these expressions for the velocity components are substituted in the irrotational flow condition given by Eq. (8-186), the result is the two-dimensional form of Laplace's equation or

$$\frac{\partial^2 \psi}{\partial x^2} + \frac{\partial^2 \psi}{\partial y^2} = 0$$

(8-215)

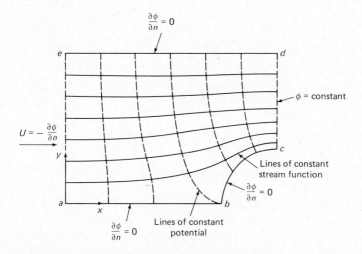

Figure 8-28 Only one-quarter of the flow field needs to be modeled because of the double symmetry. The boundary conditions shown apply to the velocity potential formulation.

It should be noted that lines of constant ψ are really streamlines, and lines of constant potential ϕ must always be orthogonal or perpendicular to these lines.

Using the Galerkin method on an element basis gives

$$\int_{A^e} \mathbf{N}^T \left[\frac{\partial^2 \psi}{\partial x^2} + \frac{\partial^2 \psi}{\partial y^2} \right] dx \, dy = \mathbf{0} \qquad \textbf{(8-216)}$$

where the shape function matrix $\mathbf{N}$ is dependent on the type of element used. Following the approach taken in the velocity potential formulation, the reader should be able to show that Eq. (8-216) can be written

$$\mathbf{K}^e \mathbf{a}^e = \mathbf{f}^e \qquad \textbf{(8-217)}$$

where

$$\mathbf{K}^e = \mathbf{K}_{xx}^e + \mathbf{K}_{yy}^e \qquad \textbf{(8-218)}$$

and

$$\mathbf{f}^e = \mathbf{f}_{uB}^e + \mathbf{f}_{vB}^e \qquad \textbf{(8-219)}$$

The element stiffness matrices are now defined by

$$\mathbf{K}_{xx}^e = \int_{A^e} \frac{\partial \mathbf{N}^T}{\partial x} \frac{\partial \mathbf{N}}{\partial x} dx \, dy \qquad \textbf{(8-220a)}$$

and

$$\mathbf{K}_{yy}^e = \int_{A^e} \frac{\partial \mathbf{N}^T}{\partial y} \frac{\partial \mathbf{N}}{\partial y} dx \, dy \qquad \textbf{(8-220b)}$$

and the element nodal force vectors by

$$\mathbf{f}_{uB}^e = \int_{C^e} \mathbf{N}^T u n_y \, dC \qquad \textbf{(8-221a)}$$

or

$$\mathbf{f}_{vB}^e = - \int_{C^e} \mathbf{N}^T v n_x \, dC \qquad \textbf{(8-221b)}$$

The evaluation of these element stiffness matrices and nodal force vectors for typical two-dimensional problems is left as an exercise. For the triangular element, the direction cosines can be evaluated with the help of Eqs. (8-208).

Application: Flow around a Long Cylinder

Let us reconsider the example from the velocity potential formulation: a long cylinder placed transversely in a flow field between two flat plates as shown in Fig. 8-27(a). Figure 8-27(b) shows the streamlines and lines of constant velocity potential. The streamlines actually represent lines of constant stream function ψ. Because of the two-axis symmetry, only one-fourth of the flow field needs to be modeled as shown in Fig. 8-29. The line a-b-c represents one streamline and hence

ψ is a constant on this line. Let us denote this constant value of ψ as ψ_1. The line d-e represents a different streamline. Let us denote the value of the stream function along d-e as ψ_2. Only one of these values is arbitrary, with the other value to be determined from physical considerations as shown below. Let us assume that the inlet velocity U on face a-e is constant. The velocity U can be related to ψ by Eq. (8-214a) or

$$U = \frac{\partial \psi}{\partial y} \qquad \text{(8-222a)}$$

Since U is constant, we may write

$$U = \frac{\psi_2 - \psi_1}{y_e - y_a} \qquad \text{(8-222b)}$$

If ψ_1 is taken arbitrarily to be zero, then ψ_2 is given as

$$\psi_2 = (y_e - y_a)U \qquad \text{(8-222c)}$$

where y_e and y_a denote the y coordinates of points e and a, respectively. Since there is no tangential velocity along faces a-e and c-d, it follows that

$$\mathbf{f}_{vB}^e = 0 \qquad \text{(8-223a)}$$

on these faces. Furthermore, since $n_y = 0$ on these faces, we also have

$$\mathbf{f}_{uB}^e = 0 \qquad \text{(8-223b)}$$

The conditions expressed by Eqs. (8-223) are equivalent to the condition that the streamlines are normal to the boundary. Hence

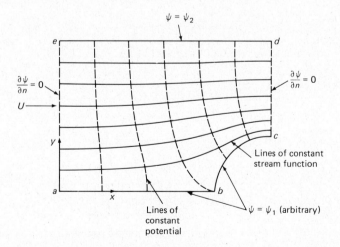

Figure 8-29   Only one-quarter of the flow field needs to be modeled because of the double symmetry. The boundary conditions shown apply to the stream function formulation.

$$\frac{\partial \psi}{\partial n} = 0 \qquad\qquad (8\text{-}224)$$

where n is the coordinate in the direction normal to the boundary. Clearly we have natural boundary conditions along faces $a\text{-}e$ and $c\text{-}d$.

8-12 CONVECTION DOMINATED FLOWS

The governing equations for heat conduction that we have seen in previous sections did not allow for any type of fluid motion within the region being analyzed. In general, fluid motion is very effective in transporting energy, usually much more effective than conduction alone. The purpose of this section is to provide an introduction on how the finite element method may be used to model such problems. Although a particular case is described and the FEM formulation given, the same basic approach may be applied to other similar problems. In this section, a particular velocity profile will be assumed for the purpose of illustrating how the temperature distribution may be obtained.

Let us consider the case of forced flow of a Newtonian, constant property, incompressible fluid between two flat plates as shown in Fig. 8-30. It is assumed that the flow is laminar and hydrodynamically fully developed, so that a parabolic velocity profile results. In this case the thermal energy equation alone needs to be solved for the temperature distribution. Various boundary conditions may be imposed as shown in Fig. 8-30. The governing equation to such a problem is given by

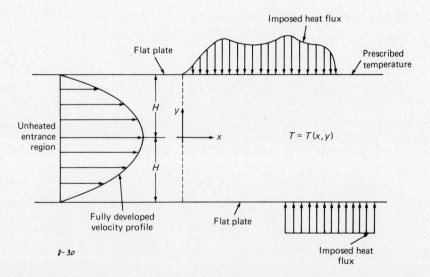

Figure 8-30 Thermal entrance region of a hydrodynamically, fully developed laminar flow of a Newtonian fluid between two long flat plates.

$$\rho c u(y) \frac{\partial T}{\partial x} = \frac{\partial}{\partial x}\left(k\frac{\partial T}{\partial x}\right) + \frac{\partial}{\partial y}\left(k\frac{\partial T}{\partial y}\right) \qquad \text{(8-225a)}$$

where T is the temperature of the fluid at any point (x,y), u is the x component of velocity, ρ is the fluid density, c is the specific heat (at constant pressure), and k is the thermal conductivity. Because of the hydrodynamically, fully developed flow assumption, the y component of velocity is assumed to be zero and, thus, is not present in Eq. (8-225a). The velocity u is given by

$$u(y) = \tfrac{3}{2}\bar{u}\left[1 - \left(\frac{y}{H}\right)^{2}\right] \qquad \text{(8-225b)}$$

where $\bar{u}$ is average velocity of the fluid and H is half of the plate separation distance as shown in Fig. 8-30.

In this formulation, the properties are assumed to be constant. However, we would proceed in much the same manner if c and k were functions of temperature. The solution (and the problem), however, would be nonlinear and the direct iteration method from Sec. 8-2 could be used to obtain the solution for the nodal temperatures. The density ρ may *not* be temperature-dependent because this would not result in a parabolic velocity profile. In fact, in this case the continuity, Navier-Stokes, and energy equations would have to be solved simultaneously. Because the formulation is two-dimensional, Eq. (8-225a) assumes that the two plates are infinitely long perpendicular to the paper.

It is interesting to note that, except for liquid metals, axial conduction may be neglected for Peclet numbers greater than 100 [7]. For liquid metals, axial conduction may be neglected for Peclet numbers greater than 160. The Peclet number is denoted as Pe and is defined to be the ratio of energy transport by convection to energy transport by conduction. If U represents the average velocity in the x direction, D_h is the hydraulic diameter, and α is the thermal diffusivity, the Peclet number is defined by

$$\text{Pe} = \frac{UD_h}{\alpha} \qquad \text{(8-226)}$$

where the thermal diffusivity is given by

$$\alpha = \frac{k}{\rho c} \qquad \text{(8-227)}$$

The hydraulic diameter D_h is defined by

$$D_h = \frac{4 \times \text{flow area}}{\text{wetted perimeter}} \qquad \text{(8-228)}$$

Note that the Peclet number is equal to the product of the Reynolds and Prandtl numbers, or

$$\text{Pe} = \text{Re Pr} \qquad \text{(8-229)}$$

where the Reynolds number is defined by

$$Re = \frac{UD_h}{\nu} \tag{8-230}$$

and the Prandtl number by

$$Pr = \frac{\nu}{\alpha} \tag{8-231}$$

where ν is the kinematic viscosity, related to the absolute viscosity μ by

$$\nu = \frac{\mu}{\rho} \tag{8-232}$$

In the case of two large flat plates separated by a distance $2H$, the Peclet number is given by

$$Pe = \frac{4UH}{\alpha} \tag{8-233}$$

The Galerkin method will be applied to Eq. (8-225a) in the usual manner. Interestingly it is not even possible to obtain the classical variational principle that corresponds to Eq. (8-225a). It will be seen that the Galerkin method in this case yields an unsymmetric stiffness matrix. The implications of this will be discussed briefly later. Try to make an educated guess as to which term in Eq. (8-225a) yields the unsymmetric stiffness matrix.

On an element basis, the Galerkin method requires that we write

$$\int_{A^e} \mathbf{N}^T \left[\rho c u \frac{\partial T}{\partial x} - \frac{\partial}{\partial x}\left(k\frac{\partial T}{\partial x}\right) - \frac{\partial}{\partial y}\left(k\frac{\partial T}{\partial y}\right) \right] t \, dx \, dy = 0 \tag{8-234}$$

The thickness t is introduced so that each term in Eq. (8-234) has dimensions of energy per unit time. We can take t to be unity for convenience. If the Green-Gauss theorem is applied to the two terms involving second-order derivatives, and if the boundary integrals are combined by using Eq. (8-48), we get

$$\int_{A^e} \mathbf{N}^T \rho c u \frac{\partial T}{\partial x} t \, dx \, dy + \int_{C^e} \mathbf{N}^T q_n t \, dC + \int_{A^e} \frac{\partial \mathbf{N}^T}{\partial x} kt \frac{\partial T}{\partial x} dx \, dy$$

$$+ \int_{A^e} \frac{\partial \mathbf{N}^T}{\partial y} kt \frac{\partial T}{\partial y} dx \, dy = 0 \tag{8-235}$$

If the heat fluxes q_{sB} on the global boundary are imposed in the direction toward the fluid, then we have

$$q_n = -q_{sB} \tag{8-236}$$

Furthermore, if we represent the temperature T on an element basis and write

$$T = \mathbf{N}\mathbf{a}^e \tag{8-237}$$

and note that the vector $\mathbf{a}^e$ is independent of x and y, Eq. (8-235) becomes

$$\mathbf{K}^e \mathbf{a}^e = \mathbf{f}^e \tag{8-238}$$

where the composite element stiffness matrix $\mathbf{K}^e$ is given by

$$\mathbf{K}^e = \mathbf{K}^e_{xx} + \mathbf{K}^e_{yy} + \mathbf{K}^e_u \tag{8-239}$$

and the element nodal force vector by

$$\mathbf{f}^e = \int_{C^e} \mathbf{N}^T q_{sB} t \, dC \tag{8-240}$$

The element stiffness matrices from conduction are identical to the expressions for $\mathbf{K}^e_{xx}$ and $\mathbf{K}^e_{yy}$ from Sec. 8-8 [see Eqs. (8-106)]. The element stiffness matrix·from the convective transport is given by

$$\mathbf{K}^e_u = \int_{A^e} \mathbf{N}^T \rho c u t \frac{\partial \mathbf{N}}{\partial x} \, dx \, dy \tag{8-241}$$

Note that $\mathbf{K}^e_u$ is unsymmetric, unlike $\mathbf{K}^e_{xx}$ and $\mathbf{K}^e_{yy}$. The element stiffness matrices and nodal force vector may be evaluated for particular two-dimensional elements, such as the triangular and rectangular elements. This is fairly routine and the details are omitted here.

The fact that one of the element stiffness matrices is unsymmetric deserves special attention. The assemblage of the element characteristics is done in the usual manner, except that the upper *and* lower parts of the assemblage stiffness matrix need to be stored. Recall from Sec. 6-8 that subroutine UACTCL [8] from Appendix C may be used in this case to obtain the solution for the nodal temperatures (after the prescribed temperatures have been imposed). In this subroutine, the upper triangular coefficients are stored in the array A, whereas the lower triangular coefficients are stored in the array C. Recall that unity is stored for the diagonal entries in the C array (the actual diagonal entries are stored in the A array).

Because the assemblage matrix is unsymmetric, the nodal temperatures that are calculated may or may not be correct. When the results are not correct, it is usually quite obvious—for example, the temperatures may be oscillatory with respect to the spacial coordinates. This oscillatory behavior should be expected if the elements are larger than a certain threshold size. More specifically, the threshold of oscillatory behavior is given in terms of a particular Peclet number referred to as the *cell* Peclet number defined as

$$\mathrm{Pe}_c = \frac{UL}{\alpha} \tag{8-242}$$

where L is a *characteristic element length*. Oscillatory behavior generally occurs when the cell Peclet number is greater than 2 [9] for one-dimensional models. Therefore, by reducing the size of the elements (and increasing the *number* of elements), we can generally improve the results after the threshold size is reached. Gresho and Lee [10] contend that the use of higher-order elements (see Chapter 9) will substantially improve the results. This contention has been verified further by Srivastiva [11] who did follow-up research on the study by Kundu and Stasa [12].

The Galerkin method is mathematically equivalent to a central finite difference. Convection-dominated flows generally require a backward (or upwind) difference to aid the convergence. In the finite element method, upwinding schemes also exist. References 13 and 14 should be consulted for more information on modeling convection–dominated flow problems. In these references, the use of upwinding schemes is also discussed. Further discussion of these topics is beyond the intended scope of this text.

8-13 INCOMPRESSIBLE VISCOUS FLOW

In this section, it will be shown how the finite element method may be used to solve for the velocity distribution in convection-dominated flows. Let us consider the laminar flow of a viscous fluid for the case of moderate Reynolds numbers such that the convective terms in the Navier-Stokes equations are significant. Typical examples would include external boundary-layer flows and internal flows. For example, we may want to determine the two-dimensional velocity field for a fluid moving through a duct with an obstruction in it as shown in Fig. 8-31. The flow field at the entrance of the duct may be assumed to be either fully developed or developing. Let us indicate how the finite element characteristics may be derived for such a complicated problem by starting with the two-dimensional form of the continuity and Navier-Stokes equations for a constant-property, incompressible, Newtonian fluid.

The continuity equation for an incompressible fluid is given by

$$\frac{\partial u}{\partial x} + \frac{\partial v}{\partial y} = 0 \tag{8-243}$$

where u and v are the x and y components of the fluid velocity at any point (x,y). The Navier-Stokes equations in the x and y directions are given by

$$\rho\left(u\frac{\partial u}{\partial x} + v\frac{\partial u}{\partial y}\right) = -\frac{\partial p}{\partial x} + \mu\left(\frac{\partial^2 u}{\partial x^2} + \frac{\partial^2 u}{\partial y^2}\right) + F_x \tag{8-244a}$$

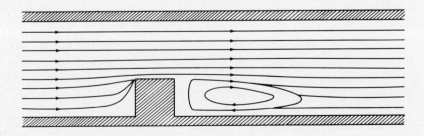

Figure 8-31 Viscous flow through an obstructed duct.

$$\rho\left(u\frac{\partial v}{\partial x} + v\frac{\partial v}{\partial y}\right) = -\frac{\partial p}{\partial y} + \mu\left(\frac{\partial^2 v}{\partial x^2} + \frac{\partial^2 v}{\partial y^2}\right) + F_y \qquad \textbf{(8-244b)}$$

where ρ is the fluid density, p is the pressure, μ is the absolute viscosity, and F_x and F_y are the body forces per unit volume in the x and y directions, respectively. In Eqs. (8-244) the convective acceleration terms [on the left-hand side of Eqs. (8-244)] make the problem to be solved nonlinear. The region to be analyzed may be discretized in the usual manner with triangular or rectangular elements. The goal is to determine the nodal values of the pressure and the velocity components, since the formulation is to be based on the so-called *primitive variables*.

If the Galerkin method is applied to Eq. (8-243), we get

$$\int_{A^e} \mathbf{N}^T\left(\frac{\partial u}{\partial x} + \frac{\partial v}{\partial y}\right) t\, dx\, dy = \mathbf{0} \qquad \textbf{(8-245)}$$

where the thickness t is introduced so that each term in Eq. (8-245) has dimensions of volume per unit time (i.e., a volumetric flow rate). For convenience, t can be taken to be unity. Let us assume

$$u = \mathbf{N}\mathbf{a}_u^e \qquad \textbf{(8-246a)}$$

and

$$v = \mathbf{N}\mathbf{a}_v^e \qquad \textbf{(8-246b)}$$

where $\mathbf{a}_u^e$ and $\mathbf{a}_v^e$ are given by

$$\mathbf{a}_u^e = [u_i \quad u_j \quad u_k]^T \qquad \textbf{(8-247a)}$$

and

$$\mathbf{a}_v^e = [v_i \quad v_j \quad v_k]^T \qquad \textbf{(8-247b)}$$

for the triangular element. In Eqs. (8-247), u_i, u_j, and u_k are the x components of the velocity at nodes i, j, and k, etc. Since $\mathbf{a}_u^e$ and $\mathbf{a}_v^e$ are independent of x and y, Eq. (8-245) can be written in the form

$$\mathbf{K}_u^e\mathbf{a}_u^e + \mathbf{K}_v^e\mathbf{a}_v^e = \mathbf{0} \qquad \textbf{(8-248)}$$

where

$$\mathbf{K}_u^e = \int_{A^e} \mathbf{N}^T \frac{\partial \mathbf{N}}{\partial x} t\, dx\, dy \qquad \textbf{(8-249)}$$

and

$$\mathbf{K}_v^e = \int_{A^e} \mathbf{N}^T \frac{\partial \mathbf{N}}{\partial y} t\, dx\, dy \qquad \textbf{(8-250)}$$

Let us now apply the Galerkin method to Eq. (8-244a) by writing

$$\int_{A^e} \mathbf{N}^T\left[\rho\left(u\frac{\partial u}{\partial x} + v\frac{\partial u}{\partial y}\right) + \frac{\partial p}{\partial x} - \mu\left(\frac{\partial^2 u}{\partial x^2} + \frac{\partial^2 u}{\partial y^2}\right) - F_x\right] t\, dx\, dy = 0 \qquad \textbf{(8-251)}$$

Note that each term in Eq. (8-251) has units of force. The Green-Guass theorem may be applied in the usual manner to the two terms involving second-order derivatives. However, the integrals around the element boundary (over C^e) need not be considered further because the velocity components are generally prescribed on the global boundary. The boundary integral terms are related to the shear stresses (which are not generally prescribed). Noting Eqs. (8-246) and representing the pressure p over the element by

$$p = \mathbf{N}\mathbf{a}_p^e \tag{8-252}$$

Eq. (8-251) may be written in the form

$$(\mathbf{K}_{uu}^e + \mathbf{K}_{vu}^e + \mathbf{K}_{ux}^e + \mathbf{K}_{uy}^e)\mathbf{a}_u^e + \mathbf{K}_{px}^e\mathbf{a}_p^e = \mathbf{f}_x^e \tag{8-253}$$

where

$$\mathbf{K}_{uu}^e = \int_{A^e} \mathbf{N}^T \rho \, \mathbf{N}\mathbf{a}_u^e \frac{\partial \mathbf{N}}{\partial x} t \, dx \, dy \tag{8-254a}$$

$$\mathbf{K}_{vu}^e = \int_{A^e} \mathbf{N}^T \rho \, \mathbf{N}\mathbf{a}_v^e \frac{\partial \mathbf{N}}{\partial y} t \, dx \, dy \tag{8-254b}$$

$$\mathbf{K}_{ux}^e = \int_{A^e} \frac{\partial \mathbf{N}^T}{\partial x} \mu \frac{\partial \mathbf{N}}{\partial x} t \, dx \, dy \tag{8-254c}$$

$$\mathbf{K}_{uy}^e = \int_{A^e} \frac{\partial \mathbf{N}^T}{\partial y} \mu \frac{\partial \mathbf{N}}{\partial y} t \, dx \, dy \tag{8-254d}$$

$$\mathbf{K}_{px}^e = \int_{A^e} \mathbf{N}^T \frac{\partial \mathbf{N}}{\partial x} t \, dx \, dy \tag{8-254e}$$

and

$$\mathbf{f}_x^e = \int_{A^e} \mathbf{N}^T F_x t \, dx \, dy \tag{8-254f}$$

In a similar fashion, it can be shown that if the Galerkin method is applied to Eq. (8-244b), the result is

$$(\mathbf{K}_{uv}^e + \mathbf{K}_{vv}^e + \mathbf{K}_{vx}^e + \mathbf{K}_{vy}^e)\mathbf{a}_v^e + \mathbf{K}_{py}^e\mathbf{a}_p^e = \mathbf{f}_y^e \tag{8-255}$$

where

$$\mathbf{K}_{uv}^e = \mathbf{K}_{uu}^e = \int_{A^e} \mathbf{N}^T \rho \, \mathbf{N}\mathbf{a}_u^e \frac{\partial \mathbf{N}}{\partial x} t \, dx \, dy \tag{8-256a}$$

$$\mathbf{K}_{vv}^e = \mathbf{K}_{vu}^e = \int_{A^e} \mathbf{N}^T \rho \, \mathbf{N}\mathbf{a}_v^e \frac{\partial \mathbf{N}}{\partial y} t \, dx \, dy \tag{8-256b}$$

$$\mathbf{K}_{vx}^e = \mathbf{K}_{ux}^e = \int_{A^e} \frac{\partial \mathbf{N}^T}{\partial x} \mu \frac{\partial \mathbf{N}}{\partial x} t \, dx \, dy \tag{8-256c}$$

$$\mathbf{K}_{vy}^e = \mathbf{K}_{uy}^e = \int_{A^e} \frac{\partial \mathbf{N}^T}{\partial y} \mu \frac{\partial \mathbf{N}}{\partial y} t \, dx \, dy \tag{8-256d}$$

$$\mathbf{K}_{py}^e = \int_{A^e} \mathbf{N}^T \frac{\partial \mathbf{N}}{\partial y} t \, dx \, dy \tag{8-256e}$$

and

$$\mathbf{f}_y^e = \int_{A^e} \mathbf{N}^T F_y t \, dx \, dy \tag{8-256f}$$

There are two different ways in which the assemblage step may be performed. The first step in both procedures is to write Eqs. (8-248), (8-253), and (8-255) in the following partitioned matrix form:

$$\left[\begin{array}{c|c|c} \mathbf{K}_u^e & \mathbf{K}_v^e & \mathbf{0} \\ \hline \mathbf{K}_{uu}^e + \mathbf{K}_{vu}^e + \mathbf{K}_{ux}^e + \mathbf{K}_{uy}^e & \mathbf{0} & \mathbf{K}_{px}^e \\ \hline \mathbf{0} & \mathbf{K}_{uv}^e + \mathbf{K}_{vv}^e + \mathbf{K}_{vx}^e + \mathbf{K}_{vy}^e & \mathbf{K}_{py}^e \end{array}\right] \left[\begin{array}{c} \mathbf{a}_u^e \\ \mathbf{a}_v^e \\ \mathbf{a}_p^e \end{array}\right] = \left[\begin{array}{c} \mathbf{0} \\ \mathbf{f}_x^e \\ \mathbf{f}_y^e \end{array}\right] \tag{8-257}$$

In the first method of assemblage, the individual terms in Eq. (8-257) are assembled such that the vector of nodal unknowns is given by

$$\mathbf{a} = [u_1 \quad u_2 \quad \cdots \quad u_n \mid v_1 \quad v_2 \quad \cdots \quad v_n \mid p_1 \quad p_2 \quad \cdots \quad p_n]^T \tag{8-258}$$

where n nodes are assumed in the discretization of the region being analyzed. This type of assemblage results in a significantly increased bandwidth of the assemblage system equations. Therefore, a more practical method of assemblage is needed, and one such method is presented below.

Let us denote the submatrices and subvectors in Eq. (8-257) as follows:

$$\left[\begin{array}{c|c|c} \mathbf{K}^{11} & \mathbf{K}^{12} & \mathbf{K}^{13} \\ \hline \mathbf{K}^{21} & \mathbf{K}^{22} & \mathbf{K}^{23} \\ \hline \mathbf{K}^{31} & \mathbf{K}^{32} & \mathbf{K}^{33} \end{array}\right] \left[\begin{array}{c} \mathbf{a}_u^e \\ \mathbf{a}_v^e \\ \mathbf{a}_p^e \end{array}\right] = \left[\begin{array}{c} \mathbf{f}^1 \\ \mathbf{f}^2 \\ \mathbf{f}^3 \end{array}\right] \tag{8-259}$$

Note the use of the numerical superscripts to indicate the relative position of each submatrix and subvector in Eq. (8-259). Let us now use subscripts to indicate the entries within each submatrix. For example, *if the three-node triangular element is used*, let us write

$$\mathbf{K}_u^e = \mathbf{K}^{11} = \begin{bmatrix} K_{11}^{11} & K_{12}^{11} & K_{13}^{11} \\ K_{21}^{11} & K_{22}^{11} & K_{23}^{11} \\ K_{31}^{11} & K_{32}^{11} & K_{33}^{11} \end{bmatrix} \tag{8-260a}$$

and

$$\mathbf{f}_x^e = \mathbf{f}^2 = \begin{bmatrix} f_1^2 \\ f_2^2 \\ f_3^2 \end{bmatrix} \tag{8-260b}$$

and so forth. Note that $\mathbf{K}^{13}$, $\mathbf{K}^{22}$, and $\mathbf{K}^{31}$ are 3×3 null matrices and $\mathbf{f}^1$ is a 3×1 null vector. Let us now do a *miniassemblage* such that the vector of nodal unknowns *on an element basis* is given by

$$\mathbf{a}^e = [u_i \quad v_i \quad p_i \mid u_j \quad v_j \quad p_j \mid u_k \quad v_k \quad p_k]^T \tag{8-261}$$

Using the nomenclature illustrated by Eqs. (8-260), Eq. (8-257) may be written

$$
\begin{bmatrix}
K^{11}_{11} & K^{12}_{11} & K^{13}_{11} & K^{11}_{12} & K^{12}_{12} & K^{13}_{12} & K^{11}_{13} & K^{12}_{13} & K^{13}_{13} \\
K^{21}_{11} & K^{22}_{11} & K^{23}_{11} & K^{21}_{12} & K^{22}_{12} & K^{23}_{12} & K^{21}_{13} & K^{22}_{13} & K^{23}_{13} \\
K^{31}_{11} & K^{32}_{11} & K^{33}_{11} & K^{31}_{12} & K^{32}_{12} & K^{33}_{12} & K^{31}_{13} & K^{32}_{13} & K^{33}_{13} \\
K^{11}_{21} & K^{12}_{21} & K^{13}_{21} & K^{11}_{22} & K^{12}_{22} & K^{13}_{22} & K^{11}_{23} & K^{12}_{23} & K^{13}_{23} \\
K^{21}_{21} & K^{22}_{21} & K^{23}_{21} & K^{21}_{22} & K^{22}_{22} & K^{23}_{22} & K^{21}_{23} & K^{22}_{23} & K^{23}_{23} \\
K^{31}_{21} & K^{32}_{21} & K^{33}_{21} & K^{31}_{22} & K^{32}_{22} & K^{33}_{22} & K^{31}_{23} & K^{32}_{23} & K^{33}_{23} \\
K^{11}_{31} & K^{12}_{31} & K^{13}_{31} & K^{11}_{32} & K^{12}_{32} & K^{13}_{32} & K^{11}_{33} & K^{12}_{33} & K^{13}_{33} \\
K^{21}_{31} & K^{22}_{31} & K^{23}_{31} & K^{21}_{32} & K^{22}_{32} & K^{23}_{32} & K^{21}_{33} & K^{22}_{33} & K^{23}_{33} \\
K^{31}_{31} & K^{32}_{31} & K^{33}_{31} & K^{31}_{32} & K^{32}_{32} & K^{33}_{32} & K^{31}_{33} & K^{32}_{33} & K^{33}_{33}
\end{bmatrix}
\begin{bmatrix}
u_i \\ v_i \\ p_i \\ u_j \\ v_j \\ p_j \\ u_k \\ v_k \\ p_k
\end{bmatrix}
=
\begin{bmatrix}
f_1^1 \\ f_1^2 \\ f_1^3 \\ f_2^1 \\ f_2^2 \\ f_2^3 \\ f_3^1 \\ f_3^2 \\ f_3^3
\end{bmatrix}
$$

$$\text{(8-262)}$$

The assemblage of Eq. (8-262) may now be done in the usual manner such that the vector of nodal unknowns is given by

$$\mathbf{a} = [u_1 \quad v_1 \quad p_1 \mid u_2 \quad v_2 \quad p_2 \mid \cdots \mid u_n \quad v_n \quad p_n]^T \qquad \text{(8-263)}$$

where again n nodes are assumed in the discretization of the region being analyzed. Note that this type of assemblage was done routinely in all stress analyses in Chapter 7 and results in a significantly smaller bandwidth than would result otherwise. Prescribed velocities and pressures would be applied by using either of the two methods from Sec. 3-2. Since the assemblage system equations are unsymmetric, subroutine UACTCL [8] from Sec. 6-8 (see Appendix C) may be used to obtain the solution for the nodal velocities and pressures.

The above formulation is based on the so-called primitive variables: the velocities and pressure. Alternate formulations exist that utilize a stream function. As in the analysis of two-dimensional potential flow problems, the stream function is defined such that the continuity equation is satisfied exactly. The three governing equations reduce to two partial differential equations, and the order of the highest derivative present increases from two to three. The interested reader should consult References 15 and 16 for more details on the stream function formulation, as well as another approach referred to as the vorticity formulation.

It has been assumed above that the same shape functions may be used for the pressure and velocity functions. This seems to give good results for rather low fluid velocities. For high fluid velocities, the elements (and hence shape functions) used for the velocity function should be one order higher than those used for pressure. Several such higher-order elements are presented in Chapter 9. Again the reader is referred to References 15 and 16 for further details.

8-14 DEVELOPMENT OF A TWO-DIMENSIONAL THERMAL ANALYSIS PROGRAM

Some helpful hints and comments are given in this section so that the reader can develop a two-dimensional thermal analysis program with further instructions from the instructor.* Specific comments are made with respect to the main program, mesh generation, data storage, variable properties, and point heat sources.

The Main Program

The main program should be kept as brief as possible. An example is the following seven lines of code:

```
PROGRAM HEAT
COMMON XL(2500)
MAX = 2500
LCONSL = 3
CALL DRIVER (MAX,LCONSL)
CALL EXIT
END
```

The variable MAX represents the length of XL array, the contents of which are described below. This variable should be assigned the value of the dimension of XL in the unlabeled COMMON. Much larger problems may be solved with the program by increasing this parameter to the memory limit of the computer being used. The variable LCONSL (as in the TRUSS program) represents the logical unit number for the console. Control is then transfered to subroutine DRIVER.

Mesh Generation

The same type of node generator discussed in Sec. 3-6 may be used in this program except that the nodes no longer need to be equally spaced. This may be accomplished with the help of two spacing factors f_x and f_y (FX and FY in the program). These factors are used as described below. This same approach is used in the stress analysis program described in Sec. 7-7. The reader who is familiar with the mesh generator in the stress analysis program may wish to skip this section.

Consider the starting node NI and the final node NF shown in Figure 8-32(a). Additional nodes can be generated between these two end nodes if a nonzero value of NG is used. The nodes need not be equally spaced, however. Without loss of generality, let us number the nodes $1, 2, \ldots, n$, as shown in Figure 8-32(b), and refer to the x and y coordinate of node 1 by x_1, y_1, etc. Then the spacing factors f_x and f_y may be defined by

---

*A two-dimensional thermal analysis program and a user's manual are included in the Instructor/Solution's manual for this text.

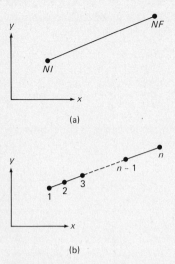

(a)

(b)

Figure 8-32   (a) Starting node *NI* and ending node *NF* are used to generate additional nodes. (b) Line along which *n* nodes are generated (not necessarily equally spaced).

$$f_x = \frac{x_3 - x_2}{x_2 - x_1} = \frac{x_4 - x_3}{x_3 - x_2} = \cdots = \frac{x_n - x_{n-1}}{x_{n-1} - x_{n-2}} \tag{8-264}$$

and

$$f_y = \frac{y_3 - y_2}{y_2 - y_1} = \frac{y_4 - y_3}{y_3 - y_2} = \cdots = \frac{y_n - y_{n-1}}{y_{n-1} - y_{n-2}} \tag{8-265}$$

Concentrating on the x direction for now, we may write

$$x_2 - x_1 = x_2 - x_1 \tag{8-266a}$$

$$x_3 - x_2 = f_x(x_2 - x_1) \tag{8-266b}$$

$$x_4 - x_3 = f_x(x_3 - x_2) = f_x^2(x_2 - x_1) \tag{8-266c}$$

$$\vdots$$

$$x_n - x_{n-1} = f_x(x_{n-1} - x_{n-2}) = f_x^{n-2}(x_2 - x_1) \tag{8-266d}$$

Adding these results gives

$$x_n - x_1 = [1 + f_x + f_x^2 + \cdots + f_x^{n-2}](x_2 - x_1) \tag{8-267}$$

from which it follows that

$$x_2 = x_1 + \frac{x_n - x_1}{1 + \sum_{i=1}^{n-2} f_x^i} \tag{8-268}$$

Since x_1 and x_n may be input (as XI and XF, respectively), and since f_x may be input (as FX), x_2 can be found from Eq. (8-268). Note that n is the total number of nodes in the generation sequence (including NI and NF). Obviously x_3 can be computed from Eq. (8-266b) or

$$x_3 = x_2 + f_x(x_2 - x_1) \tag{8-269}$$

and so forth. It should also be obvious that in the y direction, the same results hold except that each x_i is replaced by y_i and f_x is replaced by f_y. Note that if f_x (or f_y) is greater than unity, then the nodes are spaced farther apart in moving from NI to NF; if f_x (or f_y) is less than unity (but greater than zero), then the spacing between two consecutive nodes decreases in moving from NI to NF.

Data Storage

The data for the nodal coordinates, elements, boundary condition flags, material properties, boundary condition parameters, and so forth should be stored in the XL array. In addition, the assemblage stiffness matrix (in column vector form), the assemblage nodal force vector, and the diagonal pointer array (JDIAG from Sec. 6-8) should also be stored in the XL array. The partitions between each of these sections should float and if fewer nodes are used, for example, more materials may be present. The total number of storage locations required must not exceed MAX. If it does, an error message should be displayed on the console and execution should be terminated.

The numbering of the nodes is critical if the program does not renumber the nodes to minimize the bandwidth of the assemblage stiffness matrix. However, the program should store only the banded portion of this matrix, which reduces the storage requirements drastically over that of storing the full matrix. Nodes on the object being analyzed are always numbered consecutively, from 1 to the maximum number of nodes. The bandwidth is minimized when the maximum difference between any two nodes on each element is minimized. Figure 8-33 shows that this is accomplished quite simply by numbering the nodes in the direction of fewer nodes.

Note that in Fig. 8-33(a) the nodes are numbered in the direction of the smaller number of nodes. In Fig. 8-33(b) the nodes are numbered to the right and in the direction of the greater number of nodes. The storage requirements for the mesh in the latter are higher than for the mesh in the former. In Fig. 8-33(c) the nodes are numbered in an alternating sweeping fashion which more than doubles the storage requirements over that required in Fig. 8-33(a). Finally, in Fig. 8-33(d) the element on the lower left has nodes 1, 2, and 21; the implication is that the stiffness matrix is no longer banded which in turn means higher storage requirements and increased execution times. Although the program need not store the zero terms in the stiffness matrix outside the bandwidth, it should store everything within the bandwidth.

Numbering the elements is *not* critical but some sort of regular numbering scheme usually allows the use of the automatic element generation feature. Note

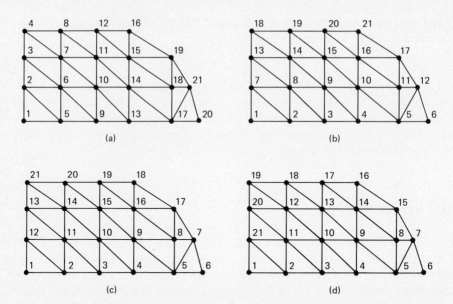

Figure 8-33   (a) Proper node numbering scheme; (b) less desirable node numbering scheme; (c) nodes should never be numbered in an alternating sweeping fashion; and (d) worst node numbering scheme (results in an unbanded stiffness matrix).

that in forming the elements shown in the lower right corner of Figure 8-33(a), the quadrilateral formed by nodes 17, 20, 21, and 18 is divided into two triangles by using the shorter diagonal (node 17 to node 21) as opposed to the longer diagonal (node 18 to node 20). This is illustrated in Fig. 8-34. Regular triangles generally give better results than obtuse or *needle-shaped* triangles.

Variable Property Routine: SUBROUTINE VPROP

It is relatively easy to provide the capability to handle both spacially dependent and temperature-dependent properties. For example, the thermal conductivity, heat transfer coefficients, etc. may be functions of either the global coordinates x and y or the temperature T. If any of the properties is a function of temperature, the problem becomes nonlinear and the direct iteration method from Sec. 8-2 is used to obtain the solution of the nodal temperatures.

The method that may be used to accomplish this is now described. If any of the material or boundary condition properties is a function of x, y, and/or T, then the user should simply enter a *unique negative integer* from -1 to -10 instead of an actual value for the property. In the discussion below, these unique negative integers are referred to as the *variable-property indicators*. For these properties, the program should be designed to call SUBROUTINE VPROP, an abridged listing of which is given below:

```
SUBROUTINE VPROP (PROP)
```

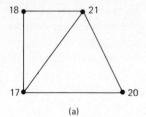

(a)

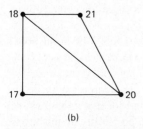

(b)

Figure 8-34   (a) Preferred way of dividing a quadrilateral into two triangles and (b) less desirable way of forming two triangles.

```
COMMON  /VPHELP/  X, Y, T
COMMON  /CONST /  PI, SIGMA, C(30)
IPROP = IABS(PROP)
GO TO (1, 2, 3, 4, 5, 6, 7, 8, 9, 10), IPROP
```
1   PROP =   definition of first variable property (may be a function of
                 X, Y, and/or T)
```
RETURN
```
2   PROP = definition of second variable property
```
RETURN

      ⋮
```
10 PROP = definition of tenth variable property
```
RETURN
END
```

The reader may want to examine SUBROUTINE PROPTY in Problem 4-78 in order to see how this may be implemented. The basic idea is that if a negative value is read for a property, the program (SUBROUTINE PROPTY in particular) should call SUBROUTINE VPROP when that property is needed and transfers control to the statement whose label is equal to the absolute value of the variable-property indicator. Note that a computed GO TO statement is used. Also note that the user may define up to 30 constants [C(1), C(2), . . . , C(30)] that may be used in the SUBROUTINE VPROP. These constants should be read by the program in the first input section. For example, let us say that the thermal conductivity is known to vary quadratically with temperature as

$$k = 32 + 10T - 0.5T^2$$

The user simply needs to decide which variable-property indicator he or she would like to use subject to the conditions that it be a unique negative integer between −1 and −10. A unique integer is one that is not used to represent any other variable property. For example, let us say that variable-property indicator 3 is not being used for any other property. In the input file, the user simply enters "−3" for the thermal conductivities of the materials for which the above equation applies. SUB-ROUTINE VPROP is modified such that the statement with the label of 3 reads:

```
3  PROP = C(1) + C(2)*T + C(3)*T*T
```

The constants C(1), C(2), and C(3) are three user-defined constants described above. It should be clear that C(1) = 32., C(2) = 10., and C(3) = −0.5. Recall that these constants should be read in *Input Section 1*. Note the use of T*T in the above expression instead of T**2, since the former executes faster than the latter.

With respect to the material property data, the reader may note that two thermal conductivities could be input for each material. These are interpreted to be the principal values of thermal conductivity in the local x and local y directions, respectively (i.e., the x' and y' directions) and allows for the possibility of anisotropic materials. Recall that anisotropic materials are materials that exhibit a direction sensitivity as shown in Fig. 8-35(a). Note that θ is the angle that the x' axis makes with the global x axis (θ is also the angle between the y' axis and the global y axis).

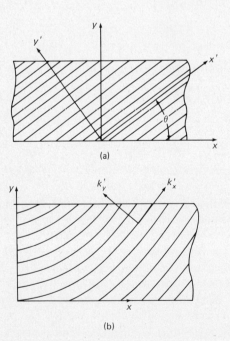

(a)

(b)

Figure 8-35 Heat conduction in anisotropic materials. (a) The principal values of thermal conductivity occur in the x' and y' directions. (b) The orientation of the local $x'y'$ coordinate system may change from point to point.

The thermal conductivities in the x' and y' directions may be denoted as TKX and TKY, respectively, in the program, and the angle θ by THETA. Note that materials such as the one depicted in Fig. 8-35(b) may be modeled with the program by defining THETA as a function of x and y (with the variable-property routine). The reader should review Problems 8-115 to 8-117 for more information on how anisotropic materials are handled with the finite element method.

Point Heat Sources

Point heat sources may be implemented in the program by specifying the value of the heat source (in units of energy per unit time and unit thickness) and the x and y coordinates of the location of the source. Any number of points sources should be present (including more than one source in an element) up to the memory limits of the computer. In the program, the element in which each source is located may be determined by checking the values of the shape functions at the location of the source. If each shape function has a value between zero and unity, then the element in which the source is located has been found. Four storage locations should be reserved in the XL array for each point source. The following information should be stored in these locations: the value of the source, the x coordinate, the y coordinate, and a flag. The flag should be equal to zero before the element is found and is equal to the element number after the proper element is found (i.e., the element that contains the source). In this way, some execution time will be saved since it is not necessary to check the locations of a source if the element in which it occurs is known. Also, this technique ensures that a source which may be on the boundary between two elements is added into the assemblage nodal force vector only once. The reader should be convinced that this approach will also yield correct results if the source happens to be at the same location as a node.

8-15   REMARKS

This chapter illustrates how the finite element method is used in steady-state thermal and fluid flow analyses. In particular, the finite element formulations of steady-state heat conduction problems in one-, two-, and three-dimensions were developed in detail. Also included was axisymmetric heat conduction. The full complement of thermal loads and boundary conditions were considered, including simple radiation to or from a large enclosure in several of the models. The extension to internodal radiation (or interelement radiation in the case of FEM) is beyond the scope of this book. Finite element formulations for steady-state, two-dimensional, incompressible potential flow were also provided. Both the velocity potential and stream function formulations were presented. Viscous effects are insignificant under the assumption of potential flow.

Various types of problems involving the flow of viscous fluids were also considered. The first type was a problem in which the energy equation including the convective energy transport term provided the governing equation. In this case the velocity profile was assumed to be given, and the main purpose of the analysis was to compute the temperature distribution in the flow field. The Galerkin method was seen to provide a relatively straightforward analysis, whereas the variational method simply could not be used (why not?). An unsymmetric stiffness matrix was seen to arise for the first time, because of the convective energy term in the governing differential equation. It was mentioned that the numerical solution may be ill-behaved unless relatively small elements (or higher-order elements) are used.

The finite element formulation was provided for the velocity and pressure distribution in a viscous fluid flowing in a two-dimensional region under steady-state conditions. The formulation was based on the Galerkin method. At first glance, the assemblage step appeared to be different from what we have seen so far. However, a closer examination revealed that the same type of assemblage procedure was used in two- and three-dimensional stress analyses in Chapter 7. Again unsymmetrical stiffness matrices arose, and again it may be necessary to use many relatively small elements (or higher-order elements) to obtain convergence. The convective acceleration terms in the Navier-Stokes equation make the formulation nonlinear. The direct iteration method from Sec. 8-2 may be used to obtain the solution for the nodal velocities and pressures. Only the formulation with the so-called primitive variables was provided in the formal development. Alternate formulations such as the stream function and vorticity formulations exist, but were not presented in detail here.

The two-dimensional, steady-state thermal analysis program was described. The program is capable of handling variable properties and allows graded meshes.

Although substructuring was introduced in Chapter 7 with an application to structural and/or stress analysis, it may also be applied to nonstructural problems. It should be recalled that substructuring makes the use of large finite element models quite practical. In fact, significant problems may be solved on microcomputers if substructuring is used. The reader should make an attempt to adapt the material in Sec. 7-6 to the analysis of two-dimensional, steady-state heat conduction with the finite element method.

REFERENCES

1. Arpaci, V. S., *Conduction Heat Transfer*, Addison-Wesley, Reading, Mass., 1966, p. 32.
2. Huebner, K. H., *The Finite Element Method for Engineers*, Wiley, New York, 1975, pp. 218–219.
3. Wilson, E. L., "Structural Analysis of Axisymmetric Solids," *AIAA Journal*, vol. 3., no. 12, pp. 2267–2274, December 1965.
4. Roberson, J. A., and C. T. Crowe, *Engineering Fluid Mechanics*, 2nd ed., Houghton Mifflin, New Jersey, 1980, pp. 437–449.
5. Shames, I. H., *Mechanics of Fluids*, McGraw-Hill, New York, 1962, pp. 173–185.

6. Currie, I. G., *Fundamental Mechanics of Fluids*, McGraw-Hill, New York, 1974, pp. 65–142.

7. Rohsenow, W. M., and J. P. Hartnett, eds., *Handbook of Heat Transfer*, McGraw-Hill, New York, 1973.

8. Zienkiewicz, O. C., *The Finite Element Method*, McGraw-Hill (UK), London, 1977, pp. 746–747.

9. Leonard, B. P., "A Survey of Finite Differences of Opinion on Numerical Muddling of the Incomprehensible Defective Confusion Equation," *Proc. ASME*, AMD-vol. 34, pp. 1–18, New York, December 2–7, 1979.

10. Gresho, P., and R. L. Lee, "Don't Suppress the Wiggles—They're Telling You Something," *Proc. ASME*, AMD-vol. 34, pp. 37–62, December 2–7, 1979.

11. Srivastava, A. K., "A Finite Element Method to Solve the Graetz Problem Using Subparametric Elements," M.S. Thesis, Florida Institute of Technology, Melbourne, Fla., July 1982.

12. Kundu, D., and F. L. Stasa, "Finite Element Solution to the Graetz Problem and Its Extensions," International Congress on Technology and Technology Exchange, Pittsburgh, May 3–6, 1982.

13. Hughes, T. J. R., ed., *Finite Elements Methods for Convection Dominated Flows*, *Proc. of the ASME*, AMD-vol. 34, December 2–7, 1979.

14. Chung, T. J., *Finite Element Analysis in Fluid Dynamics*, McGraw-Hill, New York, 1975.

15. Baker, A. J., *Finite Element Computational Fluid Mechanics*, McGraw-Hill, New York, 1983, pp. 233–305.

16. Huebner, K. H., *The Finite Element Method for Engineers*, Wiley, New York, 1975, pp. 346–361.

PROBLEMS

Note: The properties in Appendix A should be used unless indicated otherwise.

8-1 Use the direct iteration method to obtain a solution to the following nonlinear algebraic equation:

$$x^3 - 3x^2 + 6x - 4 = 0$$

Stop the solution process when three significant digits of accuracy are obtained. Use an initial guess of

a. $x = 0$ **b.** $x = 10$ **c.** $x = -5$

8-2 Use the direct iteration method to obtain a solution to the following nonlinear algebraic equation:

$$x^4 - 5x^3 + 10x^2 + 4x - 24 = 0$$

Terminate the solution process when three significant digits of accuracy are obtained. Use an initial guess of

a. $x = 0$ **b.** $x = 10$ **c.** $c = -1$

8-3 Use the direct iteration method to obtain a solution to the following nonlinear transcendental equation:

$$6x \cos \pi x = 1$$

Use an initial guess of $x = 0$ and stop the iterations when an accuracy of three significant digits is obtained. It may be verified by direct substitution that $x = \frac{1}{3}$ is also a solution of this equation. What happens if the direct iteration method is used for initial guesses of $x = 0.33$ and $x = 0.34$? What can be concluded from this?

8-4 Use the direct iteration method to obtain a solution to the following system of algebraic equations:

$$x^2 + xy^2 + 3x - 15 = 0$$
$$3x + xy - 7 = 0$$

Assume an initial guess of $x = 3$ and $y = 3$, and stop the iterations when an accuracy of three significant digits is obtained.

8-5 Use the direct iteration method to obtain a solution to the following system of algebraic equations:

$$y^2 + 5x + x^2 = 25$$
$$x^2 y = 10$$

Assume an initial guess of $x = 1$ and $y = 2$, and terminate the iterations when an accuracy of three significant digits is obtained.

8-6 With the help of the infinitesimal element of length dx shown in Fig. P8-6, derive Eq. (8-3). Proceed by performing an energy balance on the infinitesimal element. Assume that the heat flow from conduction in the x direction varies according to a first-order Taylor expansion as shown in the figure. Use Fourier's law of heat conduction (a constitutive relationship) to eliminate the heat flux q_x. Clearly state all assumptions.

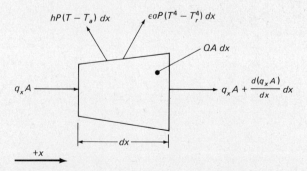

Figure P8-6

8-7 Show that the variational principle that corresponds to Eq. (8-3) is given by

$$I = \int_0^{L_f} \left[QAT - \tfrac{1}{2}hPT^2 + hPT_aT - \tfrac{1}{5}\varepsilon\sigma PT^5 \right.$$
$$\left. + \varepsilon\sigma PT_r^4T - \tfrac{1}{2}kA\left(\frac{dT}{dx}\right)^2 \right] dx$$

providing that only geometric or natural boundary conditions are allowed. Note that the one-dimensional body is assumed to have length L_f.

8-8 Determine the more general form of the variational principle in Problem 8-7 for the case of convection, radiation, and imposed heat fluxes on the boundaries as shown in Fig. 8-3.

8-9 Show that the element stiffness or conduction matrix, given by Eq. (8-20b), evaluates to

$$\mathbf{K}_{cv}^e = \frac{hPL}{6} \begin{bmatrix} 2 & 1 \\ 1 & 2 \end{bmatrix}$$

if the lineal element from Sec. 6-3 is used. Note that L is the element length. Clearly state all assumptions made in arriving at this result.

8-10 Show that the element stiffness or conduction matrix from boundary radiation, given by Eq. (8-20e), evaluates to

$$\mathbf{K}_{rB}^e = \begin{bmatrix} \varepsilon_i\sigma A_i T_i^3 & 0 \\ 0 & \varepsilon_j\sigma A_j T_j^3 \end{bmatrix}$$

Use the lineal element from Sec. 6-3. Explain how this result should be used.

8-11 Show that the element nodal force vector from lateral radiation, given by Eq. (8-21b), evaluates to

$$\mathbf{f}_r^e = \frac{\varepsilon\sigma PLT_r^4}{2} \begin{bmatrix} 1 \\ 1 \end{bmatrix}$$

where L is the element length. Assume the lineal element from Sec. 6-3. Clearly state all assumptions made in arriving at this result.

8-12 Show that the element nodal force vector from the internal heat source, given by Eq. (8-21c), evaluates to

$$\mathbf{f}_Q^e = \frac{QAL}{2} \begin{bmatrix} 1 \\ 1 \end{bmatrix}$$

where L is the element length. Assume the lineal element from Sec. 6-3. Clearly state all assumptions made in arriving at this result. Does the result seem to be intuitively correct? Please explain.

8-13 Show that the element nodal force vector from boundary radiation, given by Eq. (8-21e), evaluates to

$$\mathbf{f}_{rB}^e = \begin{bmatrix} \varepsilon_i A_i \sigma T_{ri}^4 \\ \varepsilon_j A_j \sigma T_{rj}^4 \end{bmatrix}$$

Assume the lineal element from Sec. 6-3. Clearly explain how this result should be used.

8-14 Show that the element nodal force vector from the imposed boundary heat fluxes, given by Eq. (8-21f), evaluates to

$$\mathbf{f}_{qB}^e = \begin{bmatrix} q_i A_i \\ q_j A_j \end{bmatrix}$$

Assume the lineal element from Sec. 6-3. Clearly explain how this result should be used. Note that the heat fluxes q_i and q_j are positive if directed toward the body.

8-15 Show that Eq. (8-31) holds if the element stiffness matrix from lateral convection [see Eq. (8-20b)] from a one-dimensional body is evaluated under the assumption that the perimeter P varies linearly from node i to node j over the element. Assume the lineal element from Sec. 6-3 and state all assumptions.

8-16 Consider the circular pin fin shown in Fig. P8-16. The fin is made of aluminum and has a length L_f and diameter D. The base is held a fixed temperature T_b and the tip undergoes convection. The lateral and boundary heat transfer coefficients are equal and denoted as h. The entire fin is exposed to a fluid at temperature T_a (including the tip). Discretize the fin into two equal-length elements (i.e., with three nodes). If $h = 2500$ W/m²-°C, $L_f = 5$ cm, $D = 1$ cm, $T_b = 75$°C, and $T_a = 15$°C, neglect radiation and determine:

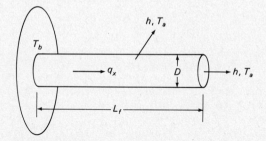

Figure P8-16

a. The temperatures at the nodal points
b. The heat removal rate from the fin (using the integration method from Sec. 4-10)
c. The fin efficiency (see Sec. 4-10)

8-17 Consider the circular pin fin shown in Fig. P8-16. The fin is made of aluminum and has a length L_f and diameter D. The base is held a fixed temperature T_b and the tip undergoes convection. The lateral and boundary heat transfer coefficients are equal and denoted as h. The entire fin is exposed to a fluid at temperature T_a (including the tip). Discretize the fin into two equal-length elements (i.e., with three nodes). If $h = 100$ Btu/hr-ft²-°F, $L_f = 3$ in., $D = 0.5$ in., $T_b = 150$°F, and $T_a = 50$°F, neglect radiation and determine:
a. The temperatures at the nodal points

b. The heat removal rate from the fin (using the integration method from Sec. 4-10)

c. The fin efficiency (see Sec. 4-10)

8-18 Consider the square pin fin of length L_f shown in Fig. P8-18. The fin is made of copper and has a cross-sectional area A. The base of the fin is held at fixed temperature T_b and the tip undergoes convection. The lateral and boundary heat transfer coefficients are equal and denoted as h. The entire fin is exposed to a fluid at temperature T_a (including the tip). Discretize the fin into two equal-length elements (i.e., with three nodes). If $h = 5000$ Btu/hr-ft²-°F, $L_f = 1$ in., $A = 1.5$ in.², $T_b = 185°F$, and $T_a = 70°F$, neglect radiation and determine:

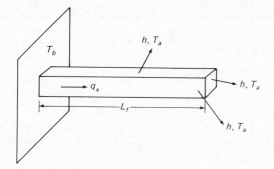

Figure P8-18

a. The temperature at each nodal point

b. The heat removal rate from the fin (see Sec. 4-10)

c. The fin efficiency (see Sec. 4-10)

8-19 Consider the square pin fin of length L_f shown in Fig. P8-18. The fin is made of copper and has a cross-sectional area A. The base of the fin is held a fixed temperature T_b and the tip undergoes convection. The lateral and boundary heat transfer coefficients are equal and denoted as h. The entire fin is exposed to a fluid at temperature T_a (including the tip). Discretize the fin into two equal-length elements (i.e., with three nodes). If $h = 10,000$ W/m²-°C, $L_f = 10$ cm, $A = 9$ cm², $T_b = 100°C$, and $T_a = 30°C$, neglect radiation and determine:

a. The temperature at each nodal point

b. The heat removal rate from the fin (see Sec. 4-10)

c. The fin efficiency (see Sec. 4-10)

8-20 Repeat part (a) of Problem 8-16 if a heat flux of 5 W/cm² is imposed on the tip of the fin (toward the fin), and

a. There is no convection from the tip of the fin.

b. There is simultaneous convection from the tip of the fin.

8-21 Repeat part (a) of Problem 8-17 if a heat flux of 500 Btu/hr-ft² is imposed on the tip of the fin (toward the fin), and

a. There is no convection from the tip of the fin.

b. There is simultaneous convection from the tip of the fin.

8-22   Repeat part (a) of Problem 8-18 if a heat flux of -7000 Btu/hr-ft^2 is imposed on the tip of the fin (away from the fin), and
   a. There is no convection from the tip of the fin.
   b. There is simultaneous convection from the tip of the fin.

8-23   Repeat part (a) of Problem 8-19 if a heat flux of -1.0 W/cm^2 is imposed on the tip of the fin (away from the fin), and
   a. There is no convection from the tip of the fin.
   b. There is simultaneous convection from the tip of the fin.

8-24   Repeat part (a) of Problem 8-16 if a uniform heat source with a strength of 10 W/cm^3 exists.

8-25   Repeat part (a) of Problem 8-17 if a uniform heat source with a strength of 120 Btu/hr-in.^{3} exists.

8-26   Repeat part (a) of Problem 8-18 if a uniform heat source with a strength of 780 Btu/hr-in.^{3} exists.

8-27   Repeat part (a) of Problem 8-19 if a uniform heat source with a strength of 0.2 W/cm^3 exists.

8-28   Reconsider the one-dimensional heat conduction model developed in Sec. 8-3. Assume that a lateral heat flux q_{lat} is imposed (along the length of the one-dimensional body) such that a fraction α is absorbed on the surface of the body. Let us denote the projected area that receives the heat flux as A_p. If the heat flux happens to be from the sun, then the parameter α is referred to as the *solar absorptivity*.
   a. Derive the integral expression for the corresponding nodal force vector. Use the Galerkin method. Note that A_p is then interpreted to be the projected area of the element.
   b. Evaluate the result for part (a) if the lineal element from Sec. 6-3 is used. Assume that α and q_{lat} are constant over the element.

8-29   Consider the annular fin shown in Fig. P8-29. Using the general nomenclature from Sec. 8-3 (k is the thermal conductivity, h is the convective heat transfer coefficient,

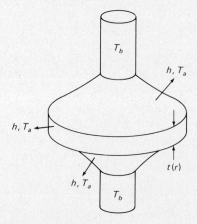

Figure P8-29

etc.), derive the expressions for the element characteristics. In particular, determine the expressions for the element stiffness matrices from conduction (in the radial direction), lateral convection, and boundary convection. In addition, determine the expressions for the element characteristics for the nodal force vectors from lateral convection, boundary convection, and boundary heat flux. Do not perform the integrations. Describe the type of element that would be appropriate here (a simple drawing showing the nodes would suffice). *Hint:* The governing equation is given in Problem 4-87 (the boundary conditions, however, are different in this problem).

8-30 Using the annular-shaped element shown in Fig. P8-30, evaluate the integrals that result in Problem 8-29. Instead of performing exact integrations, evaluate the integrands at the element centroid and treat the integrands as though they were constant. Explain how the results from the formulation in Sec. 8-3 could be used directly without reformulating the problem as required in Problem 8-29. *Hint:* What is the effective cross-sectional area for heat conduction? What is the effective perimeter?

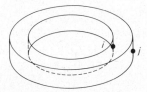

Figure P8-30

8-31 The governing equation for one-dimensional heat conduction in a sphere with a volumetric heat source is given by

$$\frac{1}{r^2} \frac{d}{dr} \left(kr^2 \frac{dT}{dr} \right) + Q = 0$$

where T is the temperature, r is the radial coordinate, k is the thermal conductivity, and Q is the strength of the heat source. What are the units of Q? Derive the expressions for the element characteristics for the case of a hollow sphere undergoing convection with heat transfer coefficients h_i and h_o on the inside and outside surfaces of the sphere, respectively, to fluids at temperatures T_i and T_o. *Hint:* When setting up the integral for the Galerkin method, integrate the product of $\mathbf{N}^T$ and the residual with respect to the elemental volume $4\pi r^2 \, dr$.

8-32 Explain how the one-dimensional heat conduction model developed in Sec. 8-3 could be used to solve for the temperature in the hollow sphere described in Problem 8-31. Be specific. *Hint:* What is the effective cross-sectional area for heat conduction? Is there any *lateral* convection?

8-33 Extend the program developed in Problem 4-78 (or the program furnished by the instructor) so that it can handle convection and imposed heat fluxes on the boundary of an arbitrary one-dimensional body as well as a uniform heat source. Boundary condition flags should be used to indicate the type of boundary conditions that exist on any given node (generally on the ends of the body). Positive integer flags should be used to denote prescribed temperature conditions, whereas negative integer flags should be used to denote convection and/or imposed heat fluxes. More specifically, three different input sections should be used in this regard as given below:

```
BOUNDARY CONDITION FLAG DATA
N  ±IBC
  (blank line)
CONVECTION AND OR HEAT FLUX DATA
IBC  QS  H  TA
  (blank line)
PRESCRIBED TEMPERATURE DATA
IBC  TEMP
  (blank line)
```

where N is the node number that has a positive IBC flag if a prescribed temperature is to be imposed, and a negative IBC flag if convection exists and/or a heat flux is to be imposed. IBC is zero (by default) if neither condition exists at the node in question. For each positive IBC used, a corresponding prescribed temperature (TEMP) must be declared in the PRESCRIBED TEMPERATURE DATA input section. For each negative IBC used, a corresponding set of imposed heat fluxes, convective heat transfer coefficients, and ambient temperatures must be present (QS, H, and TA, respectively). If only convection is present on a boundary, then the user simply sets QS to zero in the corresponding input line. If only an imposed heat flux exists on a boundary, then the user simply sets H and TA to zero in a similar fashion. Note that each of the input sections is terminated with a blank line. The program should check if N is zero (i.e., if the line is blank). If N is zero, the next input section should be read; otherwise, another set of data should be read in the same input section. Let QVOL represent the heat source strength (per unit volume) and include it in the material property data.

The variable property routine (SUBROUTINE VPROP) should be implemented as described in Problem 4-78. This approach will allow the same program to be used to solve a wide variety of one-dimensional, steady heat conduction problems.

8-34   Extend the program in Problem 8-33 (or one furnished by the instructor) to allow for lateral and boundary radiation as formulated in Sec. 8-3. Extend the material data to include the emissivity (ELAT) and receiver temperature (TRLAT) for the lateral radiation. Allow for boundary radiation by using negative boundary condition flags, and modify the next to the last input section (see Problem 8-33) to the following:

```
IMPOSED HEAT FLUX, CONVECTION, AND/OR RADIATION DATA
IBC  QS  H  TA  E  TR
  (blank line)
```

where E and TR are the boundary emissivity and receiver temperatures, respectively. If radiation is not present on a boundary, then the user simply sets E and TR to zero in the corresponding input line. Similar comments hold for the case of ELAT and TRLAT.

8-35   Solve Problems 8-16, 8-20, and 8-24 with the computer program from Problem 8-33 (or with the program furnished by the instructor). For Problem 8-16, use two, four, and eight elements to show the increased accuracy that results as the number of elements (and nodes) is increased. For Problems 8-20 and 8-24, use only eight elements.

8-36   Solve Problems 8-17, 8-21, and 8-25 with the computer program from Problem 8-33 (or with the program furnished by the instructor). For Problem 8-17, use two, four, and eight elements to show the increased accuracy that results as the number of elements (and nodes) is increased. For Problems 8-21 and 8-25, use only eight elements.

8-37   Solve Problems 8-18, 8-22, and 8-26 with the computer program from Problem 8-33 (or with the program furnished by the instructor). For Problem 8-18, use two, four, and eight elements to show the increased accuracy that results as the number of elements (and nodes) is increased. For Problems 8-22 and 8-26, use only eight elements.

8-38   Solve Problems 8-19, 8-23, and 8-27 with the computer program from Problem 8-33 (or with the program furnished by the instructor). For Problem 8-19, use two, four, and eight elements to show the increased accuracy that results as the number of elements (and nodes) is increased. For Problems 8-23 and 8-27, use only eight elements.

8-39   Use the two-dimensional form of the Green-Gauss theorem [Eqs. (8-42)] to rewrite the integrals below as a sum of two other integrals: one integral around the boundary of the two-dimensional region and the other integral over the area of the region.

a. $\displaystyle \int_A \beta \frac{\partial}{\partial x}\left(kt\frac{\partial T}{\partial x}\right) dx\, dy$

b. $\displaystyle \int_A \beta \frac{\partial}{\partial y}\left(kt\frac{\partial T}{\partial y}\right) dx\, dy$

8-40   Use the three-dimensional form of the Green-Gauss theorem [Eqs. (8-41)] to rewrite the integrals below as a sum of two other integrals: one integral over the surface of the three-dimensional region and the other integral over the volume of the region.

a. $\displaystyle \int_V \beta \frac{\partial}{\partial x}\left(k\frac{\partial T}{\partial x}\right) dx\, dy\, dz$

b. $\displaystyle \int_V \beta \frac{\partial}{\partial y}\left(k\frac{\partial T}{\partial y}\right) dx\, dy\, dz$

c. $\displaystyle \int_V \beta \frac{\partial}{\partial z}\left(k\frac{\partial T}{\partial z}\right) dx\, dy\, dz$

8-41   By performing an energy balance on the infinitesimal area element (of unit thickness) shown in Fig. P8-41, and by invoking Fourier's law of heat conduction (a constitutive relationship) to eliminate the heat fluxes from conduction (q_x and q_y), derive Eq. (8-43). Note that a heat source Q is present. Assume a uniform thickness.

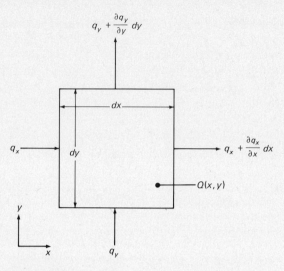

Figure P8-41

8-42   Extend Eq. (8-48) to the three-dimensional case first by inspection and then by formally extending the development in Sec. 8-5.

8-43   By performing an energy balance on the infinitesimal element of thickness *t* shown in Fig. P8-43, and by invoking Fourier's law of heat conduction (a constitutive

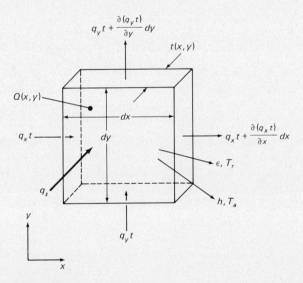

Figure P8-43

relationship) to eliminate the heat fluxes from conduction (q_x and q_y), derive Eq. (8-91). Note that lateral convection and radiation are included, as well as a laterally imposed heat flux. In addition, a heat source Q is also included.

8-44 Obtain the variational principle that corresponds to Eq. (8-91) by using the approach illustrated in Example 8-10(a). Assume that either the boundary is insulated (i.e., the natural boundary condition) or a prescribed temperature is specified on the boundary (the geometric boundary condition).

8-45 Obtain the variational principle that corresponds to Eq. (8-91) by using the approach illustrated in Example 8-10(b). Allow for the possibility of boundary convection, radiation, and imposed heat fluxes.

8-46 Show that the element stiffness matrices from conduction in a two-dimensional body are always symmetric [i.e., the element stiffness matrices given by Eqs. (8-106a) and (8-106b)].

8-47 Show that the element stiffness matrices from lateral convection and radiation from a two-dimensional body are always symmetric [i.e., the element stiffness matrices given by Eqs. (8-106c) and (8-106d)].

8-48 Show that the element stiffness matrices from boundary convection and radiation from a two-dimensional body are always symmetric [i.e., the element stiffness matrices given by Eqs. (8-106e) and (8-106f)].

8-49 Show that the element stiffness matrix from conduction in the y direction in a two-dimensional body is given by Eq. (8-109) if the three-node triangular element is used. Clearly state all assumptions made in arriving at this result.

8-50 Show that the element nodal force vector from lateral convection from a two-dimensional body is given by Eq. (8-115) if the three-node triangular element is used. Clearly state all assumptions made in arriving at this result.

8-51 Show that the element nodal force vector from lateral radiation from a two-dimensional body is given by Eq. (8-116) if the three-node triangular element is used. Clearly state all assumptions made in arriving at this result.

8-52 Show that the element nodal force vector from a laterally imposed heat flux on a two-dimensional body is given by Eq. (8-117) if the three-node triangular element is used. Clearly state all assumptions made in arriving at this result.

8-53 Show that the element nodal force vector from an internal heat source in a two-dimensional body is given by Eq. (8-118) if the three-node triangular element is used and if the heat source is uniformly distributed over the element.

8-54 Show that the element nodal force vector from boundary convection from a two-dimensional body is given by Eq. (8-121) if the three-node triangular element is used and if the convection occurs on leg ij. What is the form of this result if legs jk and ki happen to be on the global boundary?

8-55 Show that the element nodal force vector from boundary radiation from a two-dimensional body is given by Eq. (8-122) if the three-node triangular element is used and if the radiation occurs on leg ij. What is the form of this result if legs jk and ki happen to be on the global boundary?

8-56 Show that the element nodal force vector from an imposed heat flux on the boundary of a two-dimensional body is given by Eq. (8-123) if the three-node triangular element is used and if the heat flux is imposed on leg ij. What is the form of this result if legs jk and ki happen to be on the global boundary?

8-57 The expressions for the element stiffness matrices from conduction in a two-dimensional body are given by Eqs. (8-106a) and (8-106b). Since these results hold for any two-dimensional element, let us evaluate these stiffness matrices for the four-node rectangular element presented in Sec. 6-4 by proceeding as follows:

a. With the help of Eqs. (6-32), show that

$$\frac{\partial \mathbf{N}}{\partial x} = \frac{1}{a}\frac{\partial \mathbf{N}}{\partial r}$$

$$\frac{\partial \mathbf{N}}{\partial y} = \frac{1}{b}\frac{\partial \mathbf{N}}{\partial s}$$

where r and s are the serendipity coordinates.

b. Using the results from part (a), rewrite the integrals in terms of derivatives of the shape functions with respect to the serendipity coordinates r and s. Do not forget to change the limits on the integrations. Also note that $dx\ dy = ab\ dr\ ds$.

c. Evaluate the resulting integrands at the element centroid (i.e., at $r = 0$ and $s = 0$). Then treat the integrands as though they are constants and pull them through the integral. Evaluate the remaining trivial integrals. Show that the result for the element stiffness from conduction in the x direction is given by

$$\mathbf{K}_{xx}^e = \frac{ktb}{4a}\begin{bmatrix} 1 & 1 & -1 & -1 \\ 1 & 1 & -1 & -1 \\ -1 & -1 & 1 & 1 \\ -1 & -1 & 1 & 1 \end{bmatrix}$$

State the assumptions made in arriving at this result.

d. Derive the corresponding result for $\mathbf{K}_{yy}^e$? What assumptions are made?

8-58 Evaluate the element stiffness matrix from lateral convection from a two-dimensional body [i.e., Eq. (8-106c)] if the four-node rectangular element from Sec. 6-4 is used. Evaluate the integrals by first evaluating the integrands at the element centroid (i.e., at $r = 0$ and $s = 0$) and then treating the integrands as though they were constants.

8-59 Evaluate the element stiffness matrix from boundary convection from a two-dimensional body [i.e., Eq. (8-106e)] if the four-node rectangular element from Sec. 6-4 is used and if face ij happens to be on the global boundary. Evaluate the integrals by first evaluating the integrands at the centroid of face ij (i.e., at $r = +1$ and $s = 0$) and then treating the integrands as though they were constants.

8-60 Evaluate the element stiffness matrix from boundary convection from a two-dimensional body [i.e., Eq. (8-106e)] if the four-node rectangular element from Sec. 6-4 is used and if face jk happens to be on the global boundary. Evaluate the integrals by first evaluating the integrands at the centroid of face jk (i.e., at $r = 0$ and $s = +1$) and then treating the integrands as though they were constants.

8-61 Evaluate the element nodal force vector from lateral convection from a two-dimensional body [i.e., Eq. (8-107a)] if the four-node rectangular element from Sec. 6-4 is used. Evaluate the integrals by first evaluating the integrands at the element centroid (i.e., at $r = 0$ and $s = 0$) and then treating the integrands as though they were constants.

8-62 Evaluate the element nodal force vector from lateral radiation from a two-dimensional body [i.e., Eq. (8-107b)] if the four-node rectangular element from Sec. 6-4 is used. Evaluate the integrals by first evaluating the integrands at the element centroid (i.e., at $r = 0$ and $s = 0$) and then treating the integrands as though they were constants.

8-63 Evaluate the element nodal force vector from a laterally imposed heat flux on a two-dimensional body [i.e., Eq. (8-107c)] if the four-node rectangular element from Sec. 6-4 is used. Evaluate the integrals by first evaluating the integrands at the element centroid (i.e., at $r = 0$ and $s = 0$) and then treating the integrands as though they were constants.

8-64 Evaluate the element nodal force vector from an internal heat source in a two-dimensional body [i.e., Eq. (8-107d)] if the four-node rectangular element from Sec. 6-4 is used. Evaluate the integrals by first evaluating the integrands at the element centroid (i.e., at $r = 0$ and $s = 0$) and then treating the integrands as though they were constants.

8-65 Evaluate the element nodal force vector from boundary convection from a two-dimensional body [i.e., Eq. (8-107e)] if the four-node rectangular element from Sec. 6-4 is used and if face ij happens to be on the global boundary. Evaluate the integrals by first evaluating the integrands at the centroid of face ij (i.e., at $r = +1$ and $s = 0$) and then treating the integrands as though they were constants.

8-66 Evaluate the element nodal force vector from boundary convection from a two-dimensional body [i.e., Eq. (8-107e)] if the four-node rectangular element from Sec. 6-4 is used and if face jk happens to be on the global boundary. Evaluate the integrals by first evaluating the integrands at the centroid of face jk (i.e., at $r = 0$ and $s = +1$) and then treating the integrands as though they were constants.

8-67 Evaluate the element nodal force vector from boundary radiation from a two-dimensional body [i.e., Eq. (8-107f)] if the four-node rectangular element from Sec. 6-4 is used and if face km happens to be on the global boundary. Evaluate the integrals by first evaluating the integrands at the centroid of face km (i.e., at $r = -1$ and $s = 0$) and then treating the integrands as though they were constants.

8-68 Evaluate the element nodal force vector from a boundary heat flux imposed on a two-dimensional body [i.e., Eq. (8-107g)] if the four-node rectangular element from Sec. 6-4 is used and if face mi happens to be on the global boundary. Evaluate the integrals by first evaluating the integrands at the centroid of face mi (i.e., at $r = 0$ and $s = -1$) and then treating the integrands as though they were constants.

8-69 The element resultants (i.e., the heat fluxes from conduction in the x and y directions) are given by Eqs. (8-125) for the three-node triangular element. Derive the corresponding expressions for the average heat fluxes if the four-node rectangular element is used. Assume that the heat fluxes at the element centroid represent the average heat fluxes in the element. Explain why these results seem to be intuitively correct.

8-70 Consider the triangular element shown in Fig. P8-70. The plate from which the element is extracted is made of cast iron and has a thickness of 0.5 in. The nodal coordinates are $x_i = 2.0$, $y_i = 1.5$, $x_j = 1.7$, $y_j = 3.0$, $x_k = 0.6$, and $y_k = 1.8$ in.

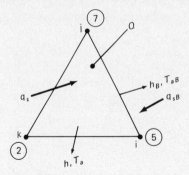

Figure P8-70

a. Determine the element stiffness matrices from conduction in the x and y directions.

b. Determine the element stiffness matrix from lateral convection if the element convects from both faces to a fluid at 35°F through a convective heat transfer coefficient of 25 Btu/hr-ft²-°F.

c. Determine the element stiffness matrix from boundary convection if leg ij happens to be on the part of the global boundary that undergoes convection to a fluid at 40°F through a convective heat transfer coefficient of 50 Btu/hr-ft²-°F.

8-71   For the element in Problem 8-70, determine the nodal force vectors
a. From lateral convection
b. From a laterally imposed heat flux of 200 Btu/hr-ft² on each face of the plate
c. From an internal heat source of 350 Btu/hr-ft³
d. From boundary convection
e. From a boundary heat flux of 425 Btu/hr-ft² imposed on leg ij (which is on the part of the global boundary also undergoing convection)

8-72   Consider the triangular element shown in Fig. P8-72. The plate from which the element is extracted is made of brass and has a thickness of 1 cm. The nodal coordinates are $x_i = 5$, $y_i = 6$, $x_j = 4$, $y_j = 4$, $x_k = 6$, and $y_k = 4$ cm.

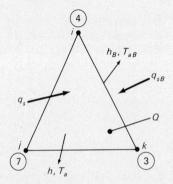

Figure P8-72

a. Determine the element stiffness matrices from conduction in the x and y directions.

b. Determine the element stiffness matrix from lateral convection if the element convects from both faces to a fluid at 20°C through a convective heat transfer coefficient of 125 W/cm²-°C.

c. Determine the element stiffness matrix from boundary convection if leg ki happens to be on the part of the global boundary that undergoes convection to a fluid at 25°C through a convective heat transfer coefficient of 250 W/cm²-°C.

8-73 For the element in Problem 8-72, determine the nodal force vectors
a. From lateral convection
b. From a laterally imposed heat flux of 300 W/cm² on each face of the plate
c. From an internal heat source of 160 W/cm³
d. From boundary convection
e. From a boundary heat flux of 235 W/cm² imposed on leg ki (which is on the part of the global boundary also undergoing convection)

8-74 Consider the triangular element shown in Fig. P8-74. The element is extracted from a thin plate of thickness 0.5 cm. The material is hot rolled, low carbon steel. The nodal coordinates are $x_i = 0, y_i = 0, x_j = 0, y_j = -1, x_k = 2$, and $y_k = -1$ cm.

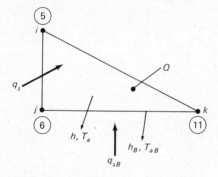

Figure P8-74

a. Determine the element stiffness matrices from conduction in the x and y directions.

b. Determine the element stiffness matrix from lateral convection if the element convects from both faces to a fluid at 40°C through convective heat transfer coefficients of 250 and 300 W/cm²-°C.

c. Determine the element stiffness matrix from boundary convection if leg jk happens to be on the part of the global boundary that undergoes convection to a fluid at 45°C through a convective heat transfer coefficient of 35 W/cm²-°C.

8-75 For the element in Problem 8-74, determine the nodal force vectors
a. From lateral convection
b. From a laterally imposed heat flux of 225 W/cm² on one face of the plate
c. From an internal heat source of 10 W/cm³
d. From boundary convection

e. From a boundary heat flux of 35 W/cm² imposed on leg *jk* (which is on the part of the global boundary also undergoing convection)

8-76 Consider the triangular element shown in Fig. P8-76. The element is extracted from a thin plate of thickness 0.75 in. The material is pure copper. The nodal coordinates are $x_i = 0$, $y_i = 0$, $x_j = 1$, $y_j = 2$, $x_k = -1$, and $y_k = 2$ in.

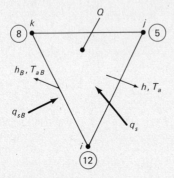

Figure P8-76

a. Determine the element stiffness matrices from conduction in the *x* and *y* directions.

b. Determine the element stiffness matrix from lateral convection if the element convects from both faces to a fluid at 50°F through a convective heat transfer coefficient of 135 Btu/hr-ft²-°F.

c. Determine the element stiffness matrix from boundary convection if leg *ki* happens to be on the part of the global boundary that undergoes convection to a fluid at 40°F through a convective heat transfer coefficient of 50 Btu/hr-ft²-°F.

8-77 For the element in Problem 8-76, determine the nodal force vectors
a. From lateral convection
b. From a laterally imposed heat flux of 150 Btu/hr-ft² on each face of the plate
c. From an internal heat source of 215 Btu/hr-ft³
d. From boundary convection
e. From a boundary heat flux of 450 Btu/hr-ft² imposed on leg *ki* (which is on the part of the global boundary also undergoing convection)

8-78 Consider the rectangular element shown in Fig. P8-78. The element is extracted from an aluminum plate with a thickness of 1.25 cm. The coordinates of the nodes

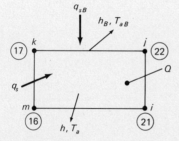

Figure P8-78

are $x_i = 4$, $y_i = 2$, $x_j = 4$, $y_j = 3$, $x_k = 2$, $y_k = 3$, $x_m = 2$, and $y_m = 2$ cm. Perform the necessary integrations by following the approaches mentioned in Problems 8-57 to 8-60, and

a. Determine the element stiffness matrices from conduction in the x and y directions.

b. Determine the element stiffness matrix from lateral convection if the element convects from both faces to a fluid at 15°C through convective heat transfer coefficients of 10 and 12 W/cm²-°C.

c. Determine the element stiffness matrix from boundary convection if face jk happens to be on the part of the global boundary that undergoes convection to a fluid at 20°C through a convective heat transfer coefficient of 35 W/cm²-°C.

8-79 Consider the element in Problem 8-78. Perform the necessary integrations by following the approaches indicated in Problems 8-61 to 8-68, and determine the nodal force vectors

a. From lateral convection

b. From a laterally imposed heat flux of 125 W/cm² on each face of the plate

c. From an internal heat source of 15 W/cm³

d. From boundary convection

e. From a boundary heat flux of 350 W/cm² imposed on face jk (which is on the part of the global boundary also undergoing convection)

8-80 Consider the rectangular element shown in Fig. P8-80. The element is extracted from a brass plate with a thickness of 0.375 in. The coordinates of the nodes are $x_i = 5$, $y_i = 2$, $x_j = 5$, $y_j = 4$, $x_k = 2$, $y_k = 4$, $x_m = 2$, and $y_m = 2$ in. Perform the necessary integrations by following the approaches mentioned in Problems 8-57 to 8-60, and

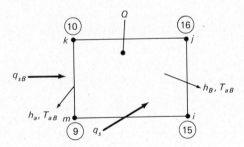

Figure P8-80

a. Determine the element stiffness matrices from conduction in the x and y directions.

b. Determine the element stiffness matrix from lateral convection if the element convects from both faces to a fluid at 60°F through a convective heat transfer coefficient of 235 Btu/hr-ft²-°F.

c. Determine the element stiffness matrix from boundary convection if face km happens to be on the part of the global boundary that undergoes convection to a fluid at 60°F through a convective heat transfer coefficient of 150 Btu/hr-ft²-°F.

8-81 Consider the element in Problem 8-80. Perform the necessary integrations by following the approaches indicated in Problems 8-61 to 8-68, and determine the nodal force vectors
  a. From lateral convection
  b. From laterally imposed heat fluxes of 350 and 400 Btu/hr-ft^2 on each face of the plate
  c. From an internal heat source of 100 Btu/hr-ft^3
  d. From boundary convection
  e. From a boundary heat flux of 340 Btu/hr-ft^2 imposed on face km (which is on the part of the global boundary also undergoing convection)

8-82 For the element in Problem 8-70, determine the element nodal force vector from a point heat source with a strength of 100 Btu/hr-in. (of thickness) if the source is located at $x_0 = 1.5$ and $y_0 = 2.0$ in.

8-83 For the element in Problem 8-72, determine the element nodal force vector from a point heat source with a strength of 120 W/cm (of thickness) if the source is located at $x_0 = 5$ and $y_0 = 5$ cm.

8-84 For the element in Problem 8-74, determine the element nodal force vector from a point heat source with a strength of 150 W/cm (of thickness) if the source is located at $x_0 = 2$ and $y_0 = -1$ cm.

8-85 For the element in Problem 8-76, determine the element nodal force vector from a point heat source with a strength of 230 Btu/hr-in. (of thickness) if the source is located at $x_0 = 1$ and $y_0 = 2$ in.

8-86 For the element in Problem 8-78, determine the element nodal force vector from a point heat source with a strength of 35 W/cm (of thickness) if the source is located at $x_0 = 3.2$ and $y_0 = 2.7$ cm.

8-87 For the element in Problem 8-78, determine the element nodal force vector from a point heat source with a strength of 45 W/cm (of thickness) if the source is located at $x_0 = 4$ and $y_0 = 2$ cm.

8-88 For the element in Problem 8-80, determine the element nodal force vector from a point heat source with a strength of 50 Btu/hr-in. (of thickness) if the source is located at $x_0 = 4$ and $y_0 = 3$ in.

8-89 For the element in Problem 8-80, determine the element nodal force vector from a point heat source with a strength of 60 Btu/hr-in. (of thickness) if the source is located at $x_0 = 2$, and $y_0 = 4$ in.

8-90 Consider the mesh shown in Fig. P8-90. The elements are defined in terms of the global node numbers as indicated in the figure. The 3×3 composite element stiffness matrix for element 1, for example, may be denoted symbolically as

$$\mathbf{K}^{(1)} = \begin{bmatrix} K_{44}^{(1)} & K_{41}^{(1)} & K_{42}^{(1)} \\ K_{14}^{(1)} & K_{11}^{(1)} & K_{12}^{(1)} \\ K_{24}^{(1)} & K_{21}^{(1)} & K_{22}^{(1)} \end{bmatrix}$$

Note that the superscript (in parentheses) denotes the element number and the subscripts are determined by the global node numbers associated with the element

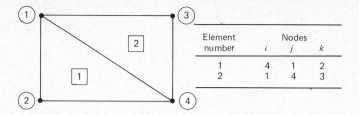

Figure P8-90

(the order is significant). With the help of this symbolic notation, give the expression for the assemblage stiffness matrix for the discretized two-dimensional region in Fig. P8-90. What is the half-bandwidth? Can the half-bandwidth be reduced? If so, explain how.

8-91 Consider the mesh shown in Fig. P8-90. The elements are defined in terms of the global node numbers as indicated in the figure. The 3×1 composite element nodal force vector for element 1, for example, may be denoted symbolically as

$$\mathbf{f}^{(1)} = \begin{bmatrix} f_4^{(1)} \\ f_1^{(1)} \\ f_2^{(1)} \end{bmatrix}$$

Note that the superscript (in parentheses) denotes the element number and the subscripts are determined by the global node numbers associated with the element (the order is significant). With the help of this symbolic notation, give the expression for the assemblage nodal force vector for the discretized two-dimensional region in Fig. P8-90.

8-92 Consider the mesh shown in Fig. P8-92. The elements are defined in terms of the global node numbers as indicated in the figure. Using the symbolic notation from Problem 8-90, give the expression for the assemblage stiffness matrix for the discretized two-dimensional region in Fig. P8-92. What is the half-bandwidth? Can the half-bandwidth be reduced? If so, explain how.

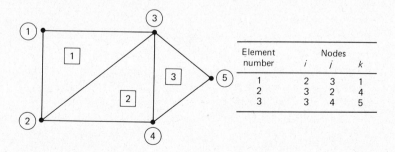

Figure P8-92

8-93 Using the symbolic notation from Problem 8-91, give the expression for the assemblage nodal force vector for the discretized two-dimensional region in Fig. P8-92.

The elements are defined in terms of the global node numbers as indicated in the figure.

8-94 Consider the mesh shown in Fig. P8-94. The elements are defined in terms of the global node numbers as indicated in the figure. Using the symbolic notation from Problem 8-90, give the expression for the assemblage stiffness matrix for the discretized two-dimensional region in Fig. P8-94. What is the half-bandwidth? Can the half-bandwidth be reduced? If so, explain how.

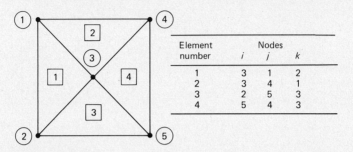

Element	Nodes		
number	i	j	k
1	3	1	2
2	3	4	1
3	2	5	3
4	5	4	3

Figure P8-94

8-95 Using the symbolic notation from Problem 8-91, give the expression for the assemblage nodal force vector for the discretized two-dimensional region in Fig. P8-94. The elements are defined in terms of the global node numbers as indicated in the figure.

8-96 Consider the mesh shown in Fig. P8-96. The elements are defined in terms of the global node numbers as indicated in the figure. By extending the symbolic notation from Problem 8-90 to the case of the rectangular element, give the expression for the assemblage stiffness matrix for the discretized two-dimensional region in Fig. P8-96. What is the half-bandwidth? Can the half-bandwidth be reduced? If so, explain how.

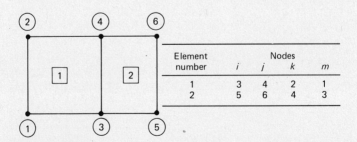

Element	Nodes			
number	i	j	k	m
1	3	4	2	1
2	5	6	4	3

Figure P8-96

8-97 By extending the symbolic notation from Problem 8-91, give the expression for the assemblage nodal force vector for the discretized two-dimensional region in Fig. P8-96. The elements are defined in terms of the global node numbers as indicated in the figure.

8-98   Consider the mesh shown in Fig. P8-98. The elements are defined in terms of the global node numbers as indicated in the figure. By extending the symbolic notation from Problem 8-90 to the case of the rectangular element, give the expression for the assemblage stiffness matrix for the discretized two-dimensional region in Fig. P8-98. What is the half-bandwidth? Can the half-bandwidth be reduced? If so, explain how.

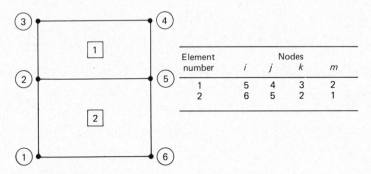

Element	Nodes			
number	i	j	k	m
1	5	4	3	2
2	6	5	2	1

Figure P8-98

8-99   By extending the symbolic notation from Problem 8-91, give the expression for the assemblage nodal force vector for the discretized two-dimensional region in Fig. P8-98. The elements are defined in terms of the global node numbers as indicated in the figure.

8-100   Consider the mesh shown in Fig. P8-100. The elements are defined in terms of the global node numbers as indicated in the figure. By extending the symbolic notation from Problem 8-90 to the case of the rectangular element, give the expression for the assemblage stiffness matrix for the discretized two-dimensional region in Fig. P8-100. Note the two different types of elements. What is the half-bandwidth? Can the half-bandwidth be reduced? If so, explain how.

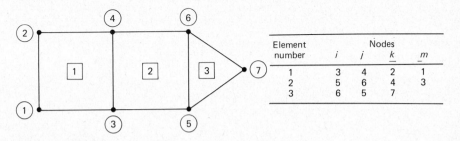

Element	Nodes			
number	i	j	k	m
1	3	4	2	1
2	5	6	4	3
3	6	5	7	

Figure P8-100

8-101   By extending the symbolic notation from Problem 8-91, give the expression for the assemblage nodal force vector for the discretized two-dimensional region in Fig. P8-100. The elements are defined in terms of the global node numbers as indicated in the figure. Note the use of the two different types of elements.

8-102 For the element in Problem 8-70 the nodal temperatures are obtained from the solution of $\mathbf{Ka} = \mathbf{f}$ and are $T_i = 175$, $T_j = 168$, and $T_k = 173°F$. Determine the average heat fluxes from conduction in the x and y directions, respectively. At what point in the element are these heat fluxes normally associated?

8-103 For the element in Problem 8-72 the nodal temperatures are obtained from the solution of $\mathbf{Ka} = \mathbf{f}$ and are $T_i = 75$, $T_j = 82$, and $T_k = 73°C$. Determine the average heat fluxes from conduction in the x and y directions, respectively. At what point in the element are these heat fluxes normally associated?

8-104 For the element in Problem 8-74 the nodal temperatures are obtained from the solution of $\mathbf{Ka} = \mathbf{f}$ and are $T_i = 94$, $T_j = 91$, and $T_k = 96°C$. Determine the average heat fluxes from conduction in the x and y directions, respectively. At what point in the element are these heat fluxes normally associated?

8-105 For the element in Problem 8-76 the nodal temperatures are obtained from the solution of $\mathbf{Ka} = \mathbf{f}$ and are $T_i = 154$, $T_j = 148$, and $T_k = 157°F$. Determine the average heat fluxes from conduction in the x and y directions, respectively. At what point in the element are these heat fluxes normally associated?

8-106 For the element in Problem 8-78 the nodal temperatures are obtained from the solution of $\mathbf{Ka} = \mathbf{f}$ and are $T_i = 92$, $T_j = 96$, $T_k = 89$, and $T_m = 95°C$. Determine the heat fluxes at the element centroid from conduction in the x and y directions, respectively. These heat fluxes may be regarded as the average conduction heat fluxes.

8-107 For the element in Problem 8-80 the nodal temperatures are obtained from the solution of $\mathbf{Ka} = \mathbf{f}$ and are $T_i = 192$, $T_j = 195$, $T_k = 189$, and $T_m = 196°C$. Determine the heat fluxes at the element centroid from conduction in the x and y directions, respectively. These heat fluxes may be regarded as the average conduction heat fluxes.

8-108 The necessary condition for the existence of an extremum of the functional

$$I = \int_A F(x,y,w,w_x,w_y,w_{xx},w_{xy},w_{yy}) \, dx \, dy$$

is that the first variation δ must vanish (i.e., equal to zero) provided that

$$\int_C \left\{ \left[\frac{\partial F}{\partial w_x} - \frac{\partial}{\partial x}\left(\frac{\partial F}{\partial w_{xx}}\right) - \frac{1}{2}\frac{\partial}{\partial y}\left(\frac{\partial F}{\partial w_{xy}}\right) \right] n_x \right. $$
$$\left. + \left[\frac{\partial F}{\partial w_y} - \frac{\partial}{\partial y}\left(\frac{\partial F}{\partial w_{yy}}\right) - \frac{1}{2}\frac{\partial}{\partial x}\left(\frac{\partial F}{\partial w_{xy}}\right) \right] n_y \right\} \delta w \, dC = 0$$

and

$$\int_C \left(\frac{\partial F}{\partial w_{xx}} n_x \, \delta w_x + \frac{\partial F}{\partial w_{yy}} n_y \, \delta w_y + \frac{1}{2}\frac{\partial F}{\partial w_{xy}} n_y \, \delta w_x + \frac{1}{2}\frac{\partial F}{\partial w_{xy}} n_x \, \delta w_y \right) dC = 0$$

on the boundary of the two-dimensional region. In the above, w is the field variable, x and y are the two independent coordinates, w_x indicates the first (partial) derivative of w with respect to x, w_{xx} indicates the second (partial) derivative of w with respect to x, etc.

a. Show that the corresponding Euler-Lagrange equation is given by

$$\frac{\partial F}{\partial w} - \frac{\partial}{\partial x}\left(\frac{\partial F}{\partial w_x}\right) - \frac{\partial}{\partial y}\left(\frac{\partial F}{\partial w_y}\right) + \frac{\partial^2}{\partial x^2}\left(\frac{\partial F}{\partial w_{xx}}\right)$$

$$+ \frac{\partial^2}{\partial x\,\partial y}\left(\frac{\partial F}{\partial w_{xy}}\right) + \frac{\partial^2}{\partial y^2}\left(\frac{\partial F}{\partial w_{yy}}\right) = 0$$

b. What order differential equation does the Euler-Lagrange equation from part (a) correspond to? Please explain.

c. What is the highest order derivative present in the functional F?

d. Try to generalize the results from parts (b) and (c). Explain why the variational formulation may be referred to as the *weak formulation*.

8-109   In Problem 8-108, identify:
a. The geometric boundary conditions
b. The natural boundary conditions

8-110   The bending of a thin isotropic plate of constant thickness t may be described by the well-known biharmonic equation

$$\frac{\partial^4 w}{\partial x^4} + 2\frac{\partial^4 w}{\partial x^2\,\partial y^2} + \frac{\partial^4 w}{\partial y^4} + \frac{12(1 - \mu^2)}{Et^3}q(x,y) = 0$$

where w is the deflection of the neutral plane of the plate (in the direction perpendicular to the xy plane), q is lateral distributed load per unit area, μ is Poisson's ratio, and E is the modulus of elasticity. Assume that the plate is rigidly supported around the boundary such that the plate has zero deflections and zero slopes on the boundary. Note that the rotations θ_x and θ_y about the x and y axes at any point in the plate are related to the deflection at the same point by

$$\theta_x = \frac{\partial w}{\partial y} \qquad \text{and} \qquad \theta_y = \frac{\partial w}{\partial x}$$

With the help of the Euler-Lagrange equation from Problem 8-108, show that the corresponding variational principle under these assumptions is given by

$$I = \int\left[\frac{1}{2}\left(\frac{\partial^2 w}{\partial x^2}\right)^2 + \left(\frac{\partial^2 w}{\partial x\,\partial y}\right)^2 + \frac{1}{2}\left(\frac{\partial^2 w}{\partial y^2}\right)^2 + \frac{12(1 - \mu^2)}{Et^3}qw\right] dx\,dy$$

8-111   Derive the variational principle for Problem 8-110 by writing

$$\int_A\left[\frac{\partial^4 w}{\partial x^4} + 2\frac{\partial^4 w}{\partial x^2\,\partial y^2} + \frac{\partial^4 w}{\partial y^4} + \frac{12(1 - \mu^2)}{Et^3}q(x,y)\right]\delta w\,dx\,dy = 0$$

and by applying the Green-Gauss theorem twice. *Hint:* The boundary integrals that arise are identically zero for the assumptions stated in Problem 8-110. Clearly explain why.

8-112   Explain what is meant by the method of quadrilateral averages. Why is the method used? Illustrate how the method is applied with a numerical example. Assume that we have two adjacent triangular elements for which the element resultants (i.e., the conduction heat fluxes in the x and y directions) are already known (from the finite element solution).

8-113 How could the method of quadrilateral averages be used in two-dimensional stress analysis (i.e., plane stress or plane strain) to improve the resulting stresses when the three-node triangular element is used? Be specific and illustrate with a numerical example.

8-114 How could the method of quadrilateral averages be used in axisymmetric stress analysis to improve the resulting stresses when the three-node triangular element is used? Be specific and illustrate with a numerical example.

8-115 In the formal development of two-dimensional, steady-state heat conduction in Sec. 8-8, only isotropic materials were considered. Let us now extend the development of that section to the case of anisotropic materials. For anisotropic materials, the thermal conductivity is dependent on direction. It is convenient to consider the two values of thermal conductivity in the so-called principal directions as shown in Fig. P8-115. Let us denote the principal values of the thermal conductivities as k_x' and k_y' in the local x and y directions, respectively. Let us denote these local directions as x' and y' (similar to the development in Chapter 3 where the transformation matrices for two- and three-dimensional trusses were developed). Note that θ is the angle between the x and x' axes (and the y and y' axes) as shown in Fig. P8-115. In this case the governing equation for two-dimensional, steady-state heat conduction must be modified to reflect the fact that there are different values of thermal conductivity in the x' and y' directions. Also, it proves to be much more convenient to express the governing equation in the local coordinate system. Only the two terms involving conduction in Eq. (8-91), i.e., the first two terms, must be modified with the resulting governing equation given by

$$\frac{\partial}{\partial x'}\left(k_x't\frac{\partial T}{\partial x'}\right) + \frac{\partial}{\partial y'}\left(k_y't\frac{\partial T}{\partial y'}\right) - h(T - T_a) - \varepsilon\sigma(T^4 - T_r^4) + q_s + Qt = 0$$

In general the local $x'y'$ axes are not in the same directions as the global xy axes as shown in Fig. P8-115. Therefore, it seems reasonable to apply the Galerkin method on an element basis where the local $x'y'$ coordinate system is used.

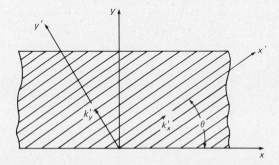

Figure P8-115

a. Show that the resulting expressions for the element stiffness matrices from conduction in the x' and y' directions are given by

$$\mathbf{K}_{xx}^e = \int_{A^e} \frac{\partial \mathbf{N}'^T}{\partial x'} k_x't \frac{\partial \mathbf{N}'}{\partial x'} \, dx' \, dy'$$

and

$$\mathbf{K}^e_{yy} = \int_{A^e} \frac{\partial \mathbf{N}'^T}{\partial y'} k'_y t \frac{\partial \mathbf{N}'}{\partial y'} dx' \, dy'$$

where $\mathbf{N}'$ indicates that the shape functions must be written in terms of the local $x'y'$ system before taking the indicated derivatives.

b. Convince yourself that since the field variable is a scalar in this problem (i.e., the temperature) that the procedure for implementing anisotropic material properties in the finite element formulation is as follows: (1) Transform the coordinates of every node for a given element from the global xy system to the $x'y'$ system with the help of Eq. (3-13), (2) compute the element stiffness matrices from the above expressions for the element stiffnesses (or conductances) using the coordinates in the local $x'y'$ system, and (3) do the remaining part of the analysis as described in Sec. 8-8. Note that it is not necessary to transform the local element stiffness matrices to global element stiffness matrices (such as in Chapter 3) because the field variable here is a scalar, not a vector. In effect, the local and global element stiffness matrices are identical in this case.

8-116 Let us apply the development of Problem 8-115 to a specific example. Consider the case of anisotropic heat conduction in a triangular element. The principal values of the thermal conductivities are denoted as k'_x and k'_y in the local x' and y' directions, respectively. The angle θ is defined to be the angle between the x and x' axes (or the y and y' axes) as shown in Fig. P8-115. The plate from which the element is extracted has a thickness of 0.5 in. The nodal coordinates are $x_i = 2.0$, $y_i = 1.5$, $x_j = 1.7$, $y_j = 3.0$, $x_k = 0.6$, and $y_k = 1.8$ in. with respect to the global xy coordinate system. If k'_x and k'_y are 120 and 60 Btu/hr-ft-°F, respectively, and if θ is 30°, determine the global element stiffness matrices for this element (from conduction only).

8-117 Repeat Problem 8-116 for an element whose nodal coordinates are $x_i = 5$, $y_i = 6$, $x_j = 4$, $y_j = 4$, $x_k = 6$, and $y_k = 4$ cm with respect to the global xy coordinate system. The plate from which the element is extracted has a thickness of 1 cm. Assume k'_x and k'_y are 200 and 110 W/m-°C, respectively, and θ is 25°.

8-118 Consider the elemental volume shown in Fig. P8-118. Perform a steady-state energy balance that includes the effects of the heat conduction in the r and z directions as well as the internal heat source Q. Assume that the conduction heat fluxes vary according to a first-order Taylor expansion as shown on the figure. Use Fourier's law of heat conduction (a constitutive relationship) to eliminate the conduction heat fluxes, thereby showing the validity of Eq. (8-126). *Hint:* Note that the effective area for heat conduction in the r direction is not constant.

8-119 Show that the variational principle that corresponds to Eq. (8-126) is given by

$$I = 2\pi \int_A \left[QT - \tfrac{1}{2} k \left(\frac{\partial T}{\partial r} \right)^2 - \tfrac{1}{2} k \left(\frac{\partial T}{\partial z} \right)^2 \right] r \, dr \, dz$$

$$+ 2\pi \int_C (q_{sB} T - \tfrac{1}{2} h_B T^2 + h_B T_{aB} T) r \, dC$$

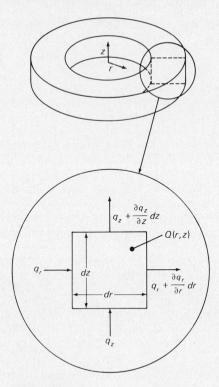

Figure P8-118

providing that imposed heat fluxes and convection are allowed on the boundary of the axisymmetric body. Note that the nomenclature from Sec. 8-9 is used and the imposed heat flux q_{sB} is *positive* if the heat flux is directed *toward* the body.

8-120 Extend the finite element formulation of the axisymmetric heat conduction problem to the case of radiation from the surface of the body to a large enclosure. The energy balance on the boundary (or surface of the body) given by Eq. (8-133a) must be modified to include the heat flux from the radiation q_{rB} which is given by

$$q_{rB} = \varepsilon_B \sigma (T^4 - T_{rB}^4)$$

Note that this assumes that the heat flux from radiation is directed away from the surface. Does this mean the formulation applies only if the radiation is from the body (being analyzed) to the receiver? Please explain. Does this boundary radiation yield another element stiffness matrix, another nodal force vector, or both? Derive them.

8-121 Show that the element stiffness matrix from conduction in the axial direction (z direction) is given by Eq. (8-152) if the three-node, axisymmetric triangular element is used. What assumptions are made in arriving at this result?

8-122   Evaluate the element stiffness matrix from boundary convection for the axisymmetric heat conduction problem [given by Eq. (8-142c)] for the case of a three-node, axisymmetric triangular element
  a. With leg jk on the global boundary
  b. With leg ki on the global boundary

8-123   Show that the element nodal force vector from an internal heat source that is assumed to be uniform over the element is given by Eq. (8-158) if the three-node, axisymmetric triangular element is used.

8-124   Show that the element nodal force vector from boundary convection from leg ij of the a three-node, axisymmetric triangular element is given by Eq. (8-159). State all assumptions made in arriving at this result.

8-125   Evaluate the element nodal force vector from boundary convection from an axisymmetric body [given by Eq. (8-143c)] if the three-node, axisymmetric triangular element is used, with leg jk on the global boundary.

8-126   Evaluate the element nodal force vector from boundary convection from an axisymmetric body [given by Eq. (8-143c)] if the three-node, axisymmetric triangular element is used, with leg ki on the global boundary.

8-127   Show that the element nodal force vector from an imposed heat flux on the boundary of an axisymmetric body is given by Eq. (8-160) if the three-node, axisymmetric triangular element is used and leg ij is on the global boundary. State all assumptions made in arriving at this result.

8-128   Evaluate the element nodal force vector from an imposed heat flux on the boundary of an axisymmetric body [given by Eq. (8-143b)] if the three-node, axisymmetric triangular element is used, with leg jk on the global boundary.

8-129   Evaluate the element nodal force vector from an imposed heat flux on the boundary of an axisymmetric body [given by Eq. (8-143b)] if the three-node, axisymmetric triangular element is used, with leg ki on the global boundary.

8-130   Consider the axisymmetric heat conduciton problem for the case of a "point" heat source. In order for the problem to be considered axisymmetric, the heat source must be uniform in the circumferential direction and is generally specified on a unit circumference basis. Consequently, it is more appropriate to refer to such a heat source as a circumferential line source. Let Q' be the heat source per unit circumference and unit time at point (r_0, z_0). Determine the corresponding element nodal force vector for the case of a three-node, axisymmetric triangular element.

8-131   Determine the expressions for the element resultants, i.e., the heat fluxes from conduction in the r and z directions if the three-node, axisymmetric triangular element is used. Note that these heat fluxes are usually considered to be the average heat fluxes over the element and are generally associated with the element centroid.

8-132   In the formal development of the axisymmetric heat conduction problem in Sec. 8-9, only isotropic materials were considered (the material may be heterogeneous, however). Recall that it is not possible to have fully anisotropic, axisymmetric materials. Why not? It is possible to have stratified materials such that the principal values of thermal conductivity, k_r and k_z (see Problems 8-115 to 8-117), occur in the r and z directions, respectively, as shown in Fig. P8-132.

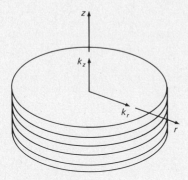

Figure P8-132

a. By using the approach indicated in Problem 8-118, show that the governing equation in this case is given by

$$\frac{1}{r}\frac{\partial}{\partial r}\left(rk_r\frac{\partial T}{\partial r}\right) + \frac{\partial}{\partial z}\left(k_z\frac{\partial T}{\partial z}\right) + Q = 0$$

b. How must the expressions for the element stiffness matrices given by Eqs. (8-142a) and (8-142b) be modified in order to account for this stratified material? Be specific.

8-133 The expressions for the element stiffness matrices from conduction in an axisymmetric body are given by Eqs. (8-142a) and (8-142b). Since these results hold for any axisymmetric element, let us evaluate these stiffness matrices for the four-node, axisymmetric rectangular element by proceeding as follows:

a. With the help of Eqs. (6-32), show that

$$\frac{\partial \mathbf{N}}{\partial x} = \frac{1}{a}\frac{\partial \mathbf{N}}{\partial r}$$

$$\frac{\partial \mathbf{N}}{\partial y} = \frac{1}{b}\frac{\partial \mathbf{N}}{\partial s}$$

b. Using the results from part (a), rewrite the integrals in terms of derivatives of the shape funcitons with respect to the serendipity coordinates r and s. Do not confuse the radius r that appears in the integral with the serendipity coordinate r (e.g., use r' to represent the radial coordinate). Do not forget to change the limits on the integrations. Also note that the elemental area $dr\ dz$ in the rz coordinate system is related to the elemental area in the rs coordinate system by $dr'\ dz = ab\ dr\ ds$.

c. Evaluate the resulting integrands at the element centroid (i.e., at $r = 0$ and $s = 0$). Then treat the integrands as though they are constant and pull them through the integral. Evaluate the remaining trivial integrals. Show that the result for the element stiffness from conduction in the radial direction is given by

$$\mathbf{K}_{rr}^e = \frac{2\pi \bar{r} k b}{4a} \begin{bmatrix} 1 & 1 & -1 & -1 \\ 1 & 1 & -1 & -1 \\ -1 & -1 & 1 & 1 \\ -1 & -1 & 1 & 1 \end{bmatrix}$$

where $\bar{r}$ is the radial coordinate at the centroid of the element. State the other assumptions made in arriving at this result.

d. Derive the corresponding result for $\mathbf{K}_{zz}^e$. What assumptions are made?

8-134 Evaluate the element stiffness matrix from boundary convection from an axisymmetric body [i.e., Eq. (8-142c)] if the four-node, axisymmetric rectangular element is used and if face ij happens to be on the global boundary. Evaluate the integrals by first evaluating the integrands at the centroid of face ij (i.e., at serendipity coordinates $r = +1$ and $s = 0$) and then treating the integrands as though they were constant. Do not confuse the radial coordinate r with the serendipity coordinate r (e.g., use r' to denote the radial coordinate).

8-135 Evaluate the element stiffness matrix from boundary convection from an axisymmetric body [i.e., Eq. (8-142c)] if the four-node, axisymmetric rectangular element is used and if face jk happens to be on the global boundary. Evaluate the integrals by first evaluating the integrands at the centroid of face jk (ie., at serendipity coordinates $r = 0$ and $s = +1$) and then treating the integrands as though they were constant. Do not confuse the radial coordinate r with the serendipity coordinate r (e.g., use r' to denote the radial coordinate).

8-136 Evaluate the element nodal force vector from an internal heat source in an axisymmetric body [i.e., Eq. (8-143a)] if the four-node, axisymmetric rectangular element is used. Evaluate the integrals by first evaluating the integrands at the element centroid (i.e., at serendipity coordinates $r = 0$ and $s = 0$) and then treating the integrands as though they were constant. Do not confuse the radial coordinate r with the serendipity coordinate r (e.g., use r' to denote the radial coordinate).

8-137 Evaluate the element nodal force vector from boundary convection from an axisymmetric body [i.e., Eq. (8-143c)] if the four-node, axisymmetric rectangular element is used and if face ij happens to be on the global boundary. Evaluate the integrals by first evaluating the integrands at the centroid of face ij (i.e., at serendipity coordinates $r = +1$ and $x = 0$) and then treating the integrands as though they were constant. Do not confuse the radial coordinate r with the serendipity coordinate r (e.g., use r' to denote the radial coordinate).

8-138 Evaluate the element nodal force vector from boundary convection from an axisymmetric body [i.e., Eq. (8-143c)] if the four-node, axisymmetric rectangular element is used and if face jk happens to be on the global boundary. Evaluate the integrals by first evaluating the integrands at the centroid of face jk (i.e., at serendipity coordinates $r = 0$ and $s = +1$) and then treating the integrands as though they were constant. Do not confuse the radial coordinate r with the serendipity coordinate r (e.g., use r' to denote the radial coordinate).

8-139 Evaluate the element nodal force vector from a boundary heat flux imposed on an axisymmetric body [i.e., Eq. (8-143b)] if the four-node, axisymmetric rectangular element is used and if face mi happens to be on the global boundary. Evaluate the integrals by first evaluating the integrands at the centroid of face mi (i.e., at serendipity coordinates $r = 0$ and $s = -1$) and then treating the integrands as though they were constant. Do not confuse the radial coordinate r with the serendipity coordinate r (e.g., use r' to denote the radial coordinate).

8-140   Derive the expressions for the average heat fluxes within an element if the four-node, axisymmetric rectangular element is used. *Hint*: Evaluate the expressions for Fourier's law of heat conduction at the element centroid and interpret these results to be the average heat fluxes from conduction in the radial and axial directions.

8-141   Consider the axisymmetric, triangular element shown in Fig. P8-141. The body from which the element is extracted is made of cast iron. The nodal coordinates are $r_i = 2.0$, $z_i = 1.5$, $r_j = 1.7$, $z_j = 3.0$, $r_k = 0.6$, and $z_k = 1.8$ in.

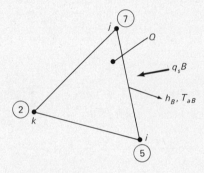

Figure P8-141

a. Determine the element stiffness matrices from conduction in the r and z directions.

b. Determine the element stiffness matrix from boundary convection if leg ij happens to be on the part of the global boundary that undergoes convection to a fluid at 58°F through a convective heat transfer coefficient of 75 Btu/hr-ft²-°F.

8-142   For the element in Problem 8-141, determine the nodal force vectors

a. From an internal heat source of 40 Btu/hr-ft³

b. From boundary convection

c. From a boundary heat flux of 325 Btu/hr-ft² imposed on leg ij (which is on the part of the global boundary also undergoing convection.)

8-143   Consider the axisymmetric, triangular element shown in Fig. P8-143. the body from which the element is extracted is made of brass. The nodal coordinates are $r_i = 5$, $z_i = 6$, $r_j = 4$, $z_j = 4$, $r_k = 6$, and $z_k = 4$ cm.

a. Determine the element stiffness matrices from conduction in the r and z directions.

b. Determine the element stiffness matrix from boundary convection if leg ki happens to be on the part of the global boundary that undergoes convection to a fluid at 55°C through a convective heat transfer coefficient of 4 W/cm²-°C.

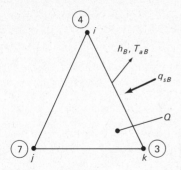

Figure P8-143

8-144   For the element in Problem 8-143, determine the nodal force vectors
   a. From an internal heat source of 80 W/cm^3
   b. From boundary convection
   c. From a boundary heat flux of 35 W/cm^2 imposed on leg ki (which is on the part of the global boundary also undergoing convection)

8-145   Consider the axisymmetric, triangular element shown in Fig. P8-145. The element is extracted from an axisymmetric body that is fabricated from hot rolled, low carbon steel. The nodal coordinates are $r_i = 0$, $z_i = 0$, $r_j = 0$, $z_j = -1$, $r_k = 2$, and $z_k = -1$ cm.

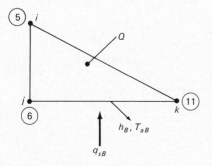

Figure P8-145

   a. Determine the element stiffness matrices from conduction in the r and z directions.
   b. Determine the element stiffness matrix from boundary convection if leg jk happens to be on the part of the global boundary that undergoes convection to a fluid at 75°C through a convective heat transfer coefficient of 5 W/cm^2-°C.

8-146   For the element in Problem 8-145, determine the nodal force vectors
   a. From an internal heat source of 15 W/cm^3
   b. From boundary convection

c. From a boundary heat flux of 50 W/cm^2 imposed on leg jk (which is on the part of the global boundary also undergoing convection)

8-147 Consider the axisymmetric, triangular element shown in Fig. P8-147. The element is extracted from an axisymmetric body that is fabricated from pure copper. The nodal coordinates are $r_i = 10$, $z_i = 10$, $r_j = 11$, $z_j = 12$, $r_k = 9$, and $z_k = 12$ in.

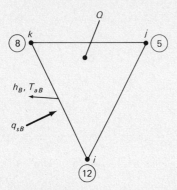

Figure P8-147

a. Determine the element stiffness matrices from conduction in the r and z directions.

b. Determine the element stiffness matrix from boundary convection if leg ki happens to be on the part of the global boundary that undergoes convection to a fluid at 56°F through a convective heat transfer coefficient of 63 Btu/hr-ft^2-°F.

8-148 For the element in Problem 8-147, determine the nodal force vectors
a. From an internal heat source of 25 Btu/hr-ft^3
b. From boundary convection
c. From a boundary heat flux of 50 Btu/hr-ft^2 imposed on leg ki (which is on the part of the global boundary also undergoing convection)

8-149 Consider the axisymmetric, rectangular element shown in Fig. P8-149. The element is extracted from an axisymmetric body that is made of aluminum. The coordinates of the nodes are $r_i = 4$, $z_i = 2$, $r_j = 4$, $z_j = 3$, $r_k = 2$, $z_k = 3$, $r_m = 2$, and $z_m = 2$ cm. Perform the necessary integrations by following the approaches mentioned in Problems 8-133 to 8-135 and

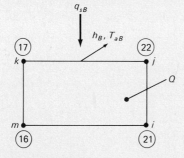

Figure P8-149

a. Determine the element stiffness matrices from conduction in the r and z directions.

b. Determine the element stiffness matrix from boundary convection if face jk happens to be on the part of the global boundary that undergoes convection to a fluid at 10°C through a convective heat transfer coefficient of 5 W/cm²-°C.

8-150 Consider the element in Problem 8-149. Perform the necessary integrations by following the approaches indicated in Problems 8-136 to 8-139, and determine the nodal force vectors

a. From an internal heat source of 5 W/cm³

b. From boundary convection

c. From a boundary heat flux of 30 W/cm² imposed on face jk (which is on the part of the global boundary also undergoing convection)

8-151 Consider the rectangular element shown in Fig. P8-151. The element is extracted from an axisymmetric body that is made of brass. The coordinates of the nodes are $r_i = 5$, $z_i = 2$, $r_j = 5$, $z_j = 4$, $r_k = 2$, $z_k = 4$, $r_m = 2$, and $z_m = 2$ in. Perform the necessary integrations by following the approaches mentioned in Problems 8-133 to 8-135 and

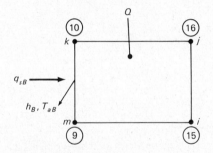

Figure P8-151

a. Determine the element stiffness matrices from conduction in the r and z directions.

b. Determine the element stiffness matrix from boundary convection if face km happens to be on the part of the global boundary that undergoes convection to a fluid at 720°F through a convective heat transfer coefficient of 75 Btu/hr-ft²-°F.

8-152 Consider the element in Problem 8-151. Perform the necessary integrations by following the approaches indicated in Problems 8-136 to 8-139, and determine the nodal force vectors

a. From an internal heat source of 10,000 Btu/hr-ft³

b. From boundary convection

c. From a boundary heat flux of 420 Btu/hr-ft² imposed on face km (which is on the part of the global boundary also undergoing convection)

8-153 For the element in Problem 8-141, determine the element nodal force vector from a circumferential line source with a strength of 50 Btu/hr per inch of circumference at the location of the source. The source is located at $r_0 = 1.5$ and $z_0 = 2.0$ in.

8-154   For the element in Problem 8-143, determine the element nodal force vector from a circumferential line source with a strength of 65 W per centimeter of circumference at the location of the source. The source is located at $r_0 = 5$ and $z_0 = 5$ cm.

8-155   For the element in Problem 8-145, determine the element nodal force vector from a circumferential line source with a strength of 40 W per centimeter of circumference at the location of the source. The source is located at $r_0 = 2$ and $z_0 = -1$ cm.

8-156   For the element in Problem 8-147, determine the element nodal force vector from a circumferential line source with a strength of 130 Btu/hr per inch of circumference at the location of the source. The source is located at $r_0 = 11$ and $z_0 = 12$ in.

8-157   For the element in Problem 8-149, determine the element nodal force vector from a circumferential line source with a strength of 30 W per centimeter of circumference at the location of the source. The source is located at $r_0 = 3.0$ and $z_0 = 2.5$ cm.

8-158   For the element in Problem 8-151, determine the element nodal force vector from a circumferential line source with strength of 30 Btu/hr per inch of circumference at the location of the source. The source is located at $r_0 = 4$ and $z_0 = 3$ in.

8-159   For the element in Problem 8-141 the nodal temperatures are obtained from the solution of $\mathbf{Ka} = \mathbf{f}$ and are $T_i = 175$, $T_j = 168$, and $T_k = 173°F$. Determine the average heat fluxes from conduction in the r and z directions, respectively. With what point in the element are these heat fluxes normally associated?

8-160   For the element in Problem 8-143 the nodal temperatures are obtained from the solution of $\mathbf{Ka} = \mathbf{f}$ and are $T_i = 75$, $T_j = 82$, and $T_k = 73°C$. Determine the average heat fluxes from conduction in the r and z directions, respectively. With what point in the element are these heat fluxes normally associated?

8-161   For the element in Problem 8-149 the nodal temperatures are obtained from the solution of $\mathbf{Ka} = \mathbf{f}$ and are $T_i = 92$, $T_j = 95$, and $T_k = 89$, $T_m = 96°C$. Determine the heat fluxes at the element centriod from conduction in the r and z directions, respectively. These heat fluxes may be regarded as the average conduction heat fluxes.

8-162   For the element in Problem 8-151 the nodal temperatures are obtained from the solution of $\mathbf{Ka} = \mathbf{f}$ and are $T_i = 192$, $T_j = 195$, $T_k = 189$, and $T_m = 196°F$. Determine the heat fluxes at the element centroid from conduction in the r and z directions, respectively. These heat fluxes may be regarded as the average conduction heat fluxes.

8-163   Consider the elemental volume $dx\ dy\ dz$ shown in Fig. P8-163. Note that the conduction heat fluxes in the x, y, and z directions are assumed to vary according to a first-order Taylor expansion. In addition, an internal heat source Q is present. By performing a steady-state energy balance and by invoking Fourier's law of heat conduction, given by Eq. (8-161), show that the governing equation for steady-state heat conduction in a heterogeneous, isotropic, three-dimensional body is given by Eq. (8-162).

8-164   Show that the variational principle that corresponds to Eq. (8-162) for three-dimensional, steady-state heat conduction in a heterogeneous, isotropic body is given by

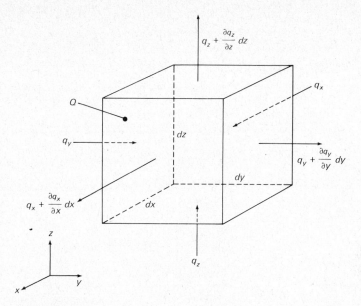

Figure P8-163

$$I = \int_V \left[QT - \tfrac{1}{2} k \left(\frac{\partial T}{\partial x}\right)^2 - \tfrac{1}{2} k \left(\frac{\partial T}{\partial y}\right)^2 - \tfrac{1}{2} k \left(\frac{\partial T}{\partial z}\right)^2 \right] dx \, dy \, dz$$

$$+ \int_S (q_{sB}T - \tfrac{1}{2} h_B T^2 + h_B T_{aB} T) \, dS$$

if imposed heat fluxes (q_{sB}) and convection from the boundary (i.e., the surface) of the three-dimensional body are taken into account. Note that the nomenclature from Sec. 8-10 is used.

8-165 Extend the formulation for three-dimensional heat conduction in Sec. 8-10 to the case of radiation from (or to) the surface of the body to (or from) a large enclosure at temperature T_{rB}. In particular, derive the expressions for the corresponding element stiffness matrix and element nodal force vector. Note that the governing equation [given by Eq. (8-162)] does not need to be modified. Why not?

8-166 Show that the element stiffness matrix from conduction in the x direction in a three-dimensional body is given by Eq. (8-177) if the four-node tetrahedral element is used. What assumptions are made in arriving at this result?

8-167 Show that the element stiffness matrix from conduction in the y direction in a three-dimensional body is given by Eq. (8-178) if the four-node tetrahedral element is used. What assumptions are made in arriving at this result?

8-168 Show that the element stiffness matrix from conduction in the z direction in a three-dimensional body is given by Eq. (8-179) if the four-node tetrahedral element is used. What assumptions are made in arriving at this result?

8-169   Evaluate the element stiffness matrix and element nodal force vector from boundary convection from a three-dimensional body [see Eqs. (8-175d) and 8-176c)] if the four-node tetrahedral element is used, and if face *jkm* of the tetrahedron happens to be on the part of the global boundary that undergoes convection.

8-170   Evaluate the element stiffness matrix and element nodal force vector from boundary convection from a three-dimensional body [see Eqs. (8-175d) and (8-176c)] if the four-node tetrahedral element is used, and if face *ikm* of the tetrahedron happens to be on the part of the global boundary that undergoes convection.

8-171   Evaluate the element stiffness matrix and element nodal force vector from boundary convection from a three-dimensional body [see Eq. (8-175d) and (8-176c)] if the four-node tetrahedral element is used, and if face *ijm* of the tetrahedron happens to be on the part of the global boundary that undergoes convection.

8-172   Evaluate the element nodal force vector from an imposed boundary heat flux on the surface of a three-dimensional body [see Eq. (8-176b)] if the four-node tetra-hedral element is used. Assume that face *jkm* of the tetrahedron happens to be on the part of the global boundary on which a heat flux is imposed.

8-173   Evaluate the element nodal force vector from an imposed boundary heat flux on the surface of a three-dimensional body [see Eq. (8-176b)] if the four-node tetra-hedral element is used. Assume that face *ijm* of the tetrahedron happens to be on the part of the global boundary on which a heat flux is imposed.

8-174   Evaluate the element nodal force vector from an imposed boundary heat flux on the surface of a three-dimensional body [see Eq. (8-176b)] if the four-node tetra-hedral element is used. Assume that face *ikm* of the tetrahedron happens to be on the part of the global boundary on which a heat flux is imposed.

8-175   Show that the element nodal force vector from an internal heat source in a three-dimensional body is given by Eq. (8-182) if the four-node tetrahedral element is used. What assumptions are made in arriving at this result?

8-176   Consider the case of a point heat source in a three-dimensional body. Let us derive an expression for the corresponding nodal force vector for such a point source. Assume the strength of the source is Q_0 (in units of Watts or Btu/hr) and the location is x_0, y_0, and z_0. Use the three-dimensional form of the delta-function (see Sec. 8-8 for the two-dimensional form of the delta-function) to represent the internal heat source in Eq. (8-176a) and evaluate the result to get

$$\mathbf{f}_Q^e = Q_0 \begin{bmatrix} N_i(x_0,y_0,z_0) \\ N_j(x_0,y_0,z_0) \\ N_k(x_0,y_0,z_0) \\ N_m(x_0,y_0,z_0) \end{bmatrix}$$

Explain the physical significance of this result.

8-177   Develop a procedure, formula, or algorithm that could be used to determine the area of a typical face of the tetrahedral element, given the coordinates of the nodes of the element. Note that this is needed in order to evaluate A_{ijk} in Eqs. (8-181), (8-183), and (8-184).

8-178   The solution of $\mathbf{Ka} = \mathbf{f}$ for the three-dimensional heat conduction problem yields the nodal temperatures. Assuming these temperatures are known, show how the average heat fluxes from conduction in the x, y, and z directions may be calculated if the four-node tetrahedral element is used.

8-179   The expression for the element stiffness matrix from conduction in the x direction in a three-dimensional body, given by Eq. (8-175a), is quite general and may be applied to any three-dimensional element. Evaluate the integral for the case of an eight-node brick element by evaluating the integrand at the element centroid (i.e., at serendipity coordinates $r = 0$, $s = 0$, and $t = 0$) and then treating the integrand as though it were constant.

8-180   Repeat Problem 8-179 for the element stiffness matrix from conduction in the y direction in a three-dimensional body, given by Eq. (8-175b).

8-181   Repeat Problem 8-179 for the element stiffness matrix from conduction in the z direction in a three-dimensional body, given by Eq. (8-175c).

8-182   The expression for the element stiffness matrix from convection from (or to) the surface of a three-dimensional body, given by Eq.(8-175d), is quite general and may be applied to any three-dimensional element. Evaluate the integral for the case of an eight-node brick element with face 1-2-3-4 on the global boundary by evaluating the integrand at the centroid of face 1-2-3-4 (what are the values of serendipity coordinates at this point?) and then treating the integrand as though it were constant. Determine the corresponding element nodal force vector in this case.

8-183   Repeat Problem 8-182 if face 1-2-5-6 of the brick element is on the part of the global boundary undergoing convection.

8-184   Repeat Problem 8-182 if face 3-4-7-8 of the brick element is on the part of the global boundary undergoing convection.

8-185   The expression for the element nodal force vector from a distributed internal heat source in a three-dimensional body, given by Eq. (8-176a), is quite general and may be applied to any three-dimensional element. Evaluate the integral for the case of an eight-node brick element by evaluating the integrand at the centroid of the element (what are the values of serendipity coordinates at this point?) and then treating the integrand as though it were constant.

8-186   The expression for the element nodal force vector from an imposed heat flux on the surface of a three-dimensional body, given by Eq. (8-176b), is quite general and may be applied to any three-dimensional element. Evaluate the integral for the case of an eight-node brick element with face 1-2-3-4 on the global boundary by evaluating the integrand at the centroid of face 1-2-3-4 (what are the values of serendipity coordinates at this point) and then treating the integrand as though it were constant.

8-187   Repeat Problem 8-186 if face 1-2-5-6 of the brick element is on the part of the global boundary over which a heat flux is imposed.

8-188   Repeat Problem 8-186 if face 3-4-7-8 of the brick element is on the part of the global boundary over which a heat flux is imposed.

8-189 The body from which a tetrahedral element is extracted is made of pure copper. The nodal coordinates of the element are $x_i = 5$, $y_i = 6$, $z_i = 0$, $x_j = 4$, $y_j = 4$, $z_j = 0$, $x_k = 5$, $y_k = 5$, $z_k = 4$, $x_m = 6$, $y_m = 4$, and $z_m = 0$ cm. Determine the element stiffness matrices from conduction in
    a. The x direction
    b. The y direction
    c. The z direction

8-190 The body from which a tetrahedral element is extracted is made of aluminum. The nodal coordinates of the element are $x_i = 2.0$, $y_i = 1.5$, $z_i = 0.0$, $x_j = 1.7$, $y_j = 3.0$, $z_j = -0.2$, $x_k = 1.5$, $y_k = 2.0$, $z_k = 1.7$, $x_m = 0.6$, $y_m = 1.8$, and $z_m = 0.1$ in. Determine the element stiffness matrices from conduction in
    a. The x direction
    b. The y direction
    c. The z direction

8-191 For the element in Problem 8-189, determine the element stiffness matrix and element nodal force vector from convection if face ijk of the tetrahedron is on the part of the global boundary that is undergoing convection. The convective heat transfer coefficient is 100 W/m²-°C and the ambient temeprature is 45°C.

8-192 For the element in Problem 8-190, determine the element stiffness matrix and element nodal force vector from convection if face ikm of the tetrahedron is on the part of the global boundary that is undergoing convection. The convective heat transfer coefficient is 50 Btu/hr-ft²-°F and the ambient temperature is 72°F.

8-193 For the element in Problem 8-189, determine the element nodal force vector from a uniform internal heat source of 50 W/cm³.

8-194 For the element in Problem 8-190, determine the element nodal force vector from a uniform internal heat source of 10,000 Btu/hr-ft³.

8-195 For the element in Problem 8-189, determine the element nodal force vector from a boundary heat flux of 200 W/cm² if face ijk is receiving the imposed flux.

8-196 For the element in Problem 8-190, determine the element nodal force vector from a boundary heat flux of 1000 Btu/hr-ft² if face ikm is receiving the imposed flux.

8-197 For the element in Problem 8-189, determine the element nodal force vector from a point heat source of 75 W at $x_0 = 5$, $y_0 = 5$, and $z_0 = 0$ cm.

8-198 For the element in Problem 8-190, determine the element nodal force vector from a point heat source of 185 Btu/hr at $x_0 = 1.6$, $y_0 = 2.1$, and $z_0 = 1.1$ in.

8-199 Consider the velocity potential formulation of the two-dimensional potential flow problem in Sec. 8-11.
    a. Show that the velocity components defined by Eqs. (8-187) satisfy the irrotational flow condition exactly.
    b. Show that Laplace's equation (8-188) results if these expressions for the velocity components are substituted into the two-dimensional continuity equation for an incompressible fluid.

8-200 Extend the two-dimensional velocity potential formulation in Sec. 8-11 to the three-dimensional case. In particular, determine the equations that correspond to Eqs.

(8-187) to (8-197). Use u, v, and w to denote the velocity components in the x, y, and z directions, respectively. Do not evaluate the integrals that appear in the expressions for the element characteristics.

8-201 Under what conditions do Eqs. (8-202) and (8-203) hold? Derive the analogous equations that would need to be used if leg jk were on the global boundary and if the velocity components u and v are assumed to vary linearly over the leg. At node j the velocity components are u_j and v_j, and at node k the velocity components are u_k and v_k.

8-202 Under what conditions do Eqs. (8-202) and (8-203) hold? Derive the analogous equations that would need to be used if leg ki is on the global boundary and if the velocity components u and v are assumed to vary linearly over the leg. At node i the velocity components are u_i and v_i, and at node k the velocity components are u_k and v_k.

8-203 The direction cosines (n_x and n_y) of the outward normal unit vector on leg ij of the triangular element may be computed with the help of Eqs. (8-208) and (8-209). Extend these equations to the case when leg jk is on the global boundary.

8-204 The direction cosines (n_x and n_y) of the outward normal unit vector on leg ij of the triangular element may be computed with the help of Eqs. (8-208) and (8-209). Derive the corresponding equations for the case of leg ki on the global boundary.

8-205 Determine the direction cosines n_x and n_y on leg jk for the element in Example 8-22 (see Fig. 8-26).

8-206 Determine the direction cosines n_x and n_y on leg ki for the element in Example 8-22 (see Fig. 8-26).

8-207 Consider the stream function formulation of the two-dimensional potential flow problem in Sec. 8-11.
   a. Show that the velocity components defined by Eqs. (8-214) satisfy the two-dimensional continuity equation exactly (for an incompressible fluid).
   b. Show that Laplace's equation (8-215) results if these expressions for the velocity components are substituted into the irrotational flow condition.

8-208 By applying the Galerkin method (on a element basis) to the governing equation given by Eq. (8-215) for two-dimensional potential flow of an incompressible fluid [see Eq. (8-216)], show that Eqs. (8-217) to (8-221) hold if the stream function formulation is used.

8-209 By performing an energy balance on the elemental area $dx \, dy$ shown in Fig. P8-209, derive the governing equation given by Eq. (8-225a). Assume that the conduction heat fluxes vary according to a first-order Taylor expansion as shown on the figure. Remember to include the effect of the fluid motion in the x direction (i.e., the energy transport by fluid motion or convection). Clearly state the conditions under which Eq. (8-225a) holds.

8-210 Show that the element stiffness matrix given by Eq. (8-241) is unsymmetric. Recall that this matrix results from the convective energy transport term in the governing equation [i.e., Eq. (8-225a)]. Evaluate the integral in Eq. (8-241) for the case of the three-node triangular element.

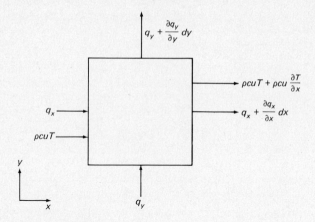

Figure P8-209

8-211 Show that the element stiffness matrix given by Eq. (8-241) is unsymmetric. Recall that this matrix results from the convective energy transport term in the governing equation [i.e., Eq. (8-225a)]. Evaluate the integral in Eq. (8-241) for the case of the four-node rectangular element by evaluating the integrand at the element centroid and then treating the integrand as though it were constant.

8-212 *Try* to obtain the variational principle that would correspond to Eq. (8-225a), and clearly explain the difficulties that are encountered.

8-213 Consider the case of a fully developed laminar flow of an incompressible Newtonian fluid in a duct of circular cross section. On the walls of the duct, either the temperature is prescribed or a heat flux is imposed (these two boundary conditions may occur on different parts of the boundary). It is assumed further that the boundary conditions are axisymmetric. It is desired to obtain the governing equation in this case. Recall from elementary fluid mechanics that if the viscosity of the fluid is assumed to be constant, the velocity profile is given by

$$u(r) = 2\bar{u} \left[1 - \left(\frac{r}{R} \right)^2 \right]$$

where $\bar{u}$ is the average fluid velocity in the duct and R is the radius of the duct. By performing an energy balance on an annular-shaped elemental volume $2\pi r \, dr \, dz$, show that the governing equation is given by

$$\rho c u(r) \frac{\partial T}{\partial z} = \frac{1}{r} \frac{\partial}{\partial r} \left(kr \frac{\partial T}{\partial r} \right) + \frac{\partial}{\partial z} \left(k \frac{\partial T}{\partial z} \right)$$

if axial conduction is not neglected. Assume a first-order Taylor expansion for the conduction heat fluxes. Include the effect of the convective energy transport in the derivation. Which term in the governing equation above results from the convective energy transport?

8-214 Obtain the expressions for the element characteristics for the situation described in Problem 8-213. Do not evaluate the resulting integrals. Identify the element stiffness matrix that is not symmetric. What is the implication of this in the solution of **Ka** = **f** for the nodal temperatures in the vector **a**?

8-215 Consider the case of laminar flow of a Newtonian fluid as described in Sec. 8-13. By beginning with Eq. (8-251) and making use of Eqs. (8-246) and (8-252), show that Eqs. (8-253) and (8-254) hold.

8-216 Identify the element stiffness matrices in Sec. 8-13 that are unsymmetric. What is the implication of this during the assemblage step? What effect do the unsymmetric stiffness matrices have on the solution for the nodal velocities and pressures? How can the results be improved?

8-217 Clearly explain the differences between the two different assemblage procedures described in Sec. 8-13 when each node has more than one degree of freedom. Which procedure results in a larger bandwidth? Please explain. Which procedure is analogous to the assemblage procedure used in stress analysis (see Chapter 7)? Please explain.

8-218 In a corrosion study to be performed on aluminum, a long specimen with rectangular cross section is to be held between two isothermal surfaces at T_1 and T_2, as shown in Fig. P8-218. The height and width of the specimen are H and W, respectively.

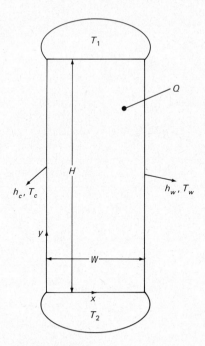

Figure P8-218

The specimen is exposed to a cold corrosive environment at a temperature T_c with a convective heat transfer coefficient h_c. The other side is exposed to relatively warm stagnant water at a temperature T_w with a convective heat transfer coefficient

h_w. A current is passed through the specimen such that a distributed internal heat source Q results at a rate given by

$$Q = C_1 \sin \frac{2\pi x}{W}$$

where x is measured in the same units as W and H. Note that the heat source is zero on the surfaces of the specimen and reaches a maximum value at $x = W/2$. The global coordinate system is shown in Fig. P8-218. The speciment is isotropic with a thermal conductivity k. A finite element solution for the steady-state temperature distribution is sought with 5 nodes in the x direction and 11 nodes in the y direction. Determine the temperature distribution for the following parameters: $W = 10$ cm, $H = 25$ cm, $T_1 = 125°C$, $T_2 = 40°C$, $T_c = 5°C$, $T_w = 25°C$, $h_c = 1000$ W/m^2-°C, $h_w = 75$ W/m^2-°C, and $C_1 = 2000$ W/m^3.

8-219 Repeat Problem 8-218 for the following parameters: $W = 0.32$ ft, $H = 0.75$ ft, $T_1 = 250°F$, $T_2 = 110°F$, $T_c = 35°F$, $T_w = 75°F$, $h_c = 250$ Btu/hr-ft^2-°F, $h_w = 15$ Btu/hr-ft^2-°F, and $C_1 = 60,000$ Btu/hr-ft^3.

8-220 Half of the outside surface of a long thick-walled boiler tube receives a uniform heat flux q_{sB} while the other half is insulated. The inside is cooled convectively with heat transfer coefficient h_B to a fluid at temperature T_{aB}. The tube is fabricated from stainless steel and copper as shown in Fig. P8-220. The inner and outer radii

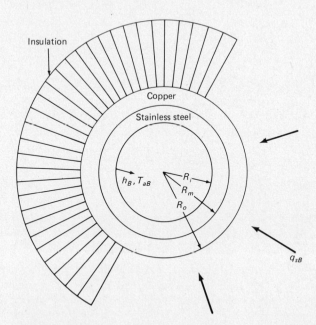

Figure P8-220

of the tube are R_i and R_o, and the interface between the stainless steel and copper is located at a radius R_m. Determine the steady-state temperature distribution in the tube for the following parameters: $q_{sB} = 20,000$ Btu/hr-ft^2, $h_B = 10,000$ Btu/hr-

$\text{ft}^2\text{-}°\text{F}$, $T_{aB} = 500°\text{F}$, $R_i = 1.0$ in., $R_m = 1.25$ in., and $R_o = 1.5$ in. Note that because of the symmetry, only one-half of the tube needs to be analyzed. Use approximately 100 three-node triangular elements.

8-221 Repeat Problem 8-220 for the following parameters: $q_{sB} = 1000$ W/cm^2, $h_B = 50{,}000$ W/m^2-°C, $T_{aB} = 250°\text{C}$, $R_i = 2.5$ cm, $R_m = 3.5$ cm, and $R_o = 4.5$ cm.

8-222 It is desired to obtain the two-dimensional steady-state temperature distribution in the thin tapered bronze fin of length L_f shown in Fig. P8-222. The thickness of the fin is t_1 at the base and tapers linearly to t_2 at the tip. The height of the fin at the base and tip are H_1 and H_2, respectively. The base is held at a temperature of T_b, while all exposed surfaces of the fin convect to a fluid at a temperature T_f with convective heat transfer coefficients of h_1, h_2 and h_3 on the lateral faces and tip, on the bottom edge, and on the top edge as shown in the figure. Use at least 80 three-node triangular elements to determine the two-dimensional temperature distribution in the fin for the following parameters: $L_f = 2.0$ cm, $t_1 = 1$ cm, $t_2 = 0.5$ cm, $H_1 = 1.5$ cm, $H_2 = 0.75$ cm, $T_b = 200°\text{C}$, $T_f = 100°\text{C}$, $h_1 = 1100$ W/m^2-°C, $h_2 = 600$ W/m^2-°C, and $h_3 = 1500$ W/m^2-°C.

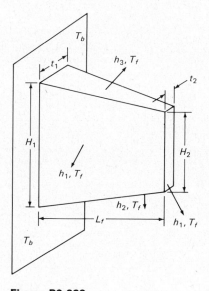

Figure P8-222

8-223 Repeat Problem 8-222 for the following parameters: $L_f = 1.0$ in., $t_1 = 0.4$ in., $t_2 = 0.25$ in., $H_1 = 0.75$ in., $H_2 = 0.375$ in., $T_b = 390°\text{F}$, $T_f = 212°\text{F}$, $h_1 = 200$ Btu/hr-ft^2-°F, $h_2 = 120$ Btu/hr-ft^2-°F, and $h_3 = 300$ Btu/hr-ft^2-°F.

9

Higher-Order Isoparametric Elements and Quadrature

9-1 INTRODUCTION

In Chapter 6 shape functions of the lowest possible order were derived for one-, two-, and three-dimensional elements. Recall that it was convenient to introduce local normalized coordinates, such as length, area, volume, and serendipity coordinates. More specifically, length, area, and volume coordinates are used with lineal, triangular, and tetrahedral elements, respectively. Serendipity coordinates are used with lineal, rectangular, and brick elements. The use of these types of coordinates is even more important when higher-order elements are used.

In this chapter several higher-order elements are described, and the shape functions are given. Numerical integration methods are also presented. Recall further from Chapter 6 that shape functions generally must meet the compatibility and completeness criteria. All the shape functions presented in this section meet these requirements. In addition, these shape functions are continuous, but their derivatives are not necessarily continuous from one element to the next at the element boundaries. In other words, the shape functions presented here have C^0-continuity only.

In this book, the only shape functions with C^1-continuity were those derived for the beam element in Sec. 7-5. When analyzing plates and shells, C^1-continuity is required because both deflection and slope continuity must be assured. The study of higher-order shape functions with C^1-continuity is beyond the scope of this text. The interested reader may want to consult the book by Zienkiewicz [1].

The letter designations for the local node numbers (i.e., i, j, k, etc.) prove to be cumbersome for the higher-order elements. Therefore, we will designate the local node numbers as 1, 2, 3, etc., and the length, area, and volume coordinates will be denoted L_1, L_2, etc.

9-2 ONE-DIMENSIONAL ELEMENTS

Two higher-order one-dimensional elements can be created by adding one or two nodes to the interior of the element as shown in Fig. 9-1. Recall that for the two-node lineal element the shape functions were linear. This is reasonable because the shape functions provide a convenient interpolation polynomial and a unique straight line may be drawn through two points. We may refer to this element as the linear-order lineal element. If we add one node to the lineal element such that it lies halfway between the two original nodes, we can obtain a quadratic-order lineal element. Note that either length or serendipity coordinates may be used. Similarly adding two nodes such that the four nodes are equidistant yields a cubic-order lineal element. The shape functions are given below in terms of the length coordinates L_1 and L_2. This is followed by the shape functions in terms of the serendipity coordinate r.

Length Coordinates

Linear order (two nodes):

$$N_1 = L_1$$
$$N_2 = L_2 \qquad\qquad (9\text{-}1)$$

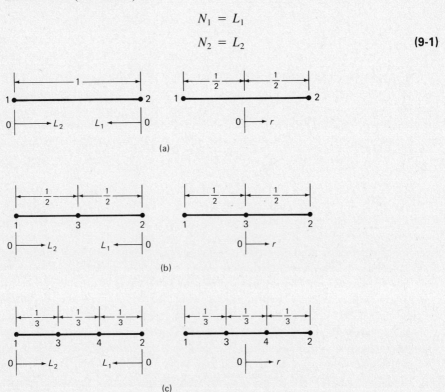

(a)

(b)

(c)

Figure 9-1 One-dimensional elements in terms of length coordinates L_1 and L_2, and serendipity coordinate r. (a) Linear order, (b) quadratic order, and (c) cubic order.

Quadratic order (three nodes):

$$N_1 = L_1(2L_1 - 1)$$
$$N_2 = L_2(2L_2 - 1)$$
$$N_3 = 4L_1L_2 \tag{9-2}$$

Cubic-order (four nodes):

$$N_1 = \tfrac{1}{2} L_1 (3L_1 - 1)(3L_1 - 2)$$
$$N_2 = \tfrac{1}{2} L_2 (3L_2 - 1)(3L_2 - 2)$$
$$N_3 = \tfrac{9}{2} L_1L_2(3L_1 - 1)$$
$$N_4 = \tfrac{9}{2} L_1L_2(3L_2 - 1) \tag{9-3}$$

Serendipity Coordinate

Linear-order (two nodes):

$$N_1 = \tfrac{1}{2}(1 - r)$$
$$N_2 = \tfrac{1}{2}(1 + r) \tag{9-4}$$

Quadratic-order (three nodes):

$$N_1 = -\tfrac{1}{2}r(1 - r)$$
$$N_2 = \tfrac{1}{2}r(1 + r)$$
$$N_3 = (1 + r)(1 - r) \tag{9-5}$$

Cubic-order (four nodes):

$$N_1 = -\tfrac{9}{144} (r - 1)(3r + 1)(3r - 1)$$
$$N_2 = \tfrac{9}{144} (3r + 1)(3r - 1)(r + 1)$$
$$N_3 = \tfrac{27}{48} (r + 1)(r - 1)(3r - 1)$$
$$N_4 = -\tfrac{27}{48} (r + 1)(r - 1)(3r + 1) \tag{9-6}$$

The shape functions for the lineal element are illustrated graphically in Fig. 9-2. Note how the shape function N_i is unity at node i. Note also that two, three, and four nodes allow linear-, quadratic-, and cubic-order interpolations. Indeed, the shape functions themselves in Fig. 9-2(a) are linear, whereas they are quadratic and cubic in Figs. 9-2(b) and (c), respectively.

The fact that the nodes must be equally spaced is significant. This restriction is relaxed later in this chapter after isoparametric elements are introduced.

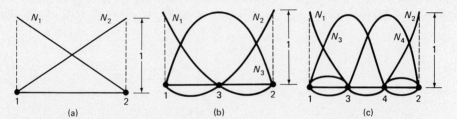

Figure 9-2 Plot of the shape functions for the (a) linear order, (b) quadratic order, and (c) cubic order one-dimensional elements.

9-3 TWO-DIMENSIONAL ELEMENTS

In this section, the triangular and rectangular elements from Chapter 6 are modified so that some of the higher-order terms are represented in the corresponding parameter functions. As in the one-dimensional higher-order element, the shape functions will be seen to be higher-order as well.

The Triangular Element

The triangular element is in the unique position of being able to include complete polynomials in the parameter function as the Pascal triangle in Fig. 9-3 shows. Recall from Chapter 6 that for the linear-order or three-node triangular element we assumed a polynomial that involves the first three terms in Fig. 9-3. A typical parameter function ϕ is represented by

$$\phi = c_1 + c_2 x + c_3 y \tag{9-7}$$

where c_1, c_2, and c_3 are constants. The position of the terms in the Pascal triangle corresponds to the position of the nodes on the triangular element. The three-node or linear-order triangular element is shown in Fig. 9-4(a).

The quadratic-order triangular element requires that the parameter function contain the first six terms from the Pascal triangle with the nodes on the element

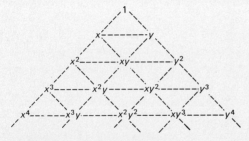

Figure 9-3 The Pascal triangle.

arranged as shown in Fig. 9-4(b). Note that nodes 4, 5, and 6 are on the midsides of legs 1-2, 2-3, and 3-1, respectively. A typical parameter function in this case is represented by

$$\phi = c_1 + c_2 x + c_3 y + c_4 x^2 + c_5 xy + c_6 y^2 \tag{9-8}$$

There are six constants in the parameter function and six nodes in the element. The quadratic-order triangular element may also be referred to as the six-node triangular element.

If Fig. 9-3 is examined carefully, it is seen that 10 terms are needed to represent a cubic-order parameter function, or

$$u = c_1 + c_2 x + c_3 y + c_4 x^2 + c_5 xy + c_6 y^2$$
$$+ c_7 x^3 + c_8 x^2 y + c_9 xy^2 + c_{10} y^3 \tag{9-9}$$

Note the position of the xy term in the Pascal triangle. Since it is in the interior of the triangle, a node must be placed in the interior of the triangular element as shown in Fig. 9-4(c). In fact, this node must be placed at the centroid of the triangle. The remaining nodes are placed on the legs of the triangle such that each leg is divided into three equal parts. Note that nodes 4 and 5 are located on leg 1-2, with node 4 closer to node 1. Similar observations may be made about the nodes on the remaining legs. Node 10 is the interior node.

The shape functions below assume the relative positions of the nodes are as shown in Fig. 9-4. In Sec. 9-6, these restrictions are relaxed.

Linear-order (three nodes):

$$N_1 = L_1$$
$$N_2 = L_2$$
$$N_3 = L_3 \tag{9-10}$$

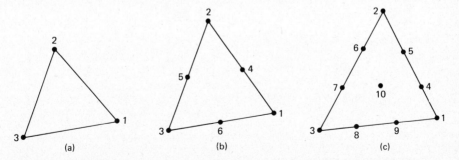

Figure 9-4 The triangular element: (a) linear order, (b) quadratic order, and (c) cubic order. Note the internal node (node 10) on the cubic-order element.

Quadratic-order (six nodes):

$$N_1 = L_1(2L_1 - 1) \qquad N_2 = L_2(2L_2 - 1)$$
$$N_3 = L_3(2L_3 - 1) \qquad N_4 = 4L_1L_2$$
$$N_5 = 4L_2L_3 \qquad\qquad N_6 = 4L_1L_3 \qquad\qquad \textbf{(9-11)}$$

Cubic-order (ten nodes):

$$N_1 = \tfrac{1}{2} L_1 (3L_1 - 1)(3L_1 - 2) \qquad N_2 = \tfrac{1}{2} L_2 (3L_2 - 1)(3L_2 - 2)$$

$$N_3 = \tfrac{1}{2} L_3 (3L_3 - 1)(3L_3 - 2) \qquad N_4 = \tfrac{9}{2} L_1L_2 (3L_1 - 1)$$

$$N_5 = \tfrac{9}{2} L_1L_2 (3L_2 - 1) \qquad N_6 = \tfrac{9}{2} L_2L_3 (3L_2 - 1)$$

$$N_7 = \tfrac{9}{2} L_2L_3 (3L_3 - 1) \qquad N_8 = \tfrac{9}{2} L_1L_3 (3L_3 - 1) \qquad \textbf{(9-12)}$$

$$N_9 = \tfrac{9}{2} L_1L_3 (3L_1 - 1) \qquad N_{10} = 27L_1L_2L_3$$

The interior node (node 10) in the cubic-order element is not shared with any other element. Therefore, before this element is assembled in the assemblage matrix $\mathbf{K}^a$, the node should be condensed by using the substructuring technique described in Sec. 7-6. Node 10 is treated simply as an interior node, and the element without this node becomes the superelement. The nodal value of the parameter function at this (and other) interior nodes could be recovered as described in Sec. 7-6.

The Rectangular Element

Several higher-order rectangular elements are shown in Fig. 9-5. Note that additional nodes are added to the sides of the rectangle. For the quadratic-order element, the midside nodes must be located midway between the corner nodes. For the cubic-order element, the midside nodes must be located such that the nodes on each side are equally spaced. In addition, the global x and y axes must be parallel to the local r and s axes, respectively. In Sec. 9-6, all these restrictions are relaxed. The shape functions are given below in terms of the serendipity coordinates r and s:

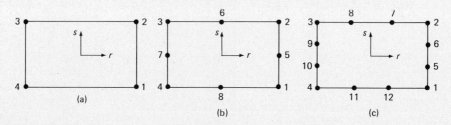

Figure 9-5 The rectangular (serendipity) element: (a) linear order, (b) quadratic order, and (c) cubic order.

Linear-order (four nodes):

$$N_1 = \tfrac{1}{4}(1 + r)(1 - s) \qquad N_2 = \tfrac{1}{4}(1 + r)(1 + s)$$

$$N_3 = \tfrac{1}{4}(1 - r)(1 + s) \qquad N_4 = \tfrac{1}{4}(1 - r)(1 - s) \qquad \text{(9-13)}$$

Quadratic-order (eight nodes):

$$N_1 = \tfrac{1}{4}(1+r)(1-s)(r-s-1) \qquad N_2 = \tfrac{1}{4}(1+r)(1+s)(r+s-1)$$

$$N_3 = \tfrac{1}{4}(1-r)(1+s)(-r+s-1) \quad N_4 = \tfrac{1}{4}(1-r)(1-s)(-r-s-1) \quad \text{(9-14)}$$

$$N_5 = \tfrac{1}{2}(1+r)(1-s^2) \qquad\qquad N_6 = \tfrac{1}{2}(1-r^2)(1+s)$$

$$N_7 = \tfrac{1}{2}(1-r)(1-s^2) \qquad\qquad N_8 = \tfrac{1}{2}(1-r^2)(1-s)$$

Cubic-order (12 nodes):

$$N_1 = \tfrac{1}{32}(1 + r)(1 - s)[-10 + 9(r^2 + s^2)]$$

$$N_2 = \tfrac{1}{32}(1 + r)(1 + s)[-10 + 9(r^2 + s^2)]$$

$$N_3 = \tfrac{1}{32}(1 - r)(1 + s)[-10 + 9(r^2 + s^2)]$$

$$N_4 = \tfrac{1}{32}(1 - r)(1 - s)[-10 + 9(r^2 + s^2)]$$

$$N_5 = \tfrac{9}{32}(1 + r)(1 - s^2)(1 - 3s)$$

$$N_6 = \tfrac{9}{32}(1 + r)(1 - s^2)(1 + 3s) \qquad \text{(9-15)}$$

$$N_7 = \tfrac{9}{32}(1 + s)(1 - r^2)(1 + 3r)$$

$$N_8 = \tfrac{9}{32}(1 + s)(1 - r^2)(1 - 3r)$$

$$N_9 = \tfrac{9}{32}(1 - r)(1 - s^2)(1 + 3s)$$

$$N_{10} = \tfrac{9}{32}(1 - r)(1 - s^2)(1 - 3s)$$

$$N_{11} = \tfrac{9}{32}(1 - s)(1 - r^2)(1 - 3r)$$

$$N_{12} = \tfrac{9}{32}(1 - s)(1 - r^2)(1 + 3r)$$

9-4 THREE-DIMENSIONAL ELEMENTS

In Chapter 6, the tetrahedral and brick elements were presented. In this section, these elements are modified so that some of the higher-order terms are represented in the corresponding parameter functions. The shape functions are presented for these elements in terms of volume coordinates for the tetrahedral element and serendipity coordinates for the brick element.

The Tetrahedral Element

The tetrahedral element is similar to the triangular element in that it is also able to include the complete polynomial in the parameter function. In Chapter 6, a typical parameter function ϕ is represented by

$$\phi = c_1 + c_2x + c_3y + c_4z \qquad (9\text{-}16)$$

for the four-node or linear-order tetrahedral element shown in Fig. 9-6(a). The quadratic-order tetrahedral element requires that the parameter function be of the form

$$\phi = c_1 + c_2x + c_3y + c_4z + c_5x^2 + c_6xy$$
$$+ c_7y^2 + c_8yz + c_9z^2 + c_{10}zx \qquad (9\text{-}17)$$

Note that 10 terms are needed, and hence 10 nodes are present in the corresponding element as shown in Fig. 9-6(b). In a similar fashion, the cubic-order tetrahedral element requires a parameter function with 20 terms, and hence the element has 20 nodes.

In the case of the quadratic-order element, nodes 5 to 10 are located midway between the respective corner nodes as shown in Fig. 9-6(b). For the cubic-order element, nodes 5 to 16 are located as shown in Fig. 9-6(c), where nodes 5 and 8 are placed such that leg 1-2 is divided into three equal segments. Similar comments hold about the remaining midside nodes. Nodes 17 to 20 are located at the centroids of faces 1-2-3, 1-3-4, etc. as shown in Fig. 9-6(c). Most of these restrictions are relaxed in Sec. 9-7, where the isoparametric tetrahedral element is presented. The shape functions are given below in terms of the volume coordinates for the tetrahedral elements.

Linear-order (four nodes):

$$N_1 = L_1 \qquad N_2 = L_2 \qquad (9\text{-}18)$$
$$N_3 = L_3 \qquad N_4 = L_4$$

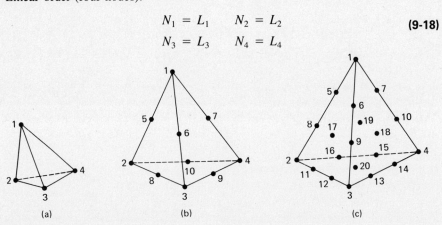

Figure 9-6 The tetrahedral element: (a) linear order, (b) quadratic order, and (c) cubic order. Note that for the cubic-order element, nodes 17, 18, 19, and 20 are located at the centroid of the respective faces of the tetrahedron.

Quadratic-order (ten nodes):

$$N_1 = L_1(2L_1 - 1) \qquad N_2 = L_2(2L_2 - 1)$$

$$N_3 = L_3(2L_3 - 1) \qquad N_4 = L_4(2L_4 - 1)$$

$$N_5 = 4L_1L_2 \qquad N_6 = 4L_1L_3$$

$$N_7 = 4L_1L_4 \qquad N_8 = 4L_2L_3 \tag{9-19}$$

$$N_9 = 4L_3L_4 \qquad N_{10} = 4L_2L_4$$

Cubic-order (twenty nodes):

$$N_1 = \tfrac{1}{2} L_1(3L_1 - 1)(3L_1 - 2) \qquad N_2 = \tfrac{1}{2} L_2(3L_2 - 1)(3L_2 - 2)$$

$$N_3 = \tfrac{1}{2} L_3(3L_3 - 1)(3L_3 - 2) \qquad N_4 = \tfrac{1}{2} L_4(3L_4 - 1)(3L_4 - 2)$$

$$N_5 = \tfrac{9}{2} L_1L_2(3L_1 - 1) \qquad N_6 = \tfrac{9}{2} L_1L_3(3L_1 - 1)$$

$$N_7 = \tfrac{9}{2} L_1L_4(3L_1 - 1) \qquad N_8 = \tfrac{9}{2} L_1L_2(3L_2 - 1)$$

$$N_9 = \tfrac{9}{2} L_1L_3(3L_3 - 1) \qquad N_{10} = \tfrac{9}{2} L_1L_4(3L_4 - 1)$$

$$N_{11} = \tfrac{9}{2} L_2L_3(3L_2 - 1) \qquad N_{12} = \tfrac{9}{2} L_2L_3(3L_3 - 1) \tag{9-20}$$

$$N_{13} = \tfrac{9}{2} L_3L_4(3L_3 - 1) \qquad N_{14} = \tfrac{9}{2} L_3L_4(3L_4 - 1)$$

$$N_{15} = \tfrac{9}{2} L_2L_4(3L_4 - 1) \qquad N_{16} = \tfrac{9}{2} L_2L_4(3L_2 - 1)$$

$$N_{17} = 27L_1L_2L_3 \qquad N_{18} = 27L_1L_3L_4$$

$$N_{19} = 27L_1L_2L_4 \qquad N_{20} = 27L_2L_3L_4$$

The Brick Element

The linear-, quadratic-, and cubic-order brick elements are shown in Fig. 9-7. Again the midside nodes must be positioned such that the edges are divided into equal segments. As mentioned in Chapter 6, the faces of the brick must line up with the global coordinate system. In Sec. 9-7, these restrictions are relaxed. The shape functions for these elements are given below in terms of the serendipity coordinates r, s, and t.

Linear-order (eight nodes):

$$N_1 = \tfrac{1}{8}(1 + r)(1 - s)(1 + t) \qquad N_2 = \tfrac{1}{8}(1 + r)(1 + s)(1 + t)$$

$$N_3 = \tfrac{1}{8}(1 - r)(1 + s)(1 + t) \qquad N_4 = \tfrac{1}{8}(1 - r)(1 - s)(1 + t) \tag{9-21}$$

$$N_5 = \tfrac{1}{8}(1 + r)(1 - s)(1 - t) \qquad N_6 = \tfrac{1}{8}(1 + r)(1 + s)(1 - t)$$

$$N_7 = \tfrac{1}{8}(1 - r)(1 + s)(1 - t) \qquad N_8 = \tfrac{1}{8}(1 - r)(1 - s)(1 - t)$$

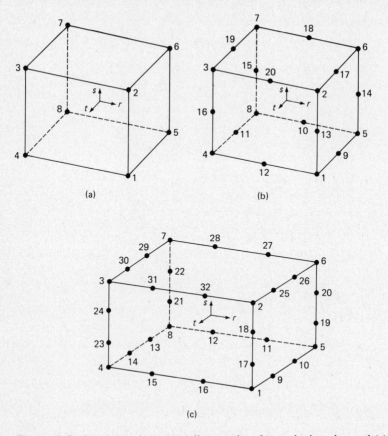

Figure 9-7   The brick element: (a) linear order, (b) quadratic order, and (c) cubic order.

Quadratic-order (20 nodes):

$$N_1 = \tfrac{1}{8}(1+r)(1-s)(1+t)(r-s+t-2)$$

$$N_2 = \tfrac{1}{8}(1+r)(1+s)(1+t)(r+s+t-2)$$

$$N_3 = \tfrac{1}{8}(1-r)(1+s)(1+t)(-r+s+t-2)$$

$$N_4 = \tfrac{1}{8}(1-r)(1-s)(1+t)(-r-s+t-2)$$

$$N_5 = \tfrac{1}{8}(1+r)(1-s)(1-t)(r-s-t-2)$$

$$N_6 = \tfrac{1}{8}(1+r)(1+s)(1-t)(r+s-t-2)$$

$$N_7 = \tfrac{1}{8}(1-r)(1+s)(1-t)(-r+s-t-2)$$

$$N_8 = \tfrac{1}{8}(1-r)(1-s)(1-t)(-r-s-t-2)$$

$$N_9 = \tfrac{1}{4}(1-t^2)(1+r)(1-s)$$

$$N_{10} = \frac{1}{4}(1-r^2)(1-s)(1-t)$$

(9-22)

$$N_{11} = \frac{1}{4}(1-t^2)(1-r)(1-s)$$

$$N_{12} = \frac{1}{4}(1-r^2)(1-s)(1+t)$$

$$N_{13} = \frac{1}{4}(1-s^2)(1+r)(1+t)$$

$$N_{14} = \frac{1}{4}(1-s^2)(1+r)(1-t)$$

$$N_{15} = \frac{1}{4}(1-s^2)(1-r)(1-t)$$

$$N_{16} = \frac{1}{4}(1-s^2)(1-r)(1+t)$$

$$N_{17} = \frac{1}{4}(1-t^2)(1+r)(1+s)$$

$$N_{18} = \frac{1}{4}(1-r^2)(1+s)(1-t)$$

$$N_{19} = \frac{1}{4}(1-t^2)(1-r)(1+s)$$

$$N_{20} = \frac{1}{4}(1-r^2)(1+s)(1+t)$$

Cubic-order (32 nodes):

$$N_1 = \frac{1}{64}(1+r)(1-s)(1+t)[9(r^2+s^2+t^2)-19]$$

$$N_2 = \frac{1}{64}(1+r)(1+s)(1+t)[9(r^2+s^2+t^2)-19]$$

$$N_3 = \frac{1}{64}(1-r)(1+s)(1+t)[9(r^2+s^2+t^2)-19]$$

$$N_4 = \frac{1}{64}(1-r)(1-s)(1+t)[9(r^2+s^2+t^2)-19]$$

$$N_5 = \frac{1}{64}(1+r)(1-s)(1-t)[9(r^2+s^2+t^2)-19]$$

$$N_6 = \frac{1}{64}(1+r)(1+s)(1-t)[9(r^2+s^2+t^2)-19]$$

$$N_7 = \frac{1}{64}(1-r)(1+s)(1-t)[9(r^2+s^2+t^2)-19]$$

$$N_8 = \frac{1}{64}(1-r)(1-s)(1-t)[9(r^2+s^2+t^2)-19]$$

$$N_9 = \frac{9}{64}(1-t^2)(1+3t)(1-s)(1+r)$$

$$N_{10} = \frac{9}{64}(1-t^2)(1-3t)(1-s)(1+r)$$

$$N_{11} = \frac{9}{64}(1-r^2)(1+3r)(1-s)(1-t)$$

$$N_{12} = \frac{9}{64}(1-r^2)(1-3r)(1-s)(1-t)$$

$$N_{13} = \frac{9}{64}(1-t^2)(1-3t)(1-s)(1-r)$$

$$N_{14} = \frac{9}{64}(1-t^2)(1+3t)(1-s)(1-r)$$

$$N_{15} = \frac{9}{64}(1-r^2)(1-3r)(1-s)(1+t)$$

$$N_{16} = \%_{64}(1-r^2)(1+3r)(1-s)(1+t) \qquad \text{(9-23)}$$

$$N_{17} = \%_{64}(1-s^2)(1-3s)(1+r)(1+t)$$

$$N_{18} = \%_{64}(1-s^2)(1+3s)(1+r)(1+t)$$

$$N_{19} = \%_{64}(1-s^2)(1-3s)(1+r)(1-t)$$

$$N_{20} = \%_{64}(1-s^2)(1+3s)(1+r)(1-t)$$

$$N_{21} = \%_{64}(1-s^2)(1-3s)(1-r)(1-t)$$

$$N_{22} = \%_{64}(1-s^2)(1+3s)(1-r)(1-t)$$

$$N_{23} = \%_{64}(1-s^2)(1-3s)(1-r)(1+t)$$

$$N_{24} = \%_{64}(1-s^2)(1+3s)(1-r)(1+t)$$

$$N_{25} = \%_{64}(1-t^2)(1+3t)(1+s)(1+r)$$

$$N_{26} = \%_{64}(1-t^2)(1-3t)(1+s)(1+r)$$

$$N_{27} = \%_{64}(1-r^2)(1+3r)(1+s)(1-t)$$

$$N_{28} = \%_{64}(1-r^2)(1-3r)(1+s)(1-t)$$

$$N_{29} = \%_{64}(1-t^2)(1-3t)(1+s)(1-r)$$

$$N_{30} = \%_{64}(1-t^2)(1+3t)(1+s)(1-r)$$

$$N_{31} = \%_{64}(1-r^2)(1-3r)(1+s)(1+t)$$

$$N_{32} = \%_{64}(1-r^2)(1+3r)(1+s)(1+t)$$

It should be noted that when $t = \pm 1$, the above shape functions reduce to the corresponding two-dimensional shape functions. Therefore, these particular elements may join a similar-order rectangular or lineal element as shown in Fig. 9-8 for the quadratic-order element.

9-5 SUBPARAMETRIC, ISOPARAMETRIC, AND SUPERPARAMETRIC ELEMENTS

As discussed below, it is relatively easy to distort elements. In particular, the distorted rectangular and brick elements can be used to accommodate practically any geometry. For example, let us take the quadratic-order rectangular element and modify it to a distorted quadrilateral element as shown in Fig. 9-9. Note that the nodes can be used for two purposes: One is to specify the locations where the parameter function is sought (i.e., the nodal displacements, temperatures, etc.),

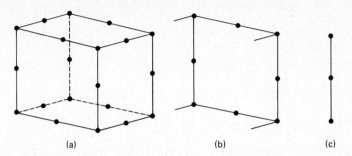

Figure 9-8 A typical face of the (a) brick element interfaces readily with the two-dimensional rectangular element in (b), which in turn interfaces with the (c) one-dimensional element.

and the other purpose is to define the geometry of the element. Note in Fig. 9-9 that a quadratic-order curve may be passed through three points (i.e., the nodes on any one side of the quadrilateral). Thus, it appears to be possible to include curved boundaries explicitly in the FEM formulations. The geometry of the boundaries is approximated by polynomials of finite order, and so the boundaries are not exactly represented in general. However, by using more nodes (or smaller elements), we can approach the actual curved boundary to a high degree of accuracy.

The isoparametric element is now defined as follows with the help of Fig. 9-10(a). When the same nodes are used to define the element geometry and the locations where the parameter function is sought, the element is said to be *isoparametric*.

When the number of nodes used to define the geometry is *less than* the number used to represent the parameter function as shown in Fig. 9-10(b), the element is said to be *subparametric*. Finally, when the number of nodes used to define the

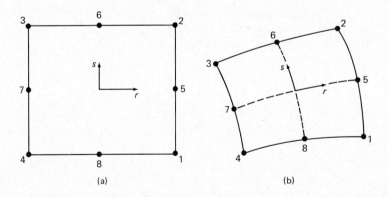

Figure 9-9 The quadratic-order rectangular element shown in (a) may be distorted into the quadratic-order quadrilateral element shown in (b). Note the curved element boundaries in (b).

geometry is *greater than* the number used to represent the parameter function as shown in Fig. 9-10(c), the element is said to be *superparametric*. Isoparametric and subparametric elements are used quite frequently in finite element analysis, whereas superparametric elements are rarely used.

Let us illustrate how we can distort an element into a more useful shape. For this purpose let us concentrate on the linear-order element shown in Fig. 9-11. Note that by distorting the rectangle in Fig. 9-11(a) into a quadrilateral, we may now

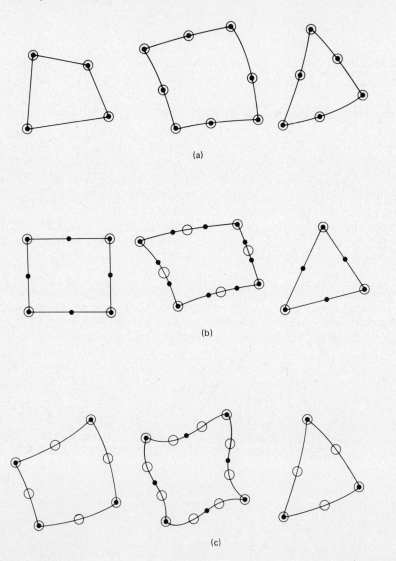

Figure 9-10 (a) Isoparametric elements, (b) subparametric elements, and (c) superparametric elements. Note that ● specifies the locations where the values of the parameter function are sought and that ○ specifies the geometry.

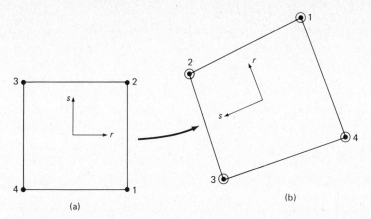

Figure 9-11   The linear-order rectangular element shown in (a) may be distorted into the linear-order quadrilateral element shown in (b).

place the nodes at more convenient locations. Thus, irregular geometries may be accommodated. Actually there are some restrictions on the placement of these nodes. These restrictions are delineated in subsequent sections.

We should be quite familiar by now with representing the parameter functions in terms of the nodal values with the help of the shape functions. For example, let us represent the temperature T within the element shown in Fig. 9-11(b) as

$$T = N_1 T_1 + N_2 T_2 + N_3 T_3 + N_4 T_4 \tag{9-24}$$

Let us also represent the two global coordinates (x,y) as follows:

$$x = N_1' x_1 + N_2' x_2 + N_3' x_3 + N_4' x_4 \tag{9-25a}$$

and

$$y = N_1' y_1 + N_2' y_2 + N_3' y_3 + N_4' y_4 \tag{9-25b}$$

where N_1', N_2', etc. represent the shape functions used to define the geometry and x_1, y_1, etc. are the nodal coordinates. Since the same nodes in Fig. 9-11(b) are being used to define the geometry and the parameter function nodal points, we have an isoparametric element. For this and all other isoparametric elements, the shape functions N_i and N_i' are equal. In other words, the same shape functions are used to define the geometry and the parameter function. Therefore, we have

$$N_i = N_i' \qquad \text{for all nodal points } i$$

It is of interest to note that if Eqs. (9-25) are evaluated at a particular node, for example, node 2, we get

$$x = x_2$$

and

$$y = y_2$$

because $N_2(x_2,y_2) = 1$, whereas $N_1(x_2,y_2) = N_3(x_2,y_2) = N_4(x_2,y_2) = 0$. Clearly this must be the result if the equations given by Eqs. (9-25) are to be meaningful.

Equations (9-25) can be generalized for all isoparametric elements by writing

$$x = \sum_{i=1}^{n} N_i x_i \qquad (9\text{-}26a)$$

$$y = \sum_{i=1}^{n} N_i y_i \qquad (9\text{-}26b)$$

where the N_i's are the shape functions, (x_i, y_i) denotes the coordinates of node i, and n is the number of nodes associated with the element.

Equations (9-26) are very powerful in that they allow us to *map* any point in the (local) coordinate system to a point in the (x,y) coordinate system. Therefore, Eqs. (9-26) are said to provide an *isoparametric mapping* from the undistorted element to the distorted element. Undistorted elements may also be referred to as *parent elements*. In subsequent sections, the conditions will be given such that we are assured of a unique or one-to-one mapping. This is to say that a point in the parent element should map into one and only one point in the distorted element (and vice versa). This will enable us to take a distorted element and map it into an undistorted element for the purpose of evaluating the integrals that naturally arise.

In Sec. 9-6, the two-dimensional isoparametric elements are developed further. In Sec. 9-7, the three-dimensional isoparametric elements are presented. In each case, the explicit form of the mapping is given.

Example 9-1.

Consider the four-node quadrilateral element defined by the following nodal coordinates: $x_1 = 5., y_1 = 7., x_2 = 1., y_2 = 4., x_3 = 2., y_3 = 1., x_4 = 8.,$ and $y_4 = 4$. Determine the global coordinates that correspond to $r = +1.0$ and $s = +0.75$ on the parent element.

Solution.

From Eqs. (9-26) and (9-13), we have

$$x = \tfrac{1}{4}(1 + r)(1 - s)x_1 + \tfrac{1}{4}(1 + r)(1 + s)x_2$$
$$+ \tfrac{1}{4}(1 - r)(1 + s)x_3 + \tfrac{1}{4}(1 - r)(1 - s)x_4$$

and

$$y = \tfrac{1}{4}(1 + r)(1 - s)y_1 + \tfrac{1}{4}(1 + r)(1 + s)y_2$$
$$+ \tfrac{1}{4}(1 - r)(1 + s)y_3 + \tfrac{1}{4}(1 - r)(1 - s)y_4$$

Substituting the values of the nodal coordinates gives

$$x = \tfrac{5}{4}(1 + r)(1 - s) + \tfrac{1}{4}(1 + r)(1 + s)$$
$$+ \tfrac{2}{4}(1 - r)(1 + s) + \tfrac{8}{4}(1 - r)(1 - s)$$

and

$$y = \tfrac{1}{4}(1 + r)(1 - s) + \tfrac{1}{4}(1 + r)(1 + s)$$
$$+ \tfrac{1}{4}(1 - r)(1 + s) + \tfrac{1}{4}(1 - r)(1 - s)$$

Now we are in a position to perform the actual mapping:

$$x = \tfrac{5}{4}(1 + 1)(1 - 0.75) + \tfrac{1}{4}(1 + 1)(1 + 0.75)$$
$$+ \tfrac{2}{4}(1 - 1)(1 + 0.75) + \tfrac{8}{4}(1 - 1)(1 - 0.75)$$

and

$$y = \tfrac{1}{4}(1 + 1)(1 - 0.75) + \tfrac{1}{4}(1 + 1)(1 + 0.75)$$
$$+ \tfrac{1}{4}(1 - 1)(1 + 0.75) + \tfrac{1}{4}(1 - 1)(1 - 0.75)$$

or

$$x = 1.500 \quad \text{and} \quad y = 4.375$$

The parent element should be plotted on a piece of graph paper in order to verify that $(1.5, 4.375)$ is on the element boundary. More specifically, this point is located between nodes 1 and 2. Why does this seem reasonable?   ■

9-6 TWO-DIMENSIONAL ISOPARAMETRIC FORMULATIONS

In Sec. 9-5 we have seen how it is possible to distort an element in order to give it a more arbitrary shape. In this section we want to show how the element stiffness matrix and element nodal force vectors are transformed for the two-dimensional isoparametric elements into those for the undistorted element. The reason for this transformation is to simplify the resulting integrations. If the integrands contain the shape functions directly (i.e., not the derivatives), then we simply represent the shape functions in terms of the appropriate normalized coordinates. No additional transformation is necessary in order to evaluate the integrals in this case (see Sec. 9-8).

Both the triangular and quadrilateral element are considered. The quadrilateral element is considered first because it is described by two independent coordinates r and s. It should be recalled that the three area coordinates used to describe the triangular element are not all independent; this will require some additional special attention.

The Quadrilateral Element

Figure 9-12 shows the isoparametric forms of the rectangular element. Both the parent (rectangular) and distorted quadrilateral elements are shown. Note that for the linear-order distorted element both the parameter function and the sides of the

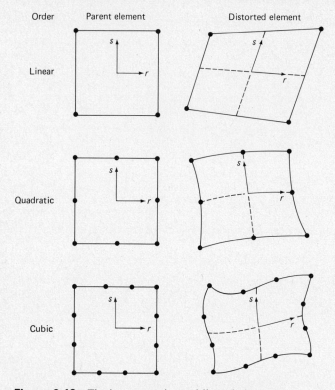

Figure 9-12   The isoparametric quadrilateral elements.

quadrilateral are linear; for the quadratic-order distorted element both the parameter function and the sides of the element are of quadratic order; for the cubic-order distorted element both the parameter function and the sides of the element are cubic order.

It should be recalled from Chapters 5 and 8 that typical, two-dimensional element stiffness matrices are given by Eqs. (5-87) and (8-106) for problems in stress analysis and heat transfer, respectively. Recall that the derivatives of the shape functions frequently appear in the integrands. If we wish to perform the integrations over the undistorted element, the integrals must be transformed into ones that contain only r and s (instead of x and y). This is accomplished as described below.

Let us first note that we have the shape functions in terms of the serendipity coordinates r and s for the parent elements as given in Sec. 9-3. Let us work with a typical shape function, for example, the ith, and note that

$$r = r(x,y)$$
$$s = s(x,y)$$

and

$$N_i = N_i(r,s)$$

Therefore, we may write the total derivatives of N_i as

$$\frac{\partial N_i}{\partial r} = \frac{\partial N_i}{\partial x}\frac{\partial x}{\partial r} + \frac{\partial N_i}{\partial y}\frac{\partial y}{\partial r} \tag{9-27a}$$

and

$$\frac{\partial N_i}{\partial s} = \frac{\partial N_i}{\partial x}\frac{\partial x}{\partial s} + \frac{\partial N_i}{\partial y}\frac{\partial y}{\partial s} \tag{9-27b}$$

Let us write these last two equations in matrix form as

$$\begin{bmatrix} \dfrac{\partial N_i}{\partial r} \\[2ex] \dfrac{\partial N_i}{\partial s} \end{bmatrix} = \begin{bmatrix} \dfrac{\partial x}{\partial r} & \dfrac{\partial y}{\partial r} \\[2ex] \dfrac{\partial x}{\partial s} & \dfrac{\partial y}{\partial s} \end{bmatrix} \begin{bmatrix} \dfrac{\partial N_i}{\partial x} \\[2ex] \dfrac{\partial N_i}{\partial y} \end{bmatrix} \tag{9-28}$$

The 2×2 matrix on the right-hand side is known as the *Jacobian matrix* and is denoted by $\mathbf{J}$, or

$$\mathbf{J} = \begin{bmatrix} \dfrac{\partial x}{\partial r} & \dfrac{\partial y}{\partial r} \\[2ex] \dfrac{\partial x}{\partial s} & \dfrac{\partial y}{\partial s} \end{bmatrix} \tag{9-29}$$

The reader may recall from calculus [2] that an infinitesimal area element $dx\, dy$ is related to an infinestimal area element in the (r,s) coordinate system by

$$dx\, dy = |\det \mathbf{J}|\, dr\, ds \tag{9-30}$$

In Eq. (9-30), the determinant of the Jacobian matrix is indicated. This determinant is referred to simply as *the Jacobian*. If Eq. (9-28) is premultiplied by $\mathbf{J}^{-1}$, we get the desired result

$$\begin{bmatrix} \dfrac{\partial N_i}{\partial x} \\[2ex] \dfrac{\partial N_i}{\partial y} \end{bmatrix} = \begin{bmatrix} \dfrac{\partial x}{\partial r} & \dfrac{\partial y}{\partial r} \\[2ex] \dfrac{\partial x}{\partial s} & \dfrac{\partial y}{\partial s} \end{bmatrix}^{-1} \begin{bmatrix} \dfrac{\partial N_i}{\partial r} \\[2ex] \dfrac{\partial N_i}{\partial s} \end{bmatrix} \tag{9-31}$$

Let us examine the Jacobian matrix more carefully. Since we have an isoparametric element, we may write [see Eqs. (9-26)]:

$$x = \sum N_j x_j \tag{9-32a}$$

and

$$y = \sum N_j y_j \tag{9-32b}$$

where the summations are made over the total number of nodes present in the element, and x_j and y_j are the coordinates of the nodes. Therefore, the Jacobian matrix becomes

$$\mathbf{J} = \begin{bmatrix} \sum \dfrac{\partial N_j}{\partial r} x_j & \sum \dfrac{\partial N_j}{\partial r} y_j \\[4mm] \sum \dfrac{\partial N_j}{\partial s} x_j & \sum \dfrac{\partial N_j}{\partial s} y_j \end{bmatrix} \tag{9-33}$$

Note that for the linear-, quadratic-, and cubic-order elements, the summations involve four, eight, and twelve terms, respectively. With the help of Eqs. (9-30), (9-31) and (9-33), every integral over the element area A^e may be transformed into an integral of the form

$$\mathbf{K}^e = \int_{-1}^{+1} \int_{-1}^{+1} \mathbf{H}(r,s) \, dr \, ds \tag{9-34a}$$

or

$$\mathbf{f}^e = \int_{-1}^{+1} \int_{-1}^{+1} \mathbf{g}(r,s) \, dr \, ds \tag{9-34b}$$

Note that the integration limits also change. In Sec. 9-8, we will see that integrals in these forms may be evaluated numerically.

It should be noted that the size of the element stiffness matrix is directly related to the order of the element as shown in Table 9-1. Note that a two-dimensional stress analysis problem analyzed with the cubic-order isoparametric, quadrilateral element has an element stiffness matrix that is of size 24×24. In this case, each node has 2 degreees of freedom and there are 12 nodes; thus the stiffness matrix is 24×24. Since the corresponding heat transfer problem has 1 degree of freedom per node, the element stiffness matrix is of size 12×12 for the cubic-order element.

Recall that integrations around the element boundary result in integrals of the form

$$\int_{S^e} \mathbf{N}^T \mathbf{s} \, dS \qquad \text{and} \qquad \int_{C^e} \mathbf{N}^T h \mathbf{N} \mathbf{t} \, dC$$

The first is recognized as the element nodal force vector from surface tractions, and the second is recognized as the element stiffness matrix from convection from the boundary of a two-dimensional body. In the case of the integral for the traction, the elemental area dS around the boundary may be expressed as

Table 9-1 Size of the Element Stiffness Matrices for the Quadrilateral Elements

Order of element	Structural (2 DOF per node)	Thermal (1 DOF per node)
Linear	8×8	4×4
Quadratic	16×16	8×8
Cubic	24×24	12×12

$$dS = t\, dC \qquad \text{(9-35)}$$

where t is the element thickness and dC is the infinitesimal arc length. From calculus, dC is given by

$$dC = \sqrt{(dx)^2 + (dy)^2} \qquad \text{(9-36)}$$

since we have $x = x(r,s)$ and $y = y(r,s)$, we may write the differentials dx and dy as:

$$dx = \frac{\partial x}{\partial r}\, dr + \frac{\partial x}{\partial s}\, ds \qquad \text{(9-37a)}$$

and

$$dy = \frac{\partial y}{\partial r}\, dr + \frac{\partial y}{\partial s}\, ds \qquad \text{(9-37b)}$$

where r and s are the serendipity coordinates. The boundary integrals need to be evaluated around the element boundary.

Let us derive the appropriate form of the expression for dC by restricting the development to the legs of the element over which the serendipity coordinate s is constant; i.e., $s = \pm 1$ on these faces. Therefore with the help of Eqs. (9-32), we may write Eqs. (9-37) as

$$dx = \frac{\partial x}{\partial r}\, dr = \sum \frac{\partial N_j}{\partial r}\, x_j\, dr \qquad \text{(9-38a)}$$

and

$$dy = \frac{\partial y}{\partial r}\, dr = \sum \frac{\partial N_j}{\partial r}\, y_j\, dr \qquad \text{(9-38b)}$$

on faces where $s = \pm 1$. A typical side (side 2-3) of a quadrilateral element for which Eqs. (9-38) apply is shown in Fig. 9-13. The elemental surface area dS or $t\, dC$ on sides of constant s becomes

$$dS = t\, dC = t\sqrt{\left(\sum \frac{\partial N_j}{\partial r}\, x_j\right)^2 + \left(\sum \frac{\partial N_j}{\partial r}\, y_j\right)^2}\, dr \qquad \text{(9-39)}$$

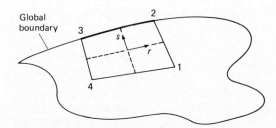

Figure 9-13   Typical quadrilateral element with leg 2-3 on the global boundary. Note that $s = +1$ on this leg.

In a similar fashion, it can be shown that on faces of constant r (i.e., faces 1-2 and 3-4), the elemental surface area dS is given

$$dS = t \, dC = t \sqrt{\left(\sum \frac{\partial N_j}{\partial s} x_j\right)^2 + \left(\sum \frac{\partial N_j}{\partial s} y_j\right)^2} \, ds \qquad \textbf{(9-40)}$$

The serendipity coordinate s (a scalar) should not be confused with the surface traction $\mathbf{s}$ (a vector). Therefore, when these expressions for dS or $t \, dC$ are substituted into typical boundary integrals, we get integrals of the form

$$\mathbf{f}^e = \int_{S^e} \mathbf{N}^T \mathbf{s} \, dS = \int_{-1}^{1} \mathbf{g}(r,s) \bigg|_{s = \pm 1} dr \qquad \textbf{(9-41)}$$

or

$$\mathbf{K}^e = \int_{C^e} \mathbf{N}^T h \mathbf{N} t \, dC = \int_{-1}^{1} \mathbf{H}(r,s) \bigg|_{s = \pm 1} dr \qquad \textbf{(9-42)}$$

on faces of constant serendipity coordinate s. Similar integrals result when faces of constant r are on the global boundary except that the integrands are evaluated at $r = \pm 1$ before integrating with respect to s.

The integrals in Eqs. (9-34), (9-41), and (9-42) look formidable. However, they may be evaluated in a relatively straightforward manner by using Gauss-Legendre quadrature as described in Sec. 9-8.

The degree of distortion that is possible before the mapping breaks down (and is no longer one-to-one) is now given. For a linear-order quadrilateral element, a one-to-one mapping is assured if the maximum angle formed by any two sides of the quadrilateral is less than 180° as shown in Fig. 9-14(a). For a quadratic-order quadrilateral element, not only must the same angle condition be satisfied but also the midside nodes must be in the middle one-third of the distorted sides. This condition is illustrated in Fig. 9-14(b). For cubic and other higher-order elements, no such simple conditions apparently exist and the necessary and sufficient condition for a one-to-one mapping is stated as follows: for a one-to-one mapping, the sign of the Jacobian (the determinant of the Jacobian matrix) must remain the same for all points in the domain mapped [3]. For cubic and higher-order elements, it is not practical to check the sign of the Jacobian at every point within the element. Instead, a few points are checked as explained later.

The Triangular Element

The isoparametric forms of the triangular element are shown in Fig. 9-15. Note that both the parent and distorted elements are shown. As in the case of the rectangular element, the number of nodes on each leg of the triangle determines the order of element (two, three, and four nodes on a leg for linear-, quadratic-, and cubic-order, respectively). Note how a quadratic-order curve may be drawn through three points (i.e., the nodes), and a cubic-order curve through four points.

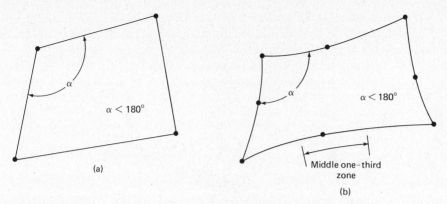

Figure 9-14 For the linear-order quadrilateral element in (a) each internal angle must be less than 180° for a unique mapping. For the quadratic-order quadrilateral element in (b) each internal angle must be less than 180° and the midside nodes must be within the middle one-third of each leg.

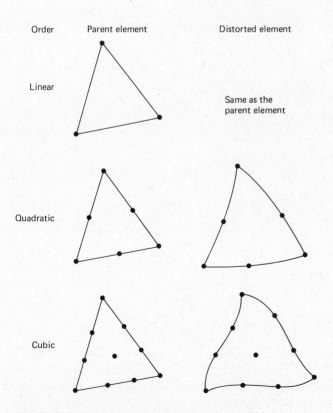

Figure 9-15 The isoparametric triangular elements.

The development in this section is very similar to that in the previous section for the quadrilateral element, except that the local, normalized coordinates are area coordinates, not serendipity coordinates. It should be recalled that the three area coordinates are not all independent because they must sum to unity [see Eq. (6-23)]. This may be taken into account by taking L_1 and L_2 to be the independent coordinates. The fact that L_3 is dependent (on L_1 and L_2) is explicitly taken into consideration later. Therefore, we may write

$$L_1 = L_1(x,y)$$

$$L_2 = L_2(x,y)$$

and

$$N_i = N_i(L_1,L_2)$$

from which it follows that

$$\frac{\partial N_i}{\partial L_1} = \frac{\partial N_i}{\partial x}\frac{\partial x}{\partial L_1} + \frac{\partial N_i}{\partial y}\frac{\partial y}{\partial L_1} \tag{9-43a}$$

and

$$\frac{\partial N_i}{\partial L_2} = \frac{\partial N_i}{\partial x}\frac{\partial x}{\partial L_2} + \frac{\partial N_i}{\partial y}\frac{\partial y}{\partial L_2} \tag{9-43b}$$

In matrix form, we have

$$\begin{bmatrix} \dfrac{\partial N_i}{\partial L_1} \\ \dfrac{\partial N_i}{\partial L_2} \end{bmatrix} = \begin{bmatrix} \dfrac{\partial x}{\partial L_1} & \dfrac{\partial y}{\partial L_1} \\ \dfrac{\partial x}{\partial L_2} & \dfrac{\partial y}{\partial L_2} \end{bmatrix} \begin{bmatrix} \dfrac{\partial N_i}{\partial x} \\ \dfrac{\partial N_i}{\partial y} \end{bmatrix} \tag{9-44}$$

Now the Jacobian matrix $\mathbf{J}$ is defined by

$$\mathbf{J} = \begin{bmatrix} \dfrac{\partial x}{\partial L_1} & \dfrac{\partial y}{\partial L_1} \\ \dfrac{\partial x}{\partial L_2} & \dfrac{\partial y}{\partial L_2} \end{bmatrix} \tag{9-45}$$

from which it follows that

$$\begin{bmatrix} \dfrac{\partial N_i}{\partial x} \\ \dfrac{\partial N_i}{\partial y} \end{bmatrix} = \begin{bmatrix} \dfrac{\partial x}{\partial L_1} & \dfrac{\partial y}{\partial L_1} \\ \dfrac{\partial x}{\partial L_2} & \dfrac{\partial y}{\partial L_2} \end{bmatrix}^{-1} \begin{bmatrix} \dfrac{\partial N_i}{\partial L_1} \\ \dfrac{\partial N_i}{\partial L_2} \end{bmatrix} \tag{9-46}$$

Note that the partial derivatives of the shape functions N_i are needed in the column vector on the right-hand side of Eq. (9-46). Recall that we have assumed N_i to be a function of L_1 and L_2 only, since L_3 is then dependent on L_1 and L_2 by Eq. (6-23). This implies that in the computations of $\partial N_i/\partial L_1$ and $\partial N_i/\partial L_2$, we must substitute

$$L_3 = 1 - L_1 - L_2 \tag{9-47}$$

into the shape functions N_i before computing these partial derivatives. For higher-order elements this approach is very tedious; fortunately, there is an easier way. Let us denote the partial derivative of N_i with respect to L_1 as

$$\left(\frac{\partial N_i}{\partial L_1}\right)_{L_2, L_3}$$

where the shape function N_i is now a function L_1, L_2, and L_3, but L_2 and L_3 are held constant in performing the differentiation. The notation

$$\frac{\partial N_i}{\partial L_1}$$

implies N_i is a function of L_1 and L_2, and L_2 is held constant [the L_2 is not written since it is not written in Eq. (9-46)]. It follows that

$$\frac{\partial N_i}{\partial L_1} = \frac{\partial L_1}{\partial L_1}\left(\frac{\partial N_i}{\partial L_1}\right)_{L_2, L_3} + \frac{\partial L_2}{\partial L_1}\left(\frac{\partial N_i}{\partial L_2}\right)_{L_1, L_3} + \frac{\partial L_3}{\partial L_1}\left(\frac{\partial N_i}{\partial L_3}\right)_{L_1, L_2} \tag{9-48}$$

and since

$$\frac{\partial L_1}{\partial L_1} = 1 \qquad \frac{\partial L_2}{\partial L_1} = 0 \qquad \frac{\partial L_3}{\partial L_1} = -1 \tag{9-49}$$

we have

$$\frac{\partial N_i}{\partial L_1} = \underbrace{\left(\frac{\partial N_i}{\partial L_1}\right)_{L_2, L_3}}_{\substack{\text{Here } N_i \text{ is taken to be a} \\ \text{function of } L_1 \text{ and } L_2 \text{ as} \\ \text{required by Eq. (9-46)}}} - \underbrace{\left(\frac{\partial N_i}{\partial L_3}\right)_{L_1, L_2}}_{\substack{\text{Here } N_i \text{ is taken to be a function of } L_1, L_2, \\ \text{and } L_3}} \tag{9-50}$$

In a completely analogous manner it follows that

$$\frac{\partial N_i}{\partial L_2} = \left(\frac{\partial N_i}{\partial L_2}\right)_{L_1, L_3} - \left(\frac{\partial N_i}{\partial L_3}\right)_{L_1, L_2} \tag{9-51}$$

Equations (9-50) and (9-51) make it unnecessary to get N_i as a function of only L_1 and L_2 before computing the partial derivatives on the right-hand side of Eq. (9-46).

The Jacobian matrix may be cast into a more usable form by writing

$$\mathbf{J} = \begin{bmatrix} \sum \dfrac{\partial N_i}{\partial L_1} x_j & \sum \dfrac{\partial N_i}{\partial L_1} y_j \\[4mm] \sum \dfrac{\partial N_i}{\partial L_2} x_j & \sum \dfrac{\partial N_i}{\partial L_2} y_j \end{bmatrix} \tag{9-52}$$

where the partial derivatives can be computed with help of Eqs. (9-50) and (9-51). Note that for the linear-, quadratic-, and cubic-order triangular elements, the summations are made over three, six, and ten terms, respectively. From elementary calculus it follows that

$$dx \, dy = |\det \mathbf{J}| \, dL_1 \, dL_2 \qquad \text{(9-53)}$$

It also follows that typical element stiffness matrices and nodal force vectors may be cast into the form

$$\mathbf{K}^e = \int_0^1 \int_0^{1-L_2} \mathbf{H}(L_1, L_2, L_3) \, dL_1 \, dL_2 \qquad \text{(9-54)}$$

and

$$\mathbf{f}^e = \int_0^1 \int_0^{1-L_2} \mathbf{g} \, (L_1, L_2, L_3) \, dL_1 \, dL_2 \qquad \text{(9-55)}$$

These integrals are in a form that may be integrated relatively easily as shown in Sec. 9-9. The resulting sizes of the element stiffness matrices for linear-, quadratic-, and cubic-order elements are given in Table 9-2. Note that the element stiffness matrices for stress analysis and heat transfer are 20×20 and 10×10, respectively, if the cubic-order element is used.

The transformation of boundary integrals (i.e., over S^e or C^e) is left to the exercises. As in the case of the quadrilateral element, the necessary and sufficient condition for a one-to-one mapping is that the sign of the Jacobian remain the same for each point in the domain mapped.

9-7 THREE-DIMENSIONAL ISOPARAMETRIC FORMULATIONS

In this section the brick and tetrahedral elements are distorted into more general shapes. However, the resulting integrations are to be done in the undistorted region. The brick is considered first because the three local normalized coordinates r, s, and t associated with such an element are all independent. In contrast, the four volume coordinates (now L_1, L_2, L_3, and L_4) associated with the tetrahedral element are not all independent. As in the two-dimensional isoparametric formulation, this will require a little more attention.

Table 9-2 Size of the Element Stiffness Matrices for the Triangular Elements

Order of element	Structural (2 DOF per node)	Thermal (1 DOF per node)
Linear	6×6	3×3
Quadratic	12×12	6×6
Cubic	20×20	10×10

The Brick Element

The isoparametric forms of the brick element are shown in Fig. 9-16. For the linear-order distorted brick, the four points used to define a face must lie in a plane. Note that the linear-, quadratic-, and cubic-order elements have two, three, and four nodes, respectively, along each of the edges.

As in the two-dimensional case, it is necessary to evaluate the element stiffness matrices and nodal force vectors for these elements. However, it is very desirable to perform the integrations in the parent element, as explained in Sec. 9-8. Let us show how the integrals for the element characteristics may be transformed into ones over the undistorted or parent regions.

Recall from Sec. 9-4 that the appropriate shape functions for the parent elements are given by Eqs. (9-21) to (9-23) in terms of the serendipity coordinates r, s, and t. Let us take a typical shape function, e.g. the ith, and denote its dependence on r, s, and t by writing

$$N_i = N_i(r,s,t)$$

But we also have

$$r = r(x,y,z)$$
$$s = s(x,y,z)$$
$$t = t(x,y,z)$$

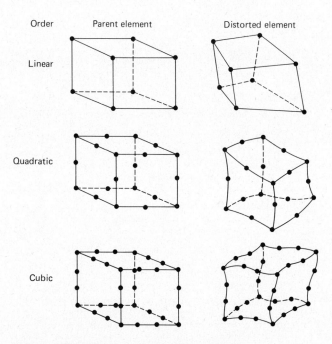

Figure 9-16 The isoparametric brick elements.

where x, y, and z are the global coordinates. The total derivatives of a typical shape function N_i are given by

$$\frac{\partial N_i}{\partial r} = \frac{\partial N_i}{\partial x}\frac{\partial x}{\partial r} + \frac{\partial N_i}{\partial y}\frac{\partial y}{\partial r} + \frac{\partial N_i}{\partial z}\frac{\partial z}{\partial r} \qquad \text{(9-56a)}$$

$$\frac{\partial N_i}{\partial s} = \frac{\partial N_i}{\partial x}\frac{\partial x}{\partial s} + \frac{\partial N_i}{\partial y}\frac{\partial y}{\partial s} + \frac{\partial N_i}{\partial z}\frac{\partial z}{\partial s} \qquad \text{(9-56b)}$$

$$\frac{\partial N_i}{\partial t} = \frac{\partial N_i}{\partial x}\frac{\partial x}{\partial t} + \frac{\partial N_i}{\partial y}\frac{\partial y}{\partial t} + \frac{\partial N_i}{\partial z}\frac{\partial z}{\partial t} \qquad \text{(9-56c)}$$

Rewriting Eqs. (9-56) in matrix form gives

$$\begin{bmatrix} \dfrac{\partial N_i}{\partial r} \\[2ex] \dfrac{\partial N_i}{\partial s} \\[2ex] \dfrac{\partial N_i}{\partial t} \end{bmatrix} = \begin{bmatrix} \dfrac{\partial x}{\partial r} & \dfrac{\partial y}{\partial r} & \dfrac{\partial z}{\partial r} \\[2ex] \dfrac{\partial x}{\partial s} & \dfrac{\partial y}{\partial s} & \dfrac{\partial z}{\partial s} \\[2ex] \dfrac{\partial x}{\partial t} & \dfrac{\partial y}{\partial t} & \dfrac{\partial z}{\partial t} \end{bmatrix} \begin{bmatrix} \dfrac{\partial N_i}{\partial x} \\[2ex] \dfrac{\partial N_i}{\partial y} \\[2ex] \dfrac{\partial N_i}{\partial z} \end{bmatrix} \qquad \text{(9-57)}$$

Now a 3×3 Jacobian matrix arises and with the help of

$$x = \sum N_j x_j \qquad y = \sum N_j y_j \qquad z = \sum N_j z_j$$

we may write the Jacobian matrix as

$$\mathbf{J} = \begin{bmatrix} \sum \dfrac{\partial N_j}{\partial r} x_j & \sum \dfrac{\partial N_j}{\partial r} y_j & \sum \dfrac{\partial N_j}{\partial r} z_j \\[3ex] \sum \dfrac{\partial N_j}{\partial s} x_j & \sum \dfrac{\partial N_j}{\partial s} y_j & \sum \dfrac{\partial N_j}{\partial s} z_j \\[3ex] \sum \dfrac{\partial N_j}{\partial t} x_j & \sum \dfrac{\partial N_j}{\partial t} y_j & \sum \dfrac{\partial N_j}{\partial t} z_j \end{bmatrix} \qquad \text{(9-58)}$$

where x_j, y_j, and z_j denote the global coordinates of the jth node. The summations in Eq. (9-58) are made over the 8, 20, and 32 nodes for the linear-, quadratic-, and cubic-order elements, respectively.

If Eq. (9-57) is premultiplied by the inverse of the Jacobian matrix, we get the desired result:

$$\begin{bmatrix} \dfrac{\partial N_i}{\partial x} \\[2ex] \dfrac{\partial N_i}{\partial y} \\[2ex] \dfrac{\partial N_i}{\partial z} \end{bmatrix} = \begin{bmatrix} \sum \dfrac{\partial N_j}{\partial r} x_j & \sum \dfrac{\partial N_j}{\partial r} y_j & \sum \dfrac{\partial N_j}{\partial r} z_j \\[3ex] \sum \dfrac{\partial N_j}{\partial s} x_j & \sum \dfrac{\partial N_j}{\partial s} y_j & \sum \dfrac{\partial N_j}{\partial s} z_j \\[3ex] \sum \dfrac{\partial N_j}{\partial t} x_j & \sum \dfrac{\partial N_j}{\partial t} y_j & \sum \dfrac{\partial N_j}{\partial t} z_j \end{bmatrix}^{-1} \begin{bmatrix} \dfrac{\partial N_i}{\partial r} \\[2ex] \dfrac{\partial N_i}{\partial s} \\[2ex] \dfrac{\partial N_i}{\partial t} \end{bmatrix} \qquad \text{(9-59)}$$

Equation (9-59) is useful because it may be used to replace derivatives of the form $\partial N_i/\partial x$, $\partial N_i/\partial y$, and $\partial N_i/\partial z$ with expressions involving $\partial N_i/\partial r$, $\partial N_i/\partial s$, and $\partial N_i/\partial t$. The elemental volume $dx\ dy\ dz$ may be written in terms of $dr\ ds\ dt$ by noting that

$$dx\ dy\ dz = |\det \mathbf{J}|\ dr\ ds\ dt \qquad \textbf{(9-60)}$$

Therefore, the element stiffness matrices and nodal force vectors may be written in the form

$$\mathbf{K}^e = \int_{-1}^{1}\int_{-1}^{1}\int_{-1}^{1} \mathbf{H}(r,s,t)\ dr\ ds\ dt \qquad \textbf{(9-61)}$$

and

$$\mathbf{f}^e = \int_{-1}^{1}\int_{-1}^{1}\int_{-1}^{1} \mathbf{g}(r,s,t)\ dr\ ds\ dt \qquad \textbf{(9-62)}$$

The simple integration limits should be noted. These integrals may be evaluated by the numerical integration technique described in Sec. 9-8.

The element stiffness matrices can become quite large for these elements as Table 9-3 shows. Note that for the cubic-order element, the element stiffness matrices are 96×96 and 32×32 for problems in stress analysis and thermal analysis, respectively. The numerical integration technique described in Sec. 9-8 makes the use of these elements practical.

Table 9-3 Size of the Element Stiffness Matrices for the Brick Elements

Order of element	Structural (3 DOF per node)	Thermal (1 DOF per node)
Linear	24×24	8×8
Quadratic	60×60	20×20
Cubic	96×96	32×32

It will now be shown how the integrals over the element surfaces S^e may be transformed into ones over the faces on the parent element. The integrands may be converted directly to functions of r, s, and t in a straightforward manner, since the shape functions are already known as a function r, s, and t. It should be noted, however, that on any given face, one of the serendipity coordinates is plus or minus unity. For example, on face 1-5-6-2, we have $r = +1$, whereas on face 5-6-7-8, we have $t = -1$, and so forth. Therefore, on face 1-5-6-2, the integrations are performed over s and t only (since $r = +1$ on this face).

The only part of the integrals over S^e that requires special attention is the elemental surface area dS. Let $d\mathbf{A}$ represent the outward normal area vector to a surface of constant serendipity coordinate, i.e., on one of the faces of the brick. For example, let us assume that face 1-2-3-4 is on the global boundary. On this face $t = +1$ and it can be shown that $d\mathbf{A}$, in this case, is given by [4]

$$dA = \left(\frac{\partial x}{\partial r}\mathbf{i} + \frac{\partial y}{\partial r}\mathbf{j} + \frac{\partial z}{\partial r}\mathbf{k}\right) \times \left(\frac{\partial x}{\partial s}\mathbf{i} + \frac{\partial y}{\partial s}\mathbf{j} + \frac{\partial z}{\partial s}\mathbf{k}\right) dr\ ds \qquad (9\text{-}63)$$

where the vector cross product is used and may be evaluated by writing it in determinant form as follows

$$dA = \det \begin{bmatrix} \mathbf{i} & \mathbf{j} & \mathbf{k} \\ \dfrac{\partial x}{\partial r} & \dfrac{\partial y}{\partial r} & \dfrac{\partial z}{\partial r} \\ \dfrac{\partial x}{\partial s} & \dfrac{\partial y}{\partial s} & \dfrac{\partial z}{\partial s} \end{bmatrix} dr\ ds \qquad (9\text{-}64)$$

The magnitude of dA is actually dS, and so we have

$$dS = \sqrt{\left(\frac{\partial y}{\partial r}\frac{\partial z}{\partial s} - \frac{\partial y}{\partial s}\frac{\partial z}{\partial r}\right)^2 + \left(\frac{\partial x}{\partial r}\frac{\partial z}{\partial s} - \frac{\partial x}{\partial s}\frac{\partial z}{\partial r}\right)^2 + \left(\frac{\partial x}{\partial r}\frac{\partial y}{\partial s} - \frac{\partial x}{\partial s}\frac{\partial y}{\partial r}\right)^2}\ dr\ ds$$

$$(9\text{-}65)$$

where we may further note that

$$\frac{\partial x}{\partial r} = \sum \frac{\partial N_j}{\partial r}x_j \qquad \frac{\partial x}{\partial s} = \sum \frac{\partial N_j}{\partial s}x_j$$

$$\frac{\partial y}{\partial r} = \sum \frac{\partial N_j}{\partial r}y_j \qquad \frac{\partial y}{\partial s} = \sum \frac{\partial N_j}{\partial s}y_j \qquad (9\text{-}66)$$

$$\frac{\partial z}{\partial r} = \sum \frac{\partial N_j}{\partial r}z_j \qquad \frac{\partial z}{\partial s} = \sum \frac{\partial N_j}{\partial s}z_j$$

Clearly dS may be written as a function of only r, s, and t and the nodal coordinates if Eqs. (9-65) and (9-66) are used. Note that Eq. (9-65) holds on surfaces of constant t (such as $t = 1$ or $t = -1$). If integrations over surfaces of constant s (or constant r) are desired, then an expression analogous to Eq. (9-65) may be written by inspection or easily derived. Therefore, integrals over a face of the element may be converted to integrals of the form

$$\mathbf{K}^e = \int_{-1}^{+1}\int_{-1}^{+1} \mathbf{H}(r,s,t)\bigg|_{t=+1} dr\ ds \qquad (9\text{-}67)$$

and

$$\mathbf{f}^e = \int_{-1}^{+1}\int_{-1}^{+1} \mathbf{g}(r,s,t)\bigg|_{t=+1} dr\ ds \qquad (9\text{-}68)$$

if the face over which $t = +1$ is on the global boundary. Similar integrals result if the other faces are on the global boundary. Again, numerical integration of these integrals is necessary as explained in Sec. 9-8.

As in the two-dimensional case, the necessary and sufficient condition for a one-to-one mapping is that the sign of the Jacobian must remain the same for all points in the domain that is mapped. Obviously it is not practical to check every point. Instead, a few select points are used as explained later.

The Tetrahedral Element

Figure 9-17 shows the isoparametric forms of the tetrahedral element. Note that the linear-, quadratic-, and cubic-order elements have 2, 3, and 4 nodes, respectively, on each edge of the tetrahedral element. Again we must convert integrals over a complicated distorted element to integrals over an undistorted parent element. This is easily accomplished if the integrals in terms of the global coordinates (x,y,z) are transformed into integrals in terms of L_1, L_2, L_3, and L_4 (the volume coordinates). The development is similar to that for the triangular element, except that an additional coordinate (L_4) needs to be considered. Now L_4 is taken to be the dependent coordinate related to the others by

$$L_4 = 1 - L_1 - L_2 - L_3 \tag{9-69}$$

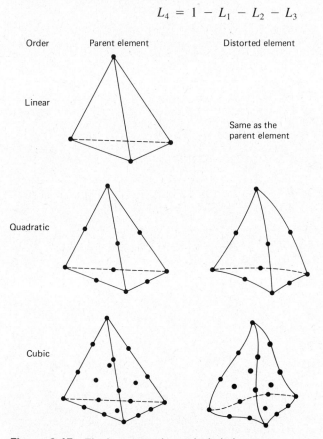

Figure 9-17   The isoparametric tetrahedral elements.

Therefore, a typical shape function, e.g., the ith, may be written as a function of L_1, L_2, and L_3, or

$$N_i = N_i(L_1, L_2, L_3) \tag{9-70}$$

But we also have

$$L_1 = L_1(x, y, z) \tag{9-71a}$$

$$L_2 = L_2(x, y, z) \tag{9-71b}$$

and

$$L_3 = L_3(x, y, z) \tag{9-71c}$$

Therefore, the total derivatives are given by

$$\frac{\partial N_i}{\partial L_1} = \frac{\partial N_i}{\partial x}\frac{\partial x}{\partial L_1} + \frac{\partial N_i}{\partial y}\frac{\partial y}{\partial L_1} + \frac{\partial N_i}{\partial z}\frac{\partial z}{\partial L_1} \tag{9-72a}$$

$$\frac{\partial N_i}{\partial L_2} = \frac{\partial N_i}{\partial x}\frac{\partial x}{\partial L_2} + \frac{\partial N_i}{\partial y}\frac{\partial y}{\partial L_2} + \frac{\partial N_i}{\partial z}\frac{\partial z}{\partial L_2} \tag{9-72b}$$

$$\frac{\partial N_i}{\partial L_3} = \frac{\partial N_i}{\partial x}\frac{\partial x}{\partial L_3} + \frac{\partial N_i}{\partial y}\frac{\partial y}{\partial L_3} + \frac{\partial N_i}{\partial z}\frac{\partial z}{\partial L_3} \tag{9-72c}$$

These three equations may be written in matrix form as

$$
\begin{bmatrix} \dfrac{\partial N_i}{\partial L_1} \\[2ex] \dfrac{\partial N_i}{\partial L_2} \\[2ex] \dfrac{\partial N_i}{\partial L_3} \end{bmatrix}
=
\begin{bmatrix} \dfrac{\partial x}{\partial L_1} & \dfrac{\partial y}{\partial L_1} & \dfrac{\partial z}{\partial L_1} \\[2ex] \dfrac{\partial x}{\partial L_2} & \dfrac{\partial y}{\partial L_2} & \dfrac{\partial z}{\partial L_2} \\[2ex] \dfrac{\partial x}{\partial L_3} & \dfrac{\partial y}{\partial L_3} & \dfrac{\partial z}{\partial L_3} \end{bmatrix}
\begin{bmatrix} \dfrac{\partial N_i}{\partial x} \\[2ex] \dfrac{\partial N_i}{\partial y} \\[2ex] \dfrac{\partial N_i}{\partial z} \end{bmatrix}
\tag{9-73}
$$

Again a 3×3 Jacobian matrix arises and with the help of

$$x = \sum N_j x_j \qquad y = \sum N_j y_j \qquad z = \sum N_j z_j$$

the Jacobian matrix is given by

$$
\mathbf{J} =
\begin{bmatrix}
\sum \dfrac{\partial N_j}{\partial L_1} x_j & \sum \dfrac{\partial N_j}{\partial L_1} y_j & \sum \dfrac{\partial N_j}{\partial L_1} z_j \\[2.5ex]
\sum \dfrac{\partial N_j}{\partial L_2} x_j & \sum \dfrac{\partial N_j}{\partial L_2} y_j & \sum \dfrac{\partial N_j}{\partial L_2} z_j \\[2.5ex]
\sum \dfrac{\partial N_j}{\partial L_3} x_j & \sum \dfrac{\partial N_j}{\partial L_3} y_j & \sum \dfrac{\partial N_j}{\partial L_3} z_j
\end{bmatrix}
\tag{9-74}
$$

where x_j, y_j, and z_j denote the global coordinates of the jth node. The summations in Eq. (9-74) are made over the 4, 10, and 20 nodes for the linear-, quadratic-, and cubic-order elements, respectively.

If Eq. (9-73) is premultiplied by the inverse of the Jacobian matrix, we get the following useful result:

$$
\begin{bmatrix} \dfrac{\partial N_i}{\partial x} \\[2ex] \dfrac{\partial N_i}{\partial y} \\[2ex] \dfrac{\partial N_i}{\partial z} \end{bmatrix} = \begin{bmatrix} \sum \dfrac{\partial N_j}{\partial L_1}x_j & \sum \dfrac{\partial N_j}{\partial L_1}y_j & \sum \dfrac{\partial N_j}{\partial L_1}z_j \\[2ex] \sum \dfrac{\partial N_j}{\partial L_2}x_j & \sum \dfrac{\partial N_j}{\partial L_2}y_j & \sum \dfrac{\partial N_j}{\partial L_2}z_j \\[2ex] \sum \dfrac{\partial N_j}{\partial L_3}x_j & \sum \dfrac{\partial N_j}{\partial L_3}y_j & \sum \dfrac{\partial N_j}{\partial L_3}z_j \end{bmatrix}^{-1} \begin{bmatrix} \dfrac{\partial N_i}{\partial L_1} \\[2ex] \dfrac{\partial N_i}{\partial L_2} \\[2ex] \dfrac{\partial N_i}{\partial L_3} \end{bmatrix}
$$

(9-75)

Again it is seen that derivatives of the shape functions must be obtained. Because we have assumed L_1, L_2, and L_3 to be independent, and L_4 to be dependent, we would have to use Eq. (9-69) to eliminate L_4 in the shape functions before taking the derivatives. This approach is not very practical and an alternate method is now presented. Let us denote the partial derivative of N_i with respect to L_1 as

$$
\left(\frac{\partial N_i}{\partial L_1} \right)_{L_2, L_3, L_4}
$$

where the shape function N_i is now a function of L_1, L_2, L_3, and L_4 (but L_2, L_3, and L_4 are held constant in performing the differentiation). The notation

$$
\frac{\partial N_i}{\partial L_1}
$$

implies N_i is a function of L_1, L_2, and L_3 only, and L_2 and L_3 are held constant [L_2 and L_3 are not written since they are not written in Eq. (9-75)]. It follows that

$$
\frac{\partial N_i}{\partial L_1} = \frac{\partial L_1}{\partial L_1}\left(\frac{\partial N_i}{\partial L_1}\right)_{L_2,L_3,L_4} + \frac{\partial L_2}{\partial L_1}\left(\frac{\partial N_i}{\partial L_2}\right)_{L_1,L_3,L_4} + \frac{\partial L_3}{\partial L_1}\left(\frac{\partial N_i}{\partial L_3}\right)_{L_1,L_2,L_4}
$$
$$
+ \frac{\partial L_4}{\partial L_1}\left(\frac{\partial N_i}{\partial L_4}\right)_{L_1,L_2,L_3}
$$

(9-76)

But

$$
\frac{\partial L_1}{\partial L_1} = 1 \qquad \frac{\partial L_2}{\partial L_1} = 0 \qquad \frac{\partial L_3}{\partial L_1} = 0 \qquad \frac{\partial L_4}{\partial L_1} = -1
$$

(9-77)

and so we have

$$
\frac{\partial N_i}{\partial L_1} = \underbrace{\left(\frac{\partial N_i}{\partial L_1}\right)_{L_2,L_3,L_4}}_{\substack{\text{Here } N_i \text{ is taken} \\ \text{to be a function} \\ \text{of only } L_1, L_2, \\ \text{and } L_3 \text{ as required} \\ \text{by Eq. (9-75)}}} - \underbrace{\left(\frac{\partial N_i}{\partial L_4}\right)_{L_1,L_2,L_3}}_{\substack{\text{Here } N_i \text{ is taken to be} \\ \text{a function of } L_1, L_2, \\ L_3, \text{ and } L_4}}
$$

(9-78)

It can similarly be shown that

$$\frac{\partial N_i}{\partial L_2} = \left(\frac{\partial N_i}{\partial L_2}\right)_{L_1,L_3,L_4} - \left(\frac{\partial N_i}{\partial L_4}\right)_{L_1,L_2,L_3} \qquad \text{(9-79)}$$

and

$$\frac{\partial N_i}{\partial L_3} = \left(\frac{\partial N_i}{\partial L_3}\right)_{L_1,L_2,L_4} - \left(\frac{\partial N_i}{\partial L_4}\right)_{L_1,L_2,L_3} \qquad \text{(9-80)}$$

Finally, the element volume $dx\,dy\,dz$ may be written as

$$dx\,dy\,dz = |\det \mathbf{J}|\,dL_1\,dL_2\,dL_3 \qquad \text{(9-80)}$$

Therefore, each integral over the element volume may be transformed into an integral of the form

$$\mathbf{K}^e = \int_0^1 \int_0^{1-L_3} \int_0^{1-L_2-L_3} \mathbf{H}(L_1,L_2,L_3)\,dL_1\,dL_2\,dL_3 \qquad \text{(9-81)}$$

or

$$\mathbf{f}^e = \int_0^1 \int_0^{1-L_3} \int_0^{1-L_2-L_3} \mathbf{g}(L_1,L_2,L_3)\,dL_1\,dL_2\,dL_3 \qquad \text{(9-82)}$$

These integrals are in forms that are suitable for numerical integration as explained in Sec. 9-9. The resulting sizes of the element stiffness matrices are given in Table 9-4. As in all other previous cases, the necessary and sufficient condition for a one-to-one mapping is that the sign of the Jacobian be the same for all points mapped.

Table 9-4 Size of the Element Stiffness Matrices for the Tetrahedral Elements

Order of element	Structural (3 DOF per node)	Thermal (1 DOF per node)
Linear	12×12	4×4
Quadratic	30×30	10×10
Cubic	60×60	20×20

9-8 NUMERICAL INTEGRATION: RECTANGULAR AND BRICK ELEMENTS

As mentioned in Secs. 9-6 and 9-7, the use of numerical integration techniques makes the isoparametric element practical. Integrals such as

$$I = \int_a^b f(x)\,dx \qquad \text{(9-83)}$$

may be evaluated approximately by writing

$$I = \int_a^b f(x) \, dx = \sum_{i=1}^{n} w_i f(x_i) \tag{9-84}$$

where the w_i are referred to as the *weights* and the x_i as the *sampling points*.

If the sampling points are chosen such that the interval $a \le x \le b$ is divided into $n - 1$ equal-length segments, the integration is referred to as *Newton-Cotes quadrature*. Quadrature is another name for numerical integration. Familiar examples of this type of numerical integration are the trapezoidal and Simpson's rules. In effect, a polynomial is used on a piecewise basis to represent $f(x)$ over the interval. The trapezoidal rule will integrate a linear function exactly, whereas Simpson's rule will integrate a quadratic function exactly. If n sampling points are used, the integration is exact if $f(x)$ is a polynomial of order $n - 1$ or less. This method is frequently used when the data to be integrated is in tabular form and equally spaced.

Integrals that arise in the finite element method have integrands that are explicit functions of the global coordinates. It was shown in Secs. 9-6 and 9-7 how these integrals could be transformed into ones in the serendipity domains for the rectangular and brick isoparametric elements. Recall, for example, Eqs. (9-34), (9-41), (9-42), etc. Note that the integration limits are also changed to reflect the fact that $-1 \le r \le +1$, $-1 \le s \le +1$, etc. Therefore, we need to integrate functions such as

$$I = \int_{-1}^{+1} f(r) \, dr \tag{9-85a}$$

$$I = \int_{-1}^{+1} \int_{-1}^{+1} f(r,s) \, dr \, ds \tag{9-85b}$$

$$I = \int_{-1}^{+1} \int_{-1}^{+1} \int_{-1}^{+1} f(r,s,t) \, dr \, ds \, dt \tag{9-85c}$$

In Eqs. (9-85), a scalar integrand is implied because the integral of a matrix is simply the matrix of integrals.

Gauss quadrature is a numerical integration method that allows the sampling points to be chosen such that the best possible accuracy may be obtained. If the sampling points and weights in Eq. (9-84) are based on Legendre polynomials, then the numerical integration is referred to as *Gauss-Legendre* quadrature. The derivation of these weights and sampling points is beyond the scope of this book. The interested reader may wish to consult reference 5. This method will integrate polynomials of order $2n - 1$ exactly, where n is the number of sampling points. Gauss-Legendre quadrature requires the integral to be in the form of Eqs. (9-85), and in one dimension we have

$$I = \int_{-1}^{+1} f(r) \, dr = \sum_{i=1}^{n} w_i f(r_i) \tag{9-86}$$

where the weights w_i and sampling points r_i are given in Table 9-5 for up to six sampling points (or $n = 6$).

Table 9-5   Sampling Point Values and Weights for Gauss-Legendre Quadrature

$$\int_{-1}^{+1} f(r) \, dr = \sum_{i=1}^{n} w_i f(r_i)$$

r_i	w_i
$n = 1$	
0.000000000000000	2.000000000000000
$n = 2$	
±0.577350269189626	1.000000000000000
$n = 3$	
0.000000000000000	0.888888888888889
±0.774596669241483	0.555555555555556
$n = 4$	
±0.339981043584856	0.652145154862546
±0.861136311594053	0.347854845137454
$n = 5$	
0.000000000000000	0.568888888888889
±0.538469310105683	0.478628670499366
±0.906179845938664	0.236926885056189
$n = 6$	
±0.238619186083197	0.467913934572691
±0.661209386466265	0.360761573048139
±0.932469514203152	0.171324492379170

Example 9-2

Evaluate the integral

$$I = \int_{-1}^{+1} (r^4 + r) \, dr$$

by using Gauss-Legendre quadrature. Perform an integration of such an order that the integral is evaluated exactly. Compare the result from the numerical evaluation with that from the analytical integration.

Solution

First, we note that the integral is in the proper form for evaluation by Gauss-Legendre quadrature. Second, since the integrand is a polynomial, we can determine the order of the quadrature that will result in an exact evaluation from

$$2n - 1 = 4$$

where the order of the polynomial is 4. Solving for n yields $n = \frac{5}{2}$, which must be rounded to give $n = 3$. Therefore, we must take three sampling points in Table

9-5. If we let $f(r) = r^4 + r$, then the calculations may be summarized in the following table:

i	r_i	w_i	$f(r_i)$	$w_i f(r_i)$
1	-0.77460	0.55556	-0.41459	-0.23033
2	0.00000	0.88889	0.00000	0.00000
3	$+0.77460$	0.55556	1.13461	$\underline{0.63033}$
				$\Sigma = 0.40000$

It can readily be shown that the exact result is $\frac{2}{5}$. Only five significant digits were carried in the numerical evaluation summarized in the table above. For all practical purposes, we have obtained the exact result by using three Gauss points. ■

Because we frequently need to integrate over rectangular and brick elements, Eq. (9-84) needs to be extended to two and three dimensions. Let us begin with Eq. (9-85b) and integrate first with respect to r by applying Eq. (9-86) to get

$$I = \int_{-1}^{+1} \int_{-1}^{+1} f(r,s) \, dr \, ds = \int_{-1}^{+1} \sum_{i=1}^{n} w_i f(r_i,s) \, ds$$

or

$$I = \int_{-1}^{+1} g(s) \, ds \tag{9-88}$$

where

$$g(s) = \sum_{i=1}^{n} w_i f(r_i,s) \tag{9-89}$$

Integrating Eq. (9-88) by applying Eq. (9-86) again gives

$$I = \sum_{j=1}^{m} w_j g(s_j) \tag{9-90}$$

But $g(s_j)$ from Eq. (9-89) is given by

$$g(s_j) = \sum_{i=1}^{n} w_i f(r_i,s_j) \tag{9-91}$$

Combining Eqs. (9-90) and (9-91) gives the desired result

$$I = \int_{-1}^{+1} \int_{-1}^{+1} f(r,s) \, dr \, ds = \sum_{j=1}^{m} \sum_{i=1}^{n} w_i w_j f(r_i,s_j) \tag{9-92}$$

Note that n sampling points are assumed in the r direction and m in the s direction. Usually m is equal to n, but this is not necessary.

Equation (9-92) is extended to three dimensions in a straightforward manner with the result

$$I = \int_{-1}^{+1}\int_{-1}^{+1}\int_{-1}^{+1} f(r,s,t) \, dr \, ds \, dt$$

$$= \sum_{k=1}^{p}\sum_{j=1}^{m}\sum_{i=1}^{n} w_i w_j w_k f(r_i, s_j, t_k) \qquad \text{(9-93)}$$

where p sampling points are assumed in the t direction. The use of these equations is demonstrated in Examples 9-3 and 9-4. ■

Example 9-3

Evaluate the integral

$$I = \int_{-1}^{+1}\int_{-1}^{+1} (r^2 + rs)s^4 \, dr \, ds$$

by using Gauss-Legendre quadrature. Compare the result with that from the exact evaluation.

Solution

Because the integrand is a polynomial, we could determine the required number of sampling points (or Gauss points) in each direction in order to get the exact value for the integral. Note that the integral is of order 2 in r and of order 5 in s. Therefore, in the r direction we have $2n - 1 = 2$ or $n = \frac{3}{2}$ and in the s direction $2m - 1 = 5$ or $m = 3$. Hence we take $n = 2$ ($\frac{3}{2}$ rounds to 2) and $m = 3$, or two Gauss points in the r direction, and three points in the s direction. It is instructive to show the six sampling points (i.e., 2×3) on a typical element as shown in Fig. 9-18.

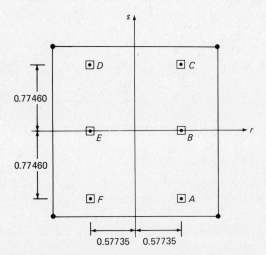

Figure 9-18 Quadrilateral element in Example 9-3. Note that ⊡ denotes a Gauss (or sampling) point.

Let us denote the integrand by $f(r,s)$, where

$$f(r,s) = (r^2 + rs)s^4$$

The calculations are summarized (to five significant digits) in the following table:

Point	r_i	s_j	w_i	w_j	$f(r_i,s_j)$	$w_i w_j f(r_i,s_j)$
A	0.57735	−0.77460	1.00000	0.55556	−0.04100	−0.02278
B	0.57735	0.00000	1.00000	0.88889	0.00000	0.00000
C	0.57735	+0.77460	1.00000	0.55556	0.21800	0.15611
D	−0.57735	+0.77460	1.00000	0.55556	−0.04100	−0.02278
E	−0.57735	0.00000	1.00000	0.88889	0.00000	0.00000
F	−0.57735	−0.77460	1.00000	0.55556	0.28100	0.15611

$$\sum = 0.26666$$

This compares favorably with the exact evaluation of $I = \frac{4}{15}$ or $I = 0.26667$ to five significant digits. The exact result is approached as the number of significant digits used is increased. ∎

Example 9-4

Evaluate the element stiffness (or conductance) matrix for heat conduction in the x direction for the element shown in Fig. 9-19, where the nodal coordinates are shown in inches. Assume a thermal conductivity k of 1 Btu/hr-in-°F and an element thickness t of 1 in.

Solution

From Eq. (8-106a), the expression for the stiffness matrix from heat conduction in the x direction is given by

$$\mathbf{K}_{xx}^e = \int_{A^e} \frac{\partial \mathbf{N}^T}{\partial x} kt \frac{\partial \mathbf{N}}{\partial x} \, dx \, dy \qquad \textbf{(8-106a)}$$

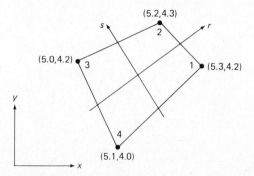

Figure 9-19   Quadrilateral element in Example 9-4.

For the element under consideration, the shape function matrix $\mathbf{N}$ is given by

$$\mathbf{N} = \begin{bmatrix} \frac{1}{4}(1 + r)(1 - s) \\ \frac{1}{4}(1 + r)(1 + s) \\ \frac{1}{4}(1 - r)(1 + s) \\ \frac{1}{4}(1 - r)(1 - s) \end{bmatrix}^T$$

Note that the four nodes are being used for two purposes: (1) to define the element geometry, and (2) to specify the locations where the values of the parameter function (in this case the temperature) are sought. Hence, the isoparametric formulation from Sec. 9-6 is directly applicable here. Recall that the Jacobian matrix $\mathbf{J}$ is given by Eq. (9-33) for this element or

$$\mathbf{J} = \begin{bmatrix} \sum_{j=1}^{4} \frac{\partial N_j}{\partial r} x_j & \sum_{j=1}^{4} \frac{\partial N_j}{\partial r} y_j \\ \sum_{j=1}^{4} \frac{\partial N_j}{\partial s} x_j & \sum_{j=1}^{4} \frac{\partial N_j}{\partial s} y_j \end{bmatrix}$$

Note that four terms are included in the summations because there are four nodes (per element). For convenience, let us denote the entries comprising $\mathbf{J}^{-1}$ as A_{11}, A_{12}, A_{21}, and A_{22}, or

$$\mathbf{J} = \begin{bmatrix} J_{11} & J_{12} \\ J_{21} & J_{22} \end{bmatrix}^{-1} = \begin{bmatrix} A_{11} & A_{12} \\ A_{21} & A_{22} \end{bmatrix}$$

The A_{ij}'s are functions of r and s. From Eq. (9-31) we may write

$$\frac{\partial N_i}{\partial x} = A_{11} \frac{\partial N_i}{\partial r} + A_{12} \frac{\partial N_i}{\partial s}$$

If all four shape functions are considered at one time, we have

$$\frac{\partial \mathbf{N}}{\partial x} = A_{11} \frac{\partial \mathbf{N}}{\partial r} + A_{12} \frac{\partial \mathbf{N}}{\partial s}$$

from which it also follows that

$$\frac{\partial \mathbf{N}^T}{\partial x} = \left(A_{11} \frac{\partial \mathbf{N}}{\partial r} + A_{12} \frac{\partial \mathbf{N}}{\partial s} \right)^T = A_{11} \frac{\partial \mathbf{N}^T}{\partial r} + A_{12} \frac{\partial \mathbf{N}^T}{\partial s}$$

Finally, we also have

$$dx \, dy = |\det \mathbf{J}| \, dr \, ds$$

and may write

$$\mathbf{K}_{xx}^e = \int_{-1}^{+1} \int_{-1}^{+1} \left(A_{11} \frac{\partial \mathbf{N}^T}{\partial r} + A_{12} \frac{\partial \mathbf{N}^T}{\partial s} \right) kt \left(A_{11} \frac{\partial \mathbf{N}}{\partial r} + A_{12} \frac{\partial \mathbf{N}}{\partial s} \right) |\det \mathbf{J}| \, dr \, ds$$

Note that the integrand is a function of r and s only. In other words, the original quadrilateral element has been mapped into the corresponding undistorted (parent) element. The global coordinates (x,y) have been eliminated in favor of the serendipity coordinates (r,s).

Let us denote the integrand by $\mathbf{H}(r,s)$ or

$$\mathbf{H}(r,s) = \mathbf{c}^T k t \mathbf{c} \, |\det \mathbf{J}|$$

where

$$\mathbf{c} = A_{11}\frac{\partial \mathbf{N}}{\partial r} + A_{12}\frac{\partial \mathbf{N}}{\partial s}$$

Clearly, we need the derivatives of the shape functions with respect to r and s, or

$$\frac{\partial \mathbf{N}}{\partial r} = \begin{bmatrix} \dfrac{\partial N_1}{\partial r} \\[2mm] \dfrac{\partial N_2}{\partial r} \\[2mm] \dfrac{\partial N_3}{\partial r} \\[2mm] \dfrac{\partial N_4}{\partial r} \end{bmatrix}^T = \begin{bmatrix} \frac{1}{4}(1 - s) \\[1mm] \frac{1}{4}(1 + s) \\[1mm] -\frac{1}{4}(1 + s) \\[1mm] -\frac{1}{4}(1 - s) \end{bmatrix}^T$$

and

$$\frac{\partial \mathbf{N}}{\partial s} = \begin{bmatrix} \dfrac{\partial N_1}{\partial s} \\[2mm] \dfrac{\partial N_2}{\partial s} \\[2mm] \dfrac{\partial N_3}{\partial s} \\[2mm] \dfrac{\partial N_4}{\partial s} \end{bmatrix}^T = \begin{bmatrix} -\frac{1}{4}(1 + r) \\[1mm] \frac{1}{4}(1 + r) \\[1mm] \frac{1}{4}(1 - r) \\[1mm] -\frac{1}{4}(1 - r) \end{bmatrix}^T$$

The Jacobian matrix $\mathbf{J}$ has four entries denoted J_{11}, J_{12}, J_{21}, and J_{22}. From Eq. (9-33), these are given by

$$J_{11} = \sum_{j=1}^{4} \frac{\partial N_j}{\partial r} x_j$$

$$= \frac{1}{4}(1 - s)x_1 + \frac{1}{4}(1 + s)x_2 - \frac{1}{4}(1 + s)x_3 - \frac{1}{4}(1 - s)x_4$$

$$J_{12} = \sum_{j=1}^{4} \frac{\partial N_j}{\partial r} y_j$$

$$= \frac{1}{4}(1 - s)y_1 + \frac{1}{4}(1 + s)y_2 - \frac{1}{4}(1 + s)y_3 - \frac{1}{4}(1 - s)y_4$$

$$J_{21} = \sum_{j=1}^{4} \frac{\partial N_j}{\partial s} x_j$$

$$= -\tfrac{1}{4}(1 + r)x_1 + \tfrac{1}{4}(1 + r)x_2 + \tfrac{1}{4}(1 - r)x_3 - \tfrac{1}{4}(1 - r)x_4$$

$$J_{22} = \sum_{j=1}^{4} \frac{\partial N_j}{\partial s} y_j$$

$$= -\tfrac{1}{4}(1 + r)y_1 + \tfrac{1}{4}(1 + r)y_2 + \tfrac{1}{4}(1 - r)y_3 - \tfrac{1}{4}(1 - r)y_4$$

Since the nodal coordinates are known, the Jacobian matrix can be readily determined for any given (r,s) and subsequently inverted to get A_{11}, A_{12}, etc. Two Gauss points in each direction will be used, since it is not possible (or desirable) to evaluate the integral exactly. Why not? The parent (undistorted) element is shown in Fig. 9-20. The calculations are summarized below for each of the four Gauss points.

Gauss point A:

$$r_2 = +0.57735027$$

$$s_1 = -0.57735027$$

$$\mathbf{J} = \begin{bmatrix} 0.10000000 & 0.08943376 \\ -0.05000000 & 0.06056624 \end{bmatrix}$$

$$\det \mathbf{J} = 0.01052831$$

$$\mathbf{J}^{-1} = \mathbf{A} = \begin{bmatrix} 5.75270206 & -8.49459588 \\ 4.74909931 & 9.49819863 \end{bmatrix}$$

$$\frac{\partial \mathbf{N}}{\partial r} = \begin{bmatrix} 0.39433757 \\ 0.10566243 \\ -0.10566243 \\ -0.39433757 \end{bmatrix}^{T} \qquad \frac{\partial \mathbf{N}}{\partial s} = \begin{bmatrix} -0.39433757 \\ 0.39433757 \\ 0.10566243 \\ -0.10566243 \end{bmatrix}^{T}$$

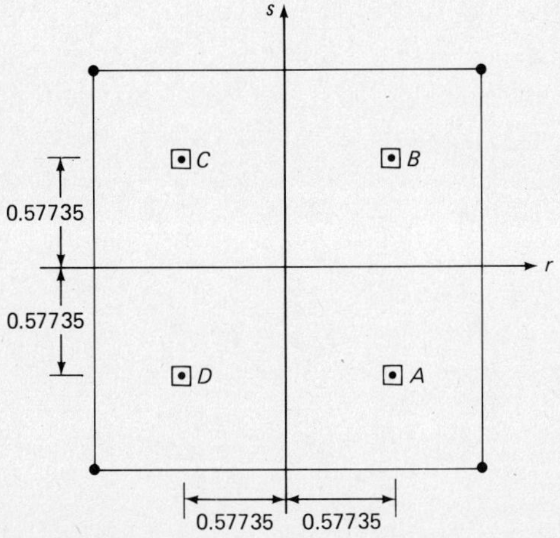

Figure 9-20 Parent element showing the 2 × 2 Gaussian quadrature for Example 9-4.

$$\mathbf{c} = \left[A_{11}\frac{\partial \mathbf{N}}{\partial r} + A_{12}\frac{\partial \mathbf{N}}{\partial s} \right] = \begin{bmatrix} 5.61824481 \\ -2.74189378 \\ -1.50540416 \\ -1.37094687 \end{bmatrix}^{T}$$

$$\mathbf{H}(r_2, s_1) = \mathbf{c}^T kt\mathbf{c} |\det \mathbf{J}|$$

$$= \begin{bmatrix} 0.33232274 & -0.16218476 & -0.08904561 & -0.08109238 \\ & 0.07915165 & 0.04345727 & 0.03957583 \\ & & 0.02385970 & 0.02172864 \\ & \text{(Symmetric)} & & 0.01978791 \end{bmatrix}$$

Gauss point B:

$$r_2 = +0.57735027$$

$$s_2 = +0.57735027$$

$$\mathbf{J} = \begin{bmatrix} 0.10000000 & 0.06056624 \\ -0.05000000 & 0.06056624 \end{bmatrix}$$

$$\det \mathbf{J} = 0.00908494$$

$$\mathbf{J}^{-1} = \mathbf{A} = \begin{bmatrix} 6.66666667 & -6.66666667 \\ 5.50361582 & 11.00723165 \end{bmatrix}$$

$$\frac{\partial \mathbf{N}}{\partial r} = \begin{bmatrix} 0.10566243 \\ 0.39433757 \\ -0.39433757 \\ -0.10566243 \end{bmatrix}^{T} \qquad \frac{\partial \mathbf{N}}{\partial s} = \begin{bmatrix} -0.39433757 \\ 0.39433757 \\ 0.10566243 \\ -0.10566243 \end{bmatrix}^{T}$$

$$\mathbf{c} = \left[A_{11}\frac{\partial \mathbf{N}}{\partial r} + A_{12}\frac{\partial \mathbf{N}}{\partial s} \right] = \begin{bmatrix} 3.33333333 \\ 0.00000000 \\ -3.33333333 \\ 0.00000000 \end{bmatrix}^{T}$$

$$\mathbf{H}(r_2, s_2) = \mathbf{c}^T kt\mathbf{c} |\det \mathbf{J}|$$

$$= \begin{bmatrix} 0.10094373 & 0.00000000 & -0.10094373 & 0.00000000 \\ & 0.00000000 & 0.00000000 & 0.00000000 \\ & & 0.10094373 & 0.00000000 \\ & \text{(Symmetric)} & & 0.00000000 \end{bmatrix}$$

Gauss point C:

$$r_1 = -0.57735027$$

$$s_2 = +0.57735027$$

$$\mathbf{J} = \begin{bmatrix} 0.10000000 & 0.06056624 \\ -0.05000000 & 0.08943376 \end{bmatrix}$$

$$\det \mathbf{J} = 0.01197169$$

$$\mathbf{J}^{-1} = \mathbf{A} = \begin{bmatrix} 7.47043843 & -5.05912314 \\ 4.17652052 & 8.35304105 \end{bmatrix}$$

$$\frac{\partial \mathbf{N}}{\partial r} = \begin{bmatrix} 0.10566243 \\ 0.39433757 \\ -0.39433757 \\ -0.10566243 \end{bmatrix}^T \qquad \frac{\partial \mathbf{N}}{\partial s} = \begin{bmatrix} -0.10566243 \\ 0.10566243 \\ 0.39433757 \\ -0.39433757 \end{bmatrix}^T$$

$$\mathbf{c} = \left[A_{11}\frac{\partial \mathbf{N}}{\partial r} + A_{12}\frac{\partial \mathbf{N}}{\partial s} \right] = \begin{bmatrix} 1.32390396 \\ 2.41131526 \\ -4.94087683 \\ 1.20565761 \end{bmatrix}^T$$

$$\mathbf{H}(r_1,s_2) = \mathbf{c}^T kt\mathbf{c}|\det \mathbf{J}|$$

$$= \begin{bmatrix} 0.02098304 & 0.03821782 & -0.07830976 & 0.01910891 \\ & 0.06960868 & -0.14263083 & 0.03480434 \\ & & 0.29225601 & -0.07131541 \\ & \text{(Symmetric)} & & 0.01740217 \end{bmatrix}$$

Gauss point D:

$$r_1 = -0.57735027$$

$$s_1 = -0.57735027$$

$$\mathbf{J} = \begin{bmatrix} 0.10000000 & 0.08943376 \\ -0.05000000 & 0.08943376 \end{bmatrix}$$

$$\det \mathbf{J} = 0.01341506$$

$$\mathbf{J}^{-1} = \mathbf{A} = \begin{bmatrix} 6.66666667 & -6.66666667 \\ 3.72715342 & 7.45430684 \end{bmatrix}$$

$$\frac{\partial \mathbf{N}}{\partial r} = \begin{bmatrix} 0.39433757 \\ 0.10566243 \\ -0.10566243 \\ -0.39433757 \end{bmatrix}^T \qquad \frac{\partial \mathbf{N}}{\partial s} = \begin{bmatrix} -0.10566243 \\ 0.10566243 \\ 0.39433757 \\ -0.39433757 \end{bmatrix}^T$$

$$\mathbf{c} = \left[A_{11}\frac{\partial \mathbf{N}}{\partial r} + A_{12}\frac{\partial \mathbf{N}}{\partial s} \right] = \begin{bmatrix} 3.33333333 \\ 0.00000000 \\ -3.33333333 \\ 0.00000000 \end{bmatrix}^T$$

$\mathbf{H}(r_1,s_1) = \mathbf{c}^T kt\mathbf{c}|\det \mathbf{J}|$

$$= \begin{bmatrix} 0.14905627 & 0.00000000 & -0.14905627 & 0.00000000 \\ & 0.00000000 & 0.00000000 & 0.00000000 \\ & & 0.14905627 & 0.00000000 \\ & \text{(Symmetric)} & & 0.00000000 \end{bmatrix}$$

These results may now be used to obtain the stiffness matrix for this element by noting that

$$\mathbf{K}_{xx}^e = \int_A \frac{\partial \mathbf{N}^T}{\partial x} kt \frac{\partial \mathbf{N}}{\partial x} \, dx \, dy = \int_{-1}^{+1} \int_{-1}^{+1} \mathbf{H}(r,s) \, dr \, ds$$

$$= \sum_{i=1}^{2} \sum_{j=1}^{2} w_i w_j \mathbf{H}(r_i,s_j)$$

where each weight (i.e., w_i and w_j) has a value of unity (see Table 9-5 for $n = 2$). The final result is given by

$$\mathbf{K}_{xx}^e = \begin{bmatrix} 0.60330578 & -0.12396694 & -0.41735537 & -0.06198347 \\ & 0.14876033 & -0.09917356 & 0.07438017 \\ & & 0.56611571 & -0.04958677 \\ & \text{(Symmetric)} & & 0.03719008 \end{bmatrix}$$

It is not possible to compare this result with that from an exact, analytical integration. Why not?   ■

9-9 NUMERICAL INTEGRATION: TRIANGULAR AND TETRAHEDRAL ELEMENTS

In Sec. 6-7 three special integration formulas were presented that could be used to evaluate integrals for the lineal, triangular, and tetrahedral elements. These formulas, given by Eqs. (6-48) to (6-50) were illustrated many times in Chapters 7 and 8. Although these formulas are applicable to the higher-order lineal, triangular, and tetrahedral elements, they are not very practical in these situations. Instead, we resort to numerical integration.

Recall from Secs. 9-6 and 9-7 that integrals with rather complicated integrands arose when higher-order triangular and tetrahedral elements were considered. For example, typical element stiffness matrices and nodal force vectors are of the forms given by Eqs. (9-54), (9-55), (9-81), and (9-82). Analytical evaluation of such integrals is impossible, and so we must resort to a numerical scheme. Formulas derived by Hammer, Marlowe, and Stroud [6] are particularly useful and easy to apply. The use of these formulas for the case of the triangle is based on the approximation given by

$$\mathbf{K}^e = \int_0^1 \int_0^{1-L_2} \mathbf{H} \; dL_1 \; dL_2 = \sum_{i=1}^n \tfrac{1}{2} w_i \mathbf{H}_i(L_{1i}, L_{2i}, L_{3i}) \tag{9-94}$$

where w_i denotes the ith *weight* and L_{1i}, L_{2i}, and L_{3i} denote the coordinates of the ith *sampling points*. Note that n sampling points are assumed. A similar formula holds for the tetrahedral element, or

$$\mathbf{K}^e = \int_0^1 \int_0^{1-L_3} \int_0^{1-L_2-L_3} \mathbf{H} \; dL_1 \; dL_2 \; dL_3$$

$$= \sum_{i=1}^n \tfrac{1}{6} w_i \mathbf{H}_i(L_{1i}, L_{2i}, L_{3i}, L_{4i}) \tag{9-95}$$

Note that the dependent area coordinate L_3 and the dependent volume coordinate L_4 were reintroduced into the integrands in Eqs. (9-94) and (9-95). The sampling points and weights are given in Figs. 9-21 and 9-22 for the triangular and tetrahedral elements, respectively. The use of these formulas is illustrated in Examples 9-5 and 9-6.

Example 9-5

By using Eq. (9-94) and Fig. 9-21, evaluate the following element stiffness matrix for the three-node triangular element

$$\mathbf{K}_{cv}^e = \int_{A^e} \mathbf{N}^T h \mathbf{N} \; dx \; dy$$

Perform a linear-order integration. Recall that this element stiffness matrix arose in Sec. 8-8 and resulted from convection from the lateral surface(s) of a plate [see Eq. (8-106c)]. The exact integration is given in Example 8-13 and will serve as a check on the numerical integrations.

Solution

In terms of the area coordinates, the shape function matrix $\mathbf{N}$ for this element is given by

$$\mathbf{N} = [L_1 \quad L_2 \quad L_3]$$

Therefore, the integrand $\mathbf{H}(L_1, L_2, L_3)$ is given by

$$\mathbf{H}(L_1, L_2, L_3) = \begin{bmatrix} L_1 \\ L_2 \\ L_3 \end{bmatrix} h[L_1 \quad L_2 \quad L_3] \, |\det \mathbf{J}|$$

The determinant of the Jacobian matrix $\mathbf{J}$ needs to be computed. From Eq. (9-50) we get

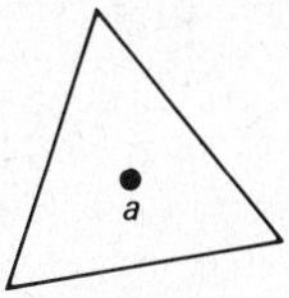

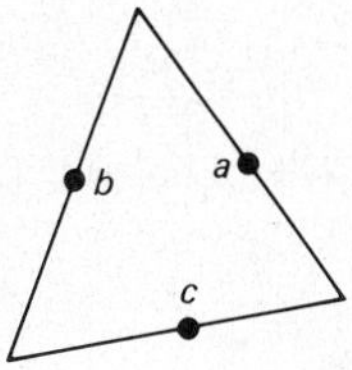

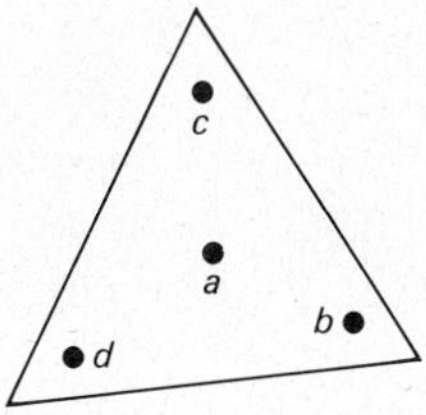

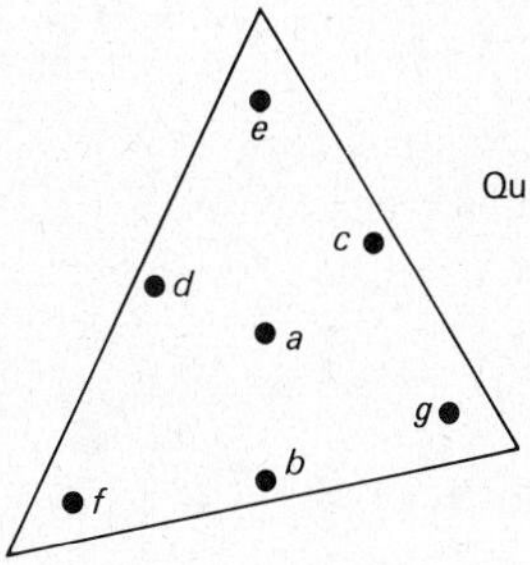

	Order	Points	Area coordinates			Weights
			L_1	L_2	L_3	w_i
	Linear	a	1/3	1/3	1/3	1
	Quadradic	a	1/2	1/2	0	
		b	1/2	0	1/2	1/3
		c	0	1/2	1/2	
	Cubic	a	1/3	1/3	1/3	$-27/48$
		b	0.6	0.2	0.2	
		c	0.2	0.6	0.2	25/48
		d	0.2	0.2	0.6	
	Quintic	a	1/3	1/3	1/3	0.225000000
		b	β	β	α	
		c	β	α	β	0.132394153
		d	α	β	β	
		e	δ	δ	γ	
		f	δ	γ	δ	0.125939180
		g	γ	δ	δ	

$$\text{where } \alpha = 0.059715872$$
$$\beta = 0.470142064$$
$$\gamma = 0.797426985$$
$$\delta = 0.101286507$$

Figure 9-21 Numerical integration formulas for triangles.

$$\frac{\partial N_1}{\partial L_1} = \left(\frac{\partial N_1}{\partial L_1}\right)_{L_2,L_3} - \left(\frac{\partial N_1}{\partial L_3}\right)_{L_1,L_2} = 1 - 0 = 1$$

$$\frac{\partial N_2}{\partial L_1} = \left(\frac{\partial N_2}{\partial L_1}\right)_{L_2,L_3} - \left(\frac{\partial N_2}{\partial L_3}\right)_{L_1,L_2} = 0 - 0 = 0$$

$$\frac{\partial N_3}{\partial L_1} = \left(\frac{\partial N_3}{\partial L_1}\right)_{L_2,L_3} - \left(\frac{\partial N_3}{\partial L_3}\right)_{L_1,L_2} = 0 - 1 = -1$$

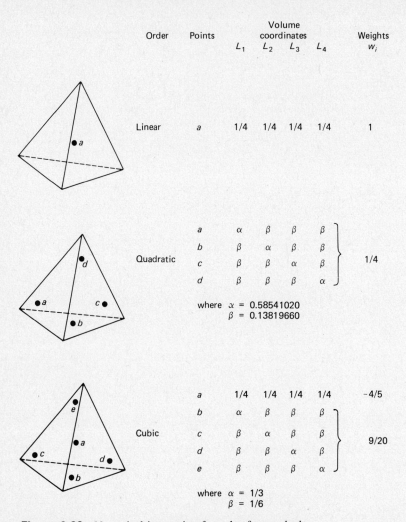

Order	Points	Volume coordinates				Weights
		L_1	L_2	L_3	L_4	w_i
Linear	a	1/4	1/4	1/4	1/4	1
Quadratic	a	α	β	β	β	1/4
	b	β	α	β	β	
	c	β	β	α	β	
	d	β	β	β	α	

where α = 0.58541020
β = 0.13819660

	a	1/4	1/4	1/4	1/4	$-4/5$
Cubic	b	α	β	β	β	9/20
	c	β	α	β	β	
	d	β	β	α	β	
	e	β	β	β	α	

where α = 1/3
β = 1/6

Figure 9-22 Numerical integration formulas for tetrahedra.

and from Eq. (9-51), we get

$$\frac{\partial N_1}{\partial L_2} = \left(\frac{\partial N_1}{\partial L_2}\right)_{L_1,L_3} - \left(\frac{\partial N_1}{\partial L_3}\right)_{L_1,L_2} = 0 - 0 = 0$$

$$\frac{\partial N_2}{\partial L_2} = \left(\frac{\partial N_2}{\partial L_2}\right)_{L_1,L_3} - \left(\frac{\partial N_2}{\partial L_3}\right)_{L_1,L_2} = 1 - 0 = 1$$

$$\frac{\partial N_3}{\partial L_2} = \left(\frac{\partial N_3}{\partial L_2}\right)_{L_1,L_3} - \left(\frac{\partial N_3}{\partial L_3}\right)_{L_1,L_2} = 0 - 1 = -1$$

Let us denote the nodal coordinates as (x_1, y_1), (x_2, y_2), and (x_3, y_3). Therefore, from the above results and Eq. (9-52), we have

$$\mathbf{J} = \begin{bmatrix} x_1 - x_3 & y_1 - y_3 \\ x_2 - x_3 & y_2 - y_3 \end{bmatrix}$$

and

$$\det \mathbf{J} = (x_1 - x_3)(y_2 - y_3) - (x_2 - x_3)(y_1 - y_3)$$
$$= x_1 y_2 - x_3 y_2 - x_1 y_3 - x_2 y_1 + x_3 y_1 + x_2 y_3 = 2A$$

where A is the area of the triangle. If the reader is not convinced the above expression is $2A$, the expression should be compared with Eq. (6-21e). Therefore, the integrand $\mathbf{H}$ is given by

$$\mathbf{H}(L_1, L_2, L_3) = 2hA \begin{bmatrix} L_1 \\ L_2 \\ L_3 \end{bmatrix} [L_1 \quad L_2 \quad L_3]$$

$$= 2hA \begin{bmatrix} L_1^2 & L_1 L_2 & L_1 L_3 \\ L_2 L_1 & L_2^2 & L_2 L_3 \\ L_3 L_1 & L_3 L_2 & L_3^2 \end{bmatrix}$$

For the linear-order integration, only one sampling point and one weight are given in Fig. 9-21. Note that the sampling point in this case corresponds to the centroid of the triangle and the weight is unity. Therefore, we have

$$\mathbf{K}_{cv}^e = \sum_{i=1}^{1} \tfrac{1}{2} w_i \mathbf{H}(L_{1_i}, L_{2_i}, L_{3_i})$$

$$= \tfrac{1}{2} 2hA \begin{bmatrix} (\tfrac{1}{3})^2 & (\tfrac{1}{3})(\tfrac{1}{3}) & (\tfrac{1}{3})(\tfrac{1}{3}) \\ (\tfrac{1}{3})(\tfrac{1}{3}) & (\tfrac{1}{3})^2 & (\tfrac{1}{3})(\tfrac{1}{3}) \\ (\tfrac{1}{3})(\tfrac{1}{3}) & (\tfrac{1}{3})(\tfrac{1}{3}) & (\tfrac{1}{3})^2 \end{bmatrix}$$

$$= \frac{hA}{9} \begin{bmatrix} 1 & 1 & 1 \\ 1 & 1 & 1 \\ 1 & 1 & 1 \end{bmatrix}$$

This result differs somewhat from that obtained in Example 8-13 from an exact integration. However, the error on the values of the nodal unknowns (i.e., the vector $\mathbf{a}$ in $\mathbf{Ka} = \mathbf{f}$) can be reduced by increasing the number of elements used. ∎

Example 9-6

Redo Example 9-5 by performing a quadratic-order quadrature. Compare the resulting stiffness matrix with that from Example 8-13, where an exact integration was performed.

Solution

The first part of Example 9-5 is applicable here, and so we may begin with

$$
\mathbf{H} = 2hA \begin{bmatrix} L_1^2 & L_1L_2 & L_1L_3 \\ L_2L_1 & L_2^2 & L_2L_3 \\ L_3L_1 & L_3L_2 & L_3^2 \end{bmatrix}
$$

Note that the quadratic-order integration requires three sampling points (and three weights). If the sampling points and weights in Fig. 9-21 are applied to the problem at hand, the result is

$$
\mathbf{K}_{cv}^e = \tfrac{1}{2}(\tfrac{1}{3})(2hA) \begin{bmatrix} (\tfrac{1}{2})^2 & (\tfrac{1}{2})(\tfrac{1}{2}) & (\tfrac{1}{2})(0) \\ (\tfrac{1}{2})(\tfrac{1}{2}) & (\tfrac{1}{2})^2 & (\tfrac{1}{2})(0) \\ (0)(\tfrac{1}{2}) & (0)(\tfrac{1}{2}) & (0)^2 \end{bmatrix}
$$

$$
+ \tfrac{1}{2}(\tfrac{1}{3})(2hA) \begin{bmatrix} (\tfrac{1}{2})^2 & (\tfrac{1}{2})(0) & (\tfrac{1}{2})(\tfrac{1}{2}) \\ (0)(\tfrac{1}{2}) & (0)^2 & (0)(\tfrac{1}{2}) \\ (\tfrac{1}{2})(\tfrac{1}{2}) & (\tfrac{1}{2})(0) & (\tfrac{1}{2})^2 \end{bmatrix}
$$

$$
+ \tfrac{1}{2}(\tfrac{1}{3})(2hA) \begin{bmatrix} (0)^2 & (0)(\tfrac{1}{2}) & (0)(\tfrac{1}{2}) \\ (\tfrac{1}{2})(0) & (\tfrac{1}{2})^2 & (\tfrac{1}{2})(\tfrac{1}{2}) \\ (\tfrac{1}{2})(0) & (\tfrac{1}{2})(\tfrac{1}{2}) & (\tfrac{1}{2})^2 \end{bmatrix}
$$

or

$$
\mathbf{K}_{cv}^e = \frac{hA}{12} \begin{bmatrix} 2 & 1 & 1 \\ 1 & 2 & 1 \\ 1 & 1 & 2 \end{bmatrix}
$$

This last result is the same as that in Example 8-13, where an exact integration was performed. Note that the integrand was of quadratic order and hence a quadratic order numerical integration gave the exact result. ■

It should be noted that the Jacobian matrix is a matrix composed solely of constants for the linear-order (or three-node) triangular element. For higher-order elements this will not necessarily be the case. In other words, the Jacobian matrix in general will be a function of the three area coordinates, in addition to the nodal coordinates. This presents no problem because the quadrature method presented here can easily accommodate this.

9-10 NUMERICAL INTEGRATION: REQUIRED ORDER

The question as to what order of integration we should use naturally arises. Zienkiewicz [7] argues that the linear-order integrations will always be convergent, but not always practical. The general guidelines may be summarized as follows. For linear-order triangles and quadrilaterals, single-point integration suffices. For quadratic-order quadrilateral and brick elements, 2×2 and $2 \times 2 \times 2$ Gauss point integrations are adequate. For quadratic-order triangular and tetrahedral elements, we should use three-point and four-point formulas from Figs. 9-21 and 9-22 (i.e., quadratic-order integrations). Cubic-order integrations should generally be performed on cubic-order elements. This implies 3×3 and $3 \times 3 \times 3$ Gauss point integrations from Table 9-5 for the cubic-order quadrilateral and brick elements, and four-point and five-point formulas from Figs. 9-21 and 9-22 for the cubic-order triangular and tetrahedral elements. Zienkiewicz's book [7] should be consulted for more information on this important aspect of finite element analysis.

REFERENCES

1. Zienkiewicz, O. C., *The Finite Element Method*, McGraw-Hill (UK), London, 1977, pp. 253–261.
2. Kaplan, W., *Advanced Calculus*, 2nd ed., Addison-Wesley, Reading, Mass., 1973, pp. 269–274.
3. Zienkiewicz, O. C., *The Finite Element Method*, McGraw-Hill, (UK) London, 1977, p. 186.
4. Zienkiewicz, O. C., *The Finite Element Method*, McGraw-Hill, (UK) London, 1977, p. 192.
5. Carnahan, B., H. A. Luther, and J. O. Wilkes, *Applied Numerical Methods*, Wiley, New York, 1969, pp. 100–116.
6. Hammer, P. C., O. P. Marlowe, and A. H. Stroud, "Numerical Integration over Simplexes and Cones," *Math. Tables Aids Comp.*, Vol. 10., pp. 130–137, 1956.
7. Zienkiewicz, O. C., *The Finite Element Method*, McGraw-Hill (UK), London, 1977, pp. 201–204.

PROBLEMS

9-1   Show that the shape function N_1 evaluates to unity at node 1 and to zero at all other nodes for the linear-, quadratic-, and cubic-order lineal element if the shape functions are given in terms of

    a. Length coordinates L_1 and L_2
    b. Serendipity coordinate r

9-2   Show that the shape function N_2 evaluates to unity at node 2 and to zero at all other nodes for the linear-, quadratic-, and cubic-order lineal element if the shape functions are given in terms of

    a. Length coordinates L_1 and L_2
    b. Serendipity coordinate r

9-3   Show that the shape function N_3 evaluates to unity at node 3 and to zero at all other nodes for the quadratic- and cubic-order lineal element if the shape functions are given in terms of

    a. Length coordinates L_1 and L_2
    b. Serendipity coordinate r

9-4 Show that the shape function N_4 evaluates to unity at node 4 and to zero at all other nodes for the cubic-order lineal element if the shape functions are given in terms of
   a. Length coordinates L_1 and L_2
   b. Serendipity coordinate r

9-5 Plot the shape functions for the lineal element on a typical element if the element is of
   a. Linear order
   b. Quadratic order
   c. Cubic order

9-6 For the quadratic-order triangular element, show that N_2 evaluates to unity at node 2. Also show that N_2 evaluates to zero if evaluated at each of the five other nodes.

9-7 For the quadratic-order triangular element, show that N_6 evaluates to unity at node 6. Also show that N_6 evaluates to zero if evaluated at each of the five other nodes.

9-8 For the cubic-order triangular element, show that N_3 evaluates to unity at node 3. Also show that N_3 evaluates to zero if evaluated at each of the nine other nodes.

9-9 For the cubic-order triangular element, show that N_7 evaluates to unity at node 7. Also show that N_7 evaluates to zero if evaluated at each of the nine other nodes.

9-10 For the cubic-order triangular element, show that N_{10} evaluates to unity at node 10. Also show that N_{10} evaluates to zero if evaluated at each of the nine other nodes.

9-11 For the quadratic-order rectangular element, show that N_3 evaluates to unity at node 3. Also show that N_3 evaluates to zero if evaluated at each of the seven other nodes.

9-12 For the quadratic-order rectangular element, show that N_6 evaluates to unity at node 6. Also show that N_6 evaluates to zero if evaluated at each of the seven other nodes.

9-13 For the cubic-order rectangular element, show that N_2 evaluates to unity at node 2. Also show that N_2 evaluates to zero if evaluated at each of the 11 other nodes.

9-14 For the cubic-order rectangular element, show that N_{10} evaluates to unity at node 10. Also show that N_{10} evaluates to zero if evaluated at each of the 11 other nodes.

9-15 For the quadratic-order tetrahedral element, show that N_2 evaluates to unity at node 2. Also show that N_2 evaluates to zero if evaluated at each of the nine other nodes.

9-16 For the quadratic-order tetrahedral element, show that N_8 evaluates to unity at node 8. Also show that N_8 evaluates to zero if evaluated at each of the nine other nodes.

9-17 For the cubic-order tetrahedral element, show that N_{11} evaluates to unity at node 11. Also show that N_{11} evaluates to zero if evaluated at each of the 19 other nodes.

9-18 For the cubic-order tetrahedral element, show that N_{19} evaluates to unity at node 19. Also show that N_{19} evaluates to zero if evaluated at each of the 19 other nodes.

9-19 For the quadratic-order brick element, show that N_5 evaluates to unity at node 5. Also show that N_5 evaluates to zero if evaluated at each of the 19 other nodes.

9-20 For the quadratic-order brick element, show that N_{14} evaluates to unity at node 14. Also show that N_{14} evaluates to zero if evaluated at each of the 19 other nodes.

9-21 For the cubic-order brick element, show that N_4 evaluates to unity at node 4. Also show that N_4 evaluates to zero if evaluated at each of the 31 other nodes.

9-22 For the cubic-order brick element, show that N_{29} evaluates to unity at node 29. Also show that N_{29} evaluates to zero if evaluated at each of the 31 other nodes.

9-23 Consider the linear-order, isoparametric quadrilateral element shown in Fig. *P9-23*. The nodal coordinates are shown on the figure. Map the point $r = +0.50$ and $s = -1.0$ on the parent element to the proper point on the distorted element (i.e., the quadrilateral). Show the distorted element on graph paper in order to see the mapping.

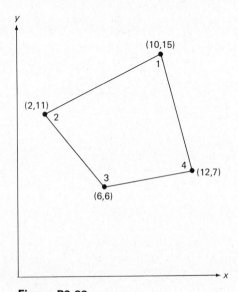

Figure P9-23

9-24 For the element in Problem 9-23, map the point $r = +0.5$ and $s = -0.75$ on the parent element to the proper point on the distorted element (i.e., the quadrilateral). Show the distorted element on graph paper in order to see the mapping.

9-25 Consider the quadratic-order, isoparametric quadrilateral element shown in Fig. *P9-25*. The nodal coordinates are shown on the figure. Map the point $r = -0.35$ and $s = +1.0$ on the parent element to the proper point on the distorted element. Show the distorted element on graph paper in order to see the mapping.

9-26 For the element in Problem 9-25, map the point $r = +0.75$ and $s = -0.25$ on the parent element to the proper point on the distorted element. Show the distorted element on graph paper in order to see the mapping.

9-27 Consider the cubic-order, isoparametric quadrilateral element shown in Fig. *P9-27*. The nodal coordinates are shown on the figure. Map the point $r = +0.50$ and $s = -1.0$ on the parent element to the proper point on the distorted element Show the distorted element on graph paper in order to see the mapping.

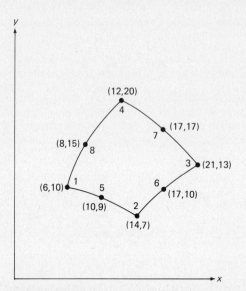

Figure P9-25

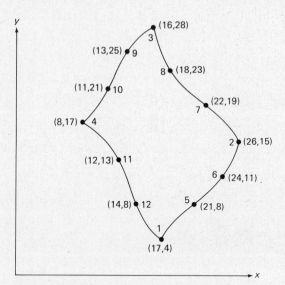

Figure P9-27

9-28 For the element in Problem 9-27, map the point $r = -0.45$ and $s = +0.75$ on the parent element to the proper point on the distorted element. Show the distorted element on graph paper in order to see the mapping.

9-29 Consider the quadratic-order, isoparametric triangular element shown in Fig. *P9-29*. The nodal coordinates are shown on the figure. Map the point $L_1 = 0.5$ and $L_2 = 0.5$ on the parent element to the proper point on the distorted element. Show the

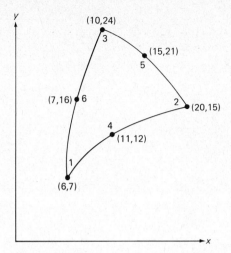

Figure P9-29

distorted element on graph paper in order to see the mapping. What is the value of L_3? Please explain.

9-30   For the element in Problem 9-29, map the point $L_1 = 0.25$ and $L_3 = 0.35$ on the parent element to the proper point on the distorted element. Show the distorted element on graph paper in order to see the mapping. What is the value of L_2? Please explain.

9-31   Consider the cubic-order, isoparametric triangular element shown in Fig. *P9-31*. The nodal coordinates are shown on the figure. Map the point $L_2 = 0.5$ and $L_3 = 0.5$ on the parent element to the proper point on the distorted element. Show the distorted element on graph paper in order to see the mapping. What is the value of L_1? Please explain.

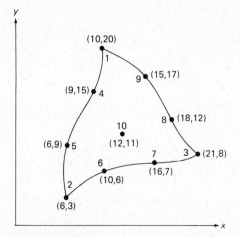

Figure P9-31

9-32   For the element in Problem 9-31, map the point $L_2 = \frac{1}{3}$ and $L_3 = \frac{1}{3}$ on the parent element to the proper point on the distorted element. Show the distorted element on graph paper in order to see the mapping. What is the value of L_1? Please explain.

9-33   For the isoparametric, quadrilateral element in Fig. P9-33, determine the Jacobian matrix.

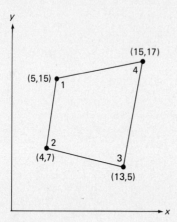

Figure P9-33

9-34   Determine the Jacobian matrix for the isoparametric, quadrilateral element shown in Fig. P9-34.

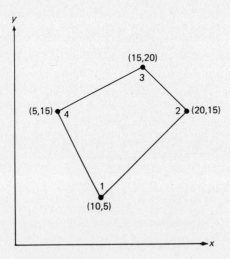

Figure P9-34

9-35   Recall from Sec. 8-3 that the element stiffness matrix from one-dimensional heat conduction is given by

$$\mathbf{K}_x^e = \int_{l^e} \frac{d\mathbf{N}^T}{dx} kA \frac{d\mathbf{N}}{dx} dx$$

where k is the thermal conductivity, A is the cross-sectional area (of the one-dimensional body in the direction x), $\mathbf{N}$ is the shape function matrix, and l^e denotes that the integration is to be performed over the length of the element. Let us consider a quadratic-order, lineal element *where the nodes are not necessarily equally spaced.* In order to evaluate $\mathbf{K}_x^e$ for this element, we must first perform an isoparametric mapping.

a. Show that the Jacobian matrix in this case is a scalar and is given by

$$J = \frac{dx}{dr} = \sum \frac{dN_j}{dr} x_j$$

How many terms are included in the summation? Please explain.

b. Show that $\mathbf{K}_x^e$ may be written

$$\mathbf{K}_x^e = \int_{-1}^{+1} \frac{d\mathbf{N}^T}{dr} \frac{kA}{\sum \dfrac{dN_j}{dr} x_j} \frac{d\mathbf{N}}{dr} dr$$

c. How are the results from parts (a) and (b) affected if the cubic-order, lineal element is used?

9-36 Consider a one-dimensional body in a two-dimensional space such as the one shown in Fig. P9-36(a). The body may be discretized as shown in Fig. P9-36(b). Note that lineal elments of any order may be used [although Fig. P9-36(b) shows only linear-order elements]. Note further that two global coordinates must be used to define the location of each node. Let us denote the global coordinates of the jth node as (x_j, y_j). Let us also use the coordinate l as shown in Fig. P9-36(c) to represent the direction that is always tangential to the one-dimensional body. In other words, the coordinate l is measured along the body. It then follows that

$$N_i = N_i(r)$$

$$r = r(l)$$

and

$$l = l(x, y)$$

where r is the serendipity coordinate, and N_i is the shape function for node i.

a. Show that the element stiffness matrix from heat conduction in the l direction (i.e., from one-dimensional heat conduction along the length of the body) is given by

$$\mathbf{K}_l^e = \int_{l^e} \frac{d\mathbf{N}^T}{dl} kA \frac{d\mathbf{N}}{dl} dl$$

where k is the thermal conductivity, A is the cross-sectional area (in the l direction), and l^e denotes that the integration is to be performed over the length of the element.

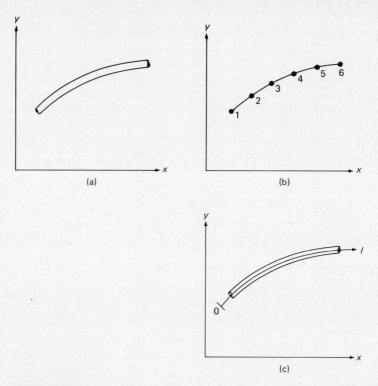

Figure P9-36

b. Show that the Jacobian matrix is a scalar in this case and is given by

$$J = \sqrt{\left(\sum \frac{dN_j}{dr} x_j\right)^2 + \left(\sum \frac{dN_j}{dr} y_j\right)^2}$$

c. Show that $\mathbf{K}_f^e$ becomes

$$\mathbf{K}_f^e = \int_{-1}^{+1} \frac{d\mathbf{N}^T}{dr} \frac{kA}{\sqrt{\left(\sum \frac{dN_j}{dr} x_j\right)^2 + \left(\sum \frac{dN_j}{dr} y_j\right)^2}} \frac{d\mathbf{N}}{dr} dr$$

d. Extend the results from parts (b) and (c) to the case of a one-dimensional body in a *three-dimensional space*.

9-37   For the isoparametric, triangular element in Fig. P9-37, determine the Jacobian matrix.

9-38   Determine the Jacobian matrix for the isoparametric, triangular element shown in Fig. P9-38.

9-39   Derive the expressions that correspond to Eqs. (9-39) and (9-40) for the isoparametric, triangular element. In other words, derive the expressions to be used in the trans-

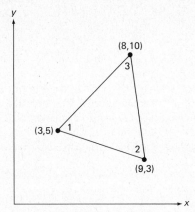

Figure P9-37

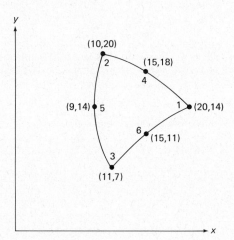

Figure P9-38

formation of dS (the elemental surface area around the boundary of the two-dimensional element) to the local coordinate system (in terms of the area coordinates). Note that the area coordinates degenerate to length coordinates in this case. Why?

9-40 Compute the following derivatives for the quadratic-order, isoparametric triangular element:

    a. $\dfrac{\partial N_2}{\partial L_1}$ **b.** $\dfrac{\partial N_5}{\partial L_2}$

9-41 Compute the following derivatives for the quadratic-order, isoparametric triangular element:

    a. $\dfrac{\partial N_3}{\partial L_1}$ **b.** $\dfrac{\partial N_6}{\partial L_2}$

9-42   Compute the following derivatives for the cubic-order, isoparametric triangular element:

   a. $\dfrac{\partial N_1}{\partial L_1}$ **b.** $\dfrac{\partial N_7}{\partial L_2}$

9-43   Derive the expressions that correspond to Eq. (9-65) for the isoparametric brick element if faces of constant r (i.e., $r = \pm 1$) happen to be on the global boundary. In other words, derive the expression to be used in the transformation of dS on faces of constant r (the elemental surface area on the boundary of the three-dimensional element) to the local coordinate system (in terms of the serendipity coordinates s and t).

9-44   Derive the expressions that correspond to Eq. (9-65) for the isoparametric brick element if faces of constant s (i.e., $s = \pm 1$) happen to be on the global boundary. In other words, derive the expression to be used in the transformation of dS on faces of constant s (the elemental surface area on the boundary of the three-dimensional element) to the local coordinate system (in terms of the serendipity coordinates r and t).

9-45   Derive the expressions that correspond to Eq. (9-65) for the isoparametric tetrahedral element. In other words, derive the expression to be used in the transformation of dS (the elemental surface area on the boundary of the three-dimensional element) to the local coordinate system (in terms of the volume coordinates). Note that the volume coordinates degenerate to area coordinates in this case. Why?

9-46   Compute the following derivatives for the quadratic-order, isoparametric tetrahedral element:

   a. $\dfrac{\partial N_2}{\partial L_1}$ **b.** $\dfrac{\partial N_8}{\partial L_2}$

9-47   Compute the following derivatives for the cubic-order, isoparametric tetrahedral element:

   a. $\dfrac{\partial N_4}{\partial L_1}$ **b.** $\dfrac{\partial N_{17}}{\partial L_2}$

9-48   Evaluate the integral given below by using Gauss-Legendre quadrature of such an order that the integral is evaluated exactly. Compare this result with that from the analytical evaluation.

$$I = \int_{-1}^{+1} (r + 3r^3)\, dr$$

9-49   Evaluate the integral given below by using Gauss-Legendre quadrature of such an order that the integral is evaluated exactly. Compare this result with that from the analytical evaluation.

$$I = \int_{-1}^{+1} (2r^3 + 3r^4)\, dr$$

9-50   Evaluate the integral given below by using two-point Guass-Legendre quadrature. Compare this result with that from the analytical evaluation.

$$I = \int_{-1}^{+1} \sin^2 \pi r \, dr$$

9-51   Repeat Problem 9-50 by performing a three-point Gauss-Legendre quadrature.

9-52   Evaluate the integral given below by using two-point Gauss-Legendre quadrature. Compare this result with that from the analytical evaluation.

$$I = \int_{-1}^{+1} \cos^2 \pi r \, dr$$

9-53   Repeat Problem 9-52 by performing a three-point Gauss-Legendre quadrature.

9-54   Evaluate the integral given below by using Gauss-Legendre quadrature of such an order that the integral is evaluated exactly. Compare this result with that from the analytical evaluation.

$$I = \int_{-1}^{+1} \int_{-1}^{+1} (2rs + 3r^2 s^3) \, dr \, ds$$

9-55   Evaluate the integral given below by using Gauss-Legendre quadrature of such an order that the integral is evaluated exactly. Compare this result with that from the analytical evaluation.

$$I = \int_{-1}^{+1} \int_{-1}^{+1} (2r^3 s^2 + 5rs^3) \, dr \, ds$$

9-56   Evaluate the element nodal force vector from a uniform body force **b** acting on a two-dimensional body if the four-node rectangular element with length $2a$ and height $2b$ is used [see Eq. (7-28)]. Perform a 2×2 Gauss-Legendre quadrature. Give the result in terms of the two components of the body force (per unit volume) b_x and b_y, the element thickness t, and the element dimensions a and b.

9-57   Evaluate the element nodal force vector from a uniform surface traction **s** acting on leg 2-3 of the four-node rectangular element with length $2a$ and height $2b$ [see Eq. (7-31). Use only two Gauss points. Give the result in terms of the two components of the surface traction s_x and s_y, the element thickness t, and the element dimensions a and b.

9-58   Evaluate the element stiffness matrix from conduction in the x direction [given by Eq. (8-106a)] for the four-node rectangular element with length $2a$ and height $2b$ by performing 2×2 Gaussian quadrature. Give the result in terms of the element dimensions a and b, the thermal conductivity k, and the element thickness t. *Hint:* See Problem 8-57 [parts (a) and (b)].

9-59   Evaluate the element stiffness matrix from conduction in the y direction [given by Eq. (8-106b)] for the four-node rectangular element with length $2a$ and height $2b$ by performing 2×2 Gaussian quadrature. Give the result in terms of the element dimensions a and b, the thermal conductivity k, and the element thickness t. *Hint:* See Problem 8-57 [parts (a) and (b)].

9-60   Evaluate the element stiffness matrix from lateral convection given by Eq. (8-106c) for the four-node rectangular element with length $2a$ and height $2b$ by performing 2×2 Gaussian quadrature. Give the result in terms of the element dimensions a and b, and the convective heat transfer coefficient h.

9-61   Evaluate the element stiffness matrix from boundary convection given by Eq. (8-106e) for the four-node rectangular element with length $2a$ and height $2b$ by assuming two Gauss points. Assume also that leg 1-2 is on the global boundary (and undergoes convection). Give the results in terms of the element dimensions a and b, the convective heat transfer coefficient h_B, and the element thickness t.

9-62   Evaluate the element stiffness matrix from boundary convection given by Eq. (8-106e) for the four-node rectangular element with length $2a$ and height $2b$ by assuming two Gauss points. Assume also that leg 4-1 is on the global boundary (and undergoes convection). Give the results in terms of the element dimensions a and b, the convective heat transfer coefficient h_B, and the element thickness t.

9-63   Evaluate the element nodal force vector from lateral convection given by Eq. (8-107a) for the four-node rectangular element with length $2a$ and height $2b$ by performing 2×2 Gaussian quadrature. Give the result in terms of the element dimensions a and b, the convective heat transfer coefficient h, and the ambient temperature T_a.

9-64   Evaluate the element nodal force vector from a lateral heat flux given by Eq. (8-107c) for the four-node rectangular element with length $2a$ and height $2b$ by performing 2×2 Gaussian quadrature. Give the result in terms of the element dimensions a and b, and the imposed heat flux q_s.

9-65   Evaluate the element nodal force vector from a distributed heat source given by Eq. (8-107d) for the four-node rectangular element with length $2a$ and height $2b$ by performing 2×2 Gaussian quadrature. Give the result in terms of the element dimensions a and b, the heat source strength Q, and the element thickness t.

9-66   Evaluate the element nodal force vector from boundary convection given by Eq. (8-107e) for the four-node rectangular element with length $2a$ and height $2b$ by assuming two Gauss points. Also assume that leg 2-3 is the global boundary (and undergoes convection). Give the result in terms of the element dimensions a and b, the convective heat transfer coefficient h_B, the ambient temperature T_{aB}, and the element thickness t.

9-67   Evaluate the element nodal force vector from a heat flux imposed on leg 3-4 of the four-node rectangular element with length $2a$ and height $2b$ by assuming two Gauss points. See Eq. (8-107g). Give the result in terms of the element dimensions a and b, the heat flux q_{sB}, and the element thickness t.

9-68   Evaluate the element nodal force vector from a uniform body force b acting on a two-dimensional body if the three-node triangular element is used [see Eq. (7-28)]. Perform a quadratic order quadrature. Give the result in terms of the two components of the body force (per unit volume) b_x and b_y, the element thickness t, and the element area A.

9-69 Evaluate the element nodal force vector from a uniform surface traction s acting on leg 2-3 of the three-node triangular element. Use only two Gauss points. Give the result in terms of the two components of the surface traction s_x and s_y, the element thickness t, and the length of leg 2-3. *Hint:* Note that N_1 is zero on leg 2-3. Replace N_2 and N_3 on leg 2-3 with the serendipity form of the shape functions for the linear element and then use Gauss-Legendre quadrature.

9-70 Evaluate the element stiffness matrix from lateral convection given by Eq. (8-106c) for the three-node triangular element by performing a quadratic order numerical integration (quadrature). Give the result in terms of the element area A and the convective heat transfer coefficient h.

9-71 Evaluate the element nodal force vector from lateral convection given by Eq. (8-107a) for the three-node triangular element by performing a quadratic-order numerical integration (quadrature). Give the result in terms of the element area A, the convective heat transfer coefficient h, and the ambient temperature T_a.

9-72 Evaluate the element nodal force vector from a lateral heat flux given by Eq. (8-107c) for the three-node triangular element by performing a quadratic-order numerical integration (quadrature). Give the result in terms of the element area A and the imposed heat flux q_s.

9-73 Evaluate the element nodal force vector from a distributed heat source given by Eq. (8-107d) for the three-node triangular element by performing a quadratic-order numerical integration (quadrature). Give the result in terms of the element area A, the heat source strength Q, and the element thickness t.

9-74 Reconsider the element in Example 9-4. Determine the element stiffness (or conductance) matrix from conduction in the y direction [see Eq. (8-106b)]. Assume 2×2 Gaussian quadrature. Take the thermal conductivity k to be 1 Btu/hr-in-°F and the element thickness to be 1 in.

9-75 Reconsider the element in Example 9-4. Determine the element nodal force vector from lateral convection [see Eq. (8-107a)]. Assume 2×2 Gaussian quadrature. Take the convective heat transfer coefficient h to be 4 Btu/hr-in^2-°F on each side of the plate and the ambient temperature to be 70°F (the same on both sides).

9-76 Reconsider the element in Example 9-4. Determine the element nodal force vector from a laterally imposed heat flux [see Eq. (8-107c)]. Assume 2×2 Gaussian quadrature. Take the heat flux q_s to be 100 Btu/hr-in^2.

9-77 Reconsider the element in Example 9-4. Determine the element nodal force vector from an imposed heat flux q_{sB} acting on leg 2-3 of the element [see Eq. (8-107g)]. Assume two Gauss points and a heat flux of 75 Btu/hr-in^2.

9-78 Consider the element shown in Fig. P9-78 where the nodal coordinates are shown in centimeters. Determine the element stiffness (or conductance) matrix from conduction in the x direction [see Eq. (8-106a)]. Assume 2×2 Gaussian quadrature. Take the thermal conductivity k to be 150 W/m-°C and the element thickness to be 2 cm.

9-79 Consider the element from Problem 9-78. Determine the element stiffness (or conductance) matrix from conduction in the y direction [see Eq. (8-106b)]. Assume

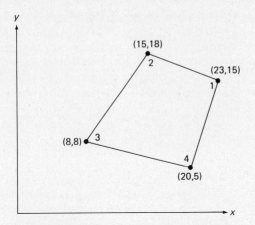

Figure P9-78

2 × 2 Gaussian quadrature. Take the thermal conductivity k to be 150 W/m-°C and the element thickness to be 2 cm.

9-80   Consider the element from Problem 9-78. Determine the element nodal force vector from lateral convection [see Eq. (8-107a)]. Assume 2 × 2 Gaussian quadrature. Take the convective heat transfer coefficient h to be 2000 W/m²-°C on each side of the plate and the ambient temperature to be 48°C (the same on both sides).

9-81   Consider the element from Problem 9-78. Determine the element nodal force vector from a laterally imposed heat flux [see Eq. (8-107c)]. Assume 2 × 2 Gaussian quadrature. Take the heat flux q_s to be 100 W/cm².

9-82   Consider the element from Problem 9-78. Determine the element nodal force vector from an imposed heat flux q_{sB} of 125 W/cm² acting on leg 3-4 of the element [see Eq. (8-107g)]. Assume two Gauss points.

9-83   Consider the problem posed in Problem 9-36. An alternate, but completely equivalent, formulation to such a problem is developed here. The main reason for the alternate formulation given below is that it may be more readily extended to the case of a two-dimensional body in a three-dimensional space (see Problem 9-84).

Consider a one-dimensional body in a two-dimensional space such as the one shown in Fig. P9-83(a). The body may be discretized as shown in Fig. P9-83(b). Note that lineal elements of any order may be used [although Fig. P9-83(b) shows only linear-order elements]. Note further that two global coordinates must be used to define the location of each node. Let us denote the global coordinates of the jth node as (x_j, y_j). Let us also use the coordinate x' as shown in Fig. P9-83(c) to represent the direction that is always tangential to the one-dimensional body. In other words, the coordinate x' is measured along the body. It then follows that

$$N_i = N_i(r)$$

$$r = r(x')$$

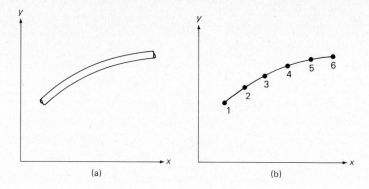

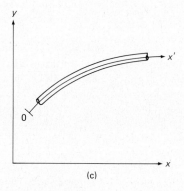

Figure P9-83

and

$$x' = x'(x,y)$$

where r is the serendipity coordinate, and N_i is the shape function for node i.

a. Show that the element stiffness matrix from heat conduction in the x' direction (i.e., from one-dimensional heat conduction along the length of the body) is given by

$$\mathbf{K}^e = \int_{l^e} \frac{d\mathbf{N}^T}{dx'} kA \frac{d\mathbf{N}}{dx'} \, dx'$$

where k is the thermal conductivity, A is the cross-sectional area (in the x' direction), and l^e denotes that the integration is to be performed over the length of the element.

b. Show that the Jacobian matrix is a scalar in this case and is now given by

$$J = \frac{\partial x'}{\partial r} = n_{11} \sum \frac{\partial N_j}{\partial r} x_j + n_{21} \sum \frac{\partial N_j}{\partial r} y_j$$

where n_{11} is the cosine of the angle between the x and x' axes, and n_{21} is the cosine of the angle between the y and the x' axes (i.e., n_{11} and n_{21} are the two direction cosines).

c. Show that $\mathbf{K}^e$ becomes

$$\mathbf{K}^e = \int_{-1}^{+1} \frac{d\mathbf{N}^T}{dr} \frac{kA}{n_{11}\sum \frac{\partial N_j}{\partial r} x_j + n_{21}\sum \frac{\partial N_j}{\partial r} y_j} \frac{d\mathbf{N}}{dr} \, dr$$

Note that the integral may be evaluated numerically with the help of Gauss-Legendre quadrature.

d. Extend the results from parts (b) and (c) to the case of a one-dimensional body in a *three*-dimensional space.

e. Show that formulation above is equivalent to that in Problem 9-36 by showing that the Jacobian J from Problem 9-36 is equivalent to the expression for J given above.

9-84   Let us extend the formulation presented in Problem 9-83 to the case of a two-dimensional body in a three-dimensional space as shown in Fig. P9-84(a). The body may be discretized into quadrilateral elements as shown in Fig. P9-84(b). Note that quadrilateral elements of any order may be used [although Fig. P9-84(b) shows only linear-order elements]. Note further that three global coordinates must be used to define the location of each node. Let us denote the global coordinates of the jth node as (x_j, y_j, z_j). Let us also use the coordinates x' and y' as shown in Fig. P9-84 to represent the local coordinate system that is in the plane of the two-dimensional body. Note that x' and y' are always perpendicular to each other (in other words, the local coordinate system and the global coordinate system both form orthogonal coordinate systems). It then follows that

$$N_i = N_i(r,s)$$

$$r = r(x',y')$$

and

$$s = s(x',y')$$

where

$$x' = x'(x,y,z)$$

and

$$y' = y'(x,y,z)$$

Note that r and s are the serendipity coordinates, and N_i is the shape function for node i. It follows that the element stiffness matrices for conduction in the x' and y' directions are given by

$$\mathbf{K}_{xx}^e = \int_{A^e} \frac{\partial \mathbf{N}^T}{\partial x'} kt \frac{\partial \mathbf{N}}{\partial x'} \, dx' \, dy'$$

and

$$\mathbf{K}_{yy}^e = \int_{A^e} \frac{\partial \mathbf{N}^T}{\partial y'} kt \frac{\partial \mathbf{N}}{\partial y'} \, dx' \, dy'$$

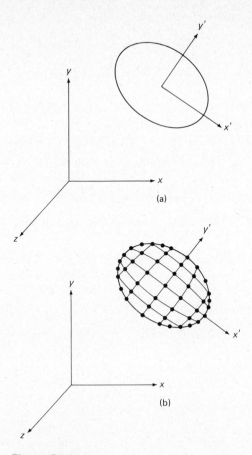

Figure P9-84

where k is the thermal conductivity, t is the thickness of the element, and A^e denotes that the integration is to be performed over the area of the element.

a. Show that the Jacobian matrix is now given by

$$
\mathbf{J} = \begin{bmatrix} \dfrac{\partial x'}{\partial r} & \dfrac{\partial y'}{\partial r} \\[2ex] \dfrac{\partial x'}{\partial s} & \dfrac{\partial y'}{\partial s} \end{bmatrix}
$$

where the entries in the Jacobian matrix are given by

$$
J_{11} = \frac{\partial x'}{\partial r} = n_{11} \sum \frac{\partial N_j}{\partial r} x_j + n_{21} \sum \frac{\partial N_j}{\partial r} y_j + n_{31} \sum \frac{\partial N_j}{\partial r} z_j
$$

$$
J_{12} = \frac{\partial y'}{\partial r} = n_{12} \sum \frac{\partial N_j}{\partial r} x_j + n_{22} \sum \frac{\partial N_j}{\partial r} y_j + n_{32} \sum \frac{\partial N_j}{\partial r} z_j
$$

$$J_{21} = \frac{\partial x'}{\partial s} = n_{11} \sum \frac{\partial N_j}{\partial s} x_j + n_{21} \sum \frac{\partial N_j}{\partial s} y_j + n_{31} \sum \frac{\partial N_j}{\partial s} z_j$$

$$J_{22} = \frac{\partial y'}{\partial s} = n_{12} \sum \frac{\partial N_j}{\partial s} x_j + n_{22} \sum \frac{\partial N_j}{\partial s} y_j + n_{32} \sum \frac{\partial N_j}{\partial s} z_j$$

and where n_{11} is the cosine of the angle between the x and x' axes, and n_{21} is the cosine of the angle between the y and the x' axes, etc. (i.e., n_{11}, n_{21}, etc. are the direction cosines).

b. By denoting the entries in the inverted Jacobian matrix as A_{11}, A_{12}, A_{21}, and A_{22}, determine the explicit form of the element stiffness (or conductance) matrices for conduction in the x and y directions in terms of the local, serendipity coordinates r and s [and the nodal coordinates (x_j, y_j, z_j)]. Note the integration is performed over the parent element and, thus, the integrals may be evaluated numerically with Gauss-Legendre quadrature. Do not attempt to perform the quadrature, however.

9-85 Recall that the condition for a one-to-one mapping is that the sign of the Jacobian remain the same for all points in the domain mapped. For the quadratic-order, lineal element show that this implies that the interior node must be in the middle one-half zone. *Hint:* Problem 9-35 gives the expression for the Jacobian.

10

Transient and Dynamic Analyses

10-1 INTRODUCTION

Up to now we have only considered steady-state nonstructural applications and static stress analysis applications. In this chapter the finite element method is extended to transient and dynamic analyses. The reader may recall from Chapter 4 that transient (nonstructural) analyses are frequently referred to as unsteady, time-dependent, or propagation problems. In structural and stress analyses, time-independent problems are referred to as static or equilibrium problems, whereas time-dependent problems are almost exclusively referred to as dynamic.

The chapter begins with a new notion referred to as partial discretization. In effect, this will allow us to discretize the time domain. The reader will recall that up to now we have only had to perform discretization in space.

The introduction to partial discretization is followed by a presentation of the governing equations for dynamic structural analysis and transient thermal analysis in Secs. 10-3 and 10-4, respectively. No longer will we get the familiar $\mathbf{Ka} = \mathbf{f}$, but rather several additional terms will arise. The concept of lumped and consistent mass and capacitance matrices is then introduced in Sec. 10-5 with the implications of each briefly discussed.

Solution methods are presented for transient thermal analysis problems in Secs. 10-7 and 10-8, following a brief introduction to this subject in Sec. 10-6. The development in Sec. 10-7 is based on the finite difference method, whereas that in Sec. 10-8 is based on the finite element method. It will be seen that both methods give the same two-point recurrence formula which will allow us to compute the nodal temperatures as a function of time. Indeed, it will be seen that the results from the finite difference method are actually special cases of the more general finite element method. It is emphasized that in all cases the goal is to compute the temperatures at the nodal points as a function of time.

Finally, in Secs. 10-9 and 10-10, solution methods to dynamic structural analysis problems are presented. In Sec. 10-9 a three-point recurrence scheme is derived by using suitable finite elements in time. This will allow us to compute the nodal displacements within a structure as a function of time. Both the inertia and damping within the structure will be explicitly included in the analysis. In Sec. 10-10 an introduction to modal analysis is given which will allow us to compute the undamped natural frequencies and the so-called mode shapes of a structure.

10-2 PARTIAL DISCRETIZATION

In the previous chapters, the unknown parameter function ϕ on an element basis was assumed to be of the form

$$\phi = \mathbf{N}\mathbf{a}^e \qquad \text{(10-1)}$$

where ϕ is a vector or a scalar and may represent the displacement, temperature, velocity, and/or pressure fields. Recall that the shape function matrix $\mathbf{N}$ is a function of the particular coordinates being used (i.e., global or local) and the vector $\mathbf{a}^e$ is composed strictly of constants. These constants are the values of the parameter functions at the locations of the nodes. For example, in two-dimensional stress analysis, we have assumed the displacement $\mathbf{u}$ on an element basis to be given by

$$\mathbf{u} = \mathbf{N}(x,y)\mathbf{a}^e \qquad \text{(10-2)}$$

where $\mathbf{u}$ itself is given by Eq. (5-58). Similarly in two-dimensional thermal analysis, we have assumed the temperature T on an element basis to be given by

$$T = \mathbf{N}(x,y)\mathbf{a}^e \qquad \text{(10-3)}$$

It is emphasized that $\mathbf{N}$ in both cases is a function of the global (or local) coordinates whereas $\mathbf{a}^e$ is composed of constants (albeit unknown constants).

Let us now consider the case of a dynamic structural problem and unsteady heat transfer problem. Unsteady problems may also be referred to as transient or time-dependent problems. Since the time variable τ enters into such problems, Eqs. (10-1) to (10-3) could be modified such that the shape function matrix could include the time variable in addition to the spatial coordinates. For example, Eq. (10-1) could be written as

$$\phi = \mathbf{N}(x,y,\tau)\mathbf{a}^e \qquad \text{(10-4)}$$

where a new shape function matrix $\mathbf{N}$ could be derived. The principal disadvantage of this approach is that it increases the number of dimensions by one. For example, a one-dimensional problem becomes two-dimensional, a two-dimensional problem becomes three-dimensional, and so forth.

This additional complexity may be avoided if we allow the vector of nodal unknowns $\mathbf{a}^e$ itself to be a function of time. This is referred to as *partial discretization* [1]. Equation (10-1) could then be written as

$$\phi = \mathbf{N}\mathbf{a}^e(\tau) \tag{10-5}$$

where the explicit dependence of $\mathbf{a}^e$ on τ is shown. The implication is that no longer will we get the familiar equation

$$\mathbf{K}^e\mathbf{a}^e = \mathbf{f}^e \tag{10-6}$$

Instead, two additional terms will arise in dynamic structural analysis problems, and one additional term will arise in transient thermal analysis problems. The explicit forms for these equations are given in Secs. 10-3 and 10-4 for dynamic structural and transient heat transfer problems, respectively.

10-3   DYNAMIC STRUCTURAL ANALYSIS

Recall from Chapter 5 that the principle of minimum potential energy or principle of virtual work may be used to derive Eq. (5-46). Both approaches gave virtually the same result, which is repeated here for easy reference:

$$\int_{V^e} (\delta\boldsymbol{\varepsilon})^T \boldsymbol{\sigma} \, dV = \int_{V^e} (\delta\mathbf{u})^T \mathbf{b} \, dV + \int_{S^e} (\delta\mathbf{u})^T \mathbf{s} \, dS + \sum (\delta\mathbf{u})^T \mathbf{f}_p \tag{5-46}$$

Note that the right-hand side of Eq. (5-46) is composed of three terms, each of which represents an external force acting on the structure, and hence the element. By applying D'Alembert's principle [2], we may introduce additional terms to the right-hand side from the inertia and the damping as shown below.

The inertia results from the mass of the structure, whereas the damping results from energy dissipation or friction within the structure. The *inertial force per unit volume* is given by the product of the mass density ρ and the acceleration. The *damping force per unit volume* is generally assumed to be the product of a viscous matrix $\boldsymbol{\mu}$ and the velocity vector. Let us denote the first derivative of the displacement field vector $\mathbf{u}$ (i.e., the parameter function) as $\dot{\mathbf{u}}$, and the second derivative as $\ddot{\mathbf{u}}$. Note that $\dot{\mathbf{u}}$ and $\ddot{\mathbf{u}}$ represent the velocity and acceleration fields, respectively. It follows from D'Alembert's principle that the elemental inertial and damping forces $d\mathbf{f}_I$ and $d\mathbf{f}_D$ are given by

$$d\mathbf{f}_I = -\rho\ddot{\mathbf{u}} \, dV \tag{10-7}$$

and

$$d\mathbf{f}_D = -\boldsymbol{\mu}\dot{\mathbf{u}} \, dV \tag{10-8}$$

Recall that the minus signs are necessary if these additional terms are to be regarded as stemming from external forces. It then follows from Chapter 5 that Eq. (5-46) may be written as

$$\int_{V^e} (\delta\boldsymbol{\varepsilon})^T \boldsymbol{\sigma} \; dV = \int_{V^e} (\delta\mathbf{u})^T \mathbf{b} \; dV + \int_{S^e} (\delta\mathbf{u})^T \mathbf{s} \; dS + \sum (\delta\mathbf{u})^T \mathbf{f}_p$$

$$- \int_{V^e} (\delta\mathbf{u})^T \rho\ddot{\mathbf{u}} \; dV - \int_{V^e} (\delta\mathbf{u})^T \boldsymbol{\mu}\dot{\mathbf{u}} \; dV \quad \textbf{(10-9)}$$

Note that

$$\dot{\mathbf{u}} = \frac{d\mathbf{u}}{d\tau} = \frac{d(\mathbf{N}\mathbf{a}^e)}{d\tau} = \mathbf{N}\frac{d\mathbf{a}^e}{d\tau} = \mathbf{N}\dot{\mathbf{a}}^e(\tau) \quad \textbf{(10-10)}$$

and

$$\mathbf{u} = \mathbf{N}\mathbf{a}^e(\tau) \quad \textbf{(10-11a)}$$

where, for example, $\mathbf{a}^e(\tau)$ is given by

$$\mathbf{a}^e = [u_1(\tau) \quad v_1(\tau) \; \vdots \; u_2(\tau) \quad v_2(\tau) \; \vdots \; \cdots \; \vdots \; u_n(\tau) \quad v_n(\tau)]^T \quad \textbf{(10-11b)}$$

for two-dimensional problems. In Eq. (10-11b), $u_i(\tau)$ and $v_i(\tau)$ are the x and y components of displacement of node i at time τ, and n is the number of nodes used to define element e.

Following the procedure used in Sec. 5-7 for static stress analysis formulations, we may write Eq. (10-9) in the form

$$\mathbf{M}^e \ddot{\mathbf{a}}^e + \mathbf{D}^e \dot{\mathbf{a}}^e + \mathbf{K}^e \mathbf{a}^e = \mathbf{f}^e \quad \textbf{(10-12)}$$

The element stiffness matrix $\mathbf{K}^e$ and element nodal force vector $\mathbf{f}^e$ remain the same as in Sec. 5-7 [see Eqs. (5-86) to (5-92)]. However, two additional terms and matrices arise. The matrix $\mathbf{M}^e$ may be referred to as the *element mass matrix*, whereas $\mathbf{D}^e$ may be referred to as the *element damping matrix*, both defined by

$$\mathbf{M}^e = \int_{V^e} \mathbf{N}^T \rho\mathbf{N} \; dV \quad \textbf{(10-13)}$$

and

$$\mathbf{D}^e = \int_{V^e} \mathbf{N}^T \boldsymbol{\mu}\mathbf{N} \; dV \quad \textbf{(10-14)}$$

The element damping matrix $\mathbf{D}^e$ should not be confused with the material property matrix $\mathbf{D}$ from previous chapters because the context makes the meaning clear (also the material property matrix $\mathbf{D}$ is never written with a superscript). Note that for each element, the nodal displacements, velocities, and accelerations are given by $\mathbf{a}^e$, $\dot{\mathbf{a}}^e$, and $\ddot{\mathbf{a}}^e$, respectively. Equation (10-12) reduces to the more familiar $\mathbf{K}^e \mathbf{a}^e = \mathbf{f}^e$ when the velocities and accelerations are zero or negligible.

The assemblage of the element mass matrices $\mathbf{M}^e$ to form the assemblage mass matrix $\mathbf{M}^a$ follows the procedure used to obtain the assemblage stiffness matrix $\mathbf{K}^a$ from the element stiffness matrices $\mathbf{K}^e$. An explicit form for the assemblage damping matrix $\mathbf{D}^a$ is not usually known because the viscous matrix $\boldsymbol{\mu}$ is not usually known. Therefore, the assemblage damping matrix $\mathbf{D}$ is generally assumed to be given by a linear combination of the assemblage mass and stiffness matrices, or

$$\mathbf{D} = \alpha\mathbf{M} + \beta\mathbf{K} \tag{10-15}$$

where α and β are experimentally determined constants [3]. The damping implied by Eq. (10-15) is referred to as *Rayleigh damping*. In any event, the assemblage system equation may be written in the form

$$\mathbf{M\ddot{a} + D\dot{a} + Ka = f} \tag{10-16}$$

Note that the superscript (ᵃ) that is normally used to denote *assemblage* is no longer written. The solution to Eq. (10-16) requires the specification of initial conditions on the nodal displacements and velocities. In addition, the boundary conditions on the nodal displacements must be imposed. These matters are dealt with in Sec. 10-9 where solution methods for Eq. (10-16) are presented.

Example 10-1

Determine the element mass matrix for one-dimensional, dynamic structural analysis problems. Assume the two-node, lineal element.

Solution

If length coordinates are used, we may write Eq. (10-13) as

$$\mathbf{M}^e = \int_{le} \begin{bmatrix} L_1 \\ L_2 \end{bmatrix} \rho [L_1 \quad L_2] A \, dx = \int_{le} \begin{bmatrix} L_1^2 & L_1L_2 \\ L_2L_1 & L_2^2 \end{bmatrix} \rho A \, dx$$

With the help of Eq. (6-48) each entry in the matrix may be integrated to give

$$\mathbf{M}^e = \frac{\rho AL}{6} \begin{bmatrix} 2 & 1 \\ 1 & 2 \end{bmatrix} \tag{10-17}$$

where L, the length of the element, should not be confused with the length coordinates (which are always written with subscripts). Note that the total mass of the element is given by ρAL. ∎

10-4 TRANSIENT THERMAL ANALYSIS

In Chapter 8 steady-state heat transfer problems were formulated. Recall that the formulations were quite general, but no mechanism for energy storage was included. It can be shown from the first law of thermodynamics that an additional term arises in each of the governing equations presented in Chapter 8. For example, for one-dimensional problems, the governing equation given by Eq. (8-3) becomes

$$\rho c A \frac{\partial T}{\partial \tau} = \frac{\partial}{\partial x} \left(kA \frac{\partial T}{\partial x} \right) - hP(T - T_a) - \varepsilon\sigma P(T^4 - T_r^4) + QA \tag{10-18}$$

where ρ and c are the mass density and specific heat, respectively. Let us limit the present discussion to heat conduction in solids (and quiescent liquids) so that the specific heats at constant volume and constant pressure are virtually the same.

Therefore, there is no need to distinguish between these two specific heats and, hence, no subscript on c is needed. In Eq. (10-18), the variable A represents the cross-sectional area of the object being analyzed and may be a function of x.

Similarly, for two- and three-dimensional heat conduction problems, the governing equations are given by

$$\rho c t \frac{\partial T}{\partial \tau} = \frac{\partial}{\partial x}\left(kt\frac{\partial T}{\partial x}\right) + \frac{\partial}{\partial y}\left(kt\frac{\partial T}{\partial y}\right) - h(T - T_a)$$
$$- \epsilon\sigma(T^4 - T_r^4) + q_s + Qt \qquad \text{(10-19)}$$

and

$$\rho c \frac{\partial T}{\partial \tau} = \frac{\partial}{\partial x}\left(k\frac{\partial T}{\partial x}\right) + \frac{\partial}{\partial y}\left(k\frac{\partial T}{\partial y}\right) + \frac{\partial}{\partial z}\left(k\frac{\partial T}{\partial z}\right) + Q \qquad \text{(10-20)}$$

In Eq. (10-19), the variable t is the thickness of the two-dimensional region (and hence the element thickness). Note that the variable τ (not t) is used to represent *time*. The governing equation for axisymmetric heat conduction is similarly modified, and the result is

$$\rho c \frac{\partial T}{\partial \tau} = \frac{1}{r}\frac{\partial}{\partial r}\left(rk\frac{\partial T}{\partial r}\right) + \frac{\partial}{\partial z}\left(k\frac{\partial T}{\partial z}\right) + Q \qquad \text{(10-21)}$$

Since a transient formulation is desired, we may employ partial discretization and write the temperature T within a typical element e as

$$T = \mathbf{N}\mathbf{a}^e(\tau)$$

The shape function matrix $\mathbf{N}$ is unchanged from Chapter 8 and $\mathbf{a}^e(\tau)$ is given by

$$\mathbf{a}^e(\tau) = [T_1(\tau) \quad T_2(\tau) \quad \cdots \quad T_n(\tau)]^T \qquad \text{(10-22)}$$

where $T_i(\tau)$ is the temperature of node i at time τ, and n is the number of nodes used to define element e.

The Galerkin weighted-residual method may be used to derive the finite element characteristics providing the energy storage term is represented in the governing equation and hence the residual. The energy storage term is on the left-hand side of Eqs. (10-18) to (10-21). For example, in two-dimensional problems, we begin by writing the weighted residual equation as

$$\int_{A^e} \mathbf{N}^T \left[\rho c t \frac{\partial T}{\partial \tau} - \frac{\partial}{\partial x}\left(kt\frac{\partial T}{\partial x}\right) - \frac{\partial}{\partial y}\left(kt\frac{\partial T}{\partial y}\right) + \cdots \right] dx\, dy = 0 \qquad \text{(10-23)}$$

The Green-Gauss theorem is applied in the usual manner on the terms involving second-order derivatives. Except for the energy storage term, the formulation is identical to that presented in Sec. 8-8. Because of the presence of the energy storage terms, we no longer get $\mathbf{K}^e\mathbf{a}^e = \mathbf{f}^e$ but rather

$$\mathbf{C}^e\dot{\mathbf{a}}^e + \mathbf{K}^e\mathbf{a}^e = \mathbf{f}^e \qquad \text{(10-24)}$$

The matrix $\mathbf{C}^e$ is referred to as the *element capacitance matrix* and is defined by

$$\mathbf{C}^e = \int_{A^e} \mathbf{N}^T \rho c t \mathbf{N} \, dx \, dy \qquad \textbf{(10-25)}$$

for two-dimensional problems. In a similar fashion it can be shown that the element capacitance matrix in one-dimensional, transient heat conduction is given by

$$\mathbf{C}^e = \int_{l^e} \mathbf{N}^T \rho c A \mathbf{N} \, dx \qquad \textbf{(10-26)}$$

whereas in three-dimensional problems it is given by

$$\mathbf{C}^e = \int_{V^e} \mathbf{N}^T \rho c \mathbf{N} \, dx \, dy \, dz \qquad \textbf{(10-27)}$$

Finally, for axisymmetric problems, we have

$$\mathbf{C}^e = \int_{A^e} 2\pi \mathbf{N}^T \rho c r \mathbf{N} \, dr \, dz \qquad \textbf{(10-28)}$$

Example 10-2

Show that the element capacitance matrix for one-dimensional heat conduction problems is given by Eq. (10-26).

Solution

From the weighted residual equation

$$\int_{A^e} \mathbf{N}^T \left[\rho c A \frac{\partial T}{\partial \tau} - \frac{\partial}{\partial x} \left(kA \frac{\partial T}{\partial x} \right) - \cdots \right] dx = \mathbf{0}$$

it follows that an additional term, namely, $\mathbf{C}^e \dot{\mathbf{a}}^e$, arises because we have

$$\frac{\partial T}{\partial \tau} = \frac{d(\mathbf{N} a^e)}{d\tau} = \mathbf{N} \frac{d a^e}{d\tau} = \mathbf{N} \dot{\mathbf{a}}^e$$

and

$$\int_{l^e} \mathbf{N}^T \rho c A \frac{\partial T}{\partial \tau} \, dx = \int_{l^e} \mathbf{N}^T \rho c A \, \mathbf{N} \dot{\mathbf{a}}^e \, dx = \left[\int_{l^e} \mathbf{N}^T \rho c A \mathbf{N} \, dx \right] \dot{\mathbf{a}}^e = \mathbf{C}^e \dot{\mathbf{a}}^e$$

The expression in the brackets is recognized as the element capacitance matrix defined by Eq. (10-26). ∎

Example 10-3

Evaluate the element capacitance matrix for the linear-order triangular element used in the discretization of a two-dimensional region.

Solution

The element capacitance matrix in this case is given by Eq. (10-25). If the shape functions are written in terms of the area coordinates, we get

$$
\mathbf{C}^e = \int_{A^e} \begin{bmatrix} L_1 \\ L_2 \\ L_3 \end{bmatrix} \rho c t \; [L_1 \quad L_2 \quad L_3] \; dx \, dy
$$

$$
= \int_{A^e} \begin{bmatrix} L_1^2 & L_1 L_2 & L_1 L_3 \\ L_2 L_1 & L_2^2 & L_2 L_3 \\ L_3 L_1 & L_3 L_2 & L_3^2 \end{bmatrix} \rho c t \; dx \, dy
$$

With the help of Eq. (6-49), this matrix evaluates to

$$
\mathbf{C}^e = \frac{\rho c t A}{12} \begin{bmatrix} 2 & 1 & 1 \\ 1 & 2 & 1 \\ 1 & 1 & 2 \end{bmatrix} \tag{10-29}
$$

where A is the area of the triangle. Note that the total capacitance of the element (i.e., $\rho c t A$) is distributed as given by Eq. (10-29). ■

The assemblage of the *element* capacitance matrices into the *assemblage* capacitance matrix is done in precisely the same manner as the assemblage of the element stiffness (or conductance) matrices to form the assemblage stiffness (or conductance) matrix. The assemblage system equation then takes the form

$$
\mathbf{C}\dot{\mathbf{a}} + \mathbf{K}\mathbf{a} = \mathbf{f} \tag{10-30}
$$

Before Eq. (10-30) can be solved for the nodal temperatures, the initial and boundary conditions must be imposed as shown in Sec. 10-7. Note that Eq. (10-30) is not written with the superscript (^{a}) which is generally used to indicate the assemblage matrices before application of the geometric boundary conditions. This superscript is dropped because it will unnecessarily clutter the notation in Sections 10-7 and 10-8.

10-5 LUMPED VERSUS CONSISTENT MATRICES

Lumped mass and capacitance matrices are always diagonal matrices, whereas consistent mass or capacitance matrices are not necessarily diagonal. A *diagonal matrix* is defined as a square matrix whose entries are zero everywhere but on the principal diagonal. Recall from Chapter 2 that the principal diagonal always runs from the upper left corner to the lower right corner of a matrix. The identity matrix is an example of a diagonal matrix.

Consistent Matrices

Recall that the element mass matrix in all dynamic structural analysis problems may be determined from Eq. (10-13), where dV must be taken to be $A \, dx$, $t \, dx \, dy$, or $dx\,dy\,dz$ for one-, two-, and three-dimensional problems, respectively. For dynamic axisymmetric stress analysis, we simply take dV to be $2\pi r\,dr\,dz$. When the shape functions from the previous chapters are used, the resulting element mass matrices are referred to as consistent element mass matrices.

In a similar fashion, the element capacitance matrices for one-, two-, and three-dimensional transient thermal analysis problems are given by Eqs. (10-25) to (10-27). Moreover, the element capacitance matrix for axisymmetric, transient thermal analysis is given by Eq. (10-28). When the shape functions from previous chapters are used, the resulting element capacitance matrices are referred to as consistent element capacitance matrices.

Consistent element mass and capacitance matrices are discussed in more detail below.

Consistent mass matrices

Recall from Example 10-1 that the element mass matrix for one-dimensional dynamic, structural analysis problems is given by

$$\mathbf{M}^e = \frac{\rho AL}{6} \begin{bmatrix} 2 & 1 \\ 1 & 2 \end{bmatrix} \tag{10-17}$$

providing the linear-order, lineal element is used. Note that this matrix is not diagonal. Note further that this result was derived in a consistent manner by using the appropriate shape functions and Eq. (10-13) directly. It is for this reason that this result is referred to as a consistent mass matrix. It can be shown that for the linear-order, triangular element the consistent mass matrix is given by

$$\mathbf{M}^e = \frac{\rho tA}{12} \left[\begin{array}{cc:cc:cc} 2 & 0 & 1 & 0 & 1 & 0 \\ 0 & 2 & 0 & 1 & 0 & 1 \\ \hdashline 1 & 0 & 2 & 0 & 1 & 0 \\ 0 & 1 & 0 & 2 & 0 & 1 \\ \hdashline 1 & 0 & 1 & 0 & 2 & 0 \\ 0 & 1 & 0 & 1 & 0 & 2 \end{array}\right] \tag{10-31}$$

Again it is seen that the consistent mass matrix is not a diagonal matrix. This implies that the assemblage mass matrix also will not be a diagonal matrix. Further implications of this are discussed later in this section. These results may be generalized to three-dimensional and axisymmetric problems as follows: If the element mass matrix is evaluated from Eq. (10-13), a consistent mass matrix is obtained (assuming that the shape functions from Chapters 6 and 9 are used).

Consistent capacitance matrices

Let us now consider the result from Example 10-3 where the element capacitance matrix for two-dimensional, transient heat conduction problems is given by Eq. (10-29) for the linear-order, triangular element. Note that this matrix is not diagonal and that this result was derived in a consistent manner by using the appropriate shape functions and Eq. (10-25) directly. Therefore, the element capacitance matrix given by Eq. (10-29) is referred to as a consistent capacitance matrix. This implies that the assemblage capacitance matrix also will not be diagonal. The implications of this in the solution step are discussed below. These results may be generalized to other heat conduction problems as follows: If the element capacitance matrix is evaluated from Eqs. (10-25) to (10-28), a consistent capacitance matrix is obtained (assuming that the shape functions from Chapters 6 and 9 are used).

Lumped Mass and Capacitance Matrices

In the solution for the nodal displacements and temperatures, the initial part of the solution may tend to be oscillatory about the true solution if the consistent mass or capacitance matrix is used. These oscillations generally do not occur if the so-called lumped mass or capacitance matrix is used. Recall that a lumped matrix is a diagonal matrix. In order to illustrate these trends, consider the temperature versus time curve for a typical node in a thermal analysis shown in Fig. 10-1. Note that the solution with the consistent capacitance matrix oscillates about the solution for the lumped matrix. This does not imply that the solution with the lumped matrix is the exact solution. Some researchers argue that the wiggles that arise when the consistent matrix is used are a signal to the analyst that smaller time steps should be used in the vicinity of the wiggles. Others such as Gresho and Lee [4] contend that the results from the lumped matrix are no more accurate, but since there are no wiggles, these results are erroneously accepted as correct.

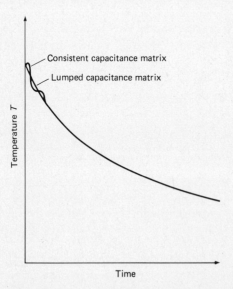

Figure 10-1 Temperature of a typical node versus time for a transient thermal analysis performed with a consistent capacitance matrix and a lumped capacitance matrix.

A rule of thumb will now be given that can be used to obtain the lumped form of the mass and capacitance matrices from the consistent matrices. The lumped mass or capacitance matrix is obtained by scaling the diagonal entries in the consistent mass or capacitance matrix such that the total mass or capacitance is preserved [5]. In general, any lumping that preserves the total mass or capacitance will lead to convergent results. If this rule is applied to the consistent mass matrices given by Eqs. (10-17) and (10-31), the corresponding lumped mass matrices are given by

$$\mathbf{M}^e = \frac{\rho A L}{2} \begin{bmatrix} 1 & \vdots & 0 \\ 0 & \vdots & 1 \end{bmatrix} \tag{10-32}$$

and

$$\mathbf{M}^e = \frac{\rho t A}{3} \begin{bmatrix} 1 & 0 & 0 & 0 & 0 & 0 \\ 0 & 1 & 0 & 0 & 0 & 0 \\ \hline 0 & 0 & 1 & 0 & 0 & 0 \\ 0 & 0 & 0 & 1 & 0 & 0 \\ \hline 0 & 0 & 0 & 0 & 1 & 0 \\ 0 & 0 & 0 & 0 & 0 & 1 \end{bmatrix} \tag{10-33}$$

respectively. Note that in both cases the total mass for the element is equally allocated to the nodes (and to each degree of freedom). Similarly, the lumped form of the capacitance matrix that corresponds to Eq. (10-29) is given by

$$\mathbf{C}^e = \frac{\rho c t A}{3} \begin{bmatrix} 1 & 0 & 0 \\ 0 & 1 & 0 \\ 0 & 0 & 1 \end{bmatrix} \tag{10-34}$$

The reader may recall that the finite difference method always yields the lumped matrices directly. One advantage of the lumped form of the capacitance matrix is that the solution for the nodal unknowns (i.e., the nodal temperatures) may be obtained in a more straightforward manner as explained in Sec. 10-7 (i.e., a so-called *explicit* solution results).

The above rule of thumb may also be applied to the higher-order elements from Chapter 9. For example, for the quadratic-order, rectangular element in the serendipity family, it can be shown that the consistent capacitance matrix is given by

$$\mathbf{C}^e = \frac{\rho c t A}{180} \begin{bmatrix} 6 & 2 & 3 & 2 & -6 & -8 & -8 & -6 \\ 2 & 6 & 2 & 3 & -6 & -6 & -8 & -8 \\ 3 & 2 & 6 & 2 & -8 & -6 & -6 & -8 \\ 2 & 3 & 2 & 6 & -8 & -8 & -6 & -6 \\ -6 & -6 & -8 & -8 & 32 & 20 & 16 & 20 \\ -8 & -6 & -6 & -8 & 20 & 32 & 20 & 16 \\ -8 & -8 & -6 & -6 & 16 & 20 & 32 & 20 \\ -6 & -8 & -8 & -6 & 20 & 16 & 20 & 32 \end{bmatrix} \tag{10-35}$$

Note that the nodes must be numbered as shown in Fig. 9-5(b). If the diagonal entries in Eq. (10-35) are scaled such that the total capacitance is preserved, the lumped capacitance matrix is given by

$$
\mathbf{C}^e = \frac{\rho ctA}{12}
\begin{bmatrix}
-1 & 0 & 0 & 0 & 0 & 0 & 0 & 0 \\
0 & -1 & 0 & 0 & 0 & 0 & 0 & 0 \\
0 & 0 & -1 & 0 & 0 & 0 & 0 & 0 \\
0 & 0 & 0 & -1 & 0 & 0 & 0 & 0 \\
0 & 0 & 0 & 0 & 4 & 0 & 0 & 0 \\
0 & 0 & 0 & 0 & 0 & 4 & 0 & 0 \\
0 & 0 & 0 & 0 & 0 & 0 & 4 & 0 \\
0 & 0 & 0 & 0 & 0 & 0 & 0 & 4
\end{bmatrix}
\tag{10-36a}
$$

This matrix was obtained by summing the entries in a given row in the consistent matrix, dividing the result by the total capacitance, and allocating this result to the diagonal entry of the row under consideration. Interestingly, the capacitance allocated to the corner nodes as given by Eq. (10-36a) is negative. Although this is known to give good results, this form of the lumped matrix is numerically inconvenient and is seldom used. Instead, the following form of the lumped capacitance matrix is frequently used:

$$
\mathbf{C}^e = \frac{\rho ctA}{36}
\begin{bmatrix}
1 & 0 & 0 & 0 & 0 & 0 & 0 & 0 \\
0 & 1 & 0 & 0 & 0 & 0 & 0 & 0 \\
0 & 0 & 1 & 0 & 0 & 0 & 0 & 0 \\
0 & 0 & 0 & 1 & 0 & 0 & 0 & 0 \\
0 & 0 & 0 & 0 & 8 & 0 & 0 & 0 \\
0 & 0 & 0 & 0 & 0 & 8 & 0 & 0 \\
0 & 0 & 0 & 0 & 0 & 0 & 8 & 0 \\
0 & 0 & 0 & 0 & 0 & 0 & 0 & 8
\end{bmatrix}
\tag{10-36b}
$$

Equation (10-36b) gives excellent results and is the recommended form of the lumped capacitance matrix in this case.

In each case, the consistent and lumped form of the matrices appear to be quite different. However, except for the very early portion of the solution, both give virtually the same results. The reader may want to consult Zienkiewicz's book [6] for more information on the subject of lumping.

10-6 SOLUTION METHODS

Equations (10-16) and (10-30) need to be solved for the nodal displacements and temperatures, respectively, as a function of time. The boundary and initial conditions also need to be imposed. The two main methods of solution that can be used are based on the finite difference and finite element methods. Both of these methods result in a two-point recurrence scheme for the solution of Eq. (10-30), whereas a three-point recurrence scheme results for the solution of Eq. (10-16). In Sec. 10-7, Eq. (10-30) is solved by using three different types of finite difference schemes. In Sec. 10-8, it is shown that these three schemes are really a special case of the more general finite element solution in time. The three-point recurrence scheme for the solution of Eq. (10-16) is derived in Sec. 10-9. The subject of numerical stability of the various solution methods is a very important one. Although a very brief introduction to this is given in Sec. 10-7, the reader should consult references 7 to 9 for more complete discussions. Recurrence schemes may also be referred to as recursion formulas. In this text, both designations are used interchangeably.

10-7 TWO-POINT RECURRENCE SCHEMES: THE FINITE DIFFERENCE METHOD

Recall that all transient thermal analyses result in the equation

$$\mathbf{C\dot{a}} + \mathbf{Ka} = \mathbf{f} \qquad (10\text{-}30)$$

In this section three different solution methods are derived for the solution of this equation by using three different types of finite differences. In particular, forward, backward, and central differences are used. It will be seen that the resulting recurrence schemes can be summarized by one all-encompassing equation. Boundary and initial conditions will be considered at the appropriate point in the development.

Forward Difference Scheme

By definition, the derivative of the $\mathbf{a}$ with respect to the time τ is given by

$$\mathbf{\dot{a}} = \frac{d\mathbf{a}}{d\tau} = \lim_{\Delta\tau \to 0} \frac{\mathbf{a}(\tau + \Delta\tau) - \mathbf{a}(\tau)}{\Delta\tau} \qquad (10\text{-}37)$$

Recall from elementary differential calculus that $\mathbf{a}(\tau)$ and $\mathbf{a}(\tau + \Delta\tau)$ denote the values of the vector $\mathbf{a}$ at times τ and $\tau + \Delta\tau$, respectively. However, the notion of a *finite difference* implies that we do not require $\Delta\tau$ to approach zero. Instead, a small but nonzero $\Delta\tau$ is used, and we approximate $\mathbf{\dot{a}}$ by

$$\mathbf{\dot{a}} \cong \frac{\mathbf{a}(\tau + \Delta\tau) - \mathbf{a}(\tau)}{\Delta\tau} \qquad (10\text{-}38)$$

The change in time, or $\Delta\tau$, is referred to as the *time step*. If Eq. (10-38) is used to eliminate $\mathbf{\dot{a}}$ in Eq. (10-30), we get

$$\mathbf{C}\frac{\mathbf{a}(\tau + \Delta\tau) - \mathbf{a}(\tau)}{\Delta\tau} + \mathbf{Ka} = \mathbf{f} \qquad (10\text{-}39)$$

Since we desire a forward difference, we must evaluate the remaining terms in Eq. (10-39) at time τ, or

$$C \frac{a(\tau + \Delta\tau) - a(\tau)}{\Delta\tau} + Ka(\tau) = f(\tau) \tag{10-40}$$

If Eq. (10-40) is multiplied by $\Delta\tau$ and written such that only the term containing $a(\tau + \Delta\tau)$ appears on the left-hand side, we get

$$Ca(\tau + \Delta\tau) = [C - K\Delta\tau]a(\tau) + f(\tau)\,\Delta\tau \tag{10-41}$$

Let us denote the vectors $a(\tau)$ and $a(\tau + \Delta\tau)$ as a_i and a_{i+1}, respectively, and $f(\tau)$ and $f(\tau + \Delta\tau)$ as f_i and f_{i+1}. Therefore, we may write Eq. (10-41) as

$$Ca_{i+1} = [C - K\,\Delta\tau]a_i + f_i\,\Delta\tau \tag{10-42}$$

Equation (10-42) is referred to as a *two-point recurrence scheme* because the right-hand side is completely known at any time τ, including $\tau = 0$ for which the initial conditions apply. Therefore, Eq. (10-42) may be applied recursively to obtain the nodal temperatures for a subsequent time given the temperatures for the preceding time. Boundary conditions may be imposed as described later in this section. The forward difference method is also known as *Euler's method*.

Note that if the lumped form of the capacitance matrix is used, the assemblage capacitance matrix C is diagonal. In this case the solution for the jth nodal temperature is given explicitly by dividing the jth row on the right-hand side of Eq. (10-42) by the jth diagonal entry in the C matrix. Hence, the solution in this case is frequently referred to as an *explicit solution*. This is the principal advantage of the forward difference (or Euler's) method. Conversely, if the consistent capacitance matrix is used, an *implicit solution* for a_{i+1} is required because C is no longer diagonal. In this case, the matrix inversion method or the active zone equation solver (i.e., subroutine ACTCOL [10]) may be used to determine the nodal temperatures at time $\tau + \Delta\tau$.

Euler's method is also convenient from the standpoint of nonlinear analyses. Recall that a thermal analysis becomes nonlinear if any of the properties is temperature-dependent or if thermal radiation is present. Since the terms in Eq. (10-30) must be evaluated at time τ and since the temperatures are always known at this time, it is *not* necessary to iterate to obtain the solution for time $\tau + \Delta\tau$ for such problems.

The principal disadvantage of this method will be illustrated numerically in Sec. 10-8 where an application is presented. Suffice it to say here that this method requires a relatively small time step for both stability and accuracy. In other words, if the time step exceeds a certain critical value for the mesh used, the solution becomes oscillatory and blows up as the time increases. Even though the time step may be below this critical value, the results may still be quite inaccurate (but stable). The accuracy of this method generally improves for successively smaller time steps. In fact, one somewhat practical way to assess the accuracy of the solution is to compare the results for two different time steps, e.g., $\Delta\tau$ and $2\,\Delta\tau$. If the results for the two different time steps are within some acceptable tolerance, a good approximation to the true solution has been obtained.

Backward Difference Scheme

If the derivative of $\mathbf{a}$ with respect to the time τ is written in the backward direction (with respect to the time) and if we do not require the time step $\Delta\tau$ to be zero, we may write

$$\dot{\mathbf{a}} \cong \frac{\mathbf{a}(\tau) - \mathbf{a}(\tau - \Delta\tau)}{\Delta\tau} \qquad \textbf{(10-43)}$$

Recognizing that the remaining terms in Eq. (10-30) should be evaluated at time τ, we get

$$\mathbf{C}\frac{\mathbf{a}(\tau) - \mathbf{a}(\tau - \Delta\tau)}{\Delta\tau} + \mathbf{Ka}(\tau) = \mathbf{f}(\tau) \qquad \textbf{(10-44)}$$

After multiplying each term by $\Delta\tau$ and rearranging such that the terms involving $\mathbf{a}(\tau)$ appear on the left-hand side, we have

$$[\mathbf{C} + \mathbf{K}\Delta\tau]\mathbf{a}(\tau) = \mathbf{Ca}(\tau - \Delta\tau) + \mathbf{f}(\tau)\,\Delta\tau \qquad \textbf{(10-45)}$$

Note that the arguments τ and $\tau - \Delta\tau$ on $\mathbf{a}$ and $\mathbf{f}$ are relative; in other words, nothing is changed if we substitute $\tau + \Delta\tau$ for τ (and τ for $\tau - \Delta\tau$) to give

$$[\mathbf{C} + \mathbf{K}\Delta\tau]\mathbf{a}(\tau + \Delta\tau) = \mathbf{Ca}(\tau) + \mathbf{f}(\tau + \Delta\tau)\,\Delta\tau \qquad \textbf{(10-46)}$$

Let us again use the subscripts i and $i + 1$ to denote times τ and $\tau + \Delta\tau$, respectively, so that the two-point recurrence scheme from the backward difference method is given by

$$[\mathbf{C} + \mathbf{K}\,\Delta\tau]\,\mathbf{a}_{i+1} = \mathbf{Ca}_i + \mathbf{f}_{i+1}\,\Delta\tau \qquad \textbf{(10-47)}$$

As in the case of the recurrence formula from the forward difference method, the right-hand side is completely known at time τ including $\mathbf{f}_{i+1}$ because it will be recalled that the vector $\mathbf{f}$ represents the forcing function for the analysis. Therefore, Eq. (10-47) may be applied recursively to obtain the nodal temperatures for a subsequent time given the temperatures for the previous time.

Note that even if the capacitance matrix $\mathbf{C}$ is diagonal, an implicit solution for the vector $\mathbf{a}_{i+1}$ must be obtained because the stiffness (or conductance) matrix $\mathbf{K}$ is never a diagonal matrix. The backward difference method is stable for all $\Delta\tau$, but the accuracy deteriorates as the time step is increased. Again the accuracy may be assessed by comparing the results for two different time steps.

Central Difference Scheme

Let us again represent the time derivative of $\mathbf{a}$ by Eq. (10-38). However instead of evaluating the other terms in Eq. (10-30) at time τ or time $\tau + \Delta\tau$, let us use the average values. In other words, let us take

$$a = \frac{a(\tau + \Delta\tau) + a(\tau)}{2}$$

and

$$f = \frac{f(\tau + \Delta\tau) + f(\tau)}{2}$$

This is equivalent to taking a *central difference* and we have

$$C\frac{a(\tau + \Delta\tau) - a(\tau)}{\Delta\tau} + K\frac{a(\tau + \Delta\tau) + a(\tau)}{2} = \frac{f(\tau + \Delta\tau) + f(\tau)}{2}$$

Multiplying through by $\Delta\tau$ and isolating the terms containing $a(\tau + \Delta\tau)$ gives

$$\left(C + \frac{K\,\Delta\tau}{2}\right)a(\tau + \Delta\tau)$$

$$= \left(C - \frac{K\,\Delta\tau}{2}\right)a(\tau) + \frac{[f(\tau) + f(\tau + \Delta\tau)]\,\Delta\tau}{2} \quad \textbf{(10-48)}$$

In terms of the subscript notation used above, Eq. (10-48) becomes

$$\left(C + \frac{K\,\Delta\tau}{2}\right)a_{i+1} = \left(C - \frac{K\,\Delta\tau}{2}\right)a_i + \frac{(f_i + f_{i+1})\,\Delta\tau}{2} \quad \textbf{(10-49)}$$

As in the backward difference scheme, the recursion formula given by Eq. (10-49) requires an *implicit solution* for the nodal temperatures at time $\tau + \Delta\tau$ given those at time τ. This method results in an oscillatory solution if the critical time step for stability is exceeded. For time steps smaller than this critical value, the accuracy of the solution improves as the time step is decreased. Not surprisingly, the central difference method is more accurate than both the forward and backward difference schemes because the central difference favors neither the temperatures at time τ nor those at time $\tau + \Delta\tau$. The central difference method is also referred to as the *Crank-Nicolson method*.

Summary and Application of the Prescribed Temperatures

The three recurrence formulas given by Eqs. (10-42), (10-47), and (10-49) may be summarized in one convenient equation as

$$[C + \theta K\,\Delta\tau]a_{i+1} = [C - (1 - \theta)K\,\Delta\tau]a_i + [(1 - \theta)f_i + \theta f_{i+1}]\,\Delta\tau \quad \textbf{(10-50)}$$

where the parameter θ takes on values 0, ½ and 1 for the forward, central, and backward difference schemes, respectively.

Strictly speaking, for values of θ other than 0, an iterative solution is required for each time step if any of the properties is temperature dependent or if thermal radiation is included in the model. In other words, the matrices K and C should really be evaluated with the nodal temperatures that correspond to the value of θ being used. The temperatures for the $(j+1)$st iteration (denoted by $\mathbf{a}^{j+1}$) may be computed from the nodal temperatures from the jth iteration with the help of the following equation:

$$\mathbf{a}^{j+1} = (1 - \theta)\mathbf{a}_i^j + \theta\mathbf{a}_{i+1}^j \tag{10-51}$$

where $\mathbf{a}_i^j$ and $\mathbf{a}_{i+1}^j$ denote the nodal temperatures at times τ and $\tau + \Delta\tau$, respectively, for the jth iteration. Note that for the forward difference or Euler's method (for which θ is zero) it is not necessary to iterate (why not?).

Recall that the initial conditions or $\mathbf{a}_0 = \mathbf{a}(0)$ are automatically incorporated in the solution process because these values of $\mathbf{a}$ (at $\tau = 0$) are used to start the recursion process. This is to say that $\mathbf{a}_0$ is taken as $\mathbf{a}(0)$ from which $\mathbf{a}_1$ may be found from Eq. (10-50). The recursion process is illustrated schematically in Fig. 10-2.

Only prescribed temperature boundary conditions need to be considered at this point because all other boundary conditions (prescribed heat flux, convection, etc.) are automatically included in the finite element formulation. Prescribed temperatures may be imposed by either of the two methods from Sec. 3-2 because Eq. (10-50) is of the form $\mathbf{K}_{\text{eff}}\mathbf{a} = \mathbf{f}_{\text{eff}}$, where $\mathbf{K}_{\text{eff}}$ is really $\mathbf{C} + \theta\mathbf{K}\,\Delta\tau$ from Eq. (10-50) and $\mathbf{f}_{\text{eff}}$ is given by the entire right-hand side of Eq. (10-50). As in the steady-state case, it is desirable to preserve symmetry because this substantially reduces the storage requirements in large problems. In summary, application of the prescribed temperature boundary conditions for transient thermal analyses is really no different from that in steady-state problems, if the prescribed temperatures are imposed on the vector $\mathbf{a}_{i+1}$ in Eq. (10-50). Note that the prescribed temperatures may possibly be a function of time.

From a practical point of view, if the numerical solution appears to oscillate when the physics of the problem would preclude this, the time step is too large for the mesh used. Reducing the time step to below that for which the instability or oscillation disappears is a very practical remedy. These comments apply in particular to the forward and central difference schemes. Accuracy may be assessed by com-

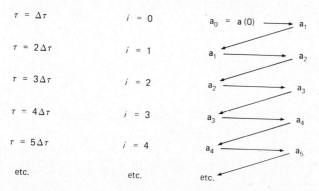

Figure 10-2 Schematic diagram of the two-point recursion process.

paring the results for two cases with different time steps: If there is a significant difference between the results, then the time step should be reduced further and the analysis repeated.

10-8   TWO-POINT RECURRENCE SCHEMES: THE FINITE ELEMENT METHOD

The finite element method itself may be applied to Eq. (10-30) as shown in this section. Because the time domain is to be discretized, the elements used may be referred to as *temporal elements* to distinguish them from the elements used to discretize the body. It will be seen that the three recursion formulas from Sec. 10-7 are really a special case of the more general finite element method. In particular Eq. (10-50) will be derived by discretizing the time domain from time τ to time $\tau + \Delta\tau$ with a suitable first-order element. One such element is shown in Fig. 10-3 where times τ and $\tau + \Delta\tau$ are denoted as τ_i and τ_{i+1}, respectively. Note the shape functions N_i and N_{i+1} are also shown. These shape functions are now a function of time. However, mathematically they are no different than those presented in previous chapters except that τ replaces x. If the weighted-residual method is used to solve Eq. (10-30), we may write

$$\int_{\tau_i}^{\tau_{i+1}} W[\mathbf{C\dot{a}} + \mathbf{Ka} - \mathbf{f}] \, d\tau = 0 \qquad \textbf{(10-52)}$$

where the weighting function W is a scalar in this case. Different choices for W will yield different two-point recurrence schemes.

With the help of the shape functions, we may represent the vector $\mathbf{a}$ in Eq. (10-52) as

$$\mathbf{a} = N_i \mathbf{a}_i + N_{i+1} \mathbf{a}_{i+1} \qquad \textbf{(10-53)}$$

because the shape functions are linear with time and evaluation of Eq. (10-53) at times τ_i and τ_{i+1} yields $\mathbf{a}_i$ and $\mathbf{a}_{i+1}$, respectively, as desired.

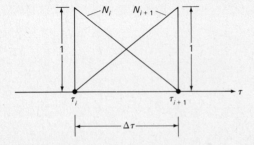

Figure 10-3   The two-node temporal element and its associated shape functions.

For convenience let us define a local normalized coordinate ξ as follows:

$$\xi = \frac{\tau - \tau_i}{\tau_{i+1} - \tau_i} = \frac{\tau - \tau_i}{\Delta\tau} \qquad \text{(10-54)}$$

such that for $\tau = \tau_i$ and $\tau = \tau_{i+1}$, we have $\xi = 0$ and $\xi = 1$, respectively. It follows that the shape functions N_i and N_{i+1} are given by

$$N_i = 1 - \xi \qquad \text{(10-55a)}$$

and

$$N_{i+1} = \xi \qquad \text{(10-55b)}$$

Figure 10-3 shows the behavior of the shape functions over the time interval from $\tau = \tau_i$ to $\tau = \tau_{i+1}$. As mentioned above, these shape functions are no different from those for the linear-order, lineal element in Chapters 6 and 9.

With the help of Eqs. (10-55), we may write Eq. (10-53) as

$$\mathbf{a} = (1 - \xi)\mathbf{a}_i + \xi\mathbf{a}_{i+1} \qquad \text{(10-56)}$$

from which it follows that

$$\dot{\mathbf{a}} = \frac{d\mathbf{a}}{d\tau} = \frac{d\mathbf{a}}{d\xi}\frac{d\xi}{d\tau} = \frac{d}{d\xi}[(1 - \xi)\mathbf{a}_i + \xi\mathbf{a}_{i+1}]\frac{d\xi}{d\tau} \qquad \text{(10-56)}$$

or

$$\dot{\mathbf{a}} = -\frac{1}{\Delta\tau}\mathbf{a}_i + \frac{1}{\Delta\tau}\mathbf{a}_{i+1} \qquad \text{(10-57)}$$

Similarly, $\mathbf{f}$ may be represented as

$$\mathbf{f} = N_i\mathbf{f}_i + N_{i+1}\mathbf{f}_{i+1} \qquad \text{(10-58)}$$

or

$$\mathbf{f} = (1 - \xi)\mathbf{f}_i + \xi\mathbf{f}_{i+1} \qquad \text{(10-59)}$$

From Eq. (10-54), it follows that $d\tau$ is related to $d\xi$ by

$$d\tau = \Delta\tau \, d\xi \qquad \text{(10-60)}$$

Therefore, Eq. (10-52) may be written as

$$\int_0^1 W\left\{\mathbf{C}\left[-\frac{1}{\Delta\tau}\mathbf{a}_i + \frac{1}{\Delta\tau}\mathbf{a}_{i+1}\right] + \mathbf{K}[(1 - \xi)\mathbf{a}_i + \xi\mathbf{a}_{i+1}]\right.$$

$$\left. - [(1 - \xi)\mathbf{f}_i + \xi\mathbf{f}_{i+1}]\right\}\Delta\tau \, d\xi = 0 \qquad \text{(10-61)}$$

where the weighting function W is still unspecified. Equation (10-61) may be put into the following more convenient form:

$$\left[\mathbf{C} \int_0^1 W \, d\xi + \mathbf{K}\Delta\tau \int_0^1 W\xi \, d\xi \right] \mathbf{a}_{i+1} = \left[\mathbf{C} \int_0^1 W \, d\xi - \mathbf{K}\Delta\tau \int_0^1 W(1 - \xi) \, d\xi \right] \mathbf{a}_i$$

$$+ \left[\Delta\tau \int_0^1 W(1 - \xi) \, d\xi \right] \mathbf{f}_i + \left[\Delta\tau \int_0^1 W\xi \, d\xi \right] \mathbf{f}_{i+1} \quad \textbf{(10-62)}$$

Dividing both sides of Eq. (10-62) by $\int_0^1 W \, d\xi$ gives the desired result:

$$[\mathbf{C} + \theta\mathbf{K}\Delta\tau]\mathbf{a}_{i+1} = [\mathbf{C} - (1 - \theta)\mathbf{K}\Delta\tau]\mathbf{a}_i + [(1 - \theta)\mathbf{f}_i + \theta\mathbf{f}_{i+1}]\,\Delta\tau \quad \textbf{(10-63)}$$

where the parameter θ is defined by

$$\theta = \frac{\displaystyle\int_0^1 W\xi \, d\xi}{\displaystyle\int_0^1 W \, d\xi} \quad \textbf{(10-64)}$$

Note that Eqs. (10-50) and (10-63) are identical. It is emphasized that Eq. (10-50) was derived and generalized from three different finite difference schemes, whereas Eq. (10-63) was derived directly via the finite element method. Different choices for the weighting function W will yield different values for the parameter θ. In particular the point collocation, subdomain collocation, and Galerkin methods from Chapter 4 will be used to determine the corresponding value of θ from Eq. (10-64).

Point Collocation

Recall from Sec. 4-6 that the weighting functions for the point collocation method are given by $\delta(\xi - \xi_j)$ such that

$$\int_0^1 \delta(\xi - \xi_j) \, d\xi = 1 \quad \text{for } \xi = \xi_j \quad \textbf{(10-66a)}$$

and

$$\int_0^1 \delta(\xi - \xi_j) \, d\xi = 0 \quad \text{for } \xi \neq \xi_j \quad \textbf{(10-66b)}$$

where ξ_j is known as a collocation point. For point collocation at time τ_i (i.e., $\tau_i = \tau_j$) where $\xi = 0$, it can be shown that $\theta = 0$. Similarly for point collocation at times $(\tau_i + \tau_{i+1})/2$ and τ_{i+1}, where $\xi = \frac{1}{2}$ and $\xi = 1$, it can be shown that $\theta = \frac{1}{2}$ and $\theta = 1$, respectively. Thus it is seen that point collocations at times τ_i, $(\tau_i + \tau_{i+1})/2$, and τ_{i+1} correspond to the forward, central, and backward difference schemes, respectively.

Example 10-4

Show that point collocation at time $(\tau_i + \tau_{i+1})/2$ gives $\theta = \frac{1}{2}$.

Solution

At time $(\tau_i + \tau_{i+1})/2$ we have $\xi = \frac{1}{2}$ and so the collocation point ξ_j must be $\frac{1}{2}$. Thus the weighting function W is given by

$$W = \delta(\xi - \frac{1}{2})$$

and Eq. (10-64) gives

$$\theta = \frac{\int_0^1 \delta(\xi - \frac{1}{2})\xi \, d\xi}{\int_0^1 \delta(\xi - \frac{1}{2}) \, d\xi} = \frac{\frac{1}{2}}{1} = \frac{1}{2}$$

Note that the numerator in the above expression for θ is zero except at the collocation point where $\xi = \frac{1}{2}$.   ∎

Subdomain Collocation

It should be recalled from Sec. 4-6 that the weighting functions for the subdomain collocation method are unity over a particular subdomain and are zero elsewhere. Since only one unknown, namely, a_{i+1}, is to be found, we must take only one weighting function or $W = 1$ over the time interval from τ to $\tau + \Delta\tau$ (i.e., from τ_i to τ_{i+1}). The value of θ for this case may be determined as follows:

$$\theta = \frac{\int_0^1 W\xi \, d\xi}{\int_0^1 W \, d\xi} = \frac{\int_0^1 (1)\xi \, d\xi}{\int_0^1 (1) \, d\xi} = \frac{\left.\frac{\xi^2}{2}\right|_0^1}{\left.\xi\right|_0^1} = \frac{1}{2}$$

Therefore, the subdomain collocation method is analogous to the central difference scheme or Crank-Nicolson method from Sec. 10-7.

Galerkin

The weighting functions for the Galerkin method are taken to be the shape functions themselves. In other words, we may take either

$$W = N_i = 1 - \xi \tag{10-67a}$$

or

$$W = N_{i+1} = \xi \tag{10-67b}$$

The shape functions are shown graphically in Fig. 10-3. It can be shown that if Eq. (10-67a) is used, we get $\theta = \frac{1}{3}$, whereas if Eq. (10-67b) is used, we get $\theta = \frac{2}{3}$. Neither of these values corresponds to any of the results from the finite difference method. However, $\theta = \frac{2}{3}$ is particularly useful because it is more accurate than the backward difference scheme ($\theta = 1$) and more stable than the central difference scheme ($\theta = \frac{1}{2}$).

Example 10-5

Show that the parameter θ takes on a value of $\frac{2}{3}$ if the weighting function is given by Eq. (10-67b).

Solution

By definition of θ, we have

$$\theta = \frac{\int_0^1 (\xi)\xi \, d\xi}{\int_0^1 (\xi) \, d\xi} = \frac{\int_0^1 \xi^2 \, d\xi}{\int_0^1 \xi \, d\xi} = \frac{\left.\dfrac{\xi^3}{3}\right|_0^1}{\left.\dfrac{\xi^2}{2}\right|_0^1} = \frac{2}{3}$$

which is the desired result. ∎

Example 10-6

Reconsider the circular pin fin from Example 4-11. For convenience, the fin is shown in Fig. 10-4(a). Recall that the fin is made of pure copper with a thermal conductivity k of 400 W/m-°C. The base is held at a temperature T_b of 85°C and

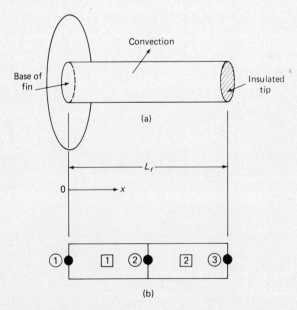

(a)

(b)

Figure 10-4 Circular pin fin (a) analyzed in Example 10-6 and (b) discretized into two elements and three nodes.

the ambient temperature T_a is maintained at 25°C. The fin length L_f is 2 cm, and the diameter D is 0.4 cm. The tip of the fin at $x = L_f$ is insulated. Determine the transient temperature distribution within the fin if the entire fin is initially at 25°C. The density ρ and specific heat c are 8900 kg/m^3 and 375 J/kg-°C, respectively. Assume the consistent form of the capacitance matrix, a time step of 0.1 sec, and $\theta = \frac{2}{3}$. Use only two elements of equal length as shown in Fig. 10-4(b).

Solution

Equation (10-50) or (10-63) provides the basis for the transient solution. Because the same fin was analyzed in Example 4-11, we may use most of the results from that example. In particular, the element stiffness matrix and nodal force vectors are unchanged. It follows that the assemblage stiffness matrix and nodal force vector are also unchanged. Therefore, from Example 4-11, we have

$$\mathbf{K} = \begin{bmatrix} 0.50893 & -0.49951 & 0 \\ -0.49951 & 1.01786 & -0.49951 \\ 0 & -0.49951 & 0.50893 \end{bmatrix}$$

and since the nodal force vector is not time-varying, we have

$$\mathbf{f}_i = \mathbf{f}_{i+1} = \mathbf{f} = \begin{bmatrix} 0.23561 \\ 0.47122 \\ 0.23561 \end{bmatrix}$$

We now need to compute the element capacitance matrix. It can be shown for the linear-order lineal element that Eq. (10-26) evaluates to

$$\mathbf{C}^e = \frac{\rho c A L}{6} \begin{bmatrix} 2 & 1 \\ 1 & 2 \end{bmatrix} \tag{10-68}$$

where A and L are the cross-sectional area and length of the element, respectively. Equation (10-68) is an expression for the consistent capacitance matrix. Let us now evaluate this matrix for the problem at hand. Noting that the two elements are identical, we may write

$$\mathbf{C}^{(1)} = \mathbf{C}^{(2)} = \frac{(8900)(375)(1.2566 \times 10^{-5})(0.01)}{6} \begin{bmatrix} 2 & 1 \\ 1 & 2 \end{bmatrix}$$

or

$$\mathbf{C}^{(1)} = \mathbf{C}^{(2)} = \begin{bmatrix} 0.13980 & 0.06990 \\ 0.06990 & 0.13980 \end{bmatrix}$$

The assemblage of the element capacitance matrices is done in precisely the same manner as the assemblage of the element stiffness matrices. The result is

$$\mathbf{C} = \begin{bmatrix} 0.13980 & 0.06990 & 0 \\ 0.06990 & 0.27960 & 0.06990 \\ 0 & 0.06990 & 0.13980 \end{bmatrix}$$

Note that $\mathbf{C}$ is not diagonal because the consistent capacitance matrix has been used. The matrix $\mathbf{K}_{\text{eff}}$ may now be computed as

$$\mathbf{K}_{\text{eff}} = \mathbf{C} + \theta\mathbf{K}\,\Delta\tau = \begin{bmatrix} 0.13980 & 0.06990 & 0 \\ 0.06990 & 0.27960 & 0.06990 \\ 0 & 0.06990 & 0.13980 \end{bmatrix}$$

$$+ \tfrac{2}{3}(0.1)\begin{bmatrix} 0.50893 & -0.49951 & 0 \\ -0.49951 & 1.01786 & -0.49951 \\ 0 & -0.49951 & 0.50893 \end{bmatrix}$$

or

$$\mathbf{K}_{\text{eff}} = \begin{bmatrix} 0.17373 & 0.03660 & 0 \\ 0.03660 & 0.34746 & 0.03660 \\ 0 & 0.03660 & 0.17373 \end{bmatrix}$$

Similarly the vector $\mathbf{f}_{\text{eff}}$ is given by the right-hand side of Eq. (10-50) [or Eq. (10-63)], or

$$\mathbf{f}_{\text{eff}} = [\mathbf{C} - (1 - \theta)\mathbf{K}\Delta\tau]\mathbf{a}_i + [(1 - \theta)\mathbf{f}_i + \theta\mathbf{f}_{i+1}]\,\Delta\tau$$

Since the force vector $\mathbf{f}$ does not change with time in this application, we should note that

$$[(1 - \theta)\mathbf{f}_i + \theta\mathbf{f}_{i+1}]\,\Delta\tau = [(1 - \theta)\mathbf{f} + \theta\mathbf{f}]\,\Delta\tau = \mathbf{f}\,\Delta\tau$$

Therefore, from

$$\mathbf{f}_{\text{eff}} = [\mathbf{C} - (1 - \theta)\mathbf{K}\,\Delta\tau]\mathbf{a}_i + \mathbf{f}\,\Delta\tau$$

and the fact that $\mathbf{a}_i$ is the vector of nodal temperatures at time $\tau = 0$ at this point in the solution, we get

$$\mathbf{f}_{\text{eff}} = \left\{ \begin{bmatrix} 0.13980 & 0.06990 & 0 \\ 0.06990 & 0.27960 & 0.06990 \\ 0 & 0.06990 & 0.13980 \end{bmatrix} \right.$$

$$\left. - (1 - \tfrac{2}{3})0.1\begin{bmatrix} 0.50893 & -0.49951 & 0 \\ -0.49951 & 1.01786 & -0.49951 \\ 0 & -0.49951 & 0.50893 \end{bmatrix} \right\}\begin{bmatrix} 25. \\ 25. \\ 25. \end{bmatrix}$$

$$+ \begin{bmatrix} 0.23561 \\ 0.47122 \\ 0.23561 \end{bmatrix}(0.1)$$

$$= \begin{bmatrix} 0.12284 & 0.08655 & 0 \\ 0.08655 & 0.24567 & 0.08655 \\ 0 & 0.08655 & 0.12284 \end{bmatrix}\begin{bmatrix} 25. \\ 25. \\ 25. \end{bmatrix} + \begin{bmatrix} 0.02356 \\ 0.04712 \\ 0.02356 \end{bmatrix}$$

or

$$\mathbf{f}_{\text{eff}} = \begin{bmatrix} 5.2582 \\ 10.5165 \\ 5.2582 \end{bmatrix}$$

The temperatures *at time* $\tau = 0.1$ *sec* are then computed from the solution of $\mathbf{K}_{\text{eff}}\mathbf{a} = \mathbf{f}_{\text{eff}}$, or

$$\begin{bmatrix} 0.17373 & 0.03660 & 0 \\ 0.03660 & 0.34746 & 0.03660 \\ 0 & 0.03660 & 0.17373 \end{bmatrix} \begin{bmatrix} T_1 \\ T_2 \\ T_3 \end{bmatrix} = \begin{bmatrix} 5.2582 \\ 10.5165 \\ 5.2582 \end{bmatrix}$$

However, the prescribed temperature of 85°C must be imposed on node 1. Using Method 1 from Sec. 3-2 (and preserving the symmetry), we get

$$\begin{bmatrix} 1.00000 & 0 & 0 \\ 0 & 0.34746 & 0.03660 \\ 0 & 0.03660 & 0.17373 \end{bmatrix} \begin{bmatrix} T_1 \\ T_2 \\ T_3 \end{bmatrix} = \begin{bmatrix} 85.0000 \\ 7.4055 \\ 5.2582 \end{bmatrix}$$

Solving this system of linear algebraic equations gives

$$T_1 = 85.0°C \qquad T_2 = 18.5°C \qquad T_3 = 26.4°C$$

for the nodal temperatures at the end of the first time step, i.e., at $\tau = 0.1$ sec. Note that the temperature of node 2 (i.e., T_2) is not physically realizable. This is a consequence of using the consistent capacitance matrix and is discussed further below.

The nodal temperatures at subsequent time steps may be computed from the two-point recurrence relation:

$$\begin{bmatrix} 0.17373 & 0.03660 & 0 \\ 0.03660 & 0.34746 & 0.03660 \\ 0 & 0.03660 & 0.17373 \end{bmatrix} \begin{bmatrix} T_1(\tau + \Delta\tau) \\ T_2(\tau + \Delta\tau) \\ T_3(\tau + \Delta\tau) \end{bmatrix}$$

$$= \begin{bmatrix} 0.12284 & 0.08655 & 0 \\ 0.08655 & 0.24567 & 0.08655 \\ 0 & 0.08655 & 0.12284 \end{bmatrix} \begin{bmatrix} T_1(\tau) \\ T_2(\tau) \\ T_3(\tau) \end{bmatrix} + \begin{bmatrix} 0.02356 \\ 0.04712 \\ 0.02356 \end{bmatrix}$$

The results are summarized in Table 10-1 for τ up to 2 sec and in Fig. 10-5(a) for τ up to 6 sec, the time at which steady-state appears to have been attained. In order to see the initial oscillations in the nodal temperatures, the results for the first five time steps are shown in Fig. 10-5(b). Recall that these initial oscillations are characteristic of the solution when the consistent capacitance matrix is used. ∎

With the help of the problem posed in Example 10-6, let us discuss the accuracy and stability issues further. Figure 10-6 shows the solution for the temperature of the tip of the fin from Example 10-6 in comparison with two other cases. Both of these other transients were obtained with a 0.5-sec time step. The first case corresponds to Euler's method ($\theta = 0$); the second case to the Crank-Nicolson method ($\theta = \frac{1}{2}$). Note that the critical time step for stability has been exceeded in the case of Euler's method. This is evidenced by the oscillatory temperature as a function of time. Note further that the amplitudes of the oscillations are growing with time (recall that this solution can be made stable if a small enough time step is used). In direct contrast the Crank-Nicolson method happens to be stable for a 0.5-sec time step. Although it is not shown here, the Crank-Nicolson method tends to give

Table 10-1   Resulting Nodal Temperatures for Example 10-6

Time, sec	T_1, °C	T_2, °C	T_3, °C
0.1	85.0	18.5	26.4
0.2	85.0	29.7	21.7
0.3	85.0	36.4	22.7
0.4	85.0	41.0	25.6
0.5	85.0	44.7	29.3
0.6	85.0	47.8	33.1
0.7	85.0	50.5	36.7
0.8	85.0	53.0	40.1
0.9	85.0	55.2	43.2
1.0	85.0	57.3	46.1
1.1	85.0	59.2	48.8
1.2	85.0	61.0	51.3
1.3	85.0	62.6	53.6
1.4	85.0	64.1	55.7
1.5	85.0	65.5	57.7
1.6	85.0	66.8	59.5
1.7	85.0	67.9	61.1
1.8	85.0	69.0	62.7
1.9	85.0	70.0	64.1
2.0	85.0	70.9	65.4

oscillatory (but stable) results if an excessively large time step is used and the accuracy of the solution deteriorates as well. Recall that the backward difference scheme is stable for all time steps, but it becomes increasingly less accurate as the time step increases. Hence, a good compromise is to carry out the solution with a value of θ of $\frac{2}{3}$.

Example 10-7

Repeat Example 10-6 with a lumped capacitance matrix in order to illustrate the lack of oscillations in the early portion of the transient solution for the nodal temperatures.

Solution

The lumped form of the capacitance matrix for the two-node, lineal element is given by

$$\mathbf{C}^e = \frac{\rho c A L}{2} \begin{bmatrix} 1 & 0 \\ 0 & 1 \end{bmatrix}$$

Therefore, numerically the element capacitance matrices are given by

$$\mathbf{C}^{(1)} = \mathbf{C}^{(2)} = \begin{bmatrix} 0.20970 & 0 \\ 0 & 0.20970 \end{bmatrix}$$

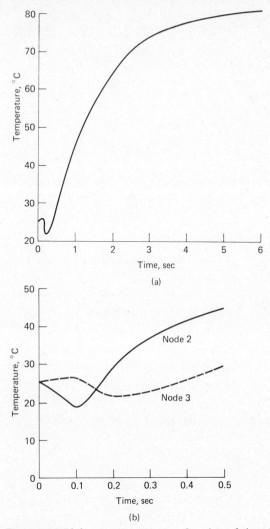

Figure 10-5(a) Temperature as a function of time for the node at the tip of the fin in Example 10-6. (b) Temperatures as a function of time of nodes 2 and 3 in Example 10-6. Note the oscillations in the early portion of the transient.

It is left as an exercise to show that the two-point recurrence relation in this case becomes

$$
\begin{bmatrix}
0.24363 & -0.03330 & 0 \\
-0.03330 & 0.48726 & -0.03330 \\
0 & -0.03330 & 0.24363
\end{bmatrix}
\begin{bmatrix}
T_1(\tau + \Delta\tau) \\
T_2(\tau + \Delta\tau) \\
T_3(\tau + \Delta\tau)
\end{bmatrix}
$$

$$
=
\begin{bmatrix}
0.19274 & 0.01665 & 0 \\
0.01665 & 0.38547 & 0.01665 \\
0 & 0.01665 & 0.19274
\end{bmatrix}
\begin{bmatrix}
T_1(\tau) \\
T_2(\tau) \\
T_3(\tau)
\end{bmatrix}
+
\begin{bmatrix}
0.02356 \\
0.04712 \\
0.02356
\end{bmatrix}
$$

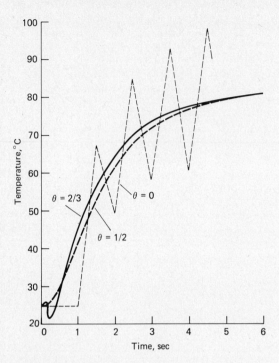

Figure 10-6 Comparison of the tip temperatures for the fin in Example 10-6 for different values of the parameter θ.

Table 10-2 shows the nodal temperatures for the first 2 sec of the transient. Note the lack of oscillations in the nodal temperatures during the first few time steps (unlike the results with the consistent capacitance matrix). Note also that the results from Example 10-6 with a consistent capacitance matrix and the results from this example are within 1°C of each other after about one second. ∎

10-9 THREE-POINT RECURRENCE SCHEMES

In this section we will develop a three-point recursion formula for the solution of Eq. (10-16). A suitable temporal element is used as shown below. In particular, the time domain will be discretized with second-order, lineal elements. One such element is shown in Fig. 10-7 where the shape functions are also shown. Note that the element runs from time $\tau - \Delta\tau$ to time $\tau + \Delta\tau$ and the center node corresponds to time τ. Times $\tau - \Delta\tau$, τ, and $\tau + \Delta\tau$ will be denoted by τ_{i-1}, τ_i, and τ_{i+1}, respectively. The shape functions for nodes $i - 1$, i, and $i + 1$ are denoted as N_{i-1}, N_i, and N_{i+1} and are given by Eq. (9-5) in terms of the serendipity coordinate r.

Table 10-2   Resulting Nodal Temperatures for Example 10-7

Time, sec	T_1, °C	T_2, °C	T_3, °C
0.1	85.0	29.1	25.6
0.2	85.0	34.6	27.0
0.3	85.0	39.1	29.2
0.4	85.0	42.9	31.7
0.5	85.0	46.2	34.4
0.6	85.0	49.1	37.2
0.7	85.0	51.6	39.9
0.8	85.0	53.9	42.6
0.9	85.0	56.0	45.1
1.0	85.0	57.9	47.5
1.1	85.0	59.7	49.8
1.2	85.0	61.3	52.0
1.3	85.0	62.7	54.0
1.4	85.0	64.1	55.8
1.5	85.0	65.4	57.6
1.6	85.0	66.5	59.2
1.7	85.0	67.6	60.7
1.8	85.0	68.6	62.1
1.9	85.0	69.6	63.5
2.0	85.0	70.4	64.7

The weighted-residual method requires that we weight the residual with a weighting function, integrate the result with respect to time over the interval from τ_{i-1} to τ_{i+1}, and set the result to zero, or

$$\int_{\tau_{i-1}}^{\tau_{i+1}} W\{\mathbf{M\ddot{a}} + \mathbf{D\dot{a}} + \mathbf{Ka} - \mathbf{f}\}\, d\tau = 0 \qquad \textbf{(10-69)}$$

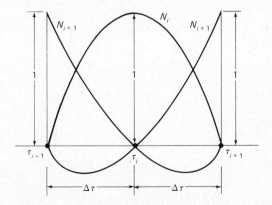

Figure 10-7   The three-node temporal element and its associated shape functions.

As in the case of the two-point recurrence formula, different choices for the scalar weighting function W will yield different three-point recurrence schemes.

With the help of the shape functions, the vector **a** in Eq. (10-69) may be written as

$$\mathbf{a} = N_{i-1}\mathbf{a}_{i-1} + N_i\mathbf{a}_i + N_{i+1}\mathbf{a}_{i+1} \tag{10-70}$$

Note that the vector **a** is assumed to vary parabolically with time such that at time τ_{i-1}, we have $\mathbf{a}_{i-1}$, etc. It is emphasized that the shape functions are functions of time and $\mathbf{a}_{i-1}$, $\mathbf{a}_i$, and $\mathbf{a}_{i+1}$ are constant vectors, where $\mathbf{a}_{i+1}$ denotes the nodal displacements at time $\tau + \Delta\tau$, etc. The second derivative of **a** [the $\ddot{\mathbf{a}}$ in Eq. (10-69)] requires that at least second-order shape functions be used; otherwise, $\ddot{\mathbf{a}}$ would be identically zero. For convenience the shape functions are given below.

$$N_1 = -\tfrac{1}{2}r(1 - r)$$

$$N_2 = \tfrac{1}{2}r(1 + r) \tag{9-5}$$

$$N_3 = (1 + r)(1 - r)$$

Recall that the midside node (at time τ_i) must be midway between the two end nodes. Note that the serendipity coordinate r is related to the time τ by

$$r = \frac{\tau - \tau_i}{\tau_{i+1} - \tau_i} = \frac{\tau - \tau_i}{\Delta\tau} \qquad \text{for} \quad \tau_{i-1} \leq \tau \leq \tau_{i+1} \tag{10-71}$$

and dr is related to $d\tau$ by

$$dr = \frac{1}{\Delta\tau}\,d\tau \tag{10-72}$$

Therefore, we may write Eq. (10-70) as

$$\mathbf{a} = -\tfrac{1}{2}r(1 - r)\mathbf{a}_{i-1} + (1 - r)(1 + r)\mathbf{a}_i + \tfrac{1}{2}r(1 + r)\mathbf{a}_{i+1} \tag{10-73}$$

For $\dot{\mathbf{a}}$, we get

$$\dot{\mathbf{a}} = \frac{d\mathbf{a}}{d\tau} = \frac{d\mathbf{a}}{dr}\frac{dr}{d\tau} = \frac{1}{\Delta\tau}\frac{d\mathbf{a}}{dr} \tag{10-74a}$$

or

$$\dot{\mathbf{a}} = \frac{1}{\Delta\tau}[(-\tfrac{1}{2} + r)\mathbf{a}_{i-1} - 2r\mathbf{a}_i + (\tfrac{1}{2} + r)\mathbf{a}_{i+1}] \tag{10-74b}$$

For $\ddot{\mathbf{a}}$, we have

$$\ddot{\mathbf{a}} = \frac{d}{d\tau}\left(\frac{d\dot{\mathbf{a}}}{d\tau}\right) = \frac{d}{dr}\left(\frac{1}{\Delta\tau}\frac{d\mathbf{a}}{dr}\right)\frac{dr}{d\tau} \tag{10-75a}$$

or

$$\ddot{\mathbf{a}} = \frac{1}{(\Delta\tau)^2}\frac{d^2\mathbf{a}}{dr^2} = \frac{1}{(\Delta\tau)^2}[\mathbf{a}_{i-1} - 2\mathbf{a}_i + \mathbf{a}_{i+1}] \tag{10-75b}$$

Similarly, the vector **f** may be represented as

$$\mathbf{f} = N_{i-1}\mathbf{f}_{i-1} + N_i\mathbf{f}_i + N_{i+1}\mathbf{f}_{i+1} \tag{10-76a}$$

or

$$\mathbf{f} = -\tfrac{1}{2}r(1-r)\mathbf{f}_{i-1} + (1-r)(1+r)\mathbf{f}_i + \tfrac{1}{2}r(1+r)\mathbf{f}_{i+1} \tag{10-76b}$$

Therefore, Eq. (10-69) may be written as

$$\int_{-1}^{+1} W\Bigg\{ \mathbf{M}\left(\frac{1}{\Delta\tau}\right)^2[\mathbf{a}_{i-1} - 2\mathbf{a}_i + \mathbf{a}_{i+1}]$$

$$+ \mathbf{D}\frac{1}{\Delta\tau}[(-\tfrac{1}{2} + r)\mathbf{a}_{i-1} - 2r\mathbf{a}_i + (\tfrac{1}{2} + r)\mathbf{a}_{i+1}]$$

$$+ \mathbf{K}[-\tfrac{1}{2}r(1-r)\mathbf{a}_{i-1} + (1-r)(1+r)\mathbf{a}_i$$

$$+ \tfrac{1}{2}r(1+r)\mathbf{a}_{i+1}] - [-\tfrac{1}{2}r(1-r)\mathbf{f}_{i-1}$$

$$+ (1-r)(1+r)\mathbf{f}_i + \tfrac{1}{2}r(1+r)\mathbf{f}_{i+1}]\Bigg\}\Delta\tau\, dr = 0 \tag{10-77}$$

It should be emphasized that the weighting function W is as yet unspecified. Let us rewrite this last equation in a more convenient form by first multiplying every term by $\Delta\tau$ and then rearranging to get

$$\left[\mathbf{M}\int_{-1}^{+1} W\, dr \quad + \mathbf{D}\,\Delta\tau \int_{-1}^{+1} W(\tfrac{1}{2} + r)\, dr\right.$$

$$\left. + \mathbf{K}(\Delta\tau)^2 \int_{-1}^{+1} \tfrac{1}{2}Wr(1 + r)\, dr\right]\mathbf{a}_{i+1}$$

$$= -\left[-2\mathbf{M}\int_{-1}^{+1} W\, dr - 2\mathbf{D}\,\Delta\tau \int_{-1}^{+1} Wr\, dr\right.$$

$$\left. + \mathbf{K}(\Delta\tau)^2 \int_{-1}^{+1} W(1 - r)(1 + r)\, dr\right]\mathbf{a}_i$$

$$- \left[\mathbf{M}\int_{-1}^{+1} W\, dr + \mathbf{D}\,\Delta\tau \int_{-1}^{+1} W(-\tfrac{1}{2} + r)\, dr\right.$$

$$\left. - \mathbf{K}(\Delta\tau)^2 \int_{-1}^{+1} \tfrac{1}{2}Wr(1 - r)\, dr\right]\mathbf{a}_{i-1}$$

$$- \left[\int_{-1}^{+1} \tfrac{1}{2}Wr(1 - r)\, dr\right](\Delta\tau)^2\mathbf{f}_{i-1}$$

$$+ \left[\int_{-1}^{+1} W(1 - r)(1 + r) \, dr \right] (\Delta\tau)^2 \mathbf{f}_i$$

$$+ \left[\int_{-1}^{+1} \tfrac{1}{2} Wr(1 + r) \, dr \right] (\Delta\tau)^2 \mathbf{f}_{i+1} \qquad \textbf{(10-78)}$$

After dividing both sides by $\int_{-1}^{+1} W \, dr$ and after some manipulation, we get [11]

$$[\mathbf{M} \quad + \gamma\mathbf{D} \, \Delta\tau + \lambda\mathbf{K}(\Delta\tau)^2]\mathbf{a}_{i+1}$$

$$= [2\mathbf{M} - (1 - 2\gamma)\mathbf{D} \, \Delta\tau - (\tfrac{1}{2} - 2\lambda + \gamma)\mathbf{K}(\Delta\tau)^2]\mathbf{a}_i$$

$$+ [-\mathbf{M} + (1 - \gamma)\mathbf{D} \, \Delta\tau - (\tfrac{1}{2} + \lambda - \gamma)\mathbf{K}(\Delta\tau)^2]\mathbf{a}_{i-1}$$

$$- (\tfrac{1}{2} + \lambda - \gamma)(\Delta\tau)^2\mathbf{f}_{i-1}$$

$$- (\tfrac{1}{2} - 2\lambda + \gamma)(\Delta\tau)^2\mathbf{f}_i - \lambda(\Delta\tau)^2\mathbf{f}_{i+1} \qquad \textbf{(10-79)}$$

where λ and γ are defined by

$$\lambda = \frac{\displaystyle\int_{-1}^{+1} \tfrac{1}{2} Wr(1 + r) \, dr}{\displaystyle\int_{-1}^{+1} W \, dr} \qquad \textbf{(10-80)}$$

and

$$\gamma = \frac{\displaystyle\int_{-1}^{+1} W(\tfrac{1}{2} + r) \, dr}{\displaystyle\int_{-1}^{+1} W \, dr} \qquad \textbf{(10-81)}$$

Special Cases

Equation (10-79) is a three-point recursion formula for the solution of Eq. (10-16). Various schemes arise depending on the choice of the weighting function W. Table 10-3 shows the values of λ and γ that correspond to the point collocation, subdomain collocation, and Galerkin weighted-residual methods. Note that λ and γ are not necessarily positive.

Point collocation

Recall that the weighting functions for the point collocation method are given by $\delta(r - r_j)$, such that

$$\int_{-1}^{+1} \delta(r - r_j) \, dr = 1 \qquad \text{for } r = r_j \qquad \textbf{(10-82a)}$$

Table 10-3   Summary of the Values of the Parameters λ and γ for Use in the Three-Point Recursion Formula

Weighted Residual Method	λ	γ
Point collocation at τ_{i-1}	0	$-\frac{1}{2}$
Point collocation at τ_i	0	$\frac{1}{2}$
Point collocation at τ_{i+1}	1	$\frac{3}{2}$
Subdomain collocation	$\frac{1}{6}$	$\frac{1}{2}$
Galerkin based on N_{i-1}	$-\frac{1}{5}$	$-\frac{1}{2}$
Galerkin based on N_i	$\frac{1}{10}$	$\frac{1}{2}$
Galerkin based on N_{i+1}	$\frac{4}{5}$	$\frac{3}{2}$

$$\int_{-1}^{+1} \delta(r - r_j)\, dr = 0 \qquad \text{for } r \neq r_j \qquad \textbf{(10-82b)}$$

where r_j is known as the collocation point. For point collocation at time τ_{i-1} where $r = -1$, it can be shown that $\lambda = 0$ and $\gamma = -\frac{1}{2}$. Similarly, for point collocation at times τ_i and τ_{i+1}, where $r = 0$ and $r = +1$, it can be shown that $\lambda = 0$ and $\gamma = \frac{1}{2}$, and $\lambda = 1$ and $\gamma = \frac{3}{2}$, respectively. These results are summarized in Table 10-3.

Example 10-8

Show that $\lambda = 0$ and $\gamma = -\frac{1}{2}$ in Eq. (10-79) for point collocation at time τ_{i-1}.

Solution

At time τ_{i-1}, we have $r = -1$ and the collocation point r_j must be -1. Thus the weighting function W is given by

$$W = \delta(r + 1)$$

and Eqs. (10-80) and (10-81) give

$$\lambda = \frac{\displaystyle\int_{-1}^{+1} \frac{1}{2}\delta(r + 1)r(1 + r)\, dr}{\displaystyle\int_{-1}^{+1} \delta(r + 1)\, dr} = \frac{0}{1} = 0$$

and

$$\gamma = \frac{\displaystyle\int_{-1}^{+1} \delta(r + 1)(\frac{1}{2} + r)\, dr}{\displaystyle\int_{-1}^{+1} \delta(r + 1)\, dr} = \frac{-\frac{1}{2}}{1} = -\frac{1}{2}$$

Subdomain collocation

The weighting function for the subdomain collocation method is given by $W = 1$ over the interval τ_{i-1} to τ_{i+1}. It is shown in Example 10-9 that $\lambda = \frac{1}{6}$ and $\gamma = \frac{1}{2}$ in this case.

Example 10-9

Show that $\lambda = \frac{1}{6}$ and $\gamma = \frac{1}{2}$ in Eq. (10-79) for subdomain collocation over the interval from τ_{i-1} to τ_{i+1}.

Solution

Using $W = 1$ in Eqs. (10-80) and (10-81) we get

$$\lambda = \frac{\int_{-1}^{+1} \frac{1}{2}(1)r(1 + r)\, dr}{\int_{-1}^{+1} (1)\, dr} = \frac{\int_{-1}^{+1} (r/2 + r^2/2)\, dr}{\int_{-1}^{+1} dr} = \frac{(r^2/4 + r^3/6)\Big|_{-1}^{+1}}{r\Big|_{-1}^{+1}} = \frac{\frac{2}{6}}{2} = \frac{1}{6}$$

and

$$\gamma = \frac{\int_{-1}^{+1} (1)(\frac{1}{2} + r)\, dr}{\int_{-1}^{+1} (1)\, dr} = \frac{(r/2 + r^2/2)\Big|_{-1}^{+1}}{r\Big|_{-1}^{+1}} = \frac{1}{2}$$

∎

Galerkin

Since the weighting functions for the Galerkin method are the shape function themselves, we may take

$$W = N_{i-1} = -\frac{1}{2}r(1 - r) \tag{10-83a}$$

$$W = N_i = (1 - r)(1 + r) \tag{10-83b}$$

or

$$W = N_{i+1} = \frac{1}{2}r(1 + r) \tag{10-83c}$$

The shape functions are shown graphically in Fig. 10-7. It can be shown that if Eq. (10-83a) is used, we get $\lambda = -\frac{1}{5}$ and $\gamma = -\frac{1}{2}$. Also, if Eq. (10-83b) is used, we get $\lambda = \frac{1}{10}$ and $\gamma = \frac{1}{2}$. Finally, if Eq. (10-83c) is used, we get $\lambda = \frac{4}{5}$ and $\gamma = \frac{3}{2}$.

Example 10-10

Show that $\lambda = -\frac{1}{5}$ and $\gamma = -\frac{1}{2}$ in Eq. (10-79) for Galerkin weighting corresponding to Eq. (10-83a).

Solution

From the definition λ, we have

$$\lambda = \frac{\int_{-1}^{+1} \frac{1}{2}[-\frac{1}{2}r(1 - r)]r(1 + r)\,dr}{\int_{-1}^{+1} -\frac{1}{2}r(1 - r)\,dr} = \frac{\frac{1}{2}\int_{-1}^{+1}(r^2 - r^4)\,dr}{\int_{-1}^{+1} r - r^2\,dr}$$

or

$$\lambda = \frac{\frac{1}{2}(r^3/3 - r^5/5)\Big|_{-1}^{+1}}{(r^2/2 - r^3/3)\Big|_{-1}^{+1}} = \frac{\frac{1}{2}(\frac{4}{15})}{-\frac{2}{3}} = -\frac{1}{5}$$

and from the definition of γ, we have

$$\gamma = \frac{\int_{-1}^{+1}[-\frac{1}{2}r(1 - r)](\frac{1}{2} + r)\,dr}{\int_{-1}^{+1} -\frac{1}{2}r(1 - r)\,dr} = \frac{\int_{-1}^{+1}(\frac{1}{2}r + \frac{1}{2}r^2 - r^3)\,dr}{\int_{-1}^{+1}(r - r^2)\,dr}$$

or

$$\gamma = \frac{(r^2/4 + r^3/6 - r^4/4)\Big|_{-1}^{+1}}{(r^2/2 - r^3/3)\Big|_{-1}^{+1}} = \frac{\frac{2}{6}}{-\frac{2}{3}} = -\frac{1}{2}$$

Initial and Boundary Conditions

We have seen how the second-order, vector-differential equation given by Eq. (10-16) may be solved by the three-point recursion formula given by Eq. (10-79). The values of λ and γ used in the recursive solution are summarized in Table 10-3. In order to start the solution process, the values of $\mathbf{a}_i$ and $\mathbf{a}_{i-1}$ are needed for $i = 1$ in order to determine $\mathbf{a}_{i+1}$. It will now be shown how the *initial conditions* on the nodal displacements and velocities can be used to provide the *starting vectors* $\mathbf{a}_0$ and $\mathbf{a}_1$. Two different methods are used; the first method is simpler but less accurate than the second. Both methods require that Eq. (10-16) be written as a set of two, first-order vector differential equations, or

$$\mathbf{M\dot{b}} + \mathbf{Db} + \mathbf{Ka} = \mathbf{f} \qquad \text{(10-84)}$$

$$\mathbf{\dot{a}} = \mathbf{b} \qquad \text{(10-85)}$$

It should be noted that since the vector $\mathbf{a}$ represents the nodal displacements, the vector $\mathbf{b}$ (defined to be $\mathbf{\dot{a}}$) must represent the nodal velocities. The *initial* displacements and velocities may be denoted as $\mathbf{a}_0$ and $\mathbf{b}_0$, respectively.

Euler starting method

The starting values for $\mathbf{a}_{i-1}$ are simply given by the initial nodal displacements $\mathbf{a}_0$. We seek the starting values for the vector $\mathbf{a}_i$. The simplest approach is to apply the forward difference scheme or Euler's method to Eq. (10-85) to get

$$\frac{\mathbf{a}(\tau + \Delta\tau) - \mathbf{a}(\tau)}{\Delta\tau} = \mathbf{b}(\tau)$$

or

$$\frac{\mathbf{a}_i - \mathbf{a}_{i-1}}{\Delta\tau} = \mathbf{b}_{i-1}$$

Solving for $\mathbf{a}_i$ gives

$$\mathbf{a}_i = \mathbf{a}_{i-1} + \mathbf{b}_{i-1}\,\Delta\tau \qquad \text{(10-86)}$$

For $i = 1$, Eq. (10-86) becomes

$$\mathbf{a}_1 = \mathbf{a}_0 + \mathbf{b}_0\,\Delta\tau \qquad \text{(10-87)}$$

Since $\mathbf{a}_0$ and $\mathbf{b}_0$ are provided by the initial conditions on the nodal displacements and velocities, respectively, Eq. (10-87) provides a means of obtaining $\mathbf{a}_1$. With $\mathbf{a}_0$ and $\mathbf{a}_1$ known, the three-point recursion formula given by Eq. (10-79) could be used to determine $\mathbf{a}_2$, and so forth. The recurrence scheme is shown schematically in Fig. 10-8.

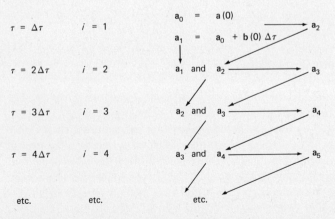

Figure 10-8 Schematic diagram of the three-point recursion process.

Crank-Nicolson starting method

While the Euler starting method is simple, it is not very accurate and a better approach results if Eqs. (10-84) and (10-85) are solved using the central difference or Crank-Nicolson method. In this case Eq. (10-84) becomes

$$\mathbf{M}\frac{\mathbf{b}_i - \mathbf{b}_{i-1}}{\Delta\tau} + \mathbf{D}\frac{\mathbf{b}_i + \mathbf{b}_{i-1}}{2} + \mathbf{K}\frac{\mathbf{a}_i + \mathbf{a}_{i-1}}{2} = \frac{\mathbf{f}_i + \mathbf{f}_{i-1}}{2} \qquad \textbf{(10-88)}$$

In a similar fashion, we may write Eq. (10-85) as

$$\frac{\mathbf{a}_i - \mathbf{a}_{i-1}}{\Delta\tau} = \frac{\mathbf{b}_i + \mathbf{b}_{i-1}}{2} \qquad \textbf{(10-89)}$$

The vector $\mathbf{b}_i$ contains the nodal velocities at the end of one time step $\Delta\tau$ and is unknown at the start of the solution. Therefore, if $\mathbf{b}_i$ is eliminated between Eqs. (10-88) and (10-89), we get the following convenient result:

$$\left[\frac{2\mathbf{M}}{\Delta\tau} + \mathbf{D} + \frac{\mathbf{K}\,\Delta\tau}{2}\right]\mathbf{a}_i$$

$$= \left[\frac{2\mathbf{M}}{\Delta\tau} + \mathbf{D} - \frac{\mathbf{K}\,\Delta\tau}{2}\right]\mathbf{a}_{i-1} + 2\mathbf{M}\mathbf{b}_{i-1} + \frac{\mathbf{f}_i\,\Delta\tau}{2} + \frac{\mathbf{f}_{i-1}\,\Delta\tau}{2} \qquad \textbf{(10-90)}$$

Equation (10-90) is useful because it provides the starting values of $\mathbf{a}_i$ when $i = 1$ for which we also have

$$\mathbf{a}_{i-1} = \mathbf{a}_0 = \mathbf{a}(0) \qquad \textbf{(10-91)}$$

$$\mathbf{b}_{i-1} = \mathbf{b}_0 = \mathbf{b}(0) \qquad \textbf{(10-92)}$$

$$\mathbf{f}_{i-1} = \mathbf{f}(0) \qquad \textbf{(10-93)}$$

$$\mathbf{f}_i = \mathbf{f}(\Delta\tau) \qquad \textbf{(10-94)}$$

Equations (10-91) and (10-92) represent the initial conditions on the nodal displacements and velocities, respectively. The vectors $\mathbf{f}(0)$ and $\mathbf{f}(\Delta\tau)$ in Eqs. (10-93) and (10-94) denote the values of $\mathbf{f}(\tau)$ at $\tau = 0$ and $\tau = \Delta\tau$, respectively.

Boundary conditions

The three-point recurrence scheme given by Eq. (10-79) is of the form $\mathbf{K}_{\text{eff}}\mathbf{a} = \mathbf{f}_{\text{eff}}$, where $\mathbf{K}_{\text{eff}}$ is given by $\mathbf{M} + \gamma\mathbf{D}\Delta\tau + \lambda\mathbf{K}(\Delta\tau)^2$ and $\mathbf{f}_{\text{eff}}$ is given by the right-hand side of Eq. (10-79). As in the case of the two-point recursion formula, we may use either Method 1 or Method 2 from Sec. 3-2 to impose the prescribed displacement boundary conditions. Since $\mathbf{K}_{\text{eff}}$ is symmetric, it is prudent to preserve symmetry if Method 1 is used. Note that an implicit solution for the nodal displacements results even if the lumped form of the mass matrix is used because Table 10-3 shows that λ and γ are never both zero.

10-10   INTRODUCTION TO MODAL ANALYSIS

In Chapter 7 the finite element method was applied to several different problems in static stress analysis. The resulting nodal displacements were then used to calculate the element strains, stresses, etc. In this chapter several sections were devoted to dynamic structural analysis. In this case, the nodal displacements are computed as a function of time for some set of initial conditions. The element resultants could be computed via the techniques from Chapter 7. In both cases, the prescribed displacements could be imposed in a routine manner.

Many times in dynamic structural analysis, the nodal displacements as a function of time are not really needed. Instead, the natural frequencies of the sustained vibrations are needed. For example, consider the case of the mass and spring system shown in Fig. 10-9(a). If the spring is assumed to be perfectly elastic and linear (i.e., linear-elastic) and massless, then the governing differential equation is given by

$$M\ddot{x} + Kx = f(t) \tag{10-95}$$

where $x(t)$ is the displacement of the mass M from the equilibrium position, K is the spring constant (i.e., the stiffness), and $f(t)$ denotes the forcing function. If damping is present as shown in Fig. 10-9(b), then the governing equation is given by

$$M\ddot{x} + D\dot{x} + Kx = f(t) \tag{10-96}$$

where D is the damping coefficient. Note that $\dot{x}$ and $\ddot{x}$ represent the velocity and acceleration of the mass, respectively. Note also that Eq. (10-96) reduces to Eq. (10-95) if the damping is negligible.

The *natural frequency* of a system is defined to be the frequency at which the system oscillates if the forcing function is identically zero. Nonzero initial conditions give rise to the oscillations in this case. If the system has only one discrete mass such as that in Fig. 10-9(a), then only one natural frequency results. However, if

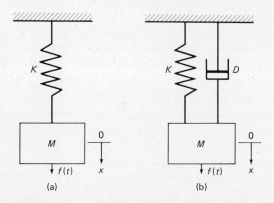

Figure 10-9   (a) A mass-spring system, and (b) a mass-spring-damper system.

the system is distributed such as a vibrating string, then an infinite number of natural frequencies exist. Because the finite element method approximates a continuous system by discrete elements (and hence masses), the stiffness-based finite element method is only capable of yielding the natural frequencies on the low end of the spectrum. In contrast, the force matrix method (not covered in this book) yields the natural frequencies on the high end of the spectrum. In both cases, the number and accuracy of the computed natural frequencies increase as the number of elements is increased.

Let us review how we obtain the natural frequencies by using Eq. (10-95) as an example. Since we desire the oscillatory or sinusoidal variations of x, we may represent $x(t)$ in complex polar form as

$$x(t) = \bar{x} e^{i\omega\tau} \tag{10-97}$$

where from *Euler's identity* we have

$$e^{i\omega\tau} = \cos \omega\tau + i \sin \omega\tau \tag{10-98}$$

Here i denotes the imaginary number defined by

$$i = \sqrt{-1} \tag{10-99}$$

and ω is the frequency of the vibration. In Eq. (10-97), $\bar{x}$ denotes the amplitude of the vibration corresponding to the frequency ω. Recall that the natural frequency is obtained by assuming a zero forcing function. It follows that

$$M\ddot{x} + Kx = 0 \tag{10-100}$$

and ω then represents the *undamped* natural frequency. From Eq. (10-97), we get

$$\dot{x} = \frac{dx}{d\tau} = i\omega\bar{x} e^{i\omega\tau}$$

and

$$\ddot{x} = \frac{d^2x}{d\tau^2} = (i\omega)^2 \bar{x} e^{i\omega\tau} = -\omega^2 \bar{x} e^{i\omega\tau} \tag{10-101}$$

Using Eqs. (10-97) and (10-101), we may write Eq. (10-100) as follows:

$$(-\omega^2 M + K)\bar{x} e^{i\omega\tau} = 0 \tag{10-102}$$

But $e^{i\omega\tau}$ is not generally zero, and we want to exclude $\bar{x} = 0$ as a solution because this would yield the trivial solution $x(\tau) = 0$. Therefore, the expression in the parenthesis must be zero, and we get

$$\omega = \sqrt{\frac{K}{M}} \tag{10-103}$$

The reader may recall this result from elementary vibrations. The natural frequency ω is frequently referred to as an *eigenvalue*. The nontrivial vector $\bar{x}$ is referred to as the eigenvector that corresponds to the frequency ω. A geometric interpretation of the eigenvector is given later in this section. The natural frequency of the mass-spring-damper system in Fig. 10-9(b) may be obtained in a similar manner by beginning with Eq. (10-96). However, the eigenvalues in this case are now complex. This implies that the oscillations decay with time.

Let us now return to Eq. (10-16) and extend this development to the finite element method. Only the *undamped* natural frequencies are to be determined here. In this case, Eq. (10-16) reduces to

$$\mathbf{M\ddot{a}} + \mathbf{Ka} = 0 \tag{10-104}$$

where the assemblage nodal force vector (the forcing function) has been set to zero. Now we represent the vector of nodal displacements as

$$\mathbf{a} = \mathbf{\bar{a}}e^{i\omega\tau} \tag{10-105}$$

from which it follows that

$$\mathbf{\ddot{a}} = (i\omega)^2\mathbf{\bar{a}}e^{i\omega\tau} = -\omega^2\mathbf{\bar{a}}e^{i\omega\tau} \tag{10-106}$$

where $\mathbf{\bar{a}}$ is now referred to as the eigenvector corresponding to the natural frequency ω. Equation (10-104) becomes

$$(-\omega^2\mathbf{M} + \mathbf{K})\mathbf{\bar{a}}e^{i\omega\tau} = 0 \tag{10-107}$$

Since the nontrival solutions for $\mathbf{\bar{a}}$ are sought, the determinant of the expression in the parenthesis must be zero, or

$$\det(-\omega^2\mathbf{M} + \mathbf{K}) = 0 \tag{10-108}$$

because this guarantees that the matrix $-\omega^2\mathbf{M} + \mathbf{K}$ is singular, thus admitting nontrivial solutions for the vector $\mathbf{\bar{a}}$. Consequently, Eq. (10-108) may be solved for the natural frequencies ω of the structure.

Note that if N nodes are used in the discretization and each node has m degrees of freedom, then Eq. (10-108) results in a polynomial of order mN and hence yields mN natural frequencies. Recall that the original (continuous) structure really has an infinite number of natural frequencies. It follows that the natural frequencies on the high end of the spectrum (i.e., for large ω's) are not accurately computed. Fortunately, the frequencies on the low end of the spectrum are quite accurate if enough elements are used. Standard FORTRAN library subroutines may be used to determine the mN values of ω^2 (and hence ω) that satisfy Eq. (10-108). The roots may be shown to be positive real numbers because the matrices $\mathbf{K}$ and $\mathbf{M}$ are always positive definite. The ω's are recognized to be eigenvalues.

Recall that the vector $\mathbf{\bar{a}}$ in Eq. (10-105) may be referred to as the eigenvector that corresponds to the frequency ω. In particular, let us define $\mathbf{\bar{a}}_j$ to be the eigenvector that corresponds to the natural frequency ω_j. It follows from Eq. (10-107) that the eigenvector $\mathbf{\bar{a}}_j$ must satisfy

$$(-\omega_j^2\mathbf{M} + \mathbf{K})\mathbf{\bar{a}}_j e^{i\omega\tau} = 0 \tag{10-109}$$

But $e^{i\omega\tau}$ is not zero, so we must have

$$(\omega_j^2 \mathbf{M} - \mathbf{K})\overline{\mathbf{a}}_j = \mathbf{0} \tag{10-110}$$

Equation (10-110) has a nontrivial solution for the eigenvector $\overline{\mathbf{a}}_j$ because the ω_j are computed such that Eq. (10-108) holds; otherwise only the trivial solution for $\overline{\mathbf{a}}_j$ exists. It should be recalled that the system of equations in Eq. (10-110) is singular. Therefore, the solution for the eigenvector is not unique and one of the entries in $\overline{\mathbf{a}}_j$ is arbitrary. Because the eigenvectors are not unique, they are generally normalized such that

$$\overline{\mathbf{a}}_j^T \mathbf{M} \overline{\mathbf{a}}_j = 1$$

Standard FORTRAN library subroutines may be used to compute the eigenvectors. The reader is referred to the book by Zienkiewicz [12] for a more detailed discussion.

The eigenvectors may be interpreted geometrically as follows. Consider the simply supported beam shown in Fig. 10-10(a). The eigevector $\overline{\mathbf{a}}_1$ that corresponds to the lowest natural frequency ω_1 results in the *mode shape* shown in Fig. 10-10(b). The eigenvector $\overline{\mathbf{a}}_2$ that corresponds to the second lowest natural frequency ω_2 results in the mode shape shown in Fig. 10-10(c). Because the beam is continuous, there are really an infinite number of natural frequencies and eigenvectors or mode shapes. The human eye sees the superposition of all mode shapes during the vibration of a structure. If a system with negligible damping is excited at one of the natural frequencies of the structure, *resonance* will ocur. The result is usually catastrophic.

Finally it should be mentioned that the lumped form of the mass matrix may be used to determine the natural frequencies and mode shapes of the structure. However, the results from the consistent mass matrix will generally be more accurate [13].

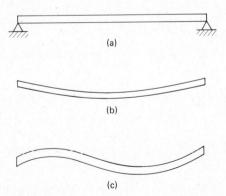

(a)

(b)

(c)

Figure 10-10   (a) A simply supported beam with (b) its first mode shape and (c) its second mode shape.

REFERENCES

1. Zienkiewicz, O. C., *The Finite Element Method*, McGraw-Hill (UK), London, 1977, pp. 60–62.
2. Hibbeler, R. C., *Engineering Mechanics: Dynamics*, Macmillan, New York, 1983, p. 80.
3. Clough, R. W., and J. Penzien, *Dynamics of Structures*, McGraw-Hill, New York, 1975.
4. Gresho, P., and R. L. Lee, "Don't Suppress the Wiggles—They're Telling You Something," *Finite Element Methods for Convection Dominated Flows*, AMD-Vol. 34, pp. 37–61, Winter Annual Meeting of the ASME, New York, Dec. 2–7, 1979.
5. Hinton, E., A. Rock, and O. C. Zienkiewicz, "A Note on Mass Lumping in Related Process in the Finite Element Method," *Int. J. Earthquake Eng. Struct. Dynam.*, Vol. 4., pp. 245–249, 1976.
6. Zienkiewicz, O. C., *The Finite Element Method*, McGraw-Hill (UK), London, 1977, pp. 535–539.
7. Zienkiewicz, O. C., *The Finite Element Method*, McGraw-Hill (UK), London, 1977, pp. 576–588.
8. Bathe, K. -J., *Finite Element Procedures in Engineering Analysis*, Prentice-Hall, Englewood Cliffs, N.J., 1982, pp. 553–554.
9. Reddy, J. N., *An Introduction to the Finite Element Method*, McGraw-Hill, New York, 1984, pp. 50–56.
10. Zienkiewicz, O. C., *The Finite Element Method*, McGraw-Hill (UK), London, 1977, pp. 740–741.
11. Zienkiewicz, O. C., *The Finite Element Method*, McGraw-Hill (UK), London, 1977, p. 582.
12. Zienkiewicz, O. C., *The Finite Element Method*, McGraw-Hill (UK), London, 1977, pp. 546–558.
13. Reddy, J. N., *An Introduction to the Finite Element Method*, McGraw-Hill, New York, 1984, p. 406.

PROBLEMS

10-1 The element mass matrix is given by Eq. (10-13). Since this expression is applicable to any element, evaluate it for the case of the three-node lineal element by using length coordinates. Use Eq. (6-48) to evaluate the integrals.

10-2 The element mass matrix is given by Eq. (10-13). Since this expression is applicable to any element, evaluate it for the case of the three-node lineal element by using the serendipity coordinate r. *Hint:* Use Gauss-Legendre quadrature of such an order that the integrals are evaluated exactly.

10-3 Consider the expression for the element mass matrix given by Eq. (10-13). Since this expression is applicable to any element, evaluate it for the case of the three-node triangular element by writing the shape functions in terms of area coordinates. *Hint:* Use the numerical integration formula from Section 9-9 of such an order that the integrals are evaluated exactly.

10-4 Consider the expression for the element mass matrix given by Eq. (10-13). Since this expression is applicable to any element, evaluate it for the case of the four-node rectangular element by writing the shape functions in terms of the serendipity coordinates r and s. *Hint:* Use Gauss-Legendre quadrature of such an order that the integrals are evaluated exactly.

10-5 The element capacitance matrix is given by Eq. (10-26) for all one-dimensional elements. Evaluate the integral for the case of the three-node lineal element by using length coordinates. Use Eq. (6-48) to evaluate the integrals.

10-6 One form of the element capacitance matrix is given by Eq. (10-26). Since this expression is applicable to any one-dimensional thermal analysis problem, evaluate it for the case of the three-node lineal element by using the serendipity coordinate r. *Hint:* Use Gauss-Legendre quadrature of such an order that the integrals are evaluated exactly.

10-7 Consider the expression for the element capacitance matrix given by Eq. (10-25). Since this expression is applicable to any two-dimensional thermal analysis problem, evaluate it for the case of the four-node rectangular element by writing the shape functions in terms of the serendipity coordinates r and s. *Hint:* Use Gauss-Legendre quadrature of such an order that the integrals are evaluated exactly.

10-8 Consider the expression for the element capacitance matrix given by Eq. (10-25). Since this expression is applicable to any two-dimensional thermal analysis problem, evaluate it for the case of the eight-node rectangular element by writing the shape functions in terms of the serendipity coordinates r and s. In other words, show that Eq. (10-35) holds in this case. *Hint:* Use Gauss-Legendre quadrature of such an order that the integrals are evaluated exactly. It may be more convenient to write a short FORTRAN program.

10-9 Determine the lumped form of the mass matrix for Problem 10-1. Clearly explain or indicate how the lumped form is obtained from the consistent form.

10-10 Determine the lumped form of the mass matrix for Problem 10-2. Clearly explain or indicate how the lumped form is obtained from the consistent form.

10-11 Determine the lumped form of the mass matrix for Problem 10-3. Clearly explain or indicate how the lumped form is obtained from the consistent form.

10-12 Determine the lumped form of the mass matrix for Problem 10-4. Clearly explain or indicate how the lumped form is obtained from the consistent form.

10-13 Determine the lumped form of the capacitance matrix for Problem 10-5. Clearly explain or indicate how the lumped form is obtained from the consistent form.

10-14 Determine the lumped form of the capacitance matrix for Problem 10-6. Clearly explain or indicate how the lumped form is obtained from the consistent form.

10-15 Determine the lumped form of the capacitance matrix for Problem 10-7. Clearly explain or indicate how the lumped form is obtained from the consistent form.

10-16 Determine the lumped form of the capacitance matrix for Problem 10-8. Clearly explain or indicate how the lumped form is obtained from the consistent form.

10-17   Conisder the following assemblage capacitance and stiffness (or conductance) matrices

$$C = \begin{bmatrix} 4 & 2 \\ 2 & 4 \end{bmatrix} \quad \text{and} \quad K = \begin{bmatrix} 8 & -5 \\ -5 & 8 \end{bmatrix}$$

and the following assemblage nodal force vector:

$$f = \begin{bmatrix} 200 \\ 200 \end{bmatrix}$$

Using the forward-difference (or Euler's) method, determine the nodal temperatures for the first four time steps if a 0.25-sec time step is used. Assume that there are no prescribed temperatures and both nodes are initially at 50°F. Explain what might happen if the time step is increased too much.

10-18   Consider the following assemblage capacitance and stiffness (or conductance) matrices

$$C = \begin{bmatrix} 8 & 4 \\ 4 & 8 \end{bmatrix} \quad \text{and} \quad K = \begin{bmatrix} 14 & -11 \\ -11 & 14 \end{bmatrix}$$

and the following assemblage nodal force vector:

$$f = \begin{bmatrix} 125 \\ 125 \end{bmatrix}$$

Using the forward-difference (or Euler's) method, determine the nodal temperatures for the first five time steps if a 0.2-sec time step is used. Assume that there are no prescribed temperatures and both nodes are initially at 25°C. Explain what might happen if the time step is increased too much.

10-19   Consider the following assemblage capacitance and stiffness (or conductance) matrices

$$C = \begin{bmatrix} 500 & 250 & 0 \\ 250 & 1000 & 250 \\ 0 & 250 & 500 \end{bmatrix} \quad \text{and} \quad K = \begin{bmatrix} 400 & -400 & 0 \\ -400 & 800 & -400 \\ 0 & -400 & 400 \end{bmatrix}$$

and the following assemblage nodal force vector:

$$f = \begin{bmatrix} 2 \\ 4 \\ 2 \end{bmatrix}$$

Using the forward-difference (or Euler's) method, determine the nodal temperatures for the first 20 time steps if a 0.1-sec time step is used. Assume that node 1 is to be prescribed at 100°C and that all nodes are initially at 10°C. Explain what might happen if the time step is increased too much.

10-20   Consider the following assemblage capacitance and stiffness (or conductance) matrices

$$C = \begin{bmatrix} 10 & 5 & 0 \\ 5 & 20 & 5 \\ 0 & 5 & 10 \end{bmatrix} \quad \text{and} \quad K = \begin{bmatrix} 100 & -100 & 0 \\ -100 & 200 & -100 \\ 0 & -100 & 100 \end{bmatrix}$$

and the following assemblage nodal force vector:

$$\mathbf{f} = \begin{bmatrix} 8 \\ 16 \\ 8 \end{bmatrix}$$

Using the forward-difference (or Euler's) method, determine the nodal temperatures for the first five time steps if a 0.05-sec time step is used. Assume that node 1 is to be prescribed at 150°F and that all nodes are initially at 75°F. Explain what might happen if the time step is increased too much.

10-21 Repeat Problem 10-17 with the backward difference scheme and a 0.5-sec time step. Explain what happens to the results as the time step is increased.

10-22 Repeat Problem 10-18 with the backward difference scheme and a 0.5-sec time step. Explain what happens to the results as the time step is increased.

10-23 Repeat Problem 10-19 for 10 time steps with the backward difference scheme and a 0.20-sec time step. Explain what happens to the results as the time step is increased.

10-24 Repeat Problem 10-20 with the backward difference scheme and a 0.05-sec time step. Explain what happens to the results as the time step is increased.

10-25 Repeat Problem 10-17 with the central difference scheme and a 0.5-sec time step. Explain what happens to the results as the time step is increased.

10-26 Repeat Problem 10-18 with the central difference scheme and a 0.5-sec time step. Explain what happens to the results as the time step is increased.

10-27 Repeat Problem 10-19 with the central difference scheme and a 0.20-sec time step. Explain what happens to the results as the time step is increased.

10-28 Repeat Problem 10-20 with the central difference scheme and a 0.05-sec time step. Explain what happens to the results as the time step is increased.

10-29 Show that point collocation at time τ_i gives $\theta = 0$. Which of the finite difference schemes does this situation correspond to? Please explain.

10-30 Show that point collocation at time τ_{i+1} gives $\theta = 1$. Which of the finite difference schemes does this situation correspond to? Please explain.

10-31 Show that the parameter θ takes on a value of $\frac{1}{3}$ if the weighting function is given by Eq. (10-67a). Is this value of θ likely to result in a stable solution for any time step $\Delta\tau$? Pleast explain the plausibility of your conclusion.

10-32 Write a small FORTRAN program that can be used to obtain the transient temperature distributions in a fin. It is recommended that the program from Problem 4-78 be used as the starting point. Allow for different values of θ. In addition, allow for either the lumped or consistent capacitance matrix (via an input parameter ILUMP, where ILUMP is 0 and 1 for a consistent and lumped capacitance matrix, respectively).

10-33 With the help of the computer program from Problem 10-32 (or one furnished by the instructor) repeat Example 10-6 for $\theta = 0$ and for both a consistent and lumped capacitance matrix. Use a 0.1-sec time step and carry out the solution until steady state is achieved (approximately 6 sec). Is the solution stable? How do the results compare with those from Example 10-6 (with $\theta = \frac{2}{3}$)?

10-34 With the help of the computer program from Problem 10-32 (or one furnished by the instructor) repeat Example 10-6 for $\theta = \frac{1}{2}$ and for both a consistent and lumped capacitance matrix. Use a 0.1-sec time step and carry out the solution until steady state is achieved (approximately 6 sec). Is the solution stable? How do the results compare with those from Example 10-6 (with $\theta = \frac{2}{3}$)?

10-35 With the help of the computer program from Problem 10-32 (or one furnished by the instructor) repeat Example 10-6 for $\theta = 1$ and for both a consistent and lumped capacitance matrix. Use a 0.1-sec time step and carry out the solution until steady state is achieved (approximately 6 sec). How do the results compare with those from Example 10-6 (with $\theta = \frac{2}{3}$)?

10-36 With the help of the computer program from Problem 10-32 (or one furnished by the instructor) repeat Example 10-6 for $\theta = \frac{1}{3}$ and for both a consistent and lumped capacitance matrix. Use a 0.1-sec time step and carry out the solution until steady state is achieved (approximately 6 sec). Is the solution stable? How do the results compare with those from Example 10-6 (with $\theta = \frac{2}{3}$)?

10-37 Show that $\lambda = 0$ and $\gamma = \frac{1}{2}$ for point collocation at τ_i.

10-38 Show that $\lambda = 1$ and $\gamma = \frac{3}{2}$ for point collocation at τ_{i+1}.

10-39 Show that $\lambda = \frac{1}{10}$ and $\gamma = \frac{1}{2}$ for Galerkin weighting corresponding to Eq. (10-83b).

10-40 Show that $\lambda = \frac{4}{5}$ and $\gamma = \frac{3}{2}$ for Galerkin weighting corresponding to Eq. (10-83c).

A

Structural and Thermal Properties

A-1 Structural Properties

English Units

	Modulus of elasticity 10^6 psi	Coefficient of thermal expansion 10^{-6} in/in-°F	Poisson's ratio —	Weight density lbf/in^3
Aluminum, 6061 alloy	10.	12.8	0.33	0.100
Brass	15.	10.5	0.33	0.300
Bronze	15.	10.0	0.33	0.300
Cast iron	15.	6.0	0.27	0.256
Copper, hard drawn	17.	9.4	0.33	0.340
Steel, hot rolled low carbon	30.	6.5	0.30	0.283
Steel, hot rolled high carbon	30.	7.0	0.30	0.283

SI Units

	Modulus of elasticity 10^{10} N/m^2	Coefficient of thermal expansion 10^{-6} m/m-°C	Poisson's ratio —	Weight density N/cm^3
Aluminum, 6061 alloy	6.90	23.0	0.33	0.0271
Brass	10.3	18.9	0.33	0.0814
Bronze	10.3	18.0	0.33	0.0814
Cast iron	10.3	10.8	0.27	0.0695
Copper, hard drawn	11.7	16.9	0.33	0.0923
Steel, hot rolled low carbon	20.7	11.7	0.30	0.0758
Steel, hot rolled high carbon	20.7	12.6	0.30	0.0758

A-2 Thermal Properties

English Units

	Thermal conductivity Btu/hr-ft-°F	Mass density lbm/ft^3	Specific heat Btu/lbm-°F
Aluminum (pure)	117.	169.	0.208
Brass	64.	530.	0.089
Bronze	15.	540.	0.079
Cast iron	30.	454.	0.100
Copper (pure)	224.	558.	0.092
Stainless steel	9.2	488.	0.107
Steel, hot rolled low carbon	31.1	488.	0.108
Steel, hot rolled high carbon	21.7	483.	0.113

SI Units

	Thermal conductivity W/m-°C	Mass density kg/m^3	Specific heat J/kg-°C
Aluminum (pure)	203.	2720.	872.
Brass	111.	8520.	373.
Bronze	26.	8670.	331.
Cast iron	52.	7305.	419.
Copper (pure)	389.	8980.	386.
Stainless steel	16.	7820.	449.
Steel, hot rolled low carbon	54.	7830.	452.
Steel, hot rolled high carbon	38.	7750.	473.

B

Program TRUSS

B-1 Listing of Source Code for Program TRUSS*

```
        PROGRAM TRUSS

        DIMENSION XCOOR(20), YCOOR(20), NODI(30), NODJ(30), MATFLG(30),
     1      DATMAT(5,2), NBCX(20), NBCY(20), DISP(5), FORCE(15), XX(2),
     2      YY(2), DIRCOS(2), SOLN(40), ASM(40,40), ANFV(40), ESM(4,4)
        COMMON NNODES, NELEM, NMATLS, NPDIS, NPLDS, LOUT, LPRINT, LIN,
     1      TITLE(20)

C....   LCONSL IS THE SCREEN DEVICE NUMBER
        LCONSL = 3
  6     WRITE (LCONSL,8)
  8     FORMAT (/////,2X,'INPUT THE NUMBER OF THE INPUT FILE (6-10): ')
        READ (LCONSL,20) LIN
        IF (LIN .EQ. 0) GO TO 9999

C       READ AN 80-COLUMN TITLE "CARD"
        READ (LIN,10) (TITLE(J), J = 1, 20)
 10     FORMAT (20A4)

C....   READ THE NUMBER OF NODES, ELEMENTS, MATERIALS, PRESCRIBED
C       DISPLACEMENTS, POINT LOADS, AND OUTPUT UNIT
        READ (LIN,20) NNODES, NELEM, NMATLS, NPDIS, NPLDS, LOUT
 20     FORMAT (6I8)
```

---

*The main program TRUSS is followed by the subroutines in the order called. Section B-2 describes the input to this program in detail. The input is also summarized in Table 3-1 in the main part of the text.

```
C....    ZERO OUT THE "BOUNDARY CONDITION" FLAGS
         DO 1000  I = 1, 20
             NBCX(I) = 0
             NBCY(I) = 0
 1000    CONTINUE

C....    READ MESH, MATERIAL, AND BOUNDARY CONDITION DATA
         CALL NODGEN (XCOOR, YCOOR)
         CALL ELEGEN (NODI, NODJ, MATFLG)
         CALL MATERL (DATMAT)
         CALL BCOND (NBCX, NBCY, FORCE, DISP)

C....    PRINT SUMMARY OF INPUT DATA
         CALL SUMMRY (XCOOR, YCOOR, NODI, NODJ, MATFLG, DATMAT, NBCX,
     1       NBCY, FORCE, DISP)

C....    ZERO OUT THE ASSEMBLAGE NODAL FORCE VECTOR AND STIFFNESS MATRIX

         DO 2000  I = 1, 40
             ANFV(I) = 0.0
         DO 2000  J = 1, 40
             ASM(I,J) = 0.0
 2000    CONTINUE

C....    GENERATE THE GLOBAL ELEMENT AND ASSEMBLAGE STIFFNESS MATRIX
         DO 2500  L = 1, NELEM
             CALL COORDS (L, NODI, NODJ, XCOOR, YCOOR, XX, YY)
             CALL LENGTH (XX, YY, ELENTH)
             CALL PROPTY (L, MATFLG, DATMAT, AREA, ELMOD)
             CALL TRANSF (L, XX, YY, ELENTH, DIRCOS)
             CALL STIFF (AREA, ELMOD, ELENTH, DIRCOS, ESM)
             CALL ASSEMK (L, NODI, NODJ, ESM, ASM)
 2500    CONTINUE

C....    GENERATE ASSEMBLAGE NODAL FORCE VECTOR
         CALL ANFVEC (NBCX, NBCY, FORCE, ANFV)
C....    IMPOSE RESTRAINTS ON THE NODAL DISPLACEMENTS
         CALL PDBC (NBCX, NBCY, DISP, ASM, ANFV)
         N = 2*NNODES

C....    SOLVE THE EQUATIONS, PRINT THE NODAL DISPLACEMENTS, AND PRINT
C            THE ELEMENT RESULTANTS
         CALL EQSOLV (N, ASM, ANFV, SOLN)
         CALL PRINTN (SOLN)
         CALL POSTPR (NODI, NODJ, XCOOR, YCOOR, MATFLG, DATMAT, SOLN)
         GO TO 6

 9999    CONTINUE
         STOP
         END


         SUBROUTINE NODGEN (XCOOR, YCOOR)
C....    READS NODAL COORDINATE DATA AND GENERATES NODES
         DIMENSION XCOOR(20), YCOOR(20)
         COMMON NNODES, NELEM, NMATLS, NPDIS, NPLDS, LOUT, LPRINT, LIN,
     1       TITLE(20)

         READ (LIN,20)  SUBT
 20      FORMAT (A4)
```

```
   25      CONTINUE
C....      READ STARTING NODE NUMBER, INCREMENT, AND FINAL NODE NUMBER
           READ (LIN,30) NI, NG, NF
   30      FORMAT (3I8)

C....      IF NI IS ZERO, RETURN SINCE NODE SPECIFICATION AND GENERATION
C....      IS COMPLETE
           IF (NI .EQ. 0) GO TO 9999


C....      IF NG IS ZERO, NO GENERATION IS DESIRED.  NODES ARE SPECIFIED
C....      ON A ONE BY ONE BASIS.
           IF (NG .NE. 0) GO TO 40
           READ (LIN,35) XCOOR(NI), YCOOR(NI)
   35      FORMAT (4F8.0)
           GO TO 25

   40      CONTINUE
C....      NODAL GENERATION--XI,YI ARE THE STARTING COORDINATE PAIR AND
C              XF,YF ARE THE ENDING COORDINATE PAIR
           READ (LIN,35) XI, YI, XF, YF
           DIV = (NF - NI) / NG
           DX = (XF - XI) / DIV
           DY = (YF - YI) / DIV
           NDIV = DIV - 1
           NN = NI + NG
           NII = NI
           XCOOR(NI) = XI
           YCOOR(NI) = YI

           DO 1000  I = 1, NDIV
               XCOOR(NN) = XCOOR(NII) + DX
               YCOOR(NN) = YCOOR(NII) + DY
               NII = NN
               NN = NN + NG
 1000      CONTINUE

C....      FINAL COORDINATES DEFINED DIRECTLY TO AVOID ROUNDOFF ERROR
           XCOOR(NF) = XF
           YCOOR(NF) = YF
           GO TO 25

 9999      CONTINUE
           RETURN
           END


           SUBROUTINE ELEGEN (NODI, NODJ, MATFLG)
C....      READS ELEMENT DATA AND GENERATES THE ELEMENTS
           DIMENSION NODI(30), NODJ(30), MATFLG(30)
           COMMON NNODES, NELEM, NMATLS, NPDIS, NPLDS, LOUT, LPRINT, LIN,
      1        TITLE(20)

           READ (LIN,20)  SUBT
   20      FORMAT (A4)

   25      CONTINUE
C....      READS THE STARTING ELEMENT NUMBER, MATERIAL SET FLAG, ELEMENT
C              INCREMENT, FINAL ELEMENT NUMBER, AND NODAL INCREMENT
```

```
        READ (LIN,30)  LI, MS, LG, LF, NG
 30     FORMAT (5I8)

C....   IF LI IS EQUAL TO ZERO, THEN ELEMENT DATA IS COMPLETED
        IF (LI .EQ. 0) GO TO 9999

C....   READ THE GLOBAL NODE NUMBERS OF ELEMENT LI
        READ (LIN,30) NI, NJ
        NODI(LI) = NI
        NODJ(LI) = NJ
        MATFLG(LI) = MS

C....   IF LG IS ZERO, INPUT ELEMENT DATA ON AN ELEMENT-BY-ELEMENT BASIS;
C           OTHERWISE, GENERATE OTHER ELEMENT DATA
        IF (LG .EQ. 0)  GO TO 25
 50     CONTINUE

        LI = LI + LG
        NI = NI + NG
        NJ = NJ + NG

        IF (LI .GT. LF)  GO TO 25
        NODI(LI) = NI
        NODJ(LI) = NJ
        MATFLG(LI) = MS
        GO TO 50

 9999   CONTINUE
        RETURN
        END



        SUBROUTINE MATERL(DATMAT)
C....   READS IN THE MATERIAL PROPERTY DATA
        DIMENSION DATMAT(5,2)
        COMMON NNODES, NELEM, NMATLS, NPDIS, NPLDS, LOUT, LPRINT, LIN,
     1      TITLE(20)

        READ (LIN,20)  SUBT
 20     FORMAT (A4)

C....   READ FOR MATERIAL SET MSNO: CROSS-SECTIONAL AREA [DATMAT(MSNO,1)]
C           AND MODULUS OF ELASTICITY [DATMAT(MSNO,2)]
 30     CONTINUE
        READ (LIN,40)  MSNO, AREA, ELMOD
 40     FORMAT (I8, 2F8.0)
        IF (MSNO .EQ. 0 .OR. MSNO .GT. 5) GO TO 9999
            DATMAT(MSNO,1) = AREA
            DATMAT(MSNO,2) = ELMOD
            GO TO 30
 9999   CONTINUE
        RETURN
        END



        SUBROUTINE BCOND (NBCX, NBCY, FORCE, DISP)
C....   READS AND GENERATES BOUNDARY CONDITION FLAGS, NODAL FORCES, AND
C           NODAL DISPLACEMENTS
        DIMENSION NBCX(20), NBCY(20), FORCE(15), DISP(5)
```

```
         COMMON NNODES, NELEM, NMATLS, NPDIS, NPLDS, LOUT, LPRINT, LIN,
    1        TITLE(20)

C....    ZERO OUT THE NBCX AND NBCY ARRAYS
         DO 10  I = 1, 20
             NBCX(I) = 0
             NBCY(I) = 0
   10    CONTINUE

         READ (LIN,20)  SUBT
   20    FORMAT (A4)

C....    READS THE STARTING NODE NUMBER, BC FLAGS IN X- AND
C....            Y-DIRECTIONS, NODAL INCREMENT, AND ENDING NODE NUMBER
   25    READ (LIN,30) NI, IBCX, IBCY, NG, NF
   30    FORMAT (5I8)

C....    IF NI IS ZERO, BC FLAG INPUT IS COMPLETE, READ NODAL FORCES
         IF (NI .EQ. 0) GO TO 200

C....    IF NG IS ZERO, NO GENERATION OF BC'S IS DESIRED.  BC INFO IS
C....    SPECIFIED ON A NODE-BY-NODE BASIS
         IF (NG .NE. 0) GO TO 100
         NBCX(NI) = IBCX
         NBCY(NI) = IBCY
         GO TO 25

C....    GENERATION PORTION OF BC'S FLAGS
  100    CONTINUE
         NN = NI
  120    CONTINUE
         NBCX(NN) = IBCX
         NBCY(NN) = IBCY
         NN = NN + NG
         IF (NN .GT. NF)  GO TO 25
         GO TO 120

  200    CONTINUE
C....    READ APPLIED NODAL FORCES
         READ (LIN,20) SUBT

C....    READ FORCE FLAG AND FORCE
  220    CONTINUE
         READ (LIN,230) NFORCE, FORC
  230    FORMAT (I8, F8.0)
         IF (NFORCE .EQ. 0  .OR. NFORCE .GT. 15)  GO TO 300
             FORCE(NFORCE) = FORC
             GO TO 220

  300    CONTINUE
C....    READ PRESCRIBED NODAL DISPLACEMENTS
         READ (LIN,20) SUBT

C....    READ DISPLACEMENT FLAG AND DISPLACEMENT
  320    CONTINUE
         READ (LIN,230)  NDISP, DISPL
         IF (NDISP .EQ. 0  .OR. NDISP .GT. 5) GO TO 9999
             DISP(NDISP) = DISPL
             GO TO 320
```

```
9999    CONTINUE
        RETURN
        END



        SUBROUTINE SUMMRY (XCOOR, YCOOR, NODI, NODJ, MATFLG, DATMAT,
     1      NBCX, NBCY, FORCE, DISP)
        DIMENSION XCOOR(20), YCOOR(20), NODI(30), NODJ(30), MATFLG(30),
     1      DATMAT(5,2), NBCX(20), NBCY(20), DISP(5), FORCE(15)
        COMMON NNODES, NELEM, NMATLS, NPDIS, NPLDS, LOUT, LPRINT, LIN,
     1      TITLE(20)


        WRITE (LOUT, 100)  TITLE
100     FORMAT (2X, 20A4, /)

        WRITE (LOUT, 110)  NNODES, NELEM, NMATLS, NPDIS, NPLDS, LOUT
110     FORMAT (8X, 'NUMBER OF NODES:    ', I3, /,
     1          8X, 'NUMBER OF ELEMENTS: ', I3, /,
     2          8X, 'NUMBER OF MATERIALS: ', I3, /,
     3          8X, 'NUMBER OF PRES DISP: ', I3, /,
     4          8X, 'NUMBER OF PT LOADS:  ', I3, /,
     5          8X, 'OUTPUT UNIT NUMBER:  ', I3)

200     WRITE (LOUT,230)
230     FORMAT (/, 2X, 'NODE NO.', 4X, 'IBCX', 2X, 'IBCY', 5X,
     1      'X-COORD', 6X, 'Y-COORD')
        DO 250  I = 1, NNODES
        WRITE (LOUT,240)  I, NBCX(I), NBCY(I), XCOOR(I), YCOOR(I)
240     FORMAT (2X, I4, 7X, I4, 2X, I4, 4X, 2G13.5)
250     CONTINUE

300     WRITE (LOUT,320)
320     FORMAT (/, 2X, 'ELEMENT NO.', 4X, 'NODE I', 3X, 'NODE J',
     1      8X, 'MAT SET FLAG')
        DO 350  I = 1, NELEM
        WRITE (LOUT, 340) I, NODI(I), NODJ(I), MATFLG(I)
340     FORMAT (I6, 10X, I4, 5X, I4, 13X, I4)
350     CONTINUE

        WRITE (LOUT, 380)
380     FORMAT (/, 2X, 'MATERIAL', 7X, 'AREA', 7X 'ELASTIC MODULUS')
400     DO 450  I = 1, NMATLS
        WRITE (LOUT, 420) I, DATMAT(I,1), DATMAT(I,2)
420     FORMAT (4X, I2, 8X, G11.4, 4X, G11.4)
450     CONTINUE

        WRITE (LOUT, 510)
510     FORMAT (/, 2X, 'SUMMARY OF DIFFERENT EXTERNAL LOADS',/,
     1          10X, 'NUMBER', 4X, 'NODAL FORCE')
        DO 550  I = 1, NPLDS
        WRITE (LOUT,520) I, FORCE(I)
520     FORMAT (12X, I2, 5X, G13.4)
550     CONTINUE

        WRITE (LOUT,610)
610     FORMAT (/, 2X, 'SUMMARY OF PRESCRIBED NODAL DISPLACEMENTS',
     1          /, 10X, 'NUMBER', 4X, 'NODAL DISP.')
        DO 650  I = 1, NPDIS
        WRITE (LOUT,520) I, DISP(I)
```

```
650     CONTINUE

        RETURN
        END



        SUBROUTINE COORDS (L, NODI, NODJ, XCOOR, YCOOR, XX, YY)
C....    DETERMINES X- AND Y-COORDINATES FOR NODES I AND J OF ELEMENT L
        DIMENSION NODI(30), NODJ(30), XCOOR(20), YCOOR(20), XX(2),
     1          YY(2)

        NDI = NODI(L)
        NDJ = NODJ(L)
        XX(1) = XCOOR(NDI)
        XX(2) = XCOOR(NDJ)
        YY(1) = YCOOR(NDI)
        YY(2) = YCOOR(NDJ)
        RETURN
        END



        SUBROUTINE LENGTH (XX, YY, ELENTH)
C....    CALCULATES THE DISTANCE BETWEEN TWO POINTS (ELEMENT LENGTH)
        DIMENSION XX(2), YY(2)
        ELENTH = SQRT ((XX(1) - XX(2))**2 + (YY(1) - YY(2))**2)
        RETURN
        END



        SUBROUTINE PROPTY (L, MATFLG, DATMAT, AREA, ELMOD)
C....    DEFINES THE CROSS-SECTIONAL AREA AND MODULUS OF ELASTICITY FOR
C              ELEMENT L
        DIMENSION MATFLG(30), DATMAT(5,2)

        NFLAG   = MATFLG(L)
        AREA    = DATMAT(NFLAG,1)
        ELMOD   = DATMAT(NFLAG,2)
        RETURN
        END



        SUBROUTINE TRANSF (L, XX, YY, ELENTH, DIRCOS)
C....    COMPUTES THE TWO DIRECTION COSINES
        DIMENSION XX(2), YY(2), DIRCOS(2)
        DIRCOS(1) = (XX(2) - XX(1)) / ELENTH
        DIRCOS(2) = (YY(2) - YY(1)) / ELENTH
        RETURN
        END



        SUBROUTINE STIFF (AREA, ELMOD, ELENTH, DIRCOS, ESM)
C....    COMPUTES THE GLOBAL ELEMENT STIFFNESS MATRIX
        DIMENSION DIRCOS(2), ESM(4,4)

        COEF = AREA * ELMOD / ELENTH

        ESM(1,1) = COEF * DIRCOS(1) * DIRCOS(1)
        ESM(1,2) = COEF * DIRCOS(1) * DIRCOS(2)
        ESM(1,3) = -ESM(1,1)
        ESM(1,4) = -ESM(1,2)
```

```
          ESM(2,1) = ESM(1,2)
          ESM(2,2) = COEF * DIRCOS(2) * DIRCOS(2)
          ESM(2,3) = -ESM(2,1)
          ESM(2,4) = -ESM(2,2)

          ESM(3,1) = ESM(1,3)
          ESM(3,2) = ESM(2,3)
          ESM(3,3) = -ESM(3,1)
          ESM(3,4) = -ESM(3,2)

          ESM(4,1) = ESM(1,4)
          ESM(4,2) = ESM(2,4)
          ESM(4,3) = ESM(3,4)
          ESM(4,4) = ESM(2,2)

          RETURN
          END


          SUBROUTINE ASSEMK (L, NODI, NODJ, ESM, ASM)
C....     ASSEMBLES THE ASSEMBLAGE STIFFNESS MATRIX
          DIMENSION NODI(30), NODJ(30), ESM(4,4), ASM(40,40)

          NI = NODI(L)
          NJ = NODJ(L)

          ASM(2*NI-1,2*NI-1) = ASM(2*NI-1,2*NI-1) + ESM(1,1)
          ASM(2*NI-1,2*NI)   = ASM(2*NI-1,2*NI)   + ESM(1,2)
          ASM(2*NI,2*NI-1)   = ASM(2*NI,2*NI-1)   + ESM(2,1)
          ASM(2*NI,2*NI)     = ASM(2*NI,2*NI)     + ESM(2,2)

          ASM(2*NI-1,2*NJ-1) = ASM(2*NI-1,2*NJ-1) + ESM(1,3)
          ASM(2*NI-1,2*NJ)   = ASM(2*NI-1,2*NJ)   + ESM(1,4)
          ASM(2*NI,2*NJ-1)   = ASM(2*NI,2*NJ-1)   + ESM(2,3)
          ASM(2*NI,2*NJ)     = ASM(2*NI,2*NJ)     + ESM(2,4)

          ASM(2*NJ-1,2*NI-1) = ASM(2*NJ-1,2*NI-1) + ESM(3,1)
          ASM(2*NJ-1,2*NI)   = ASM(2*NJ-1,2*NI)   + ESM(3,2)
          ASM(2*NJ,2*NI-1)   = ASM(2*NJ,2*NI-1)   + ESM(4,1)
          ASM(2*NJ,2*NI)     = ASM(2*NJ,2*NI)     + ESM(4,2)

          ASM(2*NJ-1,2*NJ-1) = ASM(2*NJ-1,2*NJ-1) + ESM(3,3)
          ASM(2*NJ-1,2*NJ)   = ASM(2*NJ-1,2*NJ)   + ESM(3,4)
          ASM(2*NJ,2*NJ-1)   = ASM(2*NJ,2*NJ-1)   + ESM(4,3)
          ASM(2*NJ,2*NJ)     = ASM(2*NJ,2*NJ)     + ESM(4,4)

          RETURN
          END


          SUBROUTINE ANFVEC (NBCX, NBCY, FORCE, ANFV)
C....     GENERATE ASSEMBLAGE NODAL FORCE VECTOR DIRECTLY
          DIMENSION NBCX(20), NBCY(20), FORCE(15), ANFV(40)
          COMMON NNODES, NELEM, NMATLS, NPDIS, NPLDS, LOUT, LPRINT, LIN,
         1        TITLE(20)

          DO 100 I = 1, NNODES
              IFLAGX = NBCX(I)
              IFLAGY = NBCY(I)
```

```
            IF (IFLAGX .GE. 0) GO TO 50
            IFLAG = -IFLAGX
            ANFV(2*I-1)  = FORCE(IFLAG)
 50      CONTINUE
            IF (IFLAGY .GE. 0) GO TO 100
            IFLAG = -IFLAGY
            ANFV(2*I) = FORCE(IFLAG)
100      CONTINUE

         RETURN
         END




         SUBROUTINE PDBC (NBCX, NBCY, DISP, ASM, ANFV)
C....    APPLY RESTRAINTS ON THE NODAL DISPLACEMENTS
         DIMENSION NBCX(20), NBCY(20), DISP(5), ASM(40,40), ANFV(40)
         COMMON NNODES, NELEM, NMATLS, NPDIS, NPLDS, LOUT, LPRINT, LIN,
     1       TITLE(20)

         N2 = 2*NNODES
         DO 1000 I = 1, NNODES
            IFLAGX = NBCX(I)
            IFLAGY = NBCY(I)
            IF (IFLAGX .LE. 0) GO TO 500
               DO 200  J = 1, N2
                  ASM(2*I-1,J) = 0.
200         CONTINUE
               DO 300  J = 1, N2
                  ANFV(J) = ANFV(J) - ASM(J,2*I-1)*DISP(IFLAGX)
                  ASM(J,2*I-1) = 0.0
300         CONTINUE
               ANFV(2*I-1) = DISP(IFLAGX)
               ASM(2*I-1,2*I-1) = 1.
500         CONTINUE
            IF (IFLAGY .LE. 0) GO TO 1000
               DO 700  J = 1, N2
                  ASM(2*I,J) = 0.
700         CONTINUE
               DO 800  J = 1, N2
                  ANFV(J) = ANFV(J) - ASM(J,2*I)*DISP(IFLAGY)
                  ASM(J,2*I) = 0.0
800         CONTINUE
               ANFV(2*I) = DISP(IFLAGY)
               ASM(2*I,2*I) = 1.
1000     CONTINUE

         RETURN
         END




         SUBROUTINE EQSOLV (N, A, B, X)
C....    EQUATION SOLVER BY MATRIX INVERSION
         DIMENSION A(40,40), B(40), X(40)
         CALL INVDET (A, N, DINRN, DETM)
         CALL MATVEC (N, A, B, X)
         RETURN
         END
```

```
            SUBROUTINE INVDET (C, N, DTNRM, DETM)
C....       MATRIX INVERSION SUBROUTINE--USED BY PERMISSION OF PRENTICE-
C           HALL, INC., ENGLEWOOD CLIFFS, N.J.  FROM ROBERT W. HORNBECK,
C           NUMERICAL METHODS, 1975, P. 295.

C           INVERTS AN N BY N MATRIX C AND RETURNS THE INVERTED MATRIX BACK
C           TO THE MATRIX C;  DTNRM IS THE DETERMINANT OF THE MATRIX DIVIDED
C           BY THE EUCLIDEAN NORM; DETM IS SIMPLY THE DETERMINANT.  THE
C           DIMENSIONS OF J MUST BE AT LEAST 21 GREATER THAN THE ROW OR
C           COLUMN DIMENSION OF C.  THE ROUTINE EMPLOYS GAUSS-JORDAN ELIM-
C           INATION WITH COLUMN SHIFTING TO MAXIMIZE THE PIVOT ELEMENTS.

            DIMENSION C(40,40), J(80)
            PD = 1.
            DO 124 L = 1, N
            DD = 0.
            DO 123  K = 1, N
  123       DD = DD + C(L,K)*C(L,K)
            DD = SQRT(DD)
  124       PD = PD*DD
            DETM = 1.
            DO 125  L = 1, N
  125       J(L+20) = L
            DO 144  L = 1, N
            CC = 0.
            M = L
            DO 135  K = L, N
            IF ((ABS(CC) - ABS(C(L,K))) .GE. 0.)  GO TO 135
  126       M = K
            CC = C(L,K)
  135       CONTINUE
  127       IF (L .EQ. M)  GO TO 138
  128       K = J(M+20)
            J(M+20) = J(L+20)
            J(L+20) = K
            DO 137  K = 1, N
            S = C(K,L)
            C(K,L) = C(K,M)
  137       C(K,M) = S
  138       C(L,L) = 1.
            DETM = DETM*CC
            DO 139  M = 1, N
  139       C(L,M) = C(L,M) / CC
            DO 142  M = 1, N
            IF (L .EQ. M) GO TO 142
  129       CC = C(M,L)
            IF (ABS(CC) .LE. 1.E-10)  GO TO 142
  130       C(M,L) = 0.
            DO 141  K = 1, N
  141       C(M,K) = C(M,K) - CC*C(L,K)
  142       CONTINUE
  144       CONTINUE
            DO 143  L = 1, N
            IF (J(L+20) .EQ. L) GO TO 143
  131       M = L
  132       M = M + 1
            IF (J(M+20) .EQ. L)  GO TO 133
  136       IF (N .GT. M)  GO TO 132
```

```
133     J(M+20) = J(L+20)
        DO 163  K = 1, N
        CC = C(L,K)
        C(L,K) = C(M,K)
163     C(M,K) = CC
        J(L+20) = L
143     CONTINUE
        DETM = ABS (DETM)
        DTNRM = DETM / PD
        RETURN
        END


        SUBROUTINE MATVEC(N, A, B, X)
C....   PAGE 217 "ENTRY MATVEC" OF CARNAHAN, LUTHER, AND WILKES
        DIMENSION A(40,40), B(40), X(40)
        DO 100  I = 1, N
100     X(I) = 0.
        DO 200  I = 1, N
        DO 200  J = 1, N
200     X(I) = A(I,J) * B(J) + X(I)
        RETURN
        END


        SUBROUTINE PRINTN (ARRAY)
C....   PRINTS THE NODAL DISPLACEMENTS IN THE OUTPUT
        DIMENSION ARRAY(40)
        COMMON NNODES, NELEM, NMATLS, NPDIS, NPLDS, LOUT, LPRINT, LIN,
     1      TITLE(20)

        WRITE (LOUT,25)
25      FORMAT (/, 2X, 'SUMMARY OF NODAL DISPLACEMENTS',/,
     1          5X, 'NODE NO.', 3X, 'X-COMPONENT', 5X, 'Y-COMPONENT')

        DO 100  J = 1, NNODES
        WRITE (LOUT,50)   J, ARRAY(2*J-1), ARRAY(2*J)
50      FORMAT (4X, I5, 6X, G12.5, 4X, G12.5)
100     CONTINUE

        RETURN
        END


        SUBROUTINE POSTPR (NODI, NODJ, XCOOR, YCOOR, MATFLG,
     1      DATMAT, SOLN)
C....   POSTPROCESSOR--CALCULATES AXIAL ELONGATIONS, STRAINS, STRESSES,
C           AND FORCES
        DIMENSION NODI(30), NODJ(30), XCOOR(20), YCOOR(20),
     1      MATFLG(30), DATMAT(5,2), SOLN(40), DIRCOS(2),
     2      XX(2), YY(2)
        COMMON NNODES, NELEM, NMATLS, NPDIS, NPLDS, LOUT, LPRINT, LIN,
     1      TITLE(20)

        WRITE (LOUT,50)
50      FORMAT (/, 2X, 'SUMMARY OF ELEMENT RESULTANTS',/,
     1          3X, 'ELEMENT NO.', 4X, 'ELONGATION', 5X, 'STRAIN',
     2          5X, 'STRESS', 6X, 'FORCE')

        DO 200  L = 1, NELEM
```

```
C....    PRELIMINARY CALCULATIONS
             CALL COORDS (L, NODI, NODJ, XCOOR, YCOOR, XX, YY)
             CALL LENGTH (XX, YY, ELENTH)
             CALL PROPTY (L, MATFLG, DATMAT, AREA, ELMOD)
             CALL TRANSF (L, XX, YY, ELENTH, DIRCOS)

C....    CALCULATION OF ELEMENT ELONGATION
             NI = NODI(L)
             NJ = NODJ(L)

             UI = SOLN(2*NI - 1)
             VI = SOLN(2*NI)
             UJ = SOLN(2*NJ - 1)
             VJ = SOLN(2*NJ)

             UIPR = DIRCOS(1)*UI + DIRCOS(2)*VI
             UJPR = DIRCOS(1)*UJ + DIRCOS(2)*VJ

             DELTA = UJPR - UIPR

C....    CALCULATION OF ELEMENT STRAIN
             STRAIN = DELTA / ELENTH

C....    CALCULATION OF ELEMENT STRESS
             STRESS = ELMOD * STRAIN

C....    CALCULATION OF ELEMENT FORCE
             FORCE  = STRESS * AREA

C....    PRINT THE ELEMENT RESULTANTS
             WRITE (LOUT, 100) L, DELTA, STRAIN, STRESS, FORCE
100          FORMAT (5X, I4, 7X, G12.5, 1X, G12.5, 1X, 2G12.5)

200      CONTINUE
         RETURN
         END
```

B-2 Description of Input to Program TRUSS

The purpose of this part of the appendix is to give the details on the input parameters needed to run the TRUSS program described in Sec. 3-6. Any consistent set of units may be used. It is convenient to think of the input as being divided into seven different sections. The information provided to the program in each of these sections is summarized below. A ready reference is provided in Table 3-1 in the main text.

Section	Description of Input
1	Contains two lines of input, the first of which is an 80-column title (printed in the output) and the second contains the "master control data" which define the number of nodes, the number of elements, etc.
2	Contains nodal coordinate information; the nodes could be defined both with and without nodal coordinate generation.

Section	Description of Input
3	Contains the element data including the nodal connectivity as well as the material set specification for each element; the element data could be defined both with and without element generation.
4	Contains the two "material properties" (cross-sectional area and elastic modulus) for each "material" present.
5	Contains the "boundary condition" flags, which are defined in detail later; again the flags could be generated or specified on a node-by-node basis.
6	Contains the applied nodal forces.
7	Contains the imposed nodal displacements.

Section 1 Input

Line 1—Format 20A4
  TITLE
Line 2—Format 6I8
  NNODES   NELEM   NMATLS   NPDIS   NPLDS   LOUT

where

TITLE   = 80-column title
NNODES = total number of nodes
NELEM   = total number of elements
NMATLS = number of different materials
NPDIS   = number of different prescribed displacements (usually only 1)
NPLDS   = number of different point loads
LOUT   = output unit (for example, on the Apple II Plus microcomputer, the console is 3)

Notes:
1. Restrictions: NNODES $\leq$ 20, NELEM $\leq$ 30, NMATLS $\leq$ 5, NPDIS $\leq$ 5, and NPLDS $\leq$ 15
2. This is the only input section which is *not* terminated with a blank line.

Section 2 Input

Line 1—Format 20A4
  SUBT
Line 2, 4, 6, etc.—Format 3I8
  NI   NG   NF
Line 3, 5, 7, etc.—Format 4F8.0
  XI   YI   XF   YF

where

SUBT   = appropriate identifier or subtitle like "NODAL COORDINATE DATA"

NI = starting node number in the generation sequence (see note 2 below)
NG = nodal increment
NF = number of the final node to be generated in this sequence
XI = x coordinate of node NI,
YI = y coordinate of node NI
XF = x coordinate of node NF
YF = y coordinate of node NF

Notes:
1. Restrictions: NI $\leq$ 20, NF $\leq$ 20
2. If no generation is desired, then NG, NF, XF and YF need not be input (or zero will do).
3. A mandatory blank line must end this input section.
4. See Sec. 3-6 for more details.

Section 3 Input
Line 1—Format 20A4
    SUBT
Line 2, 4, 6, etc.—Format 5I8
    LI MS LG LF NG
Line 3, 5, 7, etc.—Format 2I8
    NI NJ

where

SUBT = appropriate identifier or subtitle like "ELEMENT DATA"
LI = starting element number in the element generation sequence (see note 2 below)
MS = material set flag to be set for all elements in this generation sequence
LG = element number increment
LF = number of the final element to be generated in the sequence
NG = nodal increment
NI = global node number corresponding to node "i" on element LI
NJ = global node number corresponding to node "j" on element LI

Notes:
1. Restrictions: LI $\leq$ 30, MS $\leq$ 5, LF $\leq$ 30, NI $\leq$ 20, NJ $\leq$ 20
2. If no generation is desired, then LG, LF and NG need not be input (or zero will do).
3. A mandatory blank line must end this input section.
4. See Sec. 3-6 for more details.

Section 4 Input
Line 1—Format 20A4
    SUBT
Line 2, 3, 4, etc.—Format I8, 2F8.0
    MSNO AREA ELMOD

where

SUBT   = appropriate identifier or subtitle such as "MATERIAL PROP-
ERTY DATA"
MSNO   = unique material set flag
AREA   = corresponding cross-sectional area
ELMOD = corresponding elastic modulus

Notes:
1. Restriction: MSNO $\leq$ 5
2. A mandatory blank line must end this input section.
3. Each material set may be defined only once.

Section 5 Input

Line 1—Format 20A4
    SUBT
Lines 2, 3, 4, etc.—Format 5I8
    NI   IBCX   IBCY   NG   NF

where

SUBT = appropriate identifier or subtitle like "BOUNDARY CONDITION
FLAG DATA"
NI   = number of the starting node in this generation sequence (see note 2
below),
IBCX = boundary condition flag on the x degree of freedom which will be
assigned to all nodes in this generation sequence; the meaning of
each of the possible values of IBCX is given below:

IBCX	Meaning
-N	The Nth force (see Section 6 Input) is applied in the x direction.
0	The node (or nodes) is neither restrained nor "loaded" in the x direction.
N	The Nth displacement (see Section 7 Input) is imposed in the x direction.

IBCY = boundary condition flag on the y degree of freedom which will be
assigned to all nodes in this generation sequence; the meaning of
each of the possible values of IBCY is identical to those for IBCX,
except the y direction is affected.
NG   = increment to be added to NI to get the next node number with the
same boundary condition flags, which is in turn incremented again
to get the next node number and so forth (see note 2 below)
NF   = final node number whose boundary condition flags are set to IBCX
and IBCY in the generation sequence.

Notes:

1. Restrictions: NI $\leq$ 20, NF $\leq$ 20, $-15 \leq$ IBCX ≤ 5, $-15 \leq$ IBCY ≤ 5
2. If no generation is desired, then NG and NF need not be input (or zero will do).
3. A mandatory blank line must end this input section.
4. If the flags are not specified for one or more nodes, both flags are automatically taken to be zero by the program.

Section 6 Input

Line 1—Format 20A4
   SUBT
Line 2, 3, 4, etc.—Format I8, F8.0
   NFORCE   FORCE

where

SUBT   = appropriate identifier or subtitle like "NODAL LOADS"
NFORCE = load identification number
FORCE   = corresponding force

Notes:

1. Restrictions: $1 \leq$ NFORCE ≤ 15
2. A mandatory blank line must end this input section.

Section 7 Input

Line 1—Format 20A4
   SUBT
Line 2, 3, 4, etc.—Format I8, F8.0
   NDISP   DISP

where

SUBT   = appropriate identifier or subtitle like "PRESCRIBED DISPLACE-
            MENTS"
NDISP = displacement identification number
DISP   = corresponding displacement

Notes:

1. Restrictions: $1 \leq$ NDISP ≤ 5
2. A mandatory blank line must end this input section.

C
Active Zone Equation Solvers

SUBROUTINEs ACTCOL and UACTCL were adapted from the subroutines of the same name in O. C. Zienkiewicz's book, *The Finite Element Method*, published by McGraw-Hill Book Company (UK), 1977. These two subroutines are listed below and require the dot-product function, FUNCTION DOT (also listed below). These subprograms are used with the written permission of McGraw-Hill.

SUBROUTINEs ACTCOL and UACTCL should be used when the assemblage stiffness matrix is symmetric and unsymmetric, respectively. Recall from Sec. 6–8 in the text that the lower triangular coefficients of an unsymmetric stiffness matrix are stored in the C array, with the diagonal entries set to unity (the actual diagonal entries are stored in the A array). See Eq. (6-72) and Fig. 6-19 for an example of the A and JDIAG arrays, and refer to Table 6-3 on page 280 for the definition of the parameters A, B, C, JDIAG, NEQ, AFAC, and BACK.

```
      SUBROUTINE ACTCOL (A, B, JDIAG, NEQ, AFAC, BACK)
      LOGICAL AFAC, BACK
      REAL JDIAG(1)
      DIMENSION A(1), B(1)
C
C           FACTOR A TO UT*D*U, REDUCE B
C
      AENGY = 0.0
      JR = 0
      DO 600  J = 1, NEQ
      JD = JDIAG(J)
      JH = JD - JR
      IS = J - JH + 2
      IF (JH - 2) 600, 300, 100
```

```
  100      IF (.NOT. AFAC) GO TO 500
           IE = J - 1
           K = JR + 2
           ID = JDIAG(IS - 1)
C
C                   REDUCE ALL EQUATIONS EXCEPT DIAGONAL
C
           DO 200  I = IS, IE
           IR = ID
           ID = JDIAG(I)
           IH = MINO (ID - IR - 1, I - IS + 1)
           IF (IH .LE. 0) GO TO 200
           KKK = K - IH
           KKL = ID - IH
           A(K) = A(K) - DOT(A(KKK), A(KKL), IH)
  200      K = K + 1
C
C                   REDUCE DIAGONAL TERM
C
  300      IF (.NOT. AFAC) GO TO 500
           IR = JR + 1
           IE = JD - 1
           K  = J - JD
           DO 400  I = IR, IE
           KKK = K + I
           ID = JDIAG(KKK)
           IF (A(ID) .EQ. 0.0) GO TO 400
           D = A(I)
           A(I) = A(I)/A(ID)
           A(JD) = A(JD) - D*A(I)
  400      CONTINUE
C
C                   REDUCE RIGHT-HAND SIDE
C
  500      IF (BACK) B(J) = B(J) -DOT (A(JR + 1),B(IS - 1), JH - 1)
  600      JR = JD
           IF (.NOT. BACK) RETURN
C
C                   DIVIDE BY DIAGONAL PIVOTS
C
           DO 700  I = 1, NEQ
           ID = JDIAG(I)
           IF (A(ID) .NE. 0.0)  B(I) = B(I)/A(ID)
  700      AENGY = AENGY + B(I)*B(I)*A(ID)
C
C                   BACK SUBSTITUTE
C
           J = NEQ
           JD = JDIAG (J)
  800      D = B(J)
           J = J - 1
           IF (J .LE. 0) RETURN
```

```
          JR = JDIAG (J)
          IF ((JD - JR) .LE. 1) GO TO 1000
          IS = J - JD + JR + 2
          K = JR - IS + 1
          DO 900  I = IS, J
          KKK = I + K
  900     B(I) = B(I) - A(KKK)*D
 1000     JD = JR
          GO TO 800
          END
          SUBROUTINE UACTCL (A, C, B, JDIAG, NEQ, AFAC, BACK)
          REAL              A(1), B(1), C(1), JDIAG(1)
          LOGICAL           AFAC, BACK
C
C                  FACTOR A TO UT*D*U, REDUCE B TO Y
C
          JR = 0
          DO 300 J = 1, NEQ
          JD = JDIAG(J)
          JH = JD - JR
          IF (JH .LE. 1) GO TO 300
          IS = J + 1 - JH
          IE = J - 1
          IF (.NOT. AFAC) GO TO 250
          K = JR + 1
          ID = 0
C
C                  REDUCE ALL EQUATIONS EXCEPT DIAGONAL
C
          DO 200 I = IS, IE
          IR = ID
          ID = JDIAG(I)
          IH = MIN (ID - IR - 1, I - IS)
          IF (IH .EQ. 0) GO TO 150
          A(K) = A(K) - DOT (A(K - IH), C(ID - IH), IH)
          C(K) = C(K) - DOT (C(K - IH), A(ID - IH), IH)
  150     IF (A(ID) .NE. 0.0) C(K) = C(K)/A(ID)
  200     K = K + 1
C
C                  REDUCE DIAGONAL TERM
C
          A(JD) = A(JD) - DOT (A(JR + 1), C(JR + 1), JH - 1)
C
C                  FORWARD REDUCE THE R.H.S.
C
  250     IF (BACK) B(J) = B(J) - DOT (C(JR + 1), B(IS), JH - 1)
  300     JR = JD
          IF (.NOT.BACK) RETURN
C
C                  BACK SUBSTITUTION
C
          J = NEQ
```

```
        JD = JDIAG(J)
500     IF (A(JD) .NE. 0.0)  B(J) = B(J)/A(JD)
        D = B(J)
        J = J - 1
        IF (J .LE. 0) RETURN
        JR = JDIAG(J)
        IF ((JD - JR) .LE. 1) GO TO 700
        IS = J  - JD + JR + 2
        K  = JR - IS + 1
        DO 600 I = IS, J
600     B(I) = B(I) - A(I + K)*D
700     JD = JR
        GO TO 500
C
        END



        FUNCTION DOT (A, B, N)
        DIMENSION A(1), B(1)
        DOT = 0.0
        DO 100  I = 1, N
          DOT = DOT + A(I)*B(I)
100     CONTINUE
        RETURN
        END
```

INDEX